Natural Antioxidants and Anticarcinogens
in Nutrition, Health and Disease

Natural Antioxidants and Anticarcinogens in Nutrition, Health and Disease

Edited by

J. T. Kumpulainen
Agricultural Research Centre of Finland, Joikionen, Finland

J.T. Salonen
University of Kuopio, Kuopio, Finland

RS•C
ROYAL SOCIETY OF CHEMISTRY

The proceedings of the Second International Conference on Natural Antioxidants and Anticarcinogens in Nutrition, Health and Disease, held on 24–27 June 1998 in Helsinki, Finland

Special Publication No. 240

ISBN 0-85404-793-X

A catalogue record for this book is available from the British Library

Published by The Royal Society of Chemistry,
Thomas Graham House, Science Park, Milton Road,
Cambridge CB4 0WF, UK

For further information see our web site at www.rsc.org

Printed by MPG Books Ltd, Bodmin, Cornwall

Preface

The Second International Conference on Natural Antioxidants and Anticarcinogens in Nutrition, Health and Disease was held in Helsinki, Finland, June 24–27, 1998. Scientists working in the medical, biomedical, nutritional and food sciences all share an increasing need to exchange ideas and discuss the many new findings on the role of antioxidants in aging and in the etiology of age-related degenerative diseases, such as cardiovascular diseases, cancers and diabetes which affect European and US populations in particular. An attempt to address that need was reflected in the programme of the NAHD'98 conference. Emerging information on experimental and human research indicates that a sufficient intake of various antioxidants is the key in balanced nutrition and has a more pronounced effect on the health and well-being of aging populations than has been previously understood. Recent findings have revealed many of the underlying mechanisms of degenerative diseases and the role of antioxidants in their prevention or modulation. The NAHD'98 Conference brought together many of the world's leading scientists in the field to present their most recent findings. The conference was attended by over 300 participants from 35 countries who presented over 200 papers.

Fundamental new information on the relationship between natural antioxidants and human health and well-being is rapidly emerging. The present conference examined the current state of the art and confronted the challenges for future research in this dynamic field. To this end, the conference brought into focus the role of natural, non-nutritive antioxidants, anticarcinogens and antioxidative micronutrients in nutrition, health and disease.

The main topics of the NAHD'98 Conference included the role of dietary intake and bioavailability of natural antioxidants, anticarcinogens and pro-oxidants and the risk of coronary heart disease and cancers, the measurement of oxidative stress in humans and desirable *vs.* potentially harmful intakes of flavonoids, antioxidant vitamins, carotenoids and ubiquinone. The following topics, including a total of 81 contributions, have been selected:

1. Oxidative stress in cardiovascular diseases and aging
2. Oxidative stress and antioxidants in diabetes
3. Dietary intake, bioavailability and antioxidative effects of flavonoids and phenolics
4. Antioxidative effects of other natural antioxidants and measurement of oxidative stress or damage
5. Natural antioxidants or pro-oxidants in foods and nutrition
6. Dietary intakes and modes of action of potentially anticarcinogenic dietary compounds
7. Antioxidants, oxidative damage and cancer

I sincerely hope that these Proceedings will increase understanding and stimulate new multidisciplinary research into the role of natural antioxidants and anticarcinogens in the prevention of such age-related degenerative diseases as cardiovascular diseases, cancers and diabetes.

I express my deepest gratitude to all of the scientists who have contributed to the success of the conference and this volume, and particularly to the members of the Scientific and Organizing Committees and invited speakers. I am also very grateful to my wife, B.A. Randi Kumpulainen, for her invaluable help in the language revision and to Mr Tauno Koivisto and Ms Sari Aalto for their excellent technical assistance.

J.T. Kumpulainen
February 1999

Contents

Oxidative Stress in Cardiovascular Diseases and Aging

Antioxidants, lipid peroxidation and cardiovascular diseases   3
   J.T. Salonen

Antiatherogenicity of antioxidants against LDL oxidation   9
   M. Aviram

Are antioxidants effective in the primary prevention of coronary heart disease ?   20
   D. Kromhout

Redox imbalance and antioxidant signalling   27
   J.M.C. Gutteridge

Involvement of mitochondria in oxidative stress in aging of intact cells   34
   J. Viña, J. Sastre, F.V. Pallardó and A. Lloret

Melatonin, nitric oxide and Alzheimer´s disease - a minireview   40
   D. Lahiri

The L-arginine-nitric oxide pathway: Role as an antioxidant mechanism of atherosclerosis   46
   R.H. Böger

Antioxidant enzyme depletion and nitric oxide alterations: relationship with the risk factors
in angiographically diagnosed coronary heart disease   52
   B. Sözmen, F. Uysal, D.Gazitepe, L. Aslan and E.Y. Sözmen

Lysosomal leakage causes apoptosis following oxidative stress, growth-factor starvation and
fas-activation   57
   U. T. Brunk

Oxidative Stress and Antioxidants in Type II Diabetes

Molecular mechanisms of cellular lipid peroxidation in diabetes   69
   S.K Jain

The biochemical basis of diabetic complications - a role of oxidative stress ?   74
   J. Aaseth and O.W. Boe

Antioxidants and diabetes mellitus   78
   P. Knekt and A. Reunanen

Body iron stores, vitamin E and the risk of type 2 diabetes   80
   J.T. Salonen, T-P. Tuomainen, K. Nyyssönen and K Punnonen

Vitamin C status and oxidative stress in diabetics   83
   D. Naidoo, O. Lux and M. Phung

Antioxidant status of insulin dependent diabetics   86
   M.A. Cser, I. Sziklai-László, M.F. Larye and I. Lombeck

Dietary Intake, Bioavailability and Antioxidative Effects of Flavonoids and Phenolics

Desirable vs. harmful levels of intake of flavonoids and phenolic acids   93
    V. Breinholt

Antiatherogenicity and antioxidative properties of polyphenolic flavonoids
against LDL oxidation   106
    M. Aviram, R. Aviram and B. Fuhrman

Antioxidant potency and mode of action of flavonoids and phenolic compounds   114
    J. Kanner

Biomarkers of flavonoid intake in humans   124
    A. Ferro-Luzzi and G. Maiani

Dietary flavonoids and antioxidant protection   137
    P. Pietta

Trolox Equivalent Antioxidant Capacity of average flavonoids intake in Finland   141
    J.T. Kumpulainen, M. Lehtonen and P. Mattila

Anthocyanins in red wines: Antioxidant activity and bioavailability in humans   151
    T. Lapidot, S. Harel, R. Granit and J. Kanner

The isoflavan glabridin inhibits LDL oxidation: Structural and mechanistic aspects   161
    B. Fuhrman, J. Vaya, P. Belinky and M. Aviram

Bioavailability and antioxidant properties of luteolin   166
    K. Shimoi, H. Okada, J. Kaneko, M.Furugori, T. Goda, S. Takase, M. Suzuki,
    Y. Hara and N. Kinae

Effects of a flavonoid-rich diet on plasma quercetin and susceptibility of LDL to oxidation 174
    G. McAnlis, J. McEneny, J. Pearce and I. Young

Comparative *in vitro* and *in vivo* free radical scavenging abilities of a novel grape seed
proanthocyanidin extract and selected antioxidants   178
    D. Bagchi, R.L. Krohn, J. Balmoori, M. Bagchi, A. Garg and S. J. Stohs

Effect of parsley intake on urinary apigenin excretion, blood antioxidant enzymes
and on biomarkers for oxidative stress in humans   188
    S.E. Nielsen, J.F. Young, B. Daneshvar, S. T. Lauridsen, P. Knuthsen
    B. Sandström and L. O. Dragsted

Effect of fruit juice intake on urinary quercetin excretion and biomarkers of antioxidative status 193
    J. Young, S. E. Nielsen, J. Haraldsdóttir, B. Daneshvar, S. T. Lauridsen,
    P. Knuthsen, A. Crozier, B. Sandström and L. O. Dragsted

Antioxidant activity of hydroxycinnamic acids on human low-density lipoprotein
oxidation *in vitro*   197
    A.S. Meyer and M.F. Andreasen

Radical scavengers and inhibitors of enzymatic lipid peroxidation from plantago major,
a medicinal plant   200
    K.P. Skari, K.E. Malterud and T. Haugli

Isoflavones and plasma lipids in young women - potential effects on HDL   203
    S. Samman, P.M. Lyons-Wall, G.S.M. Chan and S.J. Smith

Flavonoids in herbs determined by HPLC with UV-PDA and MS detection   206
    P. Knuthsen and U. Justesen

Effect of processing on content and antioxidant activity of flavonoids in apple juice   209
    A.A. van der Sluis, M. Dekker and W.M.F. Jongen

Protective effect of flavonoids on linoleic acid hydroperoxide-induced toxicity of
human endothelial cells 212
  T. Kaneko and N. Baba

Synthesis of antioxidant isoflavone fatty acid esters 215
  P. Lewis, K. Wähälä, Q-H. Meng, H. Adlercreutz and M. J. Tikkanen

**Antioxidative Effects of Other Natural Antioxidants and Measurement of
Oxidative Stress or Damage**

Carotenoids: Modes of action and bioavailability of lycopene in the human 221
  H. Sies and W. Stahl

Antiatherogenic effect of lycopene and beta-carotene: inhibition of LDL oxidation and
suppression of cellular cholesterol synthesis 226
  B. Fuhrman, A. Elis and M. Aviram

Beta-carotene normalizes oxidative damage in carotenoid-depleted women 231
  B.J. Burri, A. Clifford and Z. Dixon

Prevention of singlet oxygen damage in 2'-deoxyguanosine by lycopene
entrapped in human albumin 234
  L. Yamaguchi, M.H. G. Medeiros and P. Di Mascio

Coenzyme Q protection of mitochondrial activities in rat liver under oxidative stress 238
  G. Lenaz, M.L. Genova, M. Cavazzoni, M. D'Aurelio, B. Nardo, G. Formiggini and C. Bovina

Different antioxidant mechanisms of α-tocopherol and L-arginine result in
preserved endothelial function in hypercholesterolemic rabbits 243
  S.M. Bode-Böger, R.H. Böger and J.C. Frölich

Measurement of free radicals in humans by electron spin resonance spectroscopy 251
  A. Bini, S. Bergamini, E. Ghelfi, A. Iannone, M. Meli, M.G. Staffieri and A. Tomasi

Radioimmunoassays of isoprostanes and prostaglandins as biomarkers of
oxidative stress and inflammation 260
  S. Basu

Free radical processes in 1,2-dimethylhydrazine induced carcinogenesis in rats and
protective role of melatonin 268
  *A.V. Arutjunyan, S.O. Burmistrov, T.I. Oparina, V.M. Prokopenko, M.G. Stepanov,
  I.G. Popovich, M.A. Zabezhinsky and V.N. Anisimov*

Protective effects of "sangre de drago" from croton lechleri muell.-arg. on
spontaneous lipid peroxidation 272
  C. Desmarchelier, S.M. Barros, F. Witting Schaus, J. Coussio and G.Ciccia

Natural Antioxidants or Pro-oxidants in Foods and Nutrition

Relation of dietary intake and blood antioxidants with antioxidant capacity in healthy
non-smoking men 277
  A. Jeckel, H. Boeing, D.l. Thurnham, B. Raab and H.-J. Zunft

Drinking green tea leads to a rapid increase in plasma antioxidant potential 280
  I.F.F. Benzie, Y. T. Szeto, B. Tomlinson and J.J. Strain

The antioxidant capacity of selected foods and the potential synergisms among
their main antioxidant constituents 283
  M. Salucci, R. Lázaro, G. Maiani, F. Simone, D. Pineda, and A. Ferro-Luzzi

Flavonoid content and antioxidant properties of broccoli 291
  A. Lugasi, J. Hóvári, M.N. Gasztonyi and E. Dworschák

Examination of flavonoid content in Hungarian vegetables   296
    J. Hóvári, A. Lugasi and E. Dworschák

Avenanthramide antioxidants in oats   299
    L. Dimberg

Processing of foods containing flavonoids and glucosinolates; effects on composition
and bioactivity   303
    M. Dekker, A.A. van der Sluis, R. Verkerk and W.M.F. Jongen

Measurement of cholesterol oxides in food: Results of an interlaboratory comparison study 309
    *P.C. Dutta, M.F. Caboni, U. Diczfalusy, F. Dionisi, S. Dzeletovic, J. Kumpulainen,
    V.K Lebovics, J-M. A. Grandgirard, F. Guardiola, J. Pihlava, M.T. Rodriguez-Estrada
    and F. Ulberth*

Phytosterol oxides in some samples of pure phytosterols mixture and in a
few tablet supplement preparations in Finland   316
    P. C. Dutta

Formation of sterol oxides in edible oils   320
    V.K. Lebovics, K Neszlényi, S. Latif, L. Somogyi, J. Perédi, J. Farkas and Ö. Gaál

Tocopherols, carotenoids and cholesterol oxides in plasma from women with
varying smoking and eating habits   323
    H. Billing, O. Nyren, A. Wolk and L -A. Appelqvist

Importance of *in vitro* stability for *in vivo* effects of fish oils   326
    T. Saldeen, K. Engström, R. Jokela and R. Wallin

Enhancement of absorption of vitamin E by sesaminol -an active principle of
sesame seed   331
    K. Yamashita, Y. Iizuka and I. Ikeda

Model *in vitro* studies on the protective activity of tocochromanols with respect to beta-carotene 334
    M. Nogala-Kalucka and J. Zabielski

The effect of processing on total antioxidative capacity in strawberries   338
    U. Viberg, C Alklint, B. Åkesson, G. Önning and I. Sjöholm

Quercetin content in berry products   341
    S. Häkkinen, P. Saarnia and R Törrönen

The antiperoxidative effect of dunaliella beta-carotene isomers   343
    M. J. Werman, M. Yeshurun, A. Ben-Amotz and S. Mokady

Dietary Intakes and Modes of Action of Potentially Anticarcinogenic Dietary Compounds

Dietary phytoestrogens - mechanisms of action and possible role in the development of
hormonally dependent diseases   349
    S. Mäkelä, L. Strauss, N. Saarinen, S. Salmi, T. Streng, S. Joshi and R. Santti

Dietary intakes and levels in body fluids of lignans and isoflavonoids in various populations 356
    W. Mazur and H. Adlercreuz

Inhibition of aromatase by flavonoids in cultured jeg-3 cells   369
    S.C. Joshi, M. Ahotupa, M.L. Koshan, S.I. Mäkelä and R. Santti

The effect of phytoestrogen - rich foods on urinary output of phytoestrogen metabolites
- a pilot study   375
    J. V. Woodside, M.S. Morton and A.J.C. Leathem

Mechanisms of action of the antioxidant lycopene in cancer   377
    J. Levy, M. Danilenko, M. Karas, H. Amir, A. Nahum, Y. Giat and Y. Sharoni

Intakes and modes of action of other anticarcinogenic dietary compounds 385
    L.O. Dragsted

Inhibition of CYPIAI *in vitro* by berries with different quercetin contents 395
    L. Kansanen, H. Mykkänen and R Törrönen

Chemoprotective properties of cocoa and rosemary polyphenols 398
    E.A. Offord, T. Huynh-Ba, O. Avanti and A.A. Pfeifer

Antioxidant compound 4-nerolidylcatechol inhibits *in vitro* KB cells growth
and topoisomerase i activity 404
    E. Mongelli, A. Romano, C. Desmarchelier, J. Coussio and G. Ciccia

Phytoestrogen profiling from biological samples using HPLC with coulometric
electrode array detection 407
    T. Nurmi, P. Lewis, K. Wähälä, and H. Adlercreutz

Antioxidants, Oxidative Damage and Cancer

Dietary cancer prevention: caveats seen by a toxicologist 413
    H. Verhagen

Oxidative damage to DNA: a likely cause of cancer? 417
    A. R Collins

Increased fruit and vegetable consumption reduces indices of oxidative DNA damage in
Iymphocytes and urine 423
    A.D. Haegele and H. J. Thompson

Elevated DNA damage in lymphocytes from ankylosing spondylitis patients 433
    *M. Dusinská, K Raölová, J. Lietava, M. Somorovská, H. Petrovská P. Dobríková,
    and A. Collins*

Inhibitory effect of citrus extracts on the growth of breast cancer cells *in vitro* 437
    N. Guthrie and K. Carroll

The effect of physical processing on protective effect of broccoli in relation to
damage to DNA in colonocytes 440
    B. Ratcliffe, A. R Collins, H. J. Glass, and K. Hillman

Differential effect of a novel grape seed proanthocyanidin extract on cultured human
normal and malignant cells 443
    R.L. Krohn, X. Ye, W. Liu, S.S. Joshi, M. Bagchi, H.G. Preuss, S.J. Sohs and D. Bagchi

Reduction of the incidence of metachronous adenomas of the large bowel by
means of antioxidants: A double blind randomized trial 451
    L. Bonelli, A. Camoriano, P. Ravelli, G. Missale, P. Bruzzi and H. Ase

Inhibition of colorectal carcinoma development in min mice by flavonoids 456
    B. Raab, A. Salomon, K Schmehl, S. Sander and G. Jacobasch

Subject Index **459**

Scientific Committee

Adlercreutz H. (Helsinki)
Alfthan G. (Helsinki)
Ames B. (Berkeley)
Aro A. (Helsinki)
Gutteridge J. (London)
Korpela H. (Kuopio)
Kumpulainen J. T. (Jokioinen), Chairman
Packer L. (Los Angeles)
Salonen J. T. (Kuopio)
Santti R. (Turku)
Sevanian A. (Los Angeles)
Sies H. (Düsseldorf)
Uusitupa M. (Kuopio)
Vuori E. (Helsinki)
Ylä-Herttuala S. (Kuopio)

Organizing Committee

Alfthan G. (Helsinki)
Aro A. (Helsinki)
Koivisto T. (Jokioinen)
Korpela H. (Kuopio)
Kumpulainen J. T. (Jokioinen), Chairman
Niemelä M. (Helsinki)
Mattila P. (Jokioinen)
Salonen J. T. (Kuopio)
Santti R. (Turku)
Vuori E. (Helsinki)

Oxidative Stress in Cardiovascular Diseases and Aging

ANTIOXIDANTS, LIPID PEROXIDATION AND CARDIOVASCULAR DISEASES

J.T. Salonen

Research Institute of Public Health, University of Kuopio, P.O. Box 1627, SF 70211 Kuopio, Finland.

1 FREE RADICAL STRESS IN HUMANS

Free radicals and reactive oxygen species are either synthesized endogenously e.g. in energy metabolism and by the antimicrobial defence system of the body, or produced as reactions to exogenous exposures such as cigarette smoke, imbalanced diet, exhaustive exercise, environmental pollutants and food contaminants. Oxidative and other chemical stress may modify not only polyunsaturated lipids, but also carbohydrates, proteins and complex macromolecules, forming atherogenic, carcinogenic, diabetogenic and brain degenerating substances, depending on the target organ. Modified biomolecules also interfere with gene expression, causing metabolic disturbances.

2 LIPID PEROXIDATION AND ATHEROSCLEROSIS

According to a widely supported theory the oxidative modification of lipoproteins increases their atherogenicity.[1-5] This theory is actively studied to enable a better understanding of the development and progression of atherosclerosis, currently the leading public health problem in developed countries and in the future, possibly also in the developing countries. We observed an association between a high titre of autoantibodies against malondialdehyde (MDA)-modified low density lipoprotein (LDL) and an accelerated progression of carotid atherosclerosis.[6] Our original observation has been repeated in a number of confirmatory studies, reviewed earlier.[4,5] We have also shown that high circulating levels of oxidation products of cholesterol, especially 7-OH-cholesterol, are associated with fastened progression of early carotid atherosclerosis.[7] Autoantibodies against oxidatively modified LDL have also been observed to be associated with an increased risk of acute myocardial infarction (AMI).[8] These epidemiological observations have close to established the role of lipid peroxidation in the development of atherosclerotic plaques and, although still suggestively, in the causation of cardiovascular events.

3 TRANSITION METALS AND CORONARY HEART DISEASE

Redox-active transition metals such as iron promote oxidative stress. However, the existing epidemiologic evidence concerning the role of iron, a lipid peroxidation catalyst, in coronary heart disease (CHD) is inconsistent, as reviewed earlier.[9,10] We investigated the association of the concentration ratio of serum transferrin receptor to serum ferritin (a low TfR/ferritin ratio denotes high iron stores), with the risk of AMI in a prospective case-control study in men from eastern Finland.[11] This ratio is not affected by inflammation. Transferrin receptor assays were carried out in frozen baseline samples for 99 men who later had an AMI during average 6.4 years of follow-up and 98 control men from the same cohort, matched for age, examination year and residence of 1,931 men who had no clinical CHD. AMIs were registered prospectively. The mean TfR/ferritin ratio was 15.1 (SE 2.0) among cases and 21.3 (SE 2.2) among controls (P=0.035 for difference). In logistic regression models adjusting for other strongest risk factors for AMI and indicators of inflammation and alcohol intake, men in the lowest and second lowest thirds of the TfR/ferritin ratio (highest and second highest thirds of iron stores) had a 2.9-fold (95 % confidence interval, 1.3 to 6.6, P=0.011) and 2.0-fold (0.9 to 4.2, P=0.081) risk of AMI, compared with men in the highest third (P=0.010 for trend). These data show an association between increased body iron stores and excess risk of AMI, confirming previous epidemiologic findings. There are a number of epidemiologic studies, in which body iron stores have been estimated on the basis of less precise methods, and for this reason their results are inconsistent.[9,10] In a prospective study, we observed an association between donating blood and a reduced incidence of AMI, also supporting the role of body iron stores with regard to CHD.[12, 13]

Iron and CHD: Recent Observations.

1. Supplementing experimental animals (NZ rabbit, WHHL rabbit, LDLR-/-mouse) with either oral or peroral iron accelerates the development of aortic atherosclerotic lesions.

2. Men who donate blood regularly have a reduced risk of AMI.

3. Men with high body iron stores (reduced transferrin receptor to ferritin ratio) have an increased risk of AMI.

In addition to epidemiologic observations, also animal experiments have shown that iron loading leads to accelerated development of atherosclerotic plaques.[9,10] In a prospective follow-up study, persons with high serum ferritin levels had fastened progression of carotid atherosclerosis.[14]

Mercury, especially methylmercury can promote lipid peroxidation.[10] It is an environmental pollutant that has no physiological role in the human metabolism. The principal sources of mercury intake are diet and possibly, dental amalgam fillings. Predatory fish is the main dietary source of mercury in most industrialised countries. In fish, mercury is in organic form, mainly as methylmercury, which is well absorbed in the gut and is accumulated in epithelial tissues including hair. The hair content of mercury is an indicator of mercury intake over several months.

We observed in an epidemiologic cohort study over twofold risk of AMI and mortality from CHD and cardiovascular disease (CVD) in men in eastern Finland with an elevated hair content of mercury.[15] There also was an association between elevated titres of immune complexes containing oxidized LDL and high hair mercury content.[15] Additional epidemiologic studies concerning the role of mercury and other catalytic metals are lacking.

4 NATURAL ANTIOXIDANTS AND CORONARY HEART DISEASE: THE EPIDEMIOLOGIC EVIDENCE

Two large prospective epidemiologic studies have suggested that the self-selected use of vitamin E containing supplements is associated with reduced risk of CHD events.[16,17] In our prospective studies in men from East Finland, deficiencies in selenium and vitamin C were associated with increased risk of AMI.[18,19] A selection bias (vitamin users more health conscious, multicollinearity of nutrients) can not, however, be ruled out in non-experimental epidemiologic studies. Since plasma vitamin C concentration has been so far measured only in two prospective studies, existing data are limited.[4,10,20,21] Epidemiologic findings concerning the role of dietary flavonoids and carotenoids in CVD are inconsistent, as reviewed elsewhere in detail.[4,10,21-24]

5 NATURAL ANTIOXIDANTS IN THE PREVENTION OF CARDIOVASCULAR DISEASES

No conclusive clinical trials concerning the preventive effect of any antioxidant on either atherosclerosis or cardiovascular events have so far been reported. In three large cancer trials, high doses of synthetic ß-carotene or a combination of ß-carotene with retinol had no effect on cardiovascular deaths.[25-27] In the CHAOS study, large doses (400 or 800 IU/d) of vitamin E reduced the incidence of non-fatal MI but not CVD mortality in CHD patients.[28] In a skin cancer prevention trial in the USA, 200 µg/d of selenium reduced all-cause mortality by 17 % (ns) and total cancer mortality by 50%.[29] In the Linxian study in China, a vitamin/mineral supplement reduced hypertension by 57 % in men and 8 % in women (ns) and stroke mortality by 68 % in men and 7 % (ns) in women.[30,31]

A number of randomised, placebo-controlled antioxidant supplementation trials are under way (Table 1).[4,23,32-36] The majority of these use a high dose of vitamin E as the supplement. One of these is our "Antioxidant Supplementation in Atherosclerosis Prevention" ("ASAP") study, which is a randomised 3+3 -year 2 x 2 factorial trial testing the effect of 200 mg/d of d-α-tocopherol acetate and/or 500 mg/d of vitamin C on lipid peroxidation, the progression of carotid atherosclerosis, blood pressure, cataracts and the incidence of CHD events. First three years of supplementation are double-masked, and this period was finshed in late 1998. Results concerning the first 3-year period will be presented in 1999. During the second 3-year period, all supplemented subjects will receive the combination of both supplements and those who earlier received only placebo, will remain unsupplemented. Carotid atherosclerosis and coronary events will be monitored during this open-treatment period.

The French "SU.VI.MAX"35 trial and the British "Heart Protection Study" use a combination of several antioxidants as the supplement. Other studies are factorial.The French SU.VI.MAX study is a double-masked, randomised, placebo-controlled-trial to test the effect of a combination of antioxidants on total mortality and incidence of cancers, myocardial infarction and cataracts. The supplementation consists of 15 mg of vitamin E, 120 mg of vitamin C, 6 mg of ß-carotene, 100 µg of selenium and 20 mg of zinc daily. The original sample size goal was 5,000 men aged 45-60 and 10,000 women aged 35-60 at entry.

The British Heart Protection Study is even a larger trial in 20000 men and women aged 40-75 years, with a high risk of CHD. It has a randomised placebo-controlled 2x2 factorial design. One treatment is a LDL lowering drug and the other a combination of 600 mg of vitamin E and 250 mg of vitamin C daily. The study outcomes for the vitamin part of the study are total CHD and fatal CHD. In the U.S.A. several trials testing the preventive effect of vitamin E and C on CVD events or mortality are underway: The "Women's Health Study" concerns the effect of 600 IU of vitamin E or 30 mg of ß-carotene every other day on the incidence of AMI, stroke and mortality from CVD in 40,000 postmenopausal female nurses in the U.S.A. The "Health Professionals' Follow-up Study" uses a combination of 600 IU of vitamin E and 250 mg of vitamin every other day in 20,000 patients with previous angina, stroke, claudication or diabetes, the outcome is total mortality. Other, smaller on-going studies are shown in the table.

Table 1

On-going randomized controlled trials of antioxidant vitamins in the prevention of cardiovascular diseases.

Study	Participants	n	Vitamins (dose)	Outcome
ASAP	High-risk men and women	520	Vitamin E (200 mg/d) and/or Vitamin C (500 mg/d)	Progression of carotid atherosclerosis Hypertension Cataracts CVD events (in 6 years)
SU.VI.MAX	Healthy men and women	13 000	Vitamin E, vitamin C ß-carotene, zinc and selenium	Cardiovascular endpoints and cancers
Woman's Health Study	Healthy women 40-84 years	40 000	Vitamin E (600 IU/die) ß-carotene (50 mg/die)	AMI, stroke and CDV mortality (ß-carotene terminated 1996)
Heart Protection Study	Men and women at increased risk of future MI	18 000	Vitamin E (600 mg/d), ß-carotene (20 mg/d) and Vitamin C (250 mg/d)	CHD events and total morality (ß-carotene terminated?)
WACS	Women with CVD	8 000	Vitamin E (400 IU/die) or Beta-carotene (20 mg/die) or Vitamin C (1 g/die)	AMI, stroke coronary revascularization and CVD mortality (ß-carotene terminated?)
HOPE Study	Men and women at increased risk of future MI	9 000	Vitamin E	AMI, stroke and CDV mortality
GISSI	Men and women with recent AMI	12 000	Vitamin E (300 mg/die)	Total mortality
Health Professionals' Follow-up Study	Patients with previous angina, stroke, claudication or diabetes	20 000	Vitamin E (600 IU/die), ß-carotene (20 mg/die) and vitamin C (250 mg/die)	Total mortality (ß-carotene terminated?)

The on-going preventive trials may eventually provide the necessary information about the role of vitamin E in the prevention of atherosclerotic progression and cardiovascular events. ASAP is the only study testing effects of vitamin C as a single supplement. If studies testing the effects of antioxidant combinations will have unequivocal consistent negative results, those findings will be of some value, at least regarding the kind of persons that participated in the studies. Even if the findings of the combination studies will be positive, it will be impossible to tell which antioxidant was effective. On the other hand, the view that a combination of several co-antioxidants and cofactors for antioxidative enzymes may be physiologically more relevant and thus have a better chance to be protective, has a plausible foundation. Finally, it is also possible that none of the vitamins studied (vitamin E, C, ß-carotene) is one of those nutrients that will explain the protective effects of vegetables and fruits, but rather some other phytochemicals will in fact prove to be the compounds that are responsible for the health-enhancing effects of the plant-dominated diet.

References

1. J.L. Witztum, *Lancet*. 1994, **344**, 793.
2. B. Halliwell, *Lancet*. 1994, **344**, 721.
3. J.W. Heinecke, *Atherosclerosis*. 1998, **141**, 1.
4. J.T. Salonen, *Arch. Toxicol*. 1998, **20**, S 249.
5. S. Ylä-Herttuala, *Curr. Opin. Lipidol*. 1998, **9**, 337.
6. J.T. Salonen, S. Ylä-Herttuala, R. Yamamoto, S. Butler, H. Korpela, R. Salonen, K. Nyyssönen, W. Palinski and J.L. Witztum, *Lancet*. 1992, **339**, 883.
7. J.T. Salonen, K. Nyyssönen, R. Salonen, E. Porkkala-Sarataho, T.-P. Tuomainen, U. Diczfalusy and I. Björkhem, *Circulation*. 1997, **95**, 840.
8. M. Puurunen, M. Mänttäri, V. Manninen, L. Tenkanen, G. Alfthan, C. Ehnholm, O. Vaarala, K. Aho and T. Palosuo, *Arch. Intern. Med*. 1994, **154**, 2605.
9. J.T. Salonen, In 'L. Hallberg, N. and G. Asp (Eds.) J.L. London Press, London'. 1996, 293.
10. J.T. Salonen, In 'B. Sandström and P. Walter (Eds), *Bibl. Nutr.Dieta., Barsel, Karger*'. 1998, No **54**, 112.
11. T.-P. Tuomainen, K. Punnonen, K. Nyyssönen and J.T. Salonen, *Circulation*. 1998, **97**, 1461.
12. T.-P. Tuomainen, R. Salonen, K. Nyyssönen and J.T. Salonen, *Brit. Med. J*. 1997, **314**, 1830.
13. J.T. Salonen, T.-P. Tuomainen, R. Salonen, T.A. Lakka, K. and K. Nyyssönen, *Am. J. Epidemiol*. 1998, **148**, 445.
14. S. Kiechl, J. Willeit, G. Egger, W. Poewe and F. Oberhollenzer, *Circulation*. 1998, **96**, 3261.
15. J. T. Salonen, K. Seppänen, K. Nyyssönen, H. Korpela, J. Kauhanen, M. Kantola, J. Tuomilehto, H. Esterbauer, F. Tatzber, R. Salonen. *Circulation*. 1995, **91**, 645.
16. M. J. Stampfer, C. H. Hennekens, J. E. Manson, G. A. Coditz, B. Rosner, W. C. Willett. *N. Engl. J. Med*. 1993, **328**, 1444.
17. E. B. Rimm, M. J. Stampfer, A. Ascherio, E. Giovannucci, G. A. Colditz, W. C. Willet. *N. Engl. J. Med*. 1993, **328**, 1450.
18. J. T. Salonen, G. Alfthan, J. K. Huttunen, J. Pikkarainen, P. Puska. *Lancet*. 1982, **II**, 175.
19. K. Nyyssönen, M.T. Parviainen, R. Salonen, J. Tuomilehto, J.T. Salonen, *Brit. Med. J*. 1997, **314**, 634.
20. R.A. Jacob. *Nutr. Rev*. 1998, **56**, 334.
21. K.F. Gey, *Biofactors*. 1998, **7**, 113.
22. E.M. Lonn and S. Yusuf, *Can. J. Cardiol*. 1997, **13**, 957.

23. A. Faggiotto, A. Poli and A.L. Catapano. *Curr. Opin. Lipidol.* 1998, **9**, 541.

24. A.T. Diplock, J.L. Charleux, G. Crozier-Willi, F.J. Kok, C. Rice-Evans, M. Roberfroid, W. Stahl and J. Vina-Ribes, *Br. J. Nutr.* 1998, **80**, S77.

25. The α-tocopherol, β-Carotene Cancer Prevention Group. *N. Eng. J. Med.* 1994, **330**, 1029.

26. C.H. Hennekens, J.E. Buring, J.E. Manson, M. Stampfer, B. Rosner, N.R. Cook, C. Belanger, F. LaMotte, J.M. Gaziano, P.M. Ridker, W. Willett and R. Peto, *N. Engl. J. Med.* 1996, **334**, 1145.

27. G.S. Omenn, G.E. Goodman, M.D. Thornquist, J. Balmes, M.R. Cullen, A. Glass, J.P. Keogh, F.L. Meyskens, B. Valanis, J.H. Williams, S. Barnhart and S. Hammar, *N. Engl. J. Med.* 1996, **334**, 1150.

28. N.G. Stephens, A. Parsons, P.M. Schofield, F. Kelly F, K. Cheeseman, M.J. Mitchinson, *Lancet.* 1996, **347**, 781.

29. L.C. Clark, G.F. Jr Combs, B.W. Turnbull, E.H. Slate, D.K. Chalker, J. Chow, L.S. Davis, R.A. Glover, G.F. Graham, E.G. Gross, A. Krongrad, J.L. Jr Lesher, H.K. Park, B.B. Jr Sanders, C.L. Smith and J.R. Taylor-JR. *J.A.M.A.* 1996, **276**, 1957.

30. S.D. Mark, W. Wang, J.F. Jr Fraumeni, J.Y. Li, P.R. Taylor, G.Q. Wang, W. Guo, S.M. Dawsey, B. Li, W.J. Blot, *Am. J. Epidemiol.* 1996, **143**, 658.

31. S.D. Mark, W. Wang, J.F. Jr Fraumeni, J.Y. Li, P.R. Taylor, G.Q. Wang, S.M. Dawsey, B. Li, W.J. Blot, *Epidemiology.* 1998, **9**, 9.

32. E.M. Lonn, S. Yusuf, C.I. Doris, M.J. Sabine, V. Dzavik, K. Hutchison, W.A. Riley, J. Tucker, J. Pogue and W. Taylor, *Am. J. Cardiol.* 1996, **78**, 914.

33. J.E. Price and F.G. Fowkes, *Eur. Heart. J.* 1997, **18**, 719.

34. E.B. Rimm and M.J. Stampfer, *Curr. Opin. Cardiol.* 1997, **12**, 188.

35. S. Hercberg, P. Preziosi, S. Briancon, P. Galan, I. Triol, D. Malvy, A.M. Roussel and A. Favier, *Control. Clin. Trials.* 1998, **19**, 336.

36. C.H. Hennekens, *Nutrition.* 1998, **14**, 50.

ANTIATHEROGENICITY OF ANTIOXIDANTS AGAINST LDL OXIDATION

Michael Aviram.

The Lipid Research Laboratory, Technion Faculty of Medicine, The Rappaport Family Institute for Research in the Medical Sciences and Rambam Medical Center, Haifa, Israel.

1 INTRODUCTION

Deposition of oxidized lipids in the arterial wall is a characteristic of the atherosclerotic lesion and thus the oxidation of fatty acids (in phospholipids, triglycerides and cholesteryl ester), as well as that of cholesterol, are considered to be an important risk factor for atherosclerosis. Low density lipoprotein (LDL), the major cholesterol carrier in the human serum, becomes highly atherogenic when it is converted by oxidative stress to oxidized LDL (Ox-LDL). The atherogenicity of Ox-LDL in comparison to native LDL, involves its enhanced cellular uptake by arterial wall macrophages and smooth muscle cells,[1-5] as well as several other properties (Table 1).

Table 1

Atherogenicity of Oxidized LDL.

1.	Increases uptake by macrophages (foam cell formation).
2.	Acts as a chemoattractant for circulating monocytes.
3.	Inhibits the migration of tissue macrophages back to the circulation.
4.	Induces monocyte-macrophage differentiation.
5.	Acts as chemoattractant for T-lymphocytes.
6.	Cytotoxic to cells of the arterial wall.
7.	Alters gene expression of neighboring cells .
8.	Immunogenic, and can elicit autoantibody formation.
9.	Inhibits nitric oxide-stimulation of vasodilation.
10.	Enhances coagulation processes.
11.	Increases platelet activation.
12.	Stimulates smooth muscle cells proliferation.

2 MECHANISMS OF LDL OXIDATION

LDL oxidation by arterial wall cells involve the activation of several oxygenases including nicotinamide adenine dinucleotide phosphate (NADPH)-oxidase. Studies from our laboratory have shown[6] that LDL-induced NADPH oxidase activation under oxidative stress is required for macrophage-mediated oxidation of LDL (Fig. 1). A recent report demonstrated the localization of a constitutive active phagocyte-like NADPH oxidase in the rabbit aorta.[7]

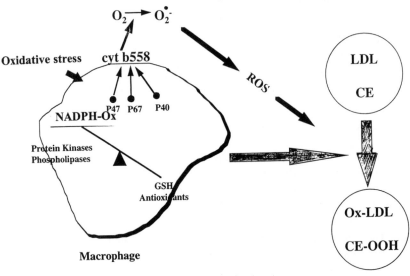

Figure 1

Possible role of macrophage NADPH oxidase in LDL oxidation. Oxidative stress can induce the translocation of the NADPH oxidased cytosolic components to the plasma membrane, where they are associated with cytochrome b-558 to produce the active enzyme. This active complex can produce superoxides from molecular oxygen, and under certain conditions the superoxides can be converted into more potent reactive oxygen species (ROS), which can oxidize LDL. Macrophage kinases and phospholipases are involved in the above pathway. The balance between NADPH oxidase activation and cellular antioxidants such as the glutathione (GSH) system determines the extent to which LDL is oxidized.

While the lipid peroxidation of LDL is under intensive investigation, little is known about lipid peroxidation of cells in the arterial wall. Macrophage lipid peroxidation (induced by iron ions, by angiotensin II or by deoxycholate) can induce cell-mediated oxidation of LDL.[8-10]

Our current knowledge on the mechanisms involved in oxidation of LDL by arterial wall macrophages is summarized in Fig. 2.

Atherogenesis: Blood and Arterial Wall Cells and Lipoproteins

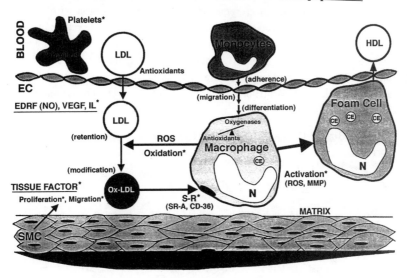

Figure 2

Macrophage foam cell formation during early atherogenesis. Monocyte derived macrophages can oxidize LDL which is retained in the arterial wall leading to the formation of Ox-LDL. Ox-LDL is taken up via the macrophage scavenger receptors (SR-A and CD-36) causing foam cell formation and cell activation. Arterial wall cells release various factors (EDRF, VEGF, IL, TF) as well as metalloproteinases (MMP), which affect the shape, size, composition and structure of the atherosclerotic lesion.

3 OXIDIZED LDL *IN VIVO*

The involvement of oxidized LDL in atherosclerosis is suggested by its presence in the atherosclerotic lesion. Oxidized LDL has been demonstrated in the lipoprotein fraction extracted from mice, rabbit and human atherosclerotic lesions.[11,12]

The demonstration of increased susceptibility to oxidation of LDL from patients with increased risk for atherosclerosis (hypertensives, hypercholesterolemics, diabetics, renal failure, xanthelasma), and from the atherosclerotic apolipoprotein E-deficient mice further supports the importance of LDL oxidation in the progression of atherosclerosis.[8] Drug therapy such as statins in hypercholesterolemia, or angiotensin II converting enzyme (ACE) inhibitors and antagonists in hypertension, results in a substantial reduction in LDL oxidizability.[8,13] In the atherosclerotic apo E deficient mice, such therapy was associated with about 70 % reduction in the progression of the atherosclerotic lesion formation.[8,14]

4 DIETARY ANTIOXIDANTS AND ATHEROSCLEROSIS

Epidemiological studies linking dietary antioxidant consumption with a reduction in cardiovascular diseases showed an inverse association between the intake of various fruits and vegetables and the mortality from coronary heart disease. Antioxidants were shown to attenuate atherogenesis in animal models, mainly by their free radicals scavenging capabilities. The serum lipid soluble, lipoprotein-associated antioxidants include vitamin E (α– and γ– tocopherols), carotenoids (mainly ß-carotene and lycopene) and also ubiquinol Q 10.

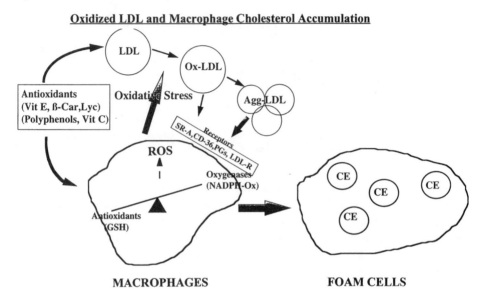

Oxidized LDL and Macrophage Cholesterol Accumulation

MACROPHAGES   **FOAM CELLS**

Figure 3
Dietary antioxidants (in the LDL and in the cells) can affect the conversion of native LDL into oxidized LDL (Ox-LDL) by macrophages [and also the subsequent conversion to aggregated LDL (Agg-LDL). The dietary antioxidants (vitamin E, ß-carotene, lycopene, polyphenolic flavonoids and vitamin C) can bind to LDL, accumulate in the serum or be taken up by arterial wall cells, including macrophages. Under increased oxidative stress Ox-LDL (and Agg-LDL) are formed, and can then be taken up by several lipoprotein receptors leading to cellular cholesterol accumulation and foam cell formation.

All these antioxidants can affect LDL oxidation (and its subsequent modifications, including aggregation) directly, and also indirectly via an effect on the arterial wall cells oxidative state and their subsequent capacity to oxidize LDL (Fig. 3). The balance between anti-oxidants and pro-oxidants in the cell, and in the lipoprotein determines the extent to which the lipoprotein is oxidized, and hence, its atherogenic potency (Fig. 4).

Oxidative Modifications of LDL

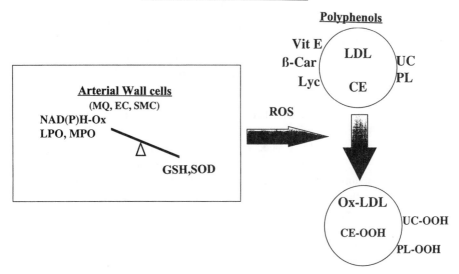

Figure 4

Role of the oxidative state in arterial wall cells and in LDL, in cell-mediated oxidation of LDL. Macrophages, endothelial cells, and smooth muscle cells (MQ, EC, SMC) oxidative state is determined by the balance between activities of oxygenases such as NAD(P)H oxidase, lipoxygenases (LPO), and myeloperoxidases (MPO) and antioxidants such as glutathione (GSH) and superoxide dismutase (SOD). LDL oxidative state is determined by the balance between its major oxidation substrate, cholesteryl ester (CE)'s and phospholipids (PL)'s polyunsaturated fatty acids, as well as it's unesterified cholesterol (UC) from one hand, and its assocated antioxidants from the other hand.

4.1. Vitamin E (tocopherols)

Vitamin E is a lipid soluble antioxidant, and most of it is LDL-associated in the plasma. Epidemiological studies have shown an association between high dietary intake, or high serum concentrations of vitamin E and lower rates of ischemic heart disease. Recently, our laboratory has shown that consumption of vitamin E by the atherosclerotic, apo-E deficient mice results in a 35 % reduction in their LDL oxidative state, and in parallel, a similar reduction was observed in the aortic lesion area.[15]

4.2. Vitamin C (ascorbic acid)

Vitamin C, unlike vitamin E, is a water soluble antioxidant and it is not bound to lipoproteins in the plasma. Vitamin C can spare vitamin E and thus, the interaction between these two vitamins (and probably also with other antioxidants such as carotenoids and polyphenolic flavonoids) is important in preventing oxidative stress. A prospective population study of men from eastern Finland recently revealed that low vitamin C plasma concentration was an important risk factor for myocardial infarction. Furthermore, vitamin C was also shown in Finnish men to be among the strongest antioxidants in human serum.[16]

4.3. Carotenoids

Carotenoids in serum are derived from plants and include ß-carotene, α-carotene, cryptoxanthin, lutein, zeaxanthin and lycopene. The most abundant carotenoids in human serum are ß-carotene (which gives carrots, apricots and peaches their color), and lycopene (which gives the red color in tomatoes). These carotenoids are lipid soluble and are potent singlet oxygen quenchers, with lycopene being most potent in this respect.

In comparison to other antioxidants, ß-carotene can inhibit LDL oxidation only to a limited degree. In diabetic patients, increased LDL oxidation state and propensity for oxidation, was significantly reduced (towards normal levels) following 3 weeks supplementation of the natural ß-carotene at a dose of 60 mg/d. Furthermore, in healthy subjects our laboratory has demonstrated an inhibitory effect on the susceptibility of LDL to oxidation by both ß-carotene and lycopene. However, we found that *not all subjects* responded to ß-carotene or lycopene supplementation by reduced *ex vivo* LDL oxidation. Comparison of the antioxidant status in "responders'" and "non-responders'" LDLs, revealed that the vitamin E content in the "responders' LDLs" was significantly higher than that found in the "non-responders' LDLs".[17] As a result of these findings, we analyzed the effect of carotenoids in combination with vitamin E, on the susceptibility of LDL to oxidation. An additive antioxidative effect against LDL oxidation was obtained when such combination of the carotenoids with vitamin E was used, instead of using the individual antioxidants separately .[17]

4.4. Polyphenolic flavonoids

Flavonoids are diphenylpropanes that commonly occur in plants, and are frequent components of the human diet. They include flavones, isoflavones and the 2,3-dihydroderivatives of flavone namely flavanones. Flavanones are tranformed to other flavonoids such as chalcones, anthocyanins and catechin.[18]

The Zutphen study clearly demonstrated an antiatherogenic property for dietary flavonoids.[19] Nutrient sources of potent antioxidants against LDL oxidation from the polyphenols family include plants such as the licorice root extract (rich with the isoflavan glabridin), red wine (rich with the grape skin's flavonol quercetin and the flavanol catechin), ginger, and grapefruit, orange, and pomegranate peels.[20] In olive oil, hydroxytyrosol was shown to contribute to it's inhibitory effect on LDL oxidation (in addition to a major effect of the oleic acid). LDL isolated from the plasma of ten normolipidemic subjects who were supplemented for a period of 2 weeks with 100 mg of licorice root extract per d, was more resistant to copper ion-induced oxidation, as well as to free radical (AAPH)-induced oxidation, by 44 % and 36 %, respectively, in comparison to LDL isolated before licorice supplementation. In E° mice, dietary supplementation of licorice (200 μg/d/mouse), or of pure glabridin (20 μg/d/mouse) for 6 weeks resulted in a 68 % and 42 % reduction in the susceptibility of their LDL to copper ion-induced oxidation, respectively. This treatment also resulted in a significant reduction in the atherosclerotic lesion area. Thus, glabridin, a polyphenol with lipophilic characteristics present in licorice ethanolic extract, is absorbed, binds to the LDL particle, and subsequently protects the LDL from oxidation in multiple modes of oxidative stress.[21]

The effect of consuming red wine with meals, on the propensity of plasma and LDL to lipid peroxidation was studied in healthy men. Red wine consumption reduced the propensity of the volunteers' LDL to copper ions-induced lipid peroxidation as determined by a 46 %, 72 % and 54 % decrement in the content of the lipoprotein-associated aldehydes, lipid peroxides, and conjugated dienes, respectively, as well as by a substantial prolongation of the lag phase required for the initiation of LDL oxidation. Thus, some phenolic substances that exist in red wine are absorbed, bind to plasma LDL and are responsible for the antioxidant properties of red wine.[22] Similar effects were recently shown for green and black tea flavonoids.[23] The effect of polyphenols-containing nutrients on LDL oxidation in humans, both *in vitro* and *in vivo*, is summarized in Table 2.

In E° mice that were supplemented with 50 μg of polyphenols/d/mouse for a period of 6 weeks, plasma LDL isolated after red wine or quercetin consumption, was less susceptible (by 50-80 %) to oxidation induced by either copper ions, by the free radical initiator AAPH, or by J-774 A.1 macrophages, in comparison to LDL isolated from placebo - treated E° mice. In parallel to these results, we found that the atherosclerotic lesion areas in E° mice that were treated with red wine or quercetin were significantly reduced, by 40 % or 38 %, in comparison to lesion areas in E° mice that were treated with placebo. Results of these studies suggest that dietary consumption by E° mice, of red wine or of its polyphenolic flavonoid quercetin, leads to reduced susceptibility of their LDL to oxidation and to attenuation in the development of atherosclerosis.[24]

Table 2

Effect of polyphenol-containing nutrients on LDL oxidation: in vitro and in vivo studies.

Nutrient	Polyphenols	Major Inhibition of LDL oxidation (%)	
		in vitro [*1]	*in vivo* [*2]
1. Red wine	Quercetin, Myricetin, Catechin	$90 \pm 7\%$	$44 \pm 5\%$
2. Olive oil	Hydroxytyrosol, Oleuropein	$72 \pm 8\%$	$28 \pm 5\%$
3. Licorice	Glabridine, Licochalcone	$92 \pm 6\%$	$44 \pm 8\%$

* 1. *LDL (200 μg of protein/ml) was incubated with $CuSO_4$ (10 μ mol/l) for 4 h at 37 °C in the presence of red wine (0.1 %, v/v), or olive oil (0.1 %, v/v) or 2 mg/L of licorice root ethanolic extract.*

*2. *LDL (200 μg of protein/ml) was obtained before and 2 weeks after the consumption of red wine (400 ml/d), olive oil (50 ml/d) or licorice (0.1 g/d), and then incubated with $CuSO_4$ (10 μ mol/l) for 4 h at 37 °C. LDL oxidation was then analyzed by the thiobarbituric acid reactive substances assay. LDL oxidation in control studies (without added nutrient) ranged between 24-36 nmoles of malondialdehyde equivalents/mg of LDL protein.*

LDL oxidation by arterial wall cells is affected not only by the lipoprotein-associated antioxidants, but also by the antioxidants which are present in the cells.

The effect of macrophage enrichment with some nutritional antioxidants on cell-mediated oxidation of LDL is illustrated in Fig. 5.

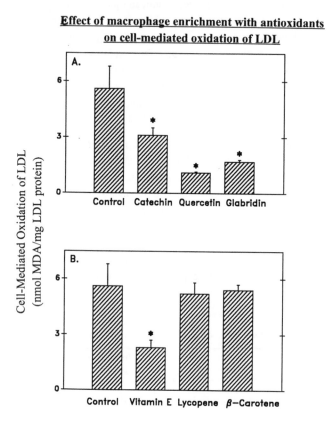

Effect of macrophage enrichment with antioxidants on cell-mediated oxidation of LDL

Figure 5

The effect of macrophage enrichment with nutritional antioxidants on cell-mediated oxidation of LDL. J-774 A.1 macrophages were incubated for 20 h at 37 °C with the indicated antioxidants concentrations: 20 µM for the polyphenols, 75 µM for vitamin E, and 50 µM for the carotenoids. LDL (100 µg of protein/ml).

4.5. Selenium and the glutathione system

Selenium is an essential trace nutrient that provides unique oxidation-reduction properties to selenoproteins. Glutathione peroxidase (GPx), a key enzyme in the glutathione system, contains an essential selenocysteine residue at its active site which catalyzes the reduction of peroxides by glutathione (GSH) and thus protects tissues against the damaging effects of peroxides. GSH is strictly conserved throughout all higher forms of aerobic life and plays an important role in reducing cellular and extracellular oxidative stress.[25]

Dietary selenium supplementation (1µg/d/mouse) to the atherosclerotic, apolipoprotein E deficient mice for a 6 months period increased GSH content and GPx activity in their peritoneal macrophages by 36 % and 30 % respectively. This effect was associated with a 46 % reduction in cell-mediated oxidation of LDL. The atherosclerotic lesion area in the aortas derived from these mice after selenium supplementation was found to be reduced by 30 %, compared to the lesion area found in non-treated mice.[26]

5 CONCLUSION

The strategies to reduce LDL oxidation which affects both cellular and lipoprotein oxidative states are summarized as follows:
1. Reduce cellular oxygenases (NADPH oxidase, myeloperoxidase, lipoxygenase).
2. Increase cellular antioxidants (glutathione, glutathione peroxidase, superoxide dismutase, Catalase).
3. Enrich LDL with monounsaturated fatty acid.
4. Enrich LDL environment with antioxidants (polyphenols, Vitamin E, ß-carotene, lycopene, Vitamin C).
5. Use angiotensin converting enzyme inhibitors and angiotensin II antagonists.
6. Increase serum paraoxonase (PON 1) activity.

References
1. Fuster V., Badimon L., Badimon J.J., et al.:The pathogenesis of coronary artery disease and the acute coronary syndrome (2). *N. Engl. J. Med.* 1992, **326**(5), 310-318.
2. Steinberg D.: LDL oxidation and its pathological significance. *J. Bio. Chem.* 1997, **272**, 20963-20966.
3. Aviram M.: LDL-platelet interaction under oxidative stress induces macrophage foam cell formation. *Thromb Haemost.* 1995, **74**(1), 560-564.
4. Brown M.S., Goldstein J.L.:A receptor-mediated pathway for cholesterol homeostasis. *Science.* 1986, **232**(4746), 34-47.
5. Aviram M.: Modified forms of low density lipoprotein and atherosclerosis. *Atherosclerosis.* 1993, **98**, 1-9.
6. Aviram M., Rosenblat M., Etzioni A. and Levy R.: Activation of NADPH oxidase is required for macrophage-mediated oxidation of low-density lipoprotein. *Metabolism.* 1996, **45**(9), 1069-1079.
7. Pagano P.J., Clark J.K., Cifuentes-Pagano M.E. et al.: Localization of a constitutively active, phagocyte-like NADPH oxidase in rabbit aortic adventitia: enhancement by angiotensin II. *Proc. Natl. Acad. Sci. USA.* 1997, **94**, 14483-14488.
8. Aviram M.: Interaction of oxidized low density lipoprotein with macrophages in atherosclerosis, and the antiatherogenicity of antioxidants. *Eur. J. Clin. Chem. Clin. Biochem.* 1996, **34**, 599-608.
9. Parthasarathy S., Santanam N.: Mechanisms of oxidation, antioxidants, and atherosclerosis. *Curr. Opin. Lipidol.* 1994, **5**(5), 371-375.
10. Stocker R.: Lipoprotein oxidation: mechanistic aspects, methodological approaches and clinical relevance. *Curr. Opin. Lipidol.* 1994, **5**(6), 422-433.

11.   Ylä-Herttuala S., Palinski W., Rosenfeld M.E. et al.: Evidence for the presence of oxidatively modified low density lipoprotein in atherosclerotic lesions of rabbit and man. *J. Clin. Invest.* 1989, **84**(4), 1086-1095.

12.   Aviram M., Maor I., Keidar S., et al.: Lesioned low density lipoprotein in atherosclerotic apolipoprotein E-deficient transgenic mice and in humans is oxidized and aggregated. *Biochem. Biophys. Res. Commun.* 1995, **216**(2), 501-513.

13.   Aviram M., Osamah H., Rosenblat M., et al.: Interactions of platelets, macrophages, and plasma lipoproteins in hypercholesterolemia: antiatherogenic effects of HMG-CoA reductase inhibitors therapy. *J.Cardiovasc. Pharmacol.* 1998, **31**, 39-45.

14.   Keidar S., Kaplan M., Shapira H., Brook J.G. and Aviram M.: Low density lipoprotein isolated from patients with essential hypertension exhibits increased propensity for oxidation and enhanced uptake by macrophages: a possible role for angiotensin II. *Atherosclerosis.* 1994, **104**, 71-84.

15.   Maor I., Hayek T., Coleman R. and Aviram M.: Plasma LDL oxidation leads to its aggregation in the atherosclerotic apolipoprotein E-deficient mice. 1996, **17**, 2995-3005.

16.   Nyyssönen K., Porkkala-Sarataho E., Kaikkonen J. and Salonen J.T.: Ascorbate and urate are the strongest determinants of plasma antioxidative capacity and seurm lipid resistance to oxidation in Finnish men. *Atherosclerosis.* 1997, **130**(1-2), 223-233.

17.   Fuhrman B., Ben-Yaish L., Attias J., Hayek T. and Aviram M.: Tomato's lycopene and β-carotene inhibit low density lipoprotein oxidation and this effect depends on the lipoprotein vitamin E content. *J. Nutr. Metab. Cardiovascular Diseases.* 1997, **7**, 433-443.

18.   Cao G., Sofic E. and Prior R.L.: Antioxidant and prooxidant behavior of flavonoids: structure-activity relationships. *Free Radic. Biol. Med.* 1997, **22**(5), 749-760.

19.   Hertog M.G., Feskens E.J., Hollman P.C., et al.: Dietary antioxidant flavonoids and risk of coronary heart disease: the Zutphen Elderly Study. *Lancet.* 1993, **342**(8878), 1007-1011.

20.   Aviram M. and Fuhrman B. :Polyphenolic flavonoids inhibit macrophage- mediated oxidation of LDL and attenuate atherogenesis. *Atherosclerosis.* 1998, in press.

21.   Fuhrman B., Buch S., Vaya J., et al.: Licorice extract and its major polyphenol glabridin protect low-density lipoprotein against lipid peroxidation: *in vitro* and *ex-vivo* studies in humans and in atherosclerotic apolipoprotein E-deficient mice. *Am. J. Clin. Nutr.* 1997, **66**, 267-275.

22.   Fuhrman B., Lavy A. and Aviram M.: Consumption of red wine with meals reduces the susceptibility of human plasma and low-density lipoprotein to lipidperoxidation. *Am. J. Clin. Nutr.* 1995, **61**, 549-554.

23.   Ishikawa T., Suzukawa M., Ito T., et al.: Effect of tea flavonoid supplementation on the susceptibility of low-density lipoprotein to oxidative modification. *Am. J. Clin. Nutr.* 1997, **66**(2), 261-266.

24.   Hayek T., Fuhrman B., Vaya J., et al.: Reduced progression of atherosclerosis in apolipoprotein E-deficient mice following consumption of red wine,or its polyphenols quercetin or catechin, is associated with reduced susceptibility of LDL to oxidation and aggregation. *Arterioscler. Thromb. Vasc. Biol.* 1997, **17**, 2744-2752.

25.   Cotgreave I.A. and Gerdes R.G.: Recent trends in glutathione biochemistry-glutathione-protein interactions: a molecular link between oxidative stress and cell proliferation? *Biochem. Biophys. Res. Commun*. 1998, **242**, 1-9.

26.   Rosenblat M. and Aviram M.: Macrophage glutathione content and glutathione peroxidase activity are inversely related to cell-mediated oxidation of LDL: *in vitro* and *in vivo* studies. *Free Radic. Bio. Med*. 1998, **24**(2), 305-317.

ARE ANTIOXIDANTS EFFECTIVE IN PRIMARY PREVENTION OF CORONARY HEART DISEASE?

D. Kromhout.

Division of Public Health Research, National Institute of Public Health and the Environment, P.O.Box 1, 3720 BA Bilthoven, The Netherlands.

1 INTRODUCTION

The importance of antioxidants in the etiology of coronary heart disease is heavily debated. The hypothesis that antioxidants play a role in the occurrence of athero-thrombotic complications like coronary heart disease, is attractive. Basic research has made clear that oxidised Low Density Lipoproteins (LDL) are of major importance in the development of atherosclerosis.[1] The LDL lipoprotein fraction can be oxidised by e.g. smoking. Oxidative damage of LDL lipoproteins can be prevented by nutritive e.g. vitamin E and non-nutritive antioxidants e.g. flavonoids.

In this paper a short overview will be given of the epidemiological evidence for the associations between dietary antioxidants and the occurrence of coronary heart disease. Thereafter the results of epidemiological studies on vegetables and fruits in relation to plasma antioxidants and coronary heart disease will be discussed. Finally the population perspective on antioxidants in primary prevention of coronary heart disease will be given.

2 ANTIOXIDANTS AND CORONARY HEART DISEASE

Research on antioxidants has traditionally focused on (pro-) vitamins with antioxidant properties e.g. tocopherols (Vitamin E), carotenoids (α-carotene) and vitamin C. Besides these so-called nutritive antioxidants there are also non-nutritive antioxidants. Non-nutritive substances are compounds in foods without nutritional value. Recently it became clear that especially plant foods contain numerous compounds with strong antioxidant properties e.g. polyphenols. Important examples are flavonoids e.g. quercetin that is present in onions and apples and catechins e.g. epigallocatechin that is present in tea. The associations between (non)-nutritive antioxidants and CHD will be summarised.

In large epidemiological studies blood samples are stored for future analyses of blood parameters. In this context associations are studied in so-called nested case-control studies. In these studies antioxidant concentrations are measured retrospectively in stored blood samples of persons who developed CHD and in controls. This is an efficient study design because only the blood samples of the persons who developed the disease and a limited number of controls have to be analysed.

In the Netherlands and Finland three nested case-control studies have been carried out studying the association between vitamin E levels in blood and cardiovascular disease.[2]

No association was observed. In a case–control study carried out in Scotland plasma vitamin E was inversely associated with angina pectoris in limited multivariate analyses.[3] However, when in the multivariate model also the linoleic acid content of adipose tissue and the eicosapentaenoic acid content of platelets was taken into account the association between plasma vitamin E and angina pectoris was no longer statistically significant. Also no association was observed between the α-tocopherol concentration in adipose tissue and myocardial infarction in the multicentre EURAMIC case-control study.[4]

In five prospective studies inverse associations were observed between vitamin E intake and CHD risk.[5-9] However in three studies this association was noted for supplement users[5-7] and in two studies for the vitamin E content in the diet.[8,9] This means that the results of these cohort studies suggest a protective effect of vitamin E on CHD risk. However it is unclear what level of intake provides protection.

In the CHAOS trial the hypothesis was tested whether vitamin E supplementation among cardiac patients could prevent the risk of re-infarction and mortality.[10] There was a reduction in non- fatal myocardial infarction in the patients supplemented with vitamin E. However no association was observed with respect to hard endpoints e.g. cardiovascular and all cause mortality. Similar results in relation to hard endpoints were observed in the ATBC trial carried out in Finland among 30,000 smokers.[11] These results suggest that vitamin E supplementation is not effective in preventing hard endpoints e.g. cardiovascular and all cause mortality.

In the EURAMIC case-control study an inverse association was found between ß-carotene concentration in adipose tissue and myocardial infarction.[4] This association was however confined to smokers. In four prospective studies no association between ß-carotene intake and CHD was observed.[6,8,9,12] An exception was the inverse association between ß-carotene intake and CHD in smokers as observed in the Health Professionals Follow-up Study.[6]

Three large intervention trials were carried out, one in Finland and two in U.S.A., in order to study the effect of ß-carotene supplementation on lung cancer and other causes of death.[11,13,14] These trials took place among smokers, asbestos workers and physicians. In spite of the hypothesised protective effect, these studies showed a detrimental effect of ß-carotene supplementation on lung cancer. In the ATBC and CARET trial ß-carotene supplementation was also associated with a higher cardiovascular and all cause mortality. In the Physical Health Trial significant associations were not observed between ß-carotene supplementation and cardiovascular and all cause mortality. The data of the Physicians Health Trial suggest that the effect of ß-carotene supplementation on myocardial infarction may be related to smoking status. An odds ratio of 0.88 (95 % CI 0.72-1.07) was observed among non- smokers and of 1.08 (95 % CI 0.80-1.48) among smokers.

In 6 prospective studies the association between vitamin C intake and CHD was studied.[6,7,8,9,12,15] No association was observed. An exception was the inverse association between vitamin C intake and CHD among non-smokers noted in the Western Electric Study.[12] Controlled trials on the association between vitamin C intake and CHD have not been carried out.

In 1993 we published a paper showing an inverse relation between the intake of flavonols and the 5-year mortality from coronary heart disease in the Zutphen Elderly Study.[16] This association was also present when the follow-up period was extended to 10 years.[17] Besides in the Netherlands this association was also studied in three prospective studies carried out in Finland, USA and Wales.[18,19,20,30-32] An inverse association was also observed in the study from Finland but not in those from the USA and Wales. It can be concluded that the association between flavonol intake and CHD is promising but more data are needed before definite conclusions on this association can be drawn.

If we try to summarise the results on the associations between the intake of different antioxidants and CHD risk we can conclude that:
- Vitamin E is related to CHD risk in prospective cohort studies but not in case-control studies and intervention trials.
- β-carotene is related to CHD in case-control studies only in smokers but not in prospective cohort studies and intervention trials.
- Vitamin C is not related to CHD risk in prospective cohort studies.
- Flavonols are related to CHD risk in some but not all prospective cohort studies.

3 FRUITS AND VEGETABLES, ANTIOXIDANTS AND CORONARY HEART DISEASE

Fruits and vegetables are important sources of carotenoids and vitamin C but not of vitamin E. The latter is mainly present in oils, margarines and nuts. A recent quantitative review concludes that in cohort studies the risk of coronary heart disease is 15 % lower at the 90 th centile of fruit and vegetable consumption compared with the 10 th centile.[21] This is equivalent to about a four fold increase in fruit consumption and a doubling of vegetable consumption. In this context it is of importance to know what the effect of such a change in fruit and vegetable consumption on the plasma level of antioxidants is.

In New Zealand a randomised controlled trial was carried out during eight weeks to study the effect of an increase of 4.7 servings of fruits and vegetables per day on plasma antioxidant levels.[22] This increase of 630 g/d was due to a 177 g/d increase in fruit, 341 g/d in juice and 104 g/day in vegetables. Such an increase of fruits and vegetables was associated with significant increases in vitamin C, α- and β-carotene and a non-significant change in α-tocopherol.

The next question to be addressed is whether an increase in fruit and vegetable consumption may reduce coronary heart disease mortality.[23] This question was addressed in a secondary prevention trial carried out in India. The cardiac patients in this trial increased their fruits and vegetable intake by about 400 g/d. They also decreased their saturated fat intake by about 3 % of energy. These changes in diet were associated with a 42 % reduction in cardiac mortality and 54 % reduction in sudden cardiac death.

These results suggest that an increase in the consumption of fruits and vegetables is associated with and increase in plasma carotenoids and vitamin C concentration. Several cohort studies and one intervention trial showed that an increased consumption of fruits and vegetables of at least 400 g/d is associated with a lowering of coronary heart disease risk.

Whether this is an effect of an increase in carotenoids and vitamin C or an effect of other bioactive compounds present in fruits and vegetables is not known.

4 ANTIOXIDANTS AND CHD FROM A PUBLIC HEALTH PERSPECTIVE

Within Europe large differences exist in mortality from coronary heart disease. Historically there was a marked North-South gradient with high rates in the North and low rates in the South. Over the last two decades CHD mortality rates decreased in most Northern European countries and the differences between Northern and Southern Europe became less pronounced.[24] Nowadays the highest CHD mortality rates are observed in Eastern Europe. From a public health point of view it is necessary to know what the major determinants of population CHD mortality rates are. If these determinants are known preventive measures can be taken in order to reduce the burden of CHD mortality in populations.

The question why CHD mortality rates differ between populations was asked in the 1950's by Prof. Ancel Keys, University of Minnesota, Minneapolis, Minnesota. He carried out pilot studies in the US, Europe and Japan in the early 1950's.[25] In 1958 he started the Seven Countries Study, a population based survey in 16 cohorts of men originally aged 40-59 years. Between 1958 and 1964 12,763 men were examined. These men were re-examined after 5 and 10 years of follow-up and were followed for mortality during 25 years. During that period about 6,000 men died and the underlying cause of death was adjudicated in a standardised way. About 1,500 men died from CHD.

Information on smoking habits was collected in all men in a standardised way in the baseline survey. Detailed dietary data were collected in small samples of the cohorts.[26] The seven days record was used as dietary survey method and duplicate portions of the foods eaten were collected and chemically analysed for fatty acids. In 1986 it was decided to recode the original dietary data of the 16 cohorts in a standardised way. These data were summarised in 16 major food groups and the average consumption for each cohort was calculated.[27] In 1987 food composites were collected representing the average food intake of the 16 cohorts in the 1960's. These food composites were chemically analysed and different nutritive and non-nutritive substances were determined.

Univariate analyses showed that the intake of saturated fatty acids was the strongest determinant of 25-year CHD mortality rates of the 16 cohorts.[28] In univariate models the dietary antioxidants vitamin E, β-carotene and vitamin C were not related to long-term CHD occurrence.[29] However, flavonoids were inversely related to 25-year CHD mortality.[30] No association was observed between cigarette smoking and CHD mortality, due to the high smoking and low CHD mortality rates in the two Japanese cohorts. However, after multivariate analysis saturated fat intake and cigarette smoking were positively related with CHD mortality and flavonoid intake inversely. These results suggest that populations characterized by diets low in saturated fat and high in flavonoids and low smoking rates have low mortality rates from CHD. In general terms this means that a diet low in animal foods and hard margarines and high in plant foods is protective in relation to CHD mortality.

The results from this cross-cultural study are consistent with the hypothesis that dietary antioxidants may protect against CHD. However, at the present time taking the results from cross-cultural, case-control, cohort and intervention studies into account it is not clear which antioxidants are responsible for this possible protective effect. The results on α-tocopherol and flavonoids are promising but far from definitive. It is therefore too early to give public health recommendations for specific dietary antioxidants. However, a general recommendation to increase the intake of fruits and vegetables to at least 400 g/day as given by WHO seems justified based on the available evidence.[31]

References

1. D. Steinberg, S. Parthasarathy, T. E. Carew, J. C. Khoo and J. L. Witzum. Beyond cholesterol. Modifications of low-density lipoprotein that increase its atherogenicity. *N. Engl. J. Med.* 1989, **320**, 915.
2. F. J. Kok. Antioxidants, lipid peroxides and coronary heart disease. *Neth. J. Cardiol.* 1992, **2**, 52.
3. R. A. Riemersma, D. A. Wood, C. C. A. MacIntyre, R. A. Elton, K. F. Gey and M. F. Oliver. Risk of angina pectoris and plasma concentrations of vitamins A, C, and carotene. *Lancet* 1991, **337**, 1.
4. A. F. M. Kardinaal, F. J. Kok, J. Ringstad et al. Antioxidants in adipose tissue and risk of myocardial infarction: the EURAMIC study. *Lancet* 1993, **342**, 1379.
5. M. J. Stampfer, C. H. Hennekens, J. E. Manson, G. A. Colditz, B. Rosner and W. C. Willett. Vitamin E consumption and the risk of coronary disease in women. *N. Engl. J. Med.* 1993, **328**, 1444.
6. E. B. Rimm, M. J. Stampfer, A. Ascherio, E. Giovanucci, G. A. Colditz and W. C. Willett. Vitamin E consumption and the risk of coronary disease in men. *N. Engl. J. Med.* 1993, **328**, 1450.
7. K. G. Losonczy, T. B. Harris and R. J. Havlik. Vitamin E and vitamin C supplement use and risk of all-cause and coronary heart disease mortality in older persons: the Establised Populations for Epidemiologic Studies of the Elderly. *Am. J. Clin. Nutr.* 1996, **64**, 190.
8. L. H. Kushi, A. R. Folsom, R. J. Prineas, P. J. Mink, Y. Wu and R. M. Bostick. Dietary antioxidant vitamins and death from coronary heart disease in postmeopausal women. *N. Engl. J. Med.* 1996, **334**, 1156.
9. P. Knekt, A. Reunanen, R. Järvinen, R. Seppänen, M. Heliövara and A. Aromaa. Antioxidant vitamin intake and coronary mortality in a longitudinal population study. *Am. J. Epidemiol.* 1994, **139**, 1180.
10. N. G. Stephens, A. Parsons, P. M. Schofield, et al. Randomised controlled trial of vitamin E in patients with coronary disease: Cambridge Heart Antioxidant Study (CHAOS). *Lancet* 1996, **347**, 781.
11. The Alpha-Tocopherol, Beta Carotene Cancer Prevention Study Group. The effect of vitamin E and beta carotene on the incidence of lung cancer and other cancers in male smokers. *N. Engl. J. Med.* 1994, **330**, 1029.
12. D. K. Pandey, R. Shelleke, B. J. Selwyn, C. Tangney and J. Stamler. Dietary vitamin C and β-carotene and risk of death in middle -aged men. The Western Electric Study. *Am. J. Epidemiol.* 1995, **142**, 1269.

13. G. S. Omenn, G. E. Goodman, M. D. Thornquist, et al. Effects of combination of beta carotene and vitamin A on lung cancer and cardiovascular disease. *N Engl J Med* 1996, **334**, 1150.

14. C. H. Hennekens, J. E. Buring, J. E. Manson, et al. Lack of effect of long-term supplementation with beta carotene on the incidence of malignant neoplasms and cardiovascular disease. *N. Engl. J. Med.* 1996, **334**, 1145.

15. C. R. Gale, C. N. Martyn, P. D. Winter, and C. Cooper. Vitamin C and risk of death from stroke and coronary heart disease in cohort of elderly people. *Brit. Med.J.* 1995, **310**, 1563.

16. M. G. L. Hertog, E. J. M. Feskens, P. C. H. Hollman, M. B. Katan and D. Kromhout. Dietary antioxidant flavonoids and risk of coronary heart disease. The Zutphen Elderly Study. *Lancet* 1993, **342**, 1007.

17. M. G. L. Hertog, E. J. M. Feskens and D. Kromhout. Antioxidant flavonols and coronary heart disease risk. Letter. *Lancet* 1997, **349**, 699.

18. P. Knekt, R. Järvinen, A. Reunanen and J. Maatela. Flavonoid intake and coronary mortality in Finland: a cohort study. *Brit. Med.J.* 1996, **312**, 478.

19. E. B. Rimm, M. B. Katan, A. Ascherio, M. J. Stampfer and W. C. Willett. Relation between intake of flavonoids and risk for coronary heart disease in male health professionals. *Ann. Intern. Med.* 1996, **125**, 384.

20. M. G. L. Hertog, P. M. Sweetman, A. M. Fehily, P. C. Elwood and D. Kromhout. Antioxidant flavonoids and ischaemic heart disease in a Welsh population of men. The Caerphilly Study. *Am. J. Clin. Nutr.* 1997, **65**, 1489.

21. M. R. Law and J. K. Morris. By how much does fruit and vegetable consumption reduce the risk of ischaemic heart disease? *Eur. J. Clin. Nutr.* 1998, **52**, 549.

22. S. Zino, M. Skeaff, S. Williams and J. Mann. Randomised controlled trial of effect of fruit and vegetable consumption on plasma concentrations of lipids and antioxidants. *Brit. Med. J.* 1997, **314**, 1787.

23. R. B. Singh, S. S. Rastogi, R. Verma, et al. Randomised controlled trial of a cardioprotective diet in patients with a recent acute myocardial infarction: results of one year follow-up. *Brit. Med.J.* 1992, **304**, 1015.

24. S. Sans, H. Kesteloot and D Kromhout D on behalf of the Task Force. The burden of cardiovascular diseases mortality in Europe. Task force of the European Society of Cardiology on Cardiovascular Mortality and Morbidity Statistics in Europe. *Eur. Heart J.* 1997, **18**, 1231.

25. D. Kromhout, A. Menotti and H. Blackburn (Eds). The Seven Countries Study. A scientific adventure in cardiovascular disease epidemiology. Brouwer Offset b.v., Utrecht, 1994.

26. A. Keys (Ed). Coronary heart disease in seven countries. *Circulation* 1970, **41**(Suppl.1), 1.

27. D. Kromhout, A. Keys, C. Aravanis, et al. Food consumption patterns in the nineteen sixties in Seven Countries. *Am. J. Clin. Nutr.* 1989, **49**, 889.

28. D. Kromhout, A. Menotti, B. Bloemberg, et al. Dietary saturated and trans fatty acids, cholesterol and 25-year mortality from coronary heart disease. The Seven Countries Study. *Prev. Med.* 1995, **24**, 308.

29. D. Kromhout, B. P. M.Bloemberg, E. J. M. Feskens, M. G. L. Hertog, A. Menotti,

and H. Blackburn, for the Seven Countries Study Group. Alcohol, fish, fiber and antioxidant vitamins do not explain population differences in coronary heart disease mortality. *Int. J. Epidemiol.* 1996, **25**, 753.

30.   M. G. L. Hertog, D. Kromhout, C. Aravanis, et al. Flavonoid intake and long-term risk of coronary heart disease and cancer in the Seven Countries Study. *Arch. Intern. Med.* 1995, **155**, 381.

31.   Report of a WHO Study Group. Diet, nutrition, and the prevention of chronic diseases. World Health Organisation Technical report Series 797, Geneva, 1990.

REDOX IMBALANCE AND ANTIOXIDANT SIGNALLING

John M. C. Gutteridge.

Oxygen Chemistry Laboratory, AICU, Royal Brompton and Harefield NHS Trust, Sydney Street, London, SW3 6NP, UK.

1 REDOX BALANCE BETWEEN ANTIOXIDANTS AND PRO-OXIDANTS

In normal health, there is a balance between the formation of oxidising chemical species and their effective removal by protective antioxidants. Antioxidants are a diverse group of molecules with diverse functions. For example, they range from large highly specific proteinaceous molecules with catalytic properties to small lipid-and water-soluble molecules with non-specific scavenging or metal-chelating properties (reviewed in 1). We might, therefore, define an antioxidant as a substance which, when present at low concentrations compared to those of the oxidisable substrate, significantly delays or inhibits oxidation of that substrate.[2] Antioxidants control the prevailing relationship between reducing or oxidising (redox) conditions in biological systems. Such control offers two major advantages, (1) the ability to remove toxic levels of oxidants before they damage critical biological molecules and (2) the ability to manipulate changes, at the subtoxic level, of molecules that can function as signal, trigger or messenger carriers.[3,4]

Plasma and tissue antioxidants operate at primary, secondary and tertiary levels mainly as constitutive molecules. However, it is becoming increasingly clear that inducible antioxidants upregulated by oxidative stress can also play key roles in body protection.

The author considers 'primary' defences to be those which prevent radical formation. The iron-binding properties of transferrin and lactoferrin fulfill such roles in extracellular fluids, since iron correctly attached to their high-affinity binding sites no longer catalyses radical formations. 'Secondary' defences are those which remove, or inactivate, formed reactive oxygen species (ROS). In some cases they may be enzymes such as the superoxide dismutases (SOD),[6,7] catalase[8] and glutathione peroxidase, or low molecular mass molecules such as vitamin E, ascorbate and glutathione (GSH). 'Tertiary' defences operate to remove and repair oxidatively damaged molecules and are particularly important for DNA.

1.1. Cellular Antioxidant defences

Oxygen metabolism occurs within cells and it is here that we expect to find antioxidants evolved to deal speedily and specifically with reactive oxidants or reductants and, as previously mentioned, these are likely to be proteinaceous catalysts, ie enzymes.

The great advantage to the cell of having constitutive catalysts as antioxidants is that, like all catalysts, they are not consumed during their normal functioning.

1.2. Membrane Antioxidant defences

Within the hydrophobic lipid interior of membranes, different types of lipophilic radicals are formed to those seen in the intracellular aqueous milieu. Lipophilic radicals require different types of antioxidants for their removal. Vitamin E (α-tocopherol), a fat-soluble vitamin, is a poor antioxidant outside a membrane bi-layer but is extremely effective when structurally incorporated into the membrane.[9] Lipid soluble antioxidants are, therefore, extremely important in protecting membrane polyunsaturated fatty acids (PUFAs) from undergoing autocatalytic free radical chain reactions known as lipid peroxidation. The peroxidation of a PUFA leads to its destruction and the formation of a plethora of oxidation products such as lipid peroxides, carbonyls and carboxylic acids. Many of these products are biologically reactive and may be used as signal molecules by the body.[10,11]

The way in which a membrane is structured from its lipids appears to play an important role in decreasing its susceptibility to oxidative damage. Thus, structural integrity requires that the correct ratios of phospholipid and cholesterol are present as well as the correct phospholipids and their fatty acid side chains.[9]

1.3. Extracellular Antioxidant defences

Blood is an effective antioxidant 'buffering' system with both plasma and red blood cells offering their own specialised protective systems. Red blood cells (RBCs) have an anion channel through which $O_2^{\cdot-}$ can enter the cell and be destroyed by CuZnSOD. Hydrogen peroxide is an uncharged molecule, which behaves much like water and can be destroyed by RBC catalase and glutathione peroxidase. Unlike plasma, RBCs contain high levels of GSH. Plasma contains both primary and secondary antioxidants and is usually a powerful inhibitor of free radical reactions (reviewed in 2). The question "what is the most important plasma antioxidant?" is often asked. To this, there is no simple answer since the experimental findings will change as the pro-oxidant used to bring about detectable oxidation is changed. Thus, vitamin E is reported to be the most important, and one of the least important, plasma antioxidants, depending on the different reaction conditions used.[12,13] When considering iron-binding proteins as primary antioxidants, it should be emphasised that they evolved for the conservation and transportation of iron in the body. An essential, but secondary, requirement is that, whilst doing so, they do not allow iron to express its powerful pro-oxidant properties.[1,2]

Plasma also contains a large number of low molecular mass molecules, many of which are redox active, that have been ascribed secondary antioxidant roles as non-specific scavengers of free radicals. These are consumed during their reactions with radicals examples being, uric acid, vitamin E, ascorbic acid and bilirubin (reviewed in 14). Several of these antioxidants are important vitamins which has, in the past, led to over-simplistic clinical intervention trials being designed with the aim of reversing life-threatening diseases.

2 SIGNALLING BY REDOX CONTROL

The oxidation-reduction (redox) potential of biological ions or molecules is a measure of their tendency to lose an electron (thereby being oxidised), and is expressed as E_0 in volts.

The more strongly reducing an ion or molecule, the more negative is its E_0. It is becoming increasingly clear that control of redox balance by antioxidants plays an important role in cellular signalling.

2.1. Reactive Iron Species (RIS)

Within cells there normally exists a pool of low molecular mass redox active iron which is essential for the synthesis of iron-requiring enzymes and proteins, and for the synthesis of DNA. This pool of iron is the target of iron chelators and is also a form of iron sensed by iron regulatory proteins. The amount, and nature of the ligands attached to this iron, remain unknown. However, a recently introduced fluorescence assay based on calcein may enhance our knowledge of intracellular iron pools.[15] In contrast to the intracellular environment, extracellular compartments do not require, or normally contain a low molecular mass iron pool. Iron-binding proteins such as transferrin and lactoferrin do not even remotely approach iron saturation in healthy subjects, indeed they retain a considerable iron-binding capacity and are able to remove mononuclear forms of iron that enter extracellular fluids. The differences between intracellular and extracellular, compartments and their requirements for low molecular mass iron deserves special comment since it is iron in this form that is the most likely catalyst of biological free radical reactions. Inside the cell low molecular mass iron need not pose a serious threat as a free radical catalyst, because the cell has specific defenses to safely and speedily remove all the $O_2{}^{\cdot-}$ and H_2O_2 and organic peroxides (such as lipid peroxides) that could react with such iron. This is achieved by intracellular enzymes such as the superoxide dismutases, catalase, and glutathione peroxidase and possibly also by thioredoxin-dependent H_2O_2 removal systems. In the extracellular space, however, we see a different pattern of protection against free radical chemistry. Here, proteins bind, conserve, transport and recycle iron and whilst doing so keep it in non- or poorly-reactive forms that do not react with H_2O_2 or organic peroxides. Proteins such as transferrin and lactoferrin bind mononuclear iron, whereas haptoglobins bind haemoglobin[16] and haemopexin binds haem.[17] In addition, plasma contains a ferrous ion oxidising protein (ferroxidase) called caeruloplasmin.[18] By keeping iron in a poorly reactive state, molecules such as $O_2{}^{\cdot-}$, H_2O_2, $NO^{\cdot}$, and lipid peroxides can survive long enough to perform important and useful functions as signal, trigger and intercellular messenger molecules.[3,4] During situations of iron-overload plasma transferrin can become fully loaded with iron (100 % iron saturation) and allow low molecular mass iron to accumulate in the plasma. Such iron, when present in micromolar concentrations, can bind to various added chelating agents such as EDTA, desferrioxamine and bleomycin that cannot abstract iron from transferrin. This non-transferrin bound iron can be associated with several ligands including citrate[19] and other organic acids, and possibly albumin. Low molecular mass ligands for iron inside the cell are also a subject of considerable debate. ATP, ADP, GTP, pyrophosphates, inositol phosphates, aminoacids and polypeptides have all been proposed.

In 1981 the author and his colleagues introduced the "Bleomycin Assay" as a first attempt to detect and measure chelatable redox active iron that could participate in free radical reactions.[20] The assay procedure is based on the ability of the metal-ion binding glycopeptide antitumour antibiotic bleomycin to degrade DNA in the presence of an iron salt, oxygen and a suitable iron reducing agent. Data obtained using the bleomycin assay in extracellular fluids for levels of RIS have recently been confirmed by quantitating such iron by its ability to activate the enzyme aconitase.[21]

2.2. Signalling through Antioxidant control of Redox Balance

It has been suggested that certain reactive oxygen (ROS), nitrogen (RNS) and (RIS), might be used as "signal, messenger and trigger molecules".[3,4,22] We are seeing increasing examples of redox regulation of gene expression, not only oxyR and NFkB but also the role of thioredoxin and of AP-1. If the redox balance is considered to play a pivotal role in signalling all functions, it then becomes highly unlikely that the body would allow short-term changes in antioxidants to influence this balance, perhaps explaining the poor record of antioxidant therapies to date.

2.2.1. Intracellular Iron Signalling. Cells normally accumulate iron via the binding of transferrin to high affinity surface receptors (TfR) followed by endocytosis. There is also a transferrin-independent pathway of cellular iron uptake that is said to involve a ferri-reductase and an Fe^{II} transmembrane transport system (reviewed in 23). When non-transferrin bound iron appears in plasma, due to iron-overload or lack of transferrin (apotransferrinaemia), it is rapidly cleared by the membrane-bound transport system constitutively present on parenchymal cells of organs, particularly those of liver, heart, pancreas and the adrenals. This latter system does not require endocytosis of a protein for iron delivery. The rate of synthesis of TfR and ferritin is regulated at the post-transcriptional level by cellular iron and co-ordinated by the iron-dependent binding of cytosolic proteins called "the iron responsive element binding proteins "(IRE-BP)" (now known as iron regulatory proteins, IRP) which bind to specific sequences on their mRNA.[24] It appears that low molecular mass iron is capable of acting as a signal to regulate ferritin and TfR synthesis in this way. Recent work has shown that IRP-I is identical to the cytosolic enzyme aconitase.[25] The protein functions as an active aconitase when it has an Fe-S cluster present or as an RNA-binding protein when iron is absent. Switching between these two forms depends on cellular iron-status.

2.2.2. Membrane Signalling. When a cell is damaged, or dies, it is highly likely that its membrane lipids will undergo peroxidation.[26] Tissue damage releases RIS and activates enzymes which catalyse peroxidation of polyunsaturated fatty acids, particularly linoleic acid, leading to a build-up of lipid peroxides.[27] Peroxidation of membrane polyunsaturated fatty acids produces a plethora of reactive primary peroxides and secondary carbonyls and it was suggested many years ago by the author that lipid oxidation products such as these resulting from cell death could act as triggers for new cell growth.[10] Through the detailed work of Hermann Esterbauer and colleagues [28] we now have clearer insights into the biological reactivity of lipid oxidation products. 4-Hydroxy-2-nonenal (HNE), a peroxidation product of (n-6) fatty acids (when RIS are present), is a potent trigger for chemotaxis, can inactive thiol-containing molecules, and activate certain enzymes (reviewed in 28). As a general rule low levels of ROS, and possibly reactive carbonyls, activate cellular process whilst higher levels turn them off. The resting cell normally has a redox potential with a reduced state, and is progressively activated as oxidation increases. Too much oxidation deposes all function[29] until eventually apoptosis or necrosis is triggered.

2.2.3. Extracellular Signalling. Human body extracellular fluids contain little, or no, catalase activity and extremely low levels of superoxide dismutase. Glutathione peroxidases in both selenium-containing and non-selenium-containing forms are present in plasma but there is little glutathione substrate in plasma (1-2 μM). "Extracellular" superoxide dismutases (EC-SOD) have been identified[7] and shown to contain copper, zinc and attached carbohydrate groups. By allowing the limited survival of $O_2^{\cdot-}$, H_2O_2 lipid peroxides (LOOH) and possibly other ROS/RNS in extracellular fluids the body can utilise these molecules, and others such as nitric oxide (NO·), as useful messenger, signal or trigger molecules.[3] A key feature of such a proposal is that $O_2^{\cdot-}$, H_2O_2, LOOH, NO· and HOCl do not meet with reactive iron or copper and that extracellular antioxidant protection has evolved to keep iron and copper in poorly or non-reactive forms.[3,30]

The major copper-containing protein of human plasma is caeruloplasmin, unique for its intense blue coloration. This protein's ferroxidase activity makes a major contribution to extracellular antioxidant protection by decreasing ferrous salt-driven lipid peroxidation and Fenton chemistry.[18]

3 IMPLICATIONS

When redox imbalance occurs in the body during illness, there are two major implications: **1)** Toxic levels of ROS, RNS and RIS can be produced which lead to molecular damage of key molecules in cells, and **2)** Production at sub-toxic levels can signal changes in cellular responses such as proliferation, apoptosis and eventually necrosis. Until the relationships between constitutive and inducible antioxidants are better understood, it seem unlikely that most current antioxidant therapies regimens will make major contributions to lifethreatening diseases.

References

1. Gutteridge JMC, Halliwell B. Antioxidants in Nutrition, Health and Disease. Oxford University Press: *Oxford*. 1994.
2. Halliwell B, Gutteridge JMC. Free Radicals in Biology and Medicine. Oxford University Press: *Oxford*. 1998, 3rd Edition.
3. Halliwell B, Gutteridge JMC. Oxygen free radicals and iron in relation to biology and medicine: some problems and concepts. *Arch. Biochem. Biophys.* 1986, **246**, 501-14.
4. Saran M, Bors W. Oxygen radicals acting as chemical messengers: A hypothesis. *Free Rad. Res. Commun.* 1989, **7**, 213-20.
5. Gutteridge JMC, Paterson SK, Segal AW, Halliwell B. Inhibition of lipid peroxidation by the iron-binding protein lactoferrin. *Biochem. J.* 1981, **199**, 259-61.
6. McCord JM, Fridovich I. Superoxide dismutase. An enzymatic function for erythrocuprein (hemocuprein). *J. Biol. Chem.* 1969, **24**, 6045-55.
7. Marklund SL, Holme E, Hellner L. Superoxide dismutase in extracellular fluids. *Clin. Chem. Acta.* 1982, **126**, 41-51.
8. Chance B, Sies H, Boveris A. Hydroperoxide metabolism in mammalian organs. *Physiol. Rev.* 1979, **59**, 527-605.

9.   Gutteridge JMC. The membrane effects of vitamin E, cholesterol and their acetates on peroxidative susceptibility. Res. Commun. Chem. Path. Pharmacol. 1978, **22**, 563-71.

10.   Gutteridge JMC, Stocks J. Peroxidation of cell lipids. *J. Med. Lab. Sci.* 1976, **53**, 281-85.

11.   Esterbauer H, Schaur RJ, Zollner H. Chemistry and biochemistry of 4-hydroxynonenal, malonaldehyde and related aldehydes. *Free Rad. Biol. Med.* 1991, **11**, 81-128.

12.   Burton GW, Joyce A, Ingold KU. First proof that vitamin E is major lipid soluble, chain-breaking antioxidant in human blood plasma. *Lancet.* 1982, 327.

13.   Stocks J, Gutteridge JMC, Sharp RJ, Dormandy TL. The inhibition of lipid auto-oxidation by human serum and its relation to serum proteins and α-tocopherol. *Clin. Sci. Mol. Med.* 1974, **47**, 223-33.

14.   Krinsky NI. Mechanism of action of biological antioxidants. *Proc. Soc. Exp. Biol. Med.* 1992, **200**, 248-54.

15.   Cabantchik ZI, Glickstein H, Milgram P, Breuer W. A fluorescence assay for assessing chelation of intracellular iron in a membrane model system and in mammalian cells. *Anal. Biochem.* 1996, **233**, 221-7.

16.   Gutteridge JMC. The antioxidant activity of haptoglobins towards haemoglobin stimulated lipid peroxidation. *Biochem. Biophys. Acta.* 1987, **917**, 219-23.

17.   Gutteridge JMC, Smith A. Antioxidant protection by haemopexin of haem-stimulated lipid peroxidation. *Biochem. J.* 1988, **256**, 261-5.

18.   Gutteridge JMC, Richmond R, Halliwell B. Oxygen free radicals and lipid peroxidation: Inhibition by the protein caeruloplasmin.*FEBS Lett.* 1980, **112**, 269-72.

19.   Grootveld M, Bell JD, Halliwell B, Aruona OI, Bomford A, Sadler P.J. Non-transferrin-bound iron in plasma or serum from patients with idiopathic hemochromatosis. *J. Biol. Chem.* 1989, **264**, 4417-22.

20.   Gutteridge JMC, Rowley DA, Halliwell B. Superoxide-dependent formation of hydroxyl radicals in the presence of iron salts. Detection of "free" iron in biological systems by using bleomycin-dependent degradation of DNA. *Biochem. J.* 1981, **199**, 263-5.

21.   Mumby S, Koizumi M. Taniguchi N, Gutteridge JMC. Reactive iron species in biological fluids activate the iron-sulphur cluster of aconitase. *Biochem. Biophys. Acta.* 1998, **1380**, 102-8.

22.   Schreck R, Albermann K, Baeuerle PA. Nuclear factor Kappa B: an oxidative stress-responsible transcription factor of eukaryotic cells (a review). *Free Rad. Res. Commun.* 1992, **17**, 221-37.

23.   De Silva DM, Askwith CC, Kaplan J. Molecular mechanisms of iron uptake in Eukaryotes. *Physiol. Rev.* 1996, **76**, 31-47.

24.   Klausner RD, Rouault TA, Harford JB. Regulating the fate of mRNA: The control of cellular iron metabolism. *Cell.* 1993, **72**, 19-28.

25. Kennedy MN, Mende-Mueller L, Blondon GA, Beinert H. Purification and characterisation of cytosolic aconitase from beef liver and its relationship to the iron-responsive element binding protein (IRE-BP). *Proc. Natl. Acad. Sci. USA.* 1992, **89**, 11730-4.

26. Halliwell B, Gutteridge JMC. Lipid peroxidation, oxygen radicals, cell damage and antioxidant therapy. *Lancet.* 1984, **1**, 1396-7.

27. Herold M, Spiteller G. Enzymic production of hydroperoxides of unsaturated fatty acids by injury of mammalian cells. *Chem. Phys. Lipids.* 1996, **79**, 113-21.

28. Esterbauer H, Schaur RJ, Zollner H. Chemistry and biochemistry of 4-hydroxynonenal, malonaldehyde and related aldehydes. *Free Rad. Biol. Med.* 1991, **11**, 81-128.

29. Burdon RH. Superoxide and H_2O_2 in relation to mammalian cell proliferation. *Free Rad. Biol. Med.* 1994, **18**, 775-94.

30. Gutteridge JMC. Signal, messenger and trigger molecules from free radical reactions, and their control by antioxidants. In: Packer L, Wirtz K, eds. Signalling Mechanisms- from transcription Factors to Oxidative Stress. Springer-Verlag: *Berlin.* 1995, 157-64.

INVOLVEMENT OF MITOCHONDRIA IN OXIDATIVE STRESS IN AGING OF INTACT CELLS

Jose Viña, Juan Sastre, Federico V. Pallardó and Ana Lloret.

Departamento de Fisiología, Facultad de Medicina, Avenida Blasco Ibáñez 17, 46010 Valencia, Spain.

1 MITOCHONDRIA ARE DAMAGED WITHIN INTACT CELLS OF OLD ANIMALS:

Reactive oxygen species (ROS), a term used for oxygen free radicals and peroxides, are generated continuously in aerobic cells and especially in the mitochondrial respiratory chain. Indeed, 1-2 % of all oxygen used by mammalian mitochondria in state 4 does not form water but oxygen activated species (ROS).

Experiments from several laboratories have shown that the rate of ROS production, and not merely the rate of oxygen consumption, appears to determine the maximal lifespan potential.

The findings mentioned above support the hypothesis that mitochondrial damage plays a key role in the aging process. However, most mitochondrial changes were found in experiments using isolated mitochondria. Thus, some of these effects could be due to an increased susceptibility of old mitochondria to the stress caused by the isolation procedure. Moreover, when intact cells were not used, the mitochondrial-cytosolic interactions were also ignored. Thus, the use of isolated mitochondria might be misleading for aging studies. Hence, whole cells, such as isolated hepatocytes, should be considered an excellent model for studies on mitochondrial aging.

Using isolated hepatocytes, we measured the rate of biochemical pathways which critically depend on mitochondrial function. Gluconeogenesis from lactate plus pyruvate, but not from glycerol or fructose, fell with aging in isolated hepatocytes. Gluconeogenesis from lactate involves mitochondria, whereas it does not from glycerol or fructose. The lower rate of gluconeogenesis from lactate plus pyruvate is due to an impaired transport of malate across mitochondrial membrane using the dicarboxylate carrier. Furthermore, post-transcriptional modifications appear to be involved in the agerelated impairment of such carrier, since its gene expression does not change with age. Oxidised proteins accumulate with cellular aging and oxidative post-traductional modifications have been reported as the main cause for the loss of some enzyme activities, such as that of malic enzyme. These findings support the hypothesis that an increased generation of oxygen free radicals may be responsible for the decline in the activity of mitochondrial membrane proteins, such as metabolite carriers. In fact, it is known that exposure of mitochondria to free radicals causes impairment of the mitochondrial inner-membrane proteins. Ketogenesis from oleate -which depends on mitochondrial performance- also decreased in hepatocytes from old animals.

Nevertheless, the fact that some mitochondrial membrane carriers are impaired upon aging does not necessarily imply that all mitochondrial functions are affected by aging. For instance, the rate of urea synthesis in hepatocytes does not change with age.

Hence, metabolic studies using whole cells show that aging affects mitochondrial function by an impairment of specific processes which involve the mitochondrial inner membrane, such as malate transport. Post-transcriptional modifications appear to be involved in the age-related loss of these carriers. These changes are likely to be due to chronic oxidative stress associated with mitochondrial aging. Metabolic pathways which do not involve mitochondria are not affected. All these findings point out the role of mitochondria as key targets of the cellular changes which occur upon the aging.

Another way of studying mitochondrial aging within intact cells of old animals is by flow cytometry, which allows a non-invasive analysis of individual cells. Thus, the effect of aging on liver mitochondrial membrane potential has been studied by flow cytometry. Aging of the liver causes a significant decrease in mitochondrial membrane potential. This may reduce the energy supply in old hepatocytes since the mitochondrial membrane potential is the driving force for ATP synthesis. Furthermore, this impairment in mitochondrial function may affect mitochondrial protein synthesis and might be involved in the age-related decrease in the level of mitochondrial transcripts.

In agreement with the age-related decrease in mitochondrial membrane potential, changes in respiratory activity have also been reported. Moreover, we have found an impairment in energization of brain and liver mitochondria under respiratory state 4.

Studies in isolated mitochondria have shown that acute oxidative stress causes an inhibition of mitochondrial respiration, which affects the mitochondrial membrane potential. Moreover, hyperoxia reduces the mitochondrial membrane potential in microvascular cells. Hence, the oxidative stress associated with aging may be responsible, at least in part, for the age-related impairment in mitochondrial membrane potential and respiratory activity. Indeed, intracellular peroxide levels increase with age in whole cells, which correlates with parallel changes in peroxide generation by isolated mitochondria. It is likely that the accumulation of peroxides in whole cells upon aging comes from the continuous peroxide generation by mitochondria throughout the cell life, although we cannot rule out that other structures, such as peroxisomes, may also have a role.

On the other hand, mitochondrial morphology is important because changes in mitochondrial ultrastructure modulate mitochondrial function. Indeed, volume-dependent regulation of matrix protein packing modulates metabolite diffusion and, in turn, mitochondrial metabolism. Enlargement, matrix vacuolization and altered cristae have been evidenced in mitochondria from old animals by flow cytometry and electron microscopy. Moreover, structural complexity of brain and liver mitochondria increased upon aging. Alterations of mitochondrial crests which occur in old mitochondria may be responsible for the age-related impairment in mitochondrial membrane potential that we have found.

In spite of changes in mitochondrial size, the number of mitochondria per cell and the mitochondrial inner membrane mass did not change with age in liver. The age-related increase in mitochondrial size can be explained by an increase in mitochondrial matrix volume, without changes in inner mitochondrial membrane mass. In fact, the electron microscopy studies showed a higher degree of vacuolization in liver mitochondria from old rats than in controls.

It is well known that acute oxidative stress causes mitochondrial swelling. Thus, age associated chronic oxidative stress may be the cause, at least in part, of mitochondrial swelling. Furthermore, a correlation between changes in mitochondrial morphology and function seems to occur upon aging.

2 OXIDATION OF MITOCHONDRIAL GLUTATHIONE IN AGING

The involvement of glutathione metabolism in aging has been known since the work of Pinto and Bartley in the late sixties. Later, different authors found age-related glutathione oxidation in several animal models and even in humans. Glutathione oxidation in aging can be due to an increased production of oxidative species, a decreased antioxidant capacity, or both factors.

So far, glutathione compartmentation during aging has not received sufficient attention. Mitochondria cannot synthesize GSH because they do not have γ-glutamylcysteine synthetase or glutathione synthetase activities. Thus, mitochondria obtain GSH by transport from the cytosol. We have recently found that glutathione oxidation increases in mitochondria from liver, kidney and brain of old rats. It is worth noting that this increase was much higher in mitochondria than in the whole tissue. These results support the idea that mitochondria are a major source of free radicals in aging as suggested by different authors, and emphasize the relevance of mitochondria as primary targets of damage associated with aging.

3 AGE-ASSOCIATED DAMAGE TO MITOCHONDRIAL DNA CORRELATES WITH OXIDATION OF MITOCHONDRIAL GLUTATHIONE

Mitochondrial DNA (mtDNA) is especially susceptible to oxidative damage and mutation because it lacks protective histones or effective repair systems. Indeed, levels of oxidative damage to mtDNA are several times higher than those in nuclear DNA, and mtDNA mutates several times more frequently than nuclear DNA.

Ames and coworkers calculated that oxygen free radicals are responsible for approximately 10,000 DNA base modifications per cell and per day. DNA-repair enzymes are able to remove most of these oxidative lesions, but not all of them. Thus, unrepaired oxidative lesions in DNA -such as 8-oxo-7,8-dihydro-2'-deoxyguanosine- accumulate with age. Most of this damage occurs in mitochondrial DNA, not in nuclear DNA.

The age-associated increase in the level of common deletions produced spontaneously in the absence of inherited cases seems to be very low (<0.1%) and may not be significant. Nevertheless, these deletions may represent only a small portion of the numerous deletions and point mutations which could accumulate with age. Indeed, several studies have found increased deletions, point mutations and aberrant forms in mtDNA of postmitotic tissues upon aging. Furthermore, since mtDNA has no introns, any mutation affects a DNA coding sequence. Thus, it was suggested that mitochondrial DNA mutations may be important contributors to aging and neurodegenerative diseases.

Oxidative damage to proteins and DNA should not be considered separately, because they can potentiate each other. Thus, accumulation of inactive forms of DNA repairing enzymes might enhance the accumulation of DNA oxidative damage and viceversa. Moreover, a loss of repairing enzymes leads to an increased spontaneous mutation rate when oxidative lesions to guanine residues are present. Therefore, oxidative lesions in DNA exhibit mutagenic potential. On these bases, oxidative damage to DNA appears to be not only involved in cell aging, but also in the pathogenesis of associated diseases, such as cancer.

Mitochondrial glutathione plays a key role in the protection against oxidative damage associated with aging. Indeed, oxidative damage to mitochondrial DNA which occurs upon aging is directly related to oxidation of mitochondrial glutathione. Similarly, glutathione oxidation may also be correlated with the oxidative damage to mitochondrial lipids and proteins related to aging. A change in the glutathione redox status would indicate that mitochondrial antioxidant systems cannot cope with the oxidant species generated throughout the cell life. Therefore, glutathione oxidation may occur prior to oxidative damage to other mitochondrial components, and it might be an early event in oxidative stress associated with mitochondrial aging. This points out the importance of maintaining a reduced glutathione status to protect cells against oxidative damage of important molecules such as DNA.

4 ANTIOXIDANTS PREVENT AGE-ASSOCIATED OXIDATIVE DAMAGE TO MITOCHONDRIA

The free radical theory of aging proposed by Harman is especially attractive because it provides a rationale for intervention, i.e. antioxidant administration may slow the aging process. Indeed, certain impairments associated with aging can be prevented by antioxidant administration. In 1979, Miquel and Economos were the first to show that administration of thiazolidine carboxylate increases the vitality and mean life span of mice. Later, Furukawa et al. reported that oral administration of glutathione protects against the age-associated decline in immune responsiveness. More recently, we found that administration of thiol-containing antioxidants protects against the age-associated glutathione depletion in mouse tissues as well as partially preventing the age-related decline in neuromuscular coordination. These antioxidants also increased the mean lifespan of *Drosophila*. In addition, we have found that some antioxidants, such as thiazolidine carboxylate derivatives or vitamins C and E, protect against mitochondrial glutathione oxidation and mtDNA oxidative damage associated with aging. Recently, we have also found that EGb 761, a standardized *Ginkgo biloba* extract with antioxidant properties, prevents the age-related changes in mitochondrial function and morphology by protecting against the age-associated oxidative damage to mitochondria. Hence, the beneficial effects of antioxidant treatment on physiological performance could be associated with the prevention of age-related oxidative stress.

The importance of an adequate dietary intake of antioxidants is corroborated by the fact that a decline in antioxidant defense, such as that obtained when dietary ascorbate is insufficient, is associated with oxidative stress evidenced by increased oxidative damage to DNA.

When using thiol-containing antioxidants such as GSH or N-acetyl cysteine, the dose must be chosen carefully to enhance the cellular antioxidant defense without increasing intracellular cysteine levels, which may be pro-oxidantive. Free cysteine undergoes spontaneous auto-oxidation and hence, it may cause cell damage. Treatment of rats with high doses of N-acetyl cysteine -administered i.p.- causes a decrease in hepatic GSH levels. Cysteine accumulates intracellularly because it is not used in glutathione synthesis when this is inhibited by physiological levels of GSH. Therefore, the dose of thiol antioxidants must be chosen carefully to prevent an excessive increase in intracellular cysteine levels.

Recent research shows that vitamin C, which can be prooxidant *in vitro* in the presence of heavy metals, acts as an antioxidant in plasma even in the presence of high levels of free iron.

The use of antioxidant supplementation in aging still requires further research, especially epidemiological studies in humans. Recent epidemiological studies suggest that dietary supplementation with antioxidants should not be applied to certain groups of the population. In fact, it was found that administration of β-carotene to smokers increases the incidence of cancer in these patients. On the other hand, vitamin E administration may increase the acute response of neutrophils during exercise in old patients. Therefore, administration of antioxidants to patients must be done carefully and should not be recommended to the overall population.

In conclusion, administration of some, but not all antioxidants may prevent the oxidative stress and the physiological impairment associated with aging. Nevertheless, further studies on dietary supplementation with antioxidants need to be carried out to establish the adequate doses which provide beneficial effect.

References

1.   Barja G., Cadenas S., Rojas C., Pérez-Campo R., and López-Torres M. Low mitochondrial free radical production per unit 02 consumption can explain the simultaneous presence of high longevity and high aerobic metabolic rate in birds. *Free Rad. Res.* 1994, **21**, 317-328.

2.   Boveris A., Oshino N., and Chance B. The cellular production of hydrogen peroxide. *Biochem. J.* 1972, **128**, 617-630.

3.   Chance B., Sies H., Boveris A. Hydroperoxide metabolism in mammalian organs. *Physiological Rev.* 1979, **59**, 527- 604.

4.   García de la Asunción J., Millán A., Plá R., Bruseghini L., Esteras A., Pallardó F.V., Sastre and J., Viña J. Mitochondrial glutathione oxidation correlates with age-associated oxidative damage to mitochondrial DNA. *FASEB J.* 1996, **10**, 333-338.

5.   Harman D. Aging: a theory based on free radical and radiation chemistry *J. Gerontol.* 1956, **11**, 298- 300.

6.   Hazelton G. A. and Lang C. A. Glutathione contents of tissues in the aging mouse.

Biochem. J. 1980, **188**, 25-30.

7. Johns D.R. Mitochondrial DNA and disease. *New Engl. J. Med.* 1995, **333**, 638-644.

8. Miquel J., Economos A.C., Fleming J., Johnson J.E. Jr. Mitochondrial role in cell aging. *Exp.Gerontol.* 1980, **15**, 579-91.

9. Orr W.C.and Sohal R.S. Extension of life-span by overexpression of superoxide dismutase and catalase in Drosophila melanogaster. *Science.* 1994, **263**, 1128-30 .

10. Sastre J., Millán A., García de la Asunción J., Plá R., Juan G., Pallardó F.V., O'Connor E., Martín J.A., Droy-Lefaix M.T., Viña J. A Ginkgo biloba extract (EGb 761) prevents mitochondrial aging by protecting against oxidative stress. *Free Rad. Biol. Med.* 1997, **24** (2), 298-304.

11. Sastre J., Pallardó F.V., Plá R., Pellín A., Juan G., O' Connor E., Estrela J.M., Miquel J. and Viña J. Aging of the liver: Age-associated mitochondrial damage in intact hepatocytes. *Hepatology.* 1996, **24**, 1199-1205.

12. Shigenaga M.K., Hagen T.M. and Ames B.N. Oxidative damage and mitochondrial decay in aging. *Proc. Natl. Acad. Sci.USA.* 1994, **91**, 10771-8.

13. Viña J., Sastre J., Anton V., Bruseghini L., Esteras A., Asensi M. Effect of aging on glutathione metabolism. Protection by antioxidants. In Free radicals and aging (Emerit I. and Chance B. eds.) pp. 136-144, Birkhauser Verlag. Basel. 1992.

MELATONIN, NITRIC OXIDE AND ALZHEIMER'S DISEASE - A MINIREVIEW

Debomoy K. Lahiri.

Laboratory of Molecular Neurogenetics, Institute of Psychiatric Research, Department of Psychiatry, Indiana University School of Medicine, 791 Union Drive, Indianapolis, IN–46202, USA.

1 INTRODUCTION

The indoleamine hormone melatonin (N-acetyl-5-methoxytryptamine) is synthesized from serotonin after a two-step biochemical sequence mediated by the enzymes N-acetyltransferase and hydroxyindole O-methyltransferase.[1] Melatonin is secreted in mammals during the dark phase of the circadian cycle. Melatonin participates in the regulation of seasonal responses to changes in the length of day in seasonally breeding mammals.[2] Melatonin is also produced by the mammalian retina and at physiological levels it inhibits Ca^{2+}-dependent dopamine release in rabbit retina. Melatonin has been implicated in the regulation of various neural and endocrine processes that are synchronized with the daily changes in photoperiod. Melatonin administered orally to humans has been used successfully to treat jet lag and some circadian-based sleep disorders. The physiological function of the pineal gland is to transform a neurochemical input, represented by noradrenaline released by the post-ganglionic sympathetic terminals, into a hormonal output constituted by melatonin.[3] Although direct light does not reach the pineal body, light inhibits melatonin synthesis through an indirect and complex pathway involving the eyes, the accessory optic tracts, SCN and the pineal sympathetic terminals.[4,5]

2 MELATONIN AND AGING

Melatonin has been suggested to be an important regulator of aging and senescence.[6] The level of melatonin in the pineal gland decreases progressively with age such that in old animals and elderly humans the level of melatonin available to the organism is a fraction of that of young individuals.[7] The age-associated decline in the production of melatonin is believed to be a consequence of an increased oxidation of its precursors. For example, increased oxidation of pineal serotonin was suggested to be responsible for reduced melatonin synthesis in the aging Djungarian hamster (*Phodopus sungorus*).[8] Aging may be as a result of accumulated free radical-mediated damage. If the drop in melatonin normally occurring in aging animals could be prevented, perhaps the aging process would also be delayed.[7] Also, supplemental administration of melatonin may be beneficial in delaying age-related degenerative conditions.

3 MELATONIN, DEMENTIA AND ALZHEIMER'S DISEASE

Melatonin may be involved in various neuropsychiatric disorders. From the measurement of the level of serum melatonin and the duration of sleep it is suggested that sleep and behavior disorders in dementia are related to decreases in the amplitude of the sleep-wake rhythm and decreases in the level of melatonin secretion.[9,10] As compared to the age-matched controls, demented patients have decreased hippocampal norepinephrine and serotonin, increased hippocampal monoamine oxidase, and decreased REM sleep.[11] Melatonin deficiency can explain these observations. Chrono-neuroendocrinological aspects of physiological aging and senile dementia have been studied.[12] Daily variation in the concentration of melatonin was analyzed in the human pineal gland with respect to the age of patients and their status of affectedness with AD.[13] On measuring melatonin in human pineals (38 controls, 16 subjects with AD) they observed that this daily variation disappeared in both the older subjects (55-92 years) and those with AD (55-89 years). In addition, the suprachiasmatic nucleus, a center for the circadian clock in the mammalian brain, undergoes significant alterations during normal aging and in AD.[14] Alzheimer's disease patients exhibit irregularities in the patterns of normal circadian (daily) rhythms. Alzheimer-type pathology has been reported in the hypothalamus and in the suprachiasmatic nuclei (SCN), the putative site of the circadian oscillator.[14,15] Moreover, a widespread axonal degeneration was also observed in the optic nerves of AD patients.[16] It was hypothesized that chronic deficiency of melatonin can cause dementia with loss of dreams.[11] Senile dementia of the Alzheimer's disease type is a result of a dysfunction of the hippocampus as suggested from neuroanatomical, neuropathological and clinical evidence.

4 MELATONIN AND ßAPP SECRETION

There is a report of a disruption in circadian regulation by brain grafts that overexpress the amyloid beta-peptide (Aß).[17] Aß is derived from a larger precursor protein termed the beta-amyloid precursor protein (ßAPP). Secreted derivatives of ßAPP (sAPP) lacking the cytoplasmic tail, transmembrane domain and a small portion of the extracellular domain generated by the proteolytic processing of full-length ßAPP have been detected in the conditioned media of several cell cultures, in human plasma and in cerebrospinal fluid.[18] In an alternate endosomal/ lysosomal pathway, proteolytic cleavages of ßAPP occur at the amino- and carboxyl-termini of the Aß sequence by 'ß'- and 'γ'-secretases, respectively, to generate soluble Aß. This pathway generates an intact Aß peptide, and thus is potentially amyloidogenic. Cultured cells secrete Aß and sAPP during normal metabolism. A relationship between melatonin, presumably added exogenously to compensate the declining level of melatonin in AD, and the metabolism of ßAPP using a cell culture system has recently been studied.[19]

Treatment with melatonin resulted in a decrease in the secretion of sAPP in the conditioned medium of PC12 and neuroblastoma cells.[19] The results from the 'reversible' experiment indicate that the treatment of cells with melatonin did not lead to a permanent change in the metabolism of ßAPP at least in these cells and melatonin was effective on the release of sAPP for as long as it was present in the cells. The possibility of whether melatonin modulates proteolytic degradation of ßAPP and/or the cholinergic receptor was also studied. The results from different drug(s)-combination experiments suggest that the effect of melatonin on the secretion of sAPP was not affected by cotreating cells with either major lysosomal proteinase inhibitors or a muscarinic agonist (Lahiri, D.K.- unpublished observations).

5 MELATONIN AND ßAPP MESSAGE

The possibility of melatonin altering the expression of the level of the ßAPP gene was studied in two cell lines. Northern blotting experiments indicate that melatonin altered the levels of three housekeeping genes such as ßAPP, ß-actin and GAPDH genes in only PC12 cells and not in the human neuroblastoma cells.[19] The decrease in the level of ßAPP message will result in a reduction in the levels of ßAPP in PC12 cells. These data taken together suggest that treatment with melatonin might reduce the pool of both amyloidogenic as well as non-amyloidogenic derivatives of ßAPP, which may influence the process of beta-amyloid deposition.

6 MELATONIN AS A LYSOSOMOTROPIC AGENT

The mechanism(s) by which melatonin regulates the secretion of sAPP is not fully understood. The function of melatonin acting as a weak base suggests that melatonin may exert lysosomotropic action leading to a decrease in the release of sAPP.[19] The lysosomotropic agent prevents the acidification of intracellular vesicles and disrupts the pH-dependent step. For example, the hormone can be taken up by cellular acidic compartments such as lysosomes; this could possibly dissipate the pH gradient there and affect the processing of ßAPP. Melatonin may accumulate in the acidic compartment of cells. At present the fate of full-length ßAPP is unclear in the melatonin-treated cells; however the proteolytic degradation of ßAPP by the lysosomal pathway or by other pathways cannot be ruled out. There is a possibility that uncleaved ßAPP molecules in treated cells are reinternalized from the cell surface to lysosomes for further degradation.

7 MELATONIN AS A FREE RADICAL SCAVENGER

The cellular and molecular actions of melatonin are poorly understood.[20] The pineal hormone melatonin is a highly efficient free radical scavenger and general antioxidant. It quenches the most toxic and damaging free radical produced in the organism, the hydroxyl radical. Melatonin acts as an intracellular scavenger of hydroxyl- and peroxyl-free radicals when administered at pharmacological doses both *in vitro* and *in vivo*.[7] However, these antioxidant effects require concentrations of melatonin of about 10^6-fold greater than the physiological concentration.[20] The potentiating effect of melatonin on NGF-mediated neuritic differentiation observed can be attributed to its antioxidant property. The antioxidant effects of melatonin may not mediate through physiologically relevant high affinity receptors, though the involvement of low affinity nuclear receptors of the retinoid Z receptor family was proposed.[21,22] The effect of melatonin on the processing of ßAPP could not be due to its effect on the melatonin receptor because of the high concentration of melatonin used here.[19] There is a possibility that massive amounts of melatonin may lead to pharmacological effects of the hormone on other hormone regulatory systems (e.g., direct action on estrogen receptors). Melatonin exerts specific neurobiological effects through high affinity receptors.[20] In vitro autoradiography has localized high affinity melatonin receptors within individual brain nuclei and in a few nonneuronal sites. These receptor sites appear to be the sites through which melatonin elicits its biological effects.[23,24] Since we observed the effect of a high concentration of melatonin on ßAPP, the involvement of a low affinity melatonin receptor or other receptors cannot be ruled out. At present the level of the melatonin receptor in PC12 and neuroblastoma cell lines is not known. However melatonin, at least in vertebrates, need not rely on the cell membrane receptor for its activity because of the fact that melatonin is highly lipophilic.

8 MELATONIN AND NITRIC OXIDE

Nitric oxide (NO), a biologically active unstable radical, is synthesized by NO synthase (NOS) when converting L-arginine to L-citrulline. We have investigated whether the treatment of cultured cells with melatonin could possibly reduce the release of NO (D.K. Lahiri and C. Ghosh-unpublished data). When neuroblastoma cells were treated with a NO donor such as sodium nitroprusside (SNP), a significant level of NO was detected in a time and dose-dependent manner in the conditioned medium as compared to the untreated cells or SNP-containing media. A significantly high level of NO release resulted a decrease in levels of sAPP in the conditioned medium of SNP-induced cells. Increased levels of NO were toxic to the cells. In neuroblastoma cells, the release of NO as mediated by SNP was significantly inhibited by treatment with melatonin. These results suggest that an elevated release of NO/ free radicals makes the cell more vulnerable to injury and that melatonin treatment could reduce the toxic effects of NO.

9 MELATONIN AND NEURONAL DIFFERENTIATION

Nerve growth factor, a neurotrophic factor for basal forebrain cholinergic neurons, is considered to be of benefit in the treatment of neurodegenerative diseases of humans. NGF interacts with a cell surface receptor on responsive neurons to initiate a series of cellular events leading to neuronal survival and/or differentiation. Melatonin can prevent the oxidation of NGF and thus increase its stability because melatonin is a highly efficient free radical scavenger and general antioxidant. Oxidation of a 2-3 tryptophan residue of NGF is known to destroy biological and immunological activity.[25] Antioxidants, such as ascorbic acid, that can prevent peroxide-mediated oxidation of recombinant human NGF have recently been described. Our results suggest that the antioxidant property of melatonin can be attributed to its potentiating effect on NGF-mediated neuronal differentiation observed here in PC12 cells. This result is consistent with the recent suggestion that growth factor (such as NGF) therapy for free-radical-mediated disease may require antioxidants (such as vitamin E) in order to be effective.[26]

It can be speculated that melatonin helps to promote the process of differentiation in sympathetic neurons and maintain them in a differentiated state. Such a role of melatonin as a differentiation-promoter can be greatly inhibited in AD due to the reduced level of melatonin in AD subjects, which can possibly be restored by adding melatonin. This suggestion is in agreement with recent reports showing that melatonin can prevent the death of neuroblastoma cells exposed to the amyloid beta-peptide and the endogenous melatonin may play a neuroprotective role.[27,28] However the extent to which we can extend our cell culture observations with *in vivo* experimental paradigms and preparations remains to be seen.

In conclusion, our present results suggest that melatonin can reduce the secretion of ßAPP in neuroblastoma and PC12 cells. Treatment with melatonin resulted in a significant decrease in expression of ßAPP message. The effect of melatonin on ßAPP may not be mediated via actions at the major melatonin receptor. The role of melatonin as a free-radical scavenger can be utilized to potentiate NGF-mediated differentiation. The study of these interactions between melatonin, free radicals and nitric oxide is relevant in the prevention and possibly treatment in relation to Alzheimer's disease.

Acknowledgements

These studies were supported by a grant from the NIH (PHS R03AG14882-01) and I sincerely thank Drs. C. Ghosh, W. Song and E. Richelson.

References

1.   J. Axelrod. *Science.* 1974, **184**, 1341-1348.
2.   J. Arendt. Melatonin and the mammalin pineal gland. Chapman and Hall, London 1995.
3.   J. Arendt and J. Broadway. *Chronobiol. Int.* 1987, **4**, 273-282.
4.   A.J. Lewy, T.A. Wehr, F.K. Goodwin, D.A. Newsome and S.P. Markey. *Science.* 1980, **210**, 1267-1269.
5.   S.M. Reppert, D.R. Weaver, S.A. Rivkees and E.G. Stopa. *Science.* 1988, **242**, 78-81.
6.   W. Pierpaoli and W. Regelson. *Proc. Natl. Acad. Sci. USA.* 1992, **91**, 787-791.
7.   R.J. Reiter. *Experiment. Gerontol.* 1995, **30**, 199-212.
8.   A. Lerchl. *Neurosci. Letts.* 1994, **176**, 25-28.
9.   P.N. Prinz, P.P. Vitaliano, M.V. Vitiello, J. Bokan, M. Raskind, E. Peskind and C. Gerber. *Neurobiol. Aging.* 1982, **3**, 361-70.
10.   K. Mishima, M. Okawa, Y. Hishikawa, S. Hozumi, H. Hori and K. Takahashi. *Acta Psychiatri. Scand.* 1994, **89**, 1-7.
11.   C.P. Maurizi.*Medical Hypotheses* 1987, **24**, 59-68.
12.   D. Dori, G. Casale, S.B. Solerte, M. Fioravanti, G. Migliorati, G. Cuzzoni, and E. Ferrari. *Chronobiol.* 1994, **21**, 121-126.
13.   D.J. Skene, B. Vivien-Roels, D.L. Sparks, J.C. Hunsaker, P. Pevet, D. Ravid and D.F. Swaab. *Brain Res.* 1990, **528**, 170-4.
14.   D.F. Swabb, E. Fliers, T.S. Partiman. *Brain Res.* 1985, **342**, 37-44.
15.   W. Witting, I.H. Kwa, P. Eikelenboom, M. Mirmiran and D.F. Swaab. *Biol. Psychiat.* 1990, **27**, 563-72.

16.   D.R. Hinton, A.A. Sadun, J.C. Blanks and C.A. Miller. *New Engl. J. Med.* 1986, **315**, 485-487.

17.   B. Tate, K.S. Aboody-Guterman, A.M. Morris, E.C. Walcott, R.E. Majocha and C.A. Marotta. *Proc. Natl. Acad. Sci. USA.* 1982, **89**, 7090-7094.

18.   D.J. Selkoe. *Science.* 1997, **275**, 630-631.

19.   W. Song and D.K. Lahiri. *J. Mol. Neurosci.* 1997, **9**, 75-92.

20.   S.M. Reppert and D. Weaver. *Cell.* 1995, **83**, 1059-1062.

21.   M. Becker-Andre, I. Wiesenberg, N. Schaeren-Wiemers, E. Andre, M. Missbach, J.H. Saurat and C. Carlberg. *J. Biol. Chem.* 1994, **269**, 28531-28534.

22.   I. Wiesenberg, M. Missbach, J.P. Kahlen, M. Schrader and C. Carlberg. *Nucl. Acids Res.* 1995, **23**, 327-333.

23.   M.L.Dubocovich. *Trends Pharmacol.Sci.* 1995, **16**, 50-56.

24.   S.M. Reppert, D.R. Weaver and T. Ebisawa. *Neuron.* 1994, **13**, 1177-1185.

25.   Y. Marushige, K. Marushige and A. Koestner *Anticancer Res.* 1992, **12**, 2069-2073.

26.   C.M. Troy and M.L. Shelanski. *Proc. Natl. Acad. Sci. USA.* 1994, **91**, 6384-6387.

27.   M.A.Pappolla, M. Sos, R.A. Omar, R.J. Bick, D.L. Hickson-Bick, R.J. Reiter, S. Efthimiopoulos and N.K. Robakis. *J. Neurosci.* 1997, **17**, 1683-1690.

28.   H. Manev, T. Uz, A. Kharlamov and J.Y. Joo. *FASEB J.* 1996, **10**, 1546-1551.

THE L-ARGININE - NITRIC OXIDE PATHWAY: ROLE AS AN ANTIOXIDANT MECHANISM IN ATHEROSCLEROSIS

Rainer H. Böger.

Institute of Clinical Pharmacology, Hannover Medical School, 30623 Hannover, Germany.

1 THE L-ARGININE- NO PATHWAY: BIOCHEMISTRY AND PHYSIOLOGY

Nitric oxide (NO) is generated from the amino acid precursor L-arginine by the activity of the enzyme NO synthase in healthy endothelium. NO plays an important role as an endothelium-derived mediator- its discovery in the 1980s was based on early knowledge of the so called "endothelium-derived relaxing factor" (EDRF) which has later been shown to be chemically almost identical to NO.[1] Besides its most prominent action - vasodilation -, NO regulates a variety of cell-cell interactions in the healthy vasculature. It inhibits platelet adhesion and aggregation, it reduces monocyte and leukocyte adhesion to the endothelium, and it inhibits proliferation of vascular smooth muscle cells. Taken together, these physiological functions of NO have led to the notion that NO acts as an "endogenous anti-atherogenic molecule".[2]

Endothelial NO synthase is activated by mediators like acetylcholine, adenosine diphosphate, or thrombin, and by mechanical forces brought about by shear stress of the blood flowing along the endothelial surface. NO then diffuses to its target cells which may be located luminally (platelets or leukocytes) or abluminally (vascular smooth muscle cells) to stimulate soluble guanylyl cyclase (Fig. 1). The intracellular second messenger cyclic GMP is responsible for many but not all of the biological effects of NO. NO is then rapidly inactivated via oxidation to the stable end products nitrite and nitrate which can be detected in plasma. Nitrate and cyclic GMP are eliminated from the body via urinary excretion. Oxidative inactivation of NO may occur even before NO has interacted with its physiological target molecule; in this case NO is converted to nitrite and nitrate without exerting its biological functions.

Many other cell types besides endothelial cells also contain nitric oxide synthases. Three major isoforms of NO synthase have been differentiated (for review cf.[3]): NOS I was isolated from neuronal cells; it generates low amounts of NO constitutively in a calcium/ calmodulin dependent manner; NO generated by this isoform of NOS functions as a neurotransmitter in non-adrenergic, non-cholinergic neurons. Another isoform of NOS was found in macrophages and named NOS II or inducible NOS. Gene expression of this isoform is induced by cytokines; it produces high amounts of NO for a short period of time which then contributes to cytotoxic effects. The activity of this isoform of NOS is independent of intracellular calcium concentration. The endothelial isoform is now called NOS III; its activity is also calcium-dependent.

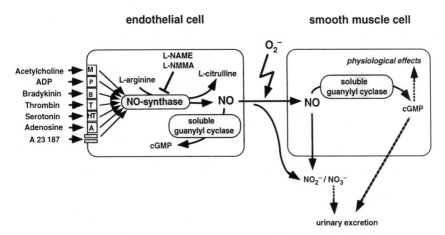

Figure 1

Schematic representation of the L-arginine-nitric oxide pathway in the vascular wall. ADP= adenosine diphosphate, A 23187= calcium ionophore A 23187, L-NMAE= NG-nitro-L-arginine methyl ester; L-NMMA= NG-monomethyl-L-arginine; cGMP= cyclic 3,5 - guanosine monophosphate. (Modified from Böger et al.[2] with permission by Elsevier Science Ltd, Ireland).

2 ROLE OF NO AND OXIDATIVE STRESS IN ATHEROSCLEROSIS

2.1. The biological activity of EDRF/NO is impaired in atherosclerosis

Endothelium-dependent relaxation is impaired in animals with experimentally induced atherosclerosis[4] and in atherosclerotic and hypercholesterolemic humans.[5] This defect is linked to decreased biological activity of endothelium-derived NO, because usually relaxation in response to exogenous NO donor drugs is not or only slightly reduced.[6] The degree of impairment of acetylcholine-induced coronary vasodilation is correlated to the number of atherosclerotic risk factors present,[7] and to the progression of atherosclerosis.[8] This defect extends beyond the coronary circulation in patients with coronary artery disease, pointing to the presence of a systemic endothelial defect in these patients. Interestingly, defective NO activity is already present in subjects who are at high risk of developing atherosclerotic cardiovascular disease, but without overt manifestations of this disease, like hypercholesterolemic subjects,[5] chronic smokers,[9] diabetics,[10] and hypertensive patients.[11] Decreased biological activity of NO is therefore supposed to be one causal factor in the development of atherosclerosis; however, the mechanism which leads to reduced activity is still controversial.

2.2. Vascular oxidative stress is increased in atherosclerosis

Several animal studies suggest that the release of oxygen-derived free radicals is increased in atherosclerotic arteries. Ohara et al.[12] found that the release of superoxide radicals by segments of rabbit aorta is increased from animals fed a cholesterol-enriched diet. A similar finding was published by Mügge et al.[13] The cellular sources of superoxide radicals may be leukocytes adhering to and invading into the hypercholesterolemic arterial wall and endothelial cells, as endothelial denudation leads to reduced superoxide release.[14] Reaction of NO with superoxide radicals is an important mechanism leading to inactivation of NO via reaction to peroxynitrite.[15] It was therefore suggested that the formation of NO by endothelial cells may be normal in hypercholesterolemia, but early oxidative inactivation may block the biological effects of NO. Minor et al.[16] even found that NO release was increased from the hypercholesterolemic vascular wall.

2.3. Functional consequences of shifted balance between NO and O_2^- in atherosclerosis

Superoxide radicals are responsible for a variety of secondary processes occurring during the pathogenesis of atherosclerosis: Oxidation of LDL lipoproteins by superoxide results in generation of oxidized LDL, which are more easily taken up by macrophages via the scavenger receptor, thereby promoting the formation of foam cells.[17] Oxidative stress is responsible for activation of oxidant-responsive genes like those coding for mononuclear cell adhesion molecules (e.g., intercellular adhesion molecule-1, vascular cell adhesion molecule-1) and chemotactic proteins like monocyte chemoattractant protein-1. These chemoattractants and adhesion factors act in concert to promote the invasion of monocytes and leukocytes into the hypercholesterolemic vascular wall and to initiate an inflammatory process which is characteristic for advanced stages of atherosclerosis. One central regulatory point for these molecules is the activation/inacitivation of the redox-sensitive transcription factor nuclear factor-kappa B (NF-κB). NF-κB is activated by oxygen-derived radicals and induces the expression of chemokines and adhesion proteins in endothelial cells.[18] NF-κB also regulates the expression of inducible NO synthase,[19] which may explain the presence of iNOS-like activity in advanced atherosclerotic lesions. NO inactivates NF-κB and thereby inhibits expression of the target genes. Thus, dysbalance between NO and redox status within the vascular wall is believed to be a major factor involved in early stages of atherogenesis.

3 POTENTIAL LINKS BETWEEN THE METABOLISM OF NO AND O_2^-

Being a radical with one free single electron itself, NO avidly interacts with other radicals. One such radical is superoxide, which has been shown as early as 1986 to react in a destructive manner with NO.[20] Looking from another perspective, one may argue that NO inactivates O_2^-. However, more recent chemical investigation revealed that the reaction between these two radicals yields peroxynitrite, an even more cytotoxic radical than O_2^-, which then rearranges to form NO_3^-.[21] Simultaneous generation of NO and O_2^-, as it occurs during leukocyte oxidative burst, is involved in unspecific host defense and tissue damage during inflammatory reactions.[22]

NO has been shown to interfere with various molecular and cellular sources of superoxide in different experimental systems. Enzymes like NADPH oxidase, cyclooxygenase, lipoxygenase, and xanthine oxidase are potential generators of superoxide. These enzymes are present in endothelial cells or in other cell types present in the atherosclerotic vascular wall. Supplementation of cholesterol-fed rabbits with the NO precursor, L-arginine, has been shown to normalize vascular superoxide release.[6,14]

Recent evidence suggests that NO synthase itself may be a source of oxygen radicals. NOS has close structural similarities with cytochrome P450 reductases.[23] The reaction catalyzed by this enzyme consists of a two-step reduction/oxidation process during which electrons are transferred from NADPH via flavins to L-arginine.[24] When the enzyme is depleted of L-arginine, less NO is produced, but at the same time generation of O_2^- occurs, indicating that the reduction and oxidation domains of the enzyme are uncoupled under these conditions, and molecular oxygen may serve as an electron acceptor. One potential mechanism by which relative substrate depletion of NOS may occur *in vivo* is by accumuation of endogenous, competitive NOS inhibitors like asymmetric dimethylarginine (ADMA). We have recently shown that elevated plasma levels of ADMA are present in hypercholesterolemia[25] and atherosclerosis[26] Further investigation is needed to study whether uncoupling of the NO synthase reaction mechanism occurs *in vivo*.

4 THERAPEUTIC IMPLICATIONS

The L-arginine/nitric oxide pathway is a major regulator of endothelial cell -blood cell interactions. Dysfunction of endothelium-dependent, NO-mediated vasodilation is a marker for the risk and the progression of atherosclerotic vascular disease *in vivo*. However, NO-mediated vasodilation is just a surrogate marker for other functions of NO which may be at least equally important during atherogenesis: inhibition of thrombocyte and leukocyte activation, and counter-regulation of oxidant stress. Endothelial dysfunction is reversed by L-arginine,[27,28] cholesterol-lowering drugs, and antioxidants like probucol, vitamin C, and vitamin E in humans. Cholesterol-lowering drugs have also been shown to improve the clinical prognosis of patients with a history of myocardial infarction.[29] Whether longer term treatment with L-arginine and antioxidants will also provide clinical benefit remains to be investigated.

References

1.   Palmer R.M.J., Ferrige A.G., Moncada S. Nitric oxide accounts for the biological activity of endothelium-derived relaxing factor. *Nature*. 1987, **327**, 524 - 526.
2.   Böger R.H., Bode-Böger S.M., Frölich J.C. The L-arginine-nitric oxide pathway: Role in atherosclerosis and therapeutic implications. *Atherosclerosis*. 1996, **127**, 1-11.
3.   Förstermann U., Closs E.I., Pollock J.S., Nakane M., Schwarz P., Gath I., Kleinert H. Nitric oxide synthase isoenzymes. Characterization, purification, molecular cloning, and functions. *Hypertension*. 1994, **23**, 1121-1131.
4.   Guerra R.jr., Brotherton A.F.A., Goodwin P.J., Clark C.R., Armstrong M.L., Harrison D.G. Mechanisms of abnormal endothelium-dependent vascular relaxation in atherosclerosis: Implications for altered autocrine and paracrine function of EDRF. *Blood Vessels*. 1989, **26**, 300-314.
5.   Creager M.A., Cooke J.P., Mendelsohn M.E., Gallagher S.J., Coleman S.M., Loscalzo J., Dzau V.J. Impaired vasodilation of forearm resistance vessels in hyper-cholesterolemic humans. *J. Clin. Invest.* 1990, **86**, 228-234.

6.   Böger R.H., Bode-Böger S.M., Mügge A., Kienke S., Brandes R., Dwenger A., Frölich   J.C. Supplementation of hypercholesterolaemic rabbits with L-arginine reduces the vascular release of superoxide anions and restores NO production. *Atherosclerosis*. 1995, **117**, 273-284.

7.   Vita J.A., Treasure C.B., Nabel E.G., Fish R.D., McLenachan J.M., Yeung A.C., Vekshtein V.I., Selwyn A.P., Ganz P. Coronary vasomotor response to acetylcholine relates to risk factors for coronary artery disease. *Circulation*. 1990, **81**, 491-497.

8.   Otsuji S., Nakajima O., Waku S., Kojima S., Hosokawa H., Kinishita I., Okubo T., Tamoto S., Takada K., Ishihara T., Osawa N. Attenuation of acetylcholine-induced vasoconstriction by L-arginine is related to the progression of atherosclerosis. *Am. Heart J.* 1995, **129**, 1094-1100.

9.   Zeiher A.M., Schächinger V., Minners J. Long-term cigarette smoking impairs endothelium-dependent coronary arterial vasodilator function. *Circulation*. 1995, **92**, 1094-1100.

10.   Johnstone M.T., Creager S.J., Scales K.M.., Cusco J.A., Le B.K., Creager M.A. Impaired endothelium-dependent vasodilation in patients with insulin-dependent diabetes mellitus. *Circulation*. 1993, **88**, 2510-2516.

11.   Panza J.A., Garcia C.E., Kilcoyne C.M., Quyyumi A.A., Cannon R.O. Impaired endothelium-dependent vasodilation in patients with essential hypertension. *Circulation*. 1995, **91**, 1732-1738.

12.   Ohara Y., Peterson T.E., Harrison D.G. Hypercholesterolemia increases endothelial superoxide anion production. *J. Clin. Invest*. 1993, **91**, 2546-2551.

13.   Mügge A., Brandes R., Böger R.H., Dwenger A., Bode-Böger S.M., Kienke S., Frölich J.C., Lichtlen P.R. Vascular release of superoxide radicals is enhanced in hypercholesterolemic rabbits. *J. Cardiovasc. Pharmacol*. 1994, **24**, 994-998.

14.   Böger R.H., Bode-Böger S.M., Brandes R.P., Phivthong-ngam L., Böhme M., Nafe R., Mügge A., Frölich J.C. Dietary L-arginine reduces the progression of atherosclerosis in cholesterol-fed rabbits -comparison with lovastatin. *Circulation*. 1997, **96**, 1282-1290.

15.   Ignarro L.J. Biosynthesis and metabolism of endothelium-derived nitric oxide. *Annu. Rev. Pharmaol. Toxicol*. 1990, **30**, 535-560.

16.   Minor R.L., Myers P.R., Guerra R., Bates J.N., Harrison D.G. Diet-induced atherosclerosis increases the release of nitrogen oxides from rabbit aorta. *J. Clin. Invest*. 1990, **86**, 2109-2116.

17.   Ross R. The pathogenesis of atherosclerosis: a perspective for the 1990s. *Nature*. 1993, **362**, 801-809.

18.   Zeiher A.M., Fisslthaler B., Schray-Utz B., Busse R. Nitric oxide modulates theexpression of monocyte chemoattractant protein-1 in cultured human endothelial cells. *Circ. Res*. 1995, **76**, 980 - 986.

19.   Kleinert H., Euchenhofer C., Ihrig-Biedert I., Förstermann U. In murine 3T3 fibroblasts, different second messenger pathways resulting in the induction of NO synthase II (iNOS) converge in the activation of transcription factor NF-κB. *J. Biol. Chem*. 1996, **271**, 6039-6044.

20. Gryglewski R.J., Palmer R.M.J., Moncada S. Superoxide anion is involved in the breakdown of endothelium-derived vascular relaxing factor. *Nature*. 1986, **320**,454 - 456.

21. Blough N.V., Zafiriou O.C. Reaction of superoxide with nitric oxide to form peroxinitrite in alkaline aqueous solution. *Inorg. Chem.* 1985, **24**, 3502-2504.

22. Miller R.A., Britigan B.E. The formation and biologic significance of phagocyte-derived oxidants. *J. Invest. Med.* 1995, **43**, 39-49.

23. White K.A., Marletta M.A. Nitric oxide synthase is a cytochrome P-450 typehemoprotein. *Biochemistry* .1992, **31**, 6627-6631.

24. Abu-Soud H.M., Feldman P.L., Clark P., Stuehr D.J. Electron transfer in the nitric oxide synthases. *J. Biol. Chem.* 1994, **269**, 32318-32326.

25. Bode-Böger S.M., Böger R.H., Kienke S., Junker W., Frölich J.C. Elevated L-arginine/ dimethylarginine ratio contributes to enhanced systemic NO production by dietary L-arginine in hypercholesterolemic rabbits. *Biochem. Biophys. Res. Commun.* 1996, **219**, 598-603.

26. Böger RH, Bode-Böger SM, Thiele W, Junker W, Alexander K, Frölich JC. Biochemical evidence for impaired nitric oxide synthesis in patients with peripheral arterial occlusive disease. *Circulation.* 1997, **95**, 2068-2074.

27. Drexler H, Zeiher AM, Meinzer K, Just H. Correction of endothelial dysfunction in coronary microcirculation of hypercholesterolaemic patients by L-arginine. *Lancet.* 1991, **338**, 1546-1550.

28. Creager M.A., Gallagher S.J., Girerd X.J., Coleman S.M., Dzau V.J., Cooke J.P. L-arginine improves endothelium-dependent vasodilation in hypercholesterolemichumans. *J. Clin. Invest.* 1992, **90**, 1248-1253.

29. Scandinavian Simvastatin Survival Study Group. Randomised trial of cholesterollowering in 4444 patients with coronary heart disease: the Scandinavian Simvastatin Survival Study (4S). *Lancet.* 1994, **344**, 1383-1389.

ANTIOXIDANT ENZYME DEPLETION AND NITRIC OXIDE ALTERATIONS: RELATIONSHIP WITH THE RISK FACTORS IN ANGIOGRAPHICALLY DIAGNOSED CORONARY HEART DISEASE

Bulent Sözmen*, Fevziye Uysal**, Dervi̦ Gazitepe*, Leyla Aslan*, Eser Yildirim Sözmen**

*Department of Internal Medicine, Atatürk State Hospital, Ye̦silyurt-Izmir, **Department of Biochemistry, Ege University Faculty of Medicine, Bornova-Izmir, Turkey.

1 INTRODUCTION

Atherosclerosis is the most common manifestation of cardiovascular diseases, however the exact mechanisms underlying its pathogenesis have not yet been completely understood. To date, a number of cardiovascular risk factors which include ageing, hypertension, hyperlipidemia, diabetes, ischemia and reperfusion have been shown to be associated with endothelial dysfunction in both experimental and clinical studies.[1]

On the other hand, published work indicates that free radicals and lipid peroxidation may also play an important role in the pathogenesis of atherosclerosis.[2] However the relationship between antioxidant enzymes and nitric oxide (NO) in this process is still obscure.

In this study, the relationship between the antioxidant enzymes and NO and some risk factors as well as the severity of coronary atherosclerosis was investigated.

2 MATERIALS AND METHODS

Patients (n=33,male) who had greater than 50 % of a major vessel stenosed (by coronary angiography) were diagnosed as coronary artery disease. Patients with previous myocardial infarction or valvular heart disease, unstable, severe or secondary hypertension, atrioventricular block exceeding first degree and not permanently paced, severe heart failure, hepatic failure, renal insufficiency or any severe or progressive disease were excluded. All patients were in synus rhythm and were not being treated with any antihypertensive drug including diuretics. All of cases had typical angina and a significant ST-segment depression in the precordial leads on their ECG during exercise or at rest. There was no important concomitant disease.

Control subjects (n=15,male) had no history of cardiac or coronary disease and hypertension, without any intercurrent diseases and any other risk factors and they presented with a normal ECG and physical examination.

All patients stopped using drugs and the diet was restricted for foods including nitrite 5 days before the analysis. Nitrite (Griess reaction) and nitrate (by the enzymatic reduction of nitrate to nitrite with nitrate reductase from Aspergillus species) that were the end products of NO in plasma,[3] superoxide dismutase (SOD)[4] and Catalase[5] in erythrocyte and the lipid profile (Total Cholesterol, LDL, Triglyceride, HDL, VLDL) in serum were detected.

Differences between the groups were evaluated using unpaired Student's t test and Kruskall Wallis nonparametric test. The software used for all statistical evaluations was SPSS V6.1 statistical package program.

3 RESULTS

Results marked an increase in cholesterol levels ($p<0.01$) as well as triglyceride ($p<0.01$) and LDL cholesterol levels ($p<0.01$) and a decrease in HDL cholesterol ($p<0.01$) and nitrate levels ($p<0.05$), SOD and catalase activities($p<0.05$, $p<0.01$, respectively) in the patient group compared with the controls (Tab. 1). One way Anova test showed that HDL (F=0,044), LDL (F=0,000), total cholesterol (F=0.000) levels and SOD (F=0.042) and catalase (F=0.017) activities of patients affected parameters changing with the number of stenosed vessels.

Table-1

The baseline characteristics of control and coronary heart disease groups. Data were given as mean SEM.

	Control group n=15	CHD group n=33
Age	55.28 ±3.31	55.54 ±1.64
TAD (mmHg)	75.00 ±1.64	80.45 ± 1.67
TAS (mmHg)	122.50 ± 2.28	132.27 ± 2.26
Triglyceride (mg/dl)	138.93 ± 7.47	199.21 ± 21.97*
Total cholesterol (mg/dl)	177.75 ± 7.20	230.91± 6.83*
LDL-cholesterol (mg/dl)	94.50 ± 4.89	145.36 ± 6.37*
HDL-cholesterol (mg/dl)	52.31 ± 2.01	40.51 ± 2.63*
SOD (U/g Hb)	2659.69 ±172.59	2089.09 ± 206.55**
Catalase (U/g Hb)	12769.31 ± 634	10511 ± 378
Nitrite (µmol/l)	7.53 ± 0.83	6.37 ± 0.64
Nitrate (µmol/l)	24.46 ±1.99	19.52 ± 1.55**
Total (nitrite+nitrate) (µmol/l)	31.73 ± 1.89	27.93 ± 1.69

* $p<0.01$, ** $p<0.05$

4 DISCUSSION

While recent investigations have focused on the NO levels in atherosclerotic tissue, no experiments on nitrite-nitrate levels in plasma were reported. Only Winlaw et al.[6] reported that there are no significant variations in nitrate levels in serum of patients with ischaemic heart disease: a data based on a study of only 5 patients. Depletion of NO in patients with risk factors for coronary heart disease, were attributed to a lower rate of synthesis of NO which may in turn result from substrate deficiency or a defect in the signal transduction pathways or in the enzyme itself. Alternatively, Quyyumi et al proposed that the depressed bioavailability of NO may be secondary to increased breakdown of normally produced NO by superoxide anions.[7] It has also been indicated that the loss of endothelium-dependent vasodilation correlates with the presence of coronary risk factors in angiographically normal coronary arteries. In our study, when correlations with age were evaluated catalase activities and total nitrite levels were found to be decreased in the patient groups compared with age-matched controls. (Fig. 1).

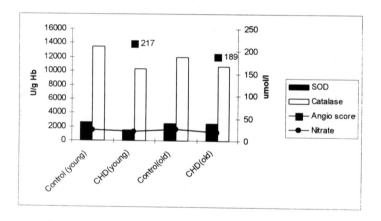

Figure 1

The relationship between antioxidant enzyme activities and angioscore and nitrate levels according to age.

We suggest that the depletion of nitrite-nitrate levels which is more expressed in the young patient group may be due to the antioxidant effect of NO, since older patients also had angio scores lower than the young patients. We found that there was no difference in nitrite-nitrate levels and antioxidant enzyme activity between the patients with high risk (serum cholesterol levels>200 mg/dl) and low risk (serum cholesterol levels<200 mg/dl) according to cholesterol levels (Fig. 2).

In addition, there was a decrease in total nitrite levels and SOD & catalase activities in patients who have low cholesterol compared to the control. Mantha et al's data[8] also showed a decrease in SOD activity in cholesterol-fed rabbits, a data that supports our findings, however their finding that catalase activity increases in cholesterol-fed rabbits-which was not fully explained in their report, was not parallel to our results. In accordance with this finding Lüscher et al.[1] proposed that the bioassayable NO release in coronary artery with hypercholesterolemia and atherosclerosis clearly is reduced. It's known that increases in the cholesterol levels, especially in LDL-cholesterol levels cause to oxidized-LDL that specifically interferes with L-arginine pathway.[1]

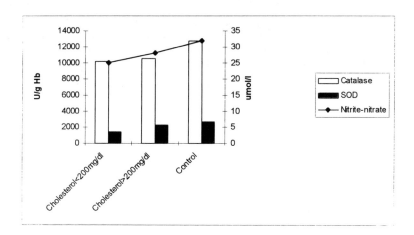

Figure 2
SOD & catalase activities and total nitrite levels in patients with low and high cholesterol levels.

In accordance with this, Ohara et al showed[2] that superoxide anion production is increased in endothelial cells from hypercholesterolemic vessels. Prasad et al.[9] suggested that hypercholesterolemia increases the level of oxygen free radicals (OFRs) which would produce endothelial damage and set the stage for development and maintenance of atherosclerosis. Increased OFRs may be due to an increased OFR-producing activity of the known sources and/or an inadequate cellular antioxidant defense systems.

It was interesting to find a negative correlation between the catalase activity and number of stenosed vessel ($r=-0.2903, p=0.05$) and between the nitrate levels and number of stenosed vessels ($r=-0.3086, p=0.081$). Based on these findings, we proposed that the insufficient antioxidant enzyme activities might be a triggerer factor to atherosclerosis; however it is not yet known if this decrease in antioxidant enzymes is a genetically determined primary factor or a secondary factor due to increased production of free radicals.

As a conclusion, it has been shown in our study, insufficient antioxidant enzyme activities and decrease in total nitrite levels might be independent trigger factors other than the well-known risk factors such as age, sex, hypercholesterolemia, hypertension for atherosclerosis: therefore prospective studies are needed to confirm the hypothesis that antioxidant enzyme activities may be used to find the subjects with risk for atherosclerosis.

References

1.   T.F. Lüscher, F.C.Tanner, M.R.Tschudi, G.Noll. *Annu Rev. Med.* 1993, **44**, 395.

2.   Y.O'hara, T.E.Peterson, D.G.Harrison. *J.Clin. Invest.* 1993, **91**, 2541.

3.   P.N.Bories, C.Bories. *Clin.Chem.* 1995, **41**(6), 904.

4.   H.P.Misra, I.Fridovich. *The J. Biol. Chem* .1972, **247**(10), 3170.

5.   H.Aebi. *Methods in Enzymol.* 1984, **105**, 121.

6.   D.S.Winlaw, G.A.Smythe, A.M.Keogh, C.G.Schyvens, P.M.Spratt, P.S.Macdonald. *The Lancet.* 1994, **344**(6), 373.

7.   A.A.Quyyumi, N.Dakak, N.P.Andrews, S.Husain, S.Arora, D.M.Gilligan, J.A.Panza, R.O.Cannon. *J. Clin. Invest.* 1995, **95**, 1747.

8.   S.V.Mantha, M.Prasad, J.Kalra, K.Prasad. *Atherosclerosis.* 1993, **101**, 135.

9.   K.Prasad, J.Kalra. *Am.. Heart. J.* 1993, **125**, 958.

LYSOSOMAL LEAKAGE CAUSES APOPTOSIS FOLLOWING OXIDATIVE STRESS, GROWTH-FACTOR STARVATION AND FAS-ACTIVATION

Ulf T. Brunk.

Division of Pathology II, Department of Neuroscience and Locomotion, Faculty of Health Sciences, Linköping University, S-581 85 Sweden.

1 INTRODUCTION

A large variety of stimuli are able to initiate apoptosis.[1-7] Some of these seem to act by oxidative mechanisms. However, given the complexity of the apoptotic process, mechanisms other than oxidative stress also might be expected to trigger apoptosis.

In earlier studies, we have found that oxidative stress will labilize lysosomal membranes. This has been observed following exposure of cells to hydrogen peroxide which induces damage to lysosomes. This damage probably occurs through iron-catalyzed oxidation mediated by the normal content of low-molecular-weight, redox-active intralysosomal iron which is almost invariably present in secondary lysosomes.[8-12]

Apoptosis due to iron-catalyzed intralysosomal oxidation is preventable by an initial exposure of the cells to the potent iron-chelator desferrioxamine (Des).[12] The endocytotic uptake of the chelator, and its distribution within the acidic vacuolar compartment, allows the conversion of the normally existing lysosomal pool of low-molecular-weight iron into a non-redox-active form.[8-12,13]

Apoptosis also is produced by lysosomal rupture arising from direct membrane photo-oxidation (using a lysosomotropic, weak base-type photosensitizer), probably by formation of singlet oxygen since it is not inhibitable by Des.[14-15] In these cases, moderate lysosomal rupture seems to initiate apoptosis, while necrosis typically follows severe oxidative stress with complete collapse of the acidic vacuolar apparatus.

These earlier observations suggested the hypothesis that partial lysosomal leak/rupture, with release of normally intralysosomal digestive enzymes into the cell cytoplasm, might transform the whole cell into sort of an autophagic vacuole within an initially intact plasma membrane. This might represent a final pathway leading to apoptosis that may be common for several different inducing mechanisms.[12,14] As further test of this hypothesis, we have here used a cell-type well characterized for its predictable behavior during apoptosis. We have investigated lysosomal stability during the induction of apoptosis by way of three commonly used stimuli: oxidative stress, growth-factor starvation, and dimerization of the Fas/APO-1/CD95 receptor. Overall, the results support the proposition that lysosomal leakage and/or rupture typically precede apoptosis regardless of the initiating stimulus.

2 MATERIAL AND METHODS

2.1. Cells and Culture Conditions

Jurkat EG.1 cells, an human T-leukemia cell line obtained from the European Collection of Animal Cell Cultures (ECACC, Salisbury, UK), were grown in suspension at 37 °C in humidified air with 5 % carbon dioxide in RPMI culture medium with 10 % fetal calf serum (FCS) and 10 mM Hepes. The cells were rapidly growing and needed to be split (1:3) three times a week.

In some experiments, cells were exposed to 1 mM Des for one hour under otherwise normal culture conditions before challenge with apoptotic stimuli.

2.2. Induction of Apoptosis

Apoptosis was induced in three different ways:

2.2.1. Growth-factor starvation. Apoptosis was induced by omitting cell splitting and medium renewal for up to seven days. This leads to growth factor starvation and is a well established technique for induction of apoptosis in various types of cells.

2.2.2. Oxidative stress. Cells (1.0×10^6/ml) were exposed to 50-750 µM hydrogen peroxide in phosphate buffered saline (PBS) for 30 min at 37 °C (controls were exposed to PBS only) and then returned to ordinary culture conditions. Cell samples were assayed directly after the oxidative stress and after another 3, 6, 15, and 24 h following return to normal culture conditions.

2.2.3. Fas/APO-1/CD95 stimulation. Cells (0.5×10^6/ml) were continuously exposed to 10-100 ng/ml of anti-human Fas/APO-1/CD95, CH-11 monoclonal antibodies (Mab) (ONCOR, Gaithersburg, MD, USA) under normal culture conditions and sampled for analyses hourly for 5 h.

2.3. Light and Transmission Electron Microscopy

Aliquots of cells were studied and counted in a Rosenthal chamber. The frequency of cells with apoptotic morphology (budding membranes and pycnotic or fragmented nuclei) was counted.

Following acridine orange staining (see below) living cells were studied in a Nikon Microphot SA microscope with a Nikon p102 photometer linked to a Macintosh computer and analyzed by static fluorometry using green exciting light (Nikon B-2A filter cube and an extra 630 barrier filter). One hundred randomly selected cells were measured, histograms constructed, and means ± SD calculated. The means of 3-6 samples were used for statistical evaluation. Cells were also analyzed and photographed by confocal laser scannning microscopy using a LMS 410 inverted microscope (Carl Zeiss, Jena, Germany) as previously described.[13] The equipment was operated in phase- and fluorescence- modes with excitation generated by a HeNe 0.35 mW laser.

Cells also were prepared for transmission electron microscopy (TEM) as previously described.[16,17] In brief, the cells were fixed in a glutaraldehyde-based fixative with an isotonic vehicle, post-fixed in osmium tetroxide, centrifuged, soaked in agar, stained *en bloc* in uranyl acetate, and embedded in Epon. Thin sections were cut with a diamond knife and studied in a Jeol 1200 TEM/STEM machine at 80 kV.

2.4. Lysosomal Stability Assay

2.4.1. Static cytofluorometry. At various time points, cells induced into apoptosis by the different treatments were exposed to the lysosomotropic weak base acridine orange (AO) as previously described.[8-14] Due to its lysosomotropic properties, this vital dye accumulates mainly in the acidic vacuolar apparatus, preferentially in secondary lysosomes, although to a minor degree also in the cytosol and nucleus.[18-20] Following excitation with blue light, AO shows a metachromatic red (high concentration in lysosomes) and green (low concentration in nuclei and cytosol) fluorescence. Following excitation with green light, however, only concentrated intralysosomal AO gives any fluorescence, being dark red. Blanks without cells (empty part of the cover slips on which the cells grew -that always gave very low values) were automatically subtracted from each measurement.

2.4.2. Flow cytofluorometry. In order to estimate fluorescence from large numbers of cells, and to relate the intactness of lysosomes to apoptotic shrunken "small" cells, intravitally AO-stained cells were also evaluated in a FACScan (Beckton-Dickinson, San José, CA, USA) flow cytofluorometer equipped with an argon laser and fitted with Lysis II software. For each evaluation 10,000 cells were registered, and the values were processed in a Macintosh computer using Cell Quest software (Becton-Dickinson, San José, CA, USA).

2.5. DNA Agarose Gel Electrophoresis

Cellular DNA was partly purified according to the technique described by Gorczyca et al.[21] and then subjected to electrophoresis on a 2 % agarose gel at 70 V and 20 mA for about 2 h. The gels were analyzed for the appearance of typical DNA laddering, arising from internucleosomal cleavage, by comparison with a DNA marker.

2.6. Statistics

The significance of differences between groups of cells was evaluated by the Mann-Whitney U-test. Values are given as mean ± SD. P-values <0.05 (*) were considered significant. $P < 0.01$ (**).

3 RESULTS

3.1. Morphology and Lysosomal Stability

3.1.1. Apoptosis induced by growth factor starvation. Cultures of cells maintained without renewal of growth medium stopped propagating after about four days of starvation, and accumulated an increasing number of shrunken cells with apoptotic morphology, as estimated by conventional light microscopy. As shown in Fig. 1A, the cultures contained close to 70 % apoptotic cells after four days of starvation. Following AO-staining, cells were studied by static and flow cytofluorometry. Cells with advanced apoptotic morphology contained greatly reduced numbers of intact (red) lysosomes, while cells in early apoptosis showed only a slight decrease in the number of normal lysosomes. Cells with reduced numbers of normal lysosomes were designated "pale" cells and identified in the flow histograms according to principles illustrated in Fig. 1B. In control cultures such "pale" cells were consistently <5 % of the total. Note that the "pale" cells constitute a quite heterogeneous group, consisting of some cells with almost normal numbers of intact lysosomes with strong red fluorescence, as well as some cells with very few such lysosomes. Since Des induces apoptosis by itself during prolonged exposure (results not shown) the iron-chelator was not applied to this group of cells.

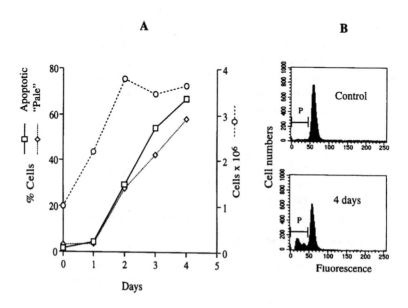

Figure 1 (A and B)
Panel A shows declining proliferation of Jurkat cells by time under growth factor starvation, as well as an increasing appearance of shrunken apoptotic cells with fragmented nuclei (registration by counting in a Rosenthal chamber), in parallel with increasing amounts of "pale cells" (cells with a reduced number of red lysosomes, identified by flow cytofluometry). The criteria for "pale" cells (P) are defined in panel B. Note that some of the "pale" cells have only a few intact lysosomes, while others are close to normal. In unstarved cultures there are <5% "pale" (P) cells, probably representing cellular fragments and the normal frequency of apoptosis. In cells starved for four days the "pale" cells had increased to about 70 %. Values are means of two experiments showing less than 10 % variation.

    3.1.2. Apoptosis induced by antibodies. As little as two hours following addition of anti-human Fas/APO-1/CD95 antibodies to the culture medium, there was a sharp increase in the number of apoptotic cells. This brisk apoptotic response was dose-dependent with the percentage of cells showing apoptotic morphology varying from about 45 to about 85 % at six h after the addition of 10 or 100 ng/ml of the antibodies, respectively (results not shown). Antibody at the concentration of 50 ng/ml resulted in about 70 % apoptotic cells after five hours and was considered suitable. Consequently, all further experiments were done using this concentration. Pretreatment with Des had little effect on the antibody-induced apoptosis, although there was a tendency towards a decline in the number of apoptotic cells. As is typically observed, the untreated control cultures contained a small, but stable, number of apoptotic cells (Fig. 2).

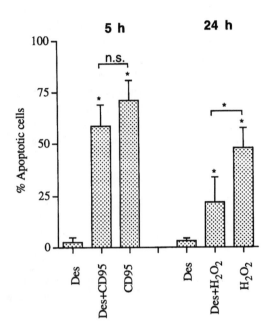

Figure 2

The effect of anti-human Fas/APO-1/CD95 antibodies and oxidative stress on the induction of apoptosis, and the effect on this process of pre-loading the lysosomes with the iron-chelator desferrioxamine (Des). Cells were exposed continuously to the antibodies for 5 h, and to oxidative stress (H_2O_2) for 30 min followed by a return to normal culture conditions for 24 h, respectively. The frequency of apoptotic cells was estimated by their typical morphology as evaluated in a Rosenthal chamber. Note the inhibitory effect by Des on the oxidative stress-induced apoptosis but not on the anti-body-induced apoptosis. Control cells showed < 5 % apoptotic cells. n=4.

In cells examined by confocal laser scanning microscopy after AO-staining, the numbers of intact lysosomes were found to decline as a function of time after exposure to the antibodies. Following prolonged incubation, most apoptotic cells contained no or only a few intact lysosomes (results not shown). By static and flow cytofluorometry the number of "pale" cells was estimated and found to increase in number time-dependently. By static fluorometry, "pale" cells were registered according to principles shown in Figure 3B. Pretreatment with Des had no significant influence (Fig. 2 and 4). The number of significantly shrunken "small" cells was registered by flow cytofluorometry (forward scatter) and found to vary in parallel with the "pale" cells, although the "small" cells were somewhat fewer than the "pale" ones (results not shown) suggesting that lysosomal loss of intactness preceded apoptotic shrinkage. On TEM of antibody-exposed cells, typical apoptotic nuclear alterations were found as was much enhanced autophagocytosis (results not shown).

3.1.3. Apoptosis induced by oxidative stress. Following oxidative stress - in the form of exposure to hydrogen peroxide (H_2O_2) in PBS - apoptosis also developed dose- and time dependently (results not shown), although more slowly than following the immunological type of induction (Fig. 3A). Exposure to 500 µM hydrogen peroxide in PBS for 30 min resulted in about 50 % apoptosis after 24 h. In contrast to Fas/APO-1/CD95 stimulation, cells exposed to oxidative stress were significantly protected by pretreatment with Des (Fig. 2 and 4). Typical apoptotic morphology was seen in the electron microscope (results not shown) and, as was observed also with cells exposed to either growth factor starvation or the Fas antibody, the number of intact lysosomes declined with time (Fig. 3A).

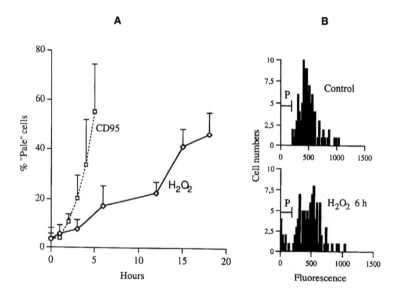

Figure 3

Pale" cells (static cytofluorometry) with decreased numbers of intact lysosomes accumulate with time in cultures exposed to both the antibody and oxidative stress (H_2O_2) (A), although at different rates. The principles for selecting "pale" (P) cells are shown in (B). Some "pale" cells are close to normal, while others have greatly reduced numbers of intact lysosomes, compare with Figure 1 (panel B). n= 5.

Immediately after the exposure to oxidative stress, there was a slight reduction in granular red fluorescence, indicating that some lysosomes rapidly had lost their membrane proton gradient. This loss of proton gradient is known to be accompanied by a release into the cytosol of lysosomal enzymes, such as cathepsin-D, and thus is likely to reflect lysosomal rupture.[9,12] When cells were returned to normal culture conditions after exposure to oxidative stress, a continuous decline in the number of intact lysosomes was found, as reflected by an increasing number of "pale" cells (Fig. 3A). This finding suggests that initial oxidative damage to lysosomal membranes is only fully expressed after some time.

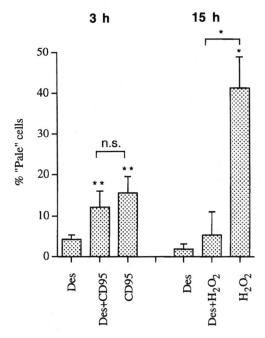

Figure 4

*Occurrence of "pale" cells (static cytofluorometry), as explained in Figure 3 **B**, following exposure to antibodies or oxidative stress (H_2O_2). Note protective effect by Des against oxidative stress but not against the immunological induction of apoptosis. n=4-5.*

As previously reported[12] pre-incubation of the cells with the iron chelator Des suppressed both lysosomal leakage/rupture and apoptosis. However, cells pre-incubated with Des alone were slightly growth-inhibited, while cells exposed to both Des and, subsequently, hydrogen peroxide, were severely growth inhibited (results not shown).

3.2. DNA Fragmentation

Separation of DNA-fragments according to size revealed typical "laddering," after all used ways of apoptosis induction. Des protected against fragmentation when oxidative stress was used as an inducer, but not when the immunological method was applied (results not shown).

4 DISCUSSION

Our hypothesis that partial rupture of the acidic vacuolar compartment - with attendant leak of lysosomal hydrolytic enzymes - might be a common final pathway in apoptosis is a new and controversial one. This concept is certainly in conflict with some older and established opinions, e.g. with those that advocate a more direct regulatory influence on death genes by calcium, oxygen-derived radicals, and other noxious agents. To try to discriminate between these different ideas we have elected to use Jurkat cells since they show an archetypal form of apoptosis, exhibiting the classic signs of cellular shrinkage, nuclear fragmentation, cytoplasmic budding, formation of apoptotic bodies, and evident DNA-laddering, while some of these phenomena may be absent during apoptosis of several other types of cells.[5] The established Jurkat cell-line is of human T4-lymphocyte origin and is well known to undergo apoptosis induced by growth factor starvation, oxidative stress or activation of the Fas/APO-1/CD95 receptor. The Jurkat cell line is thus well suited for comparative studies on lysosomal stability in relation to apoptosis induced by different modes.

As expected, and in agreement with previous findings on other cell-types,[12,14,22] moderate oxidative stress was found to rapidly induce a partial lysosomal rupture which was followed by apoptosis and further loss of intact lysosomes. Both lysosomal rupture and apoptosis are largely prevented by exposure of target cells to Des prior to oxidative stress. Des is a potent hydrophilic iron-chelator (MW=656.8) that does not readily pass the plasma membrane and is taken up primarily by endocytosis.[13,23] It is thus directed to the acidic cellular vacuolar apparatus, where it binds low-molecular-weight iron and prevents iron-mediated catalytic redox-reactions which would otherwise destabilize lysosomal membranes.[8-12] However, redox-active iron does not appear to be important in the genesis of apoptosis induced by dimerization of the Fas/APO-1/CD95 receptor, since Des has very little effect on either lysosomal rupture or apoptosis due to such induction. Nonetheless, as is the case with oxidative triggering of apoptosis, lysosomal labilization definitely precedes apoptosis caused by CD95 receptor engagement. This supports our hypothesis that release of hydrolytic enzymes (such as nucleases and proteases) from the lysosomal compartment to the cytosol is an important initiating event in the apoptotic process. However, the mechanisms behind the lysosomal labilization may be different depending upon the apoptotic stimulus applied. Oxidative stress induces iron-catalyzed intra-lysosomal oxidative reactions which destabilize the lysosomal membranes,[8-12] while such an iron-dependent mechanism is clearly not involved in apoptosis induced by antibody engagement of the Fas/APO-1/CD95 receptor, or by growth factor starvation.

We should caution, however, that the above suggestion that lysosomal leakage precedes apoptosis is not conclusively supported by our results with cells exposed to growth factor starvation. In this latter case, starved cells slowed their replicative activity

after a few days and did become apoptotic. Furthermore, the shrunken apoptotic cells had substantially decreased numbers of intact lysosomes, observations generally supporting the concept that in starvation apoptosis and lysosomal rupture occur more-or-less in parallel. However, in starved cells it is not possible to discern whether lysosomal rupture comes before apotosis or *vice versa*. The possible involvement of intralysosomal redox-active iron in starvation-induced apoptosis was not possible to test since exposure to Des for longer periods of time by itself induces apoptosis.

The stability of lysosomes in relation to cellular degeneration has been a matter of dispute ever since the discovery of these organelles by Christian de Duve, who, somewhat provocatively, nicknamed them "suicide bags", in order to emphasize their potential harmfulness to the living cell.[24,25] Today, however, the prevailing dogma is that lysosomes are rather stable organelles which break down only late during cellular degeneration and do not initiate damage of living cells by releasing their numerous hydrolytic enzymes into the cytosol.[26,27] One of the main reasons for this opinion seems to be the fact that lysosomes look ultrastructurally normal, even in cells with otherwise advanced degenerative alterations. However, lysosomes showing undamaged membranes on TEM nevertheless may have lost a substantial amount of the marker enzyme, acid phosphatase.[28] It has also been shown that minor lysosomal destabilization produces degenerative alterations that are rapidly and efficiently repaired by autophagocytotic processes in a short period of time, suggesting that minor lysosomal leakage might be a rather common and reversible occurrence.[11,15] Not until the damage to the lysosomal compartment is substantial, are the cells no longer able to survive.[8-12] These observations are in agreement with findings that one and the same type of injury may induce proliferation, apoptosis, or necrosis depending upon the size of the impact.[29] Furthermore, all lysosomes do not burst simultaneously,[30] and our results indicate that apoptosis is the result of an initial rupture of a limited number of lysosomes. Near-total loss of intact lysosomes is only observed in late post-apoptotic necrosis.

The exact mechanism by which lysosomal leak/rupture might induce apoptosis is not clear. Perhaps the most plausible explanation would be that lysosomal proteolytic enzymes, such as cathepsin-D, once released into the cytosol, act in concert with the caspases which are known to require proteolytic activation. Such a scenario would be in agreement with recent findings implying a role of cathepsin-D in apoptosis with enhanced expression of this well-known endoprotease.[31]

In summary: The results of this study suggest, in agreement with our hypothesis on lysosomal involvement in apoptosis, that partial rupture of the acidic vacuolar apparatus may be one of the final events in programmed cell death, and that such rupture may be multifactorially induced. Lysosomal enzymes, leaking out of their normal vacuolar compartment, may induce apoptosis by proteolytic activation of the caspase-family of enzymes, and, perhaps, by contributing with their own degradative activities as well.

References

1.   J.F.R. Kerr, A.H. Wyllie and A.R. Currie. *Br. J. Cancer.* 1972, **26**, 239.
2.   A.H. Wyllie. In 'Cell death in biology and pathology'. I.D. Bowen and R.A. Lockshin eds. London: Chapman & Hall, 1981 p. 9.
3.   S.V. Lennon, S.J. Martin and T.G. Cotter. *Cell Prolif.* 1991, **24**, 203.
4.   M.C. Raff. *Nature.* 1992, **356**, 397.
5.   J.J. Cohen. *Immunol. Today.* 1993, **14**, 126.
6.   T.M. Buttke and P.A. Sandstrom. *Immunol. Today.* 1994, **15**, 7.
7.   C.B. Thompson. *Science.* 1995, **267**, 1456.
8.   J.M. Zdolsek, H. Zhang, K. Roberg and U.T. Brunk. *Free Radic. Res. Commun.* 1993, **18**, 71.
9.   U.T. Brunk, H. Zhang, K. Roberg and K. Öllinger. *Redox Report.* 1995, **1**, 267.

10. K. Öllinger and U.T. Brunk. *Free Radic. Biol. Med.* 1995, **19**, 565.

11. U.T. Brunk, H. Zhang, H. Dalen and K. Öllinger. *Free Radic. Biol.Med.* 1995, **19**, 813.

12. H.B. Hellquist, I. Svensson and U.T. Brunk. *Redox Report.* 1997, **3**, 65.

13. J.B. Lloyd, H. Cable and C. Rice-Evans. *Biochem. Pharmacol.* 1991, **41**, 1361.

14. J.M. Zdolsek, G.M. Olsson and U.T. Brunk. *Photochem. Photobiol.* 1990, **51**, 67.

15. U.T. Brunk, H. Dalen, K. Roberg and H.B. Hellquist. *Free Radic. Biol. Med.* 1997, **23**, 616.

16. B. Arborgh, P. Bell, U.T. Brunk and V.P. Collins. *J. Ultrastruct. Res.* 1976, **56**, 339.

17. K. Abok, B-A. Fredriksson and U.T. Brunk. *APMIS.* 1988, **96**, 589.

18. E. Robbins and P.I. Marcus. *J. Cell Biol.* 1963, **18**, 237.

19. A.C. Allison and M.R. Young. *Life Sci.* 1964, **3**, 1407.

20. A.V. Zelenin. *Nature.* 1966, **212**, 425.

21. W. Gorczyca, J. Gong and Z. Darzynkiewicz. *Cancer Research.* 1993, **53**, 1945.

22. K. Roberg and K. Öllinger. *Am. J. Pathol.* 1998, **152**, 1151.

23. B.T. Olejnicka, K.Öllinger and U.T. Brunk. *APMIS.* 1997, **105**, 689.

24. C. de Duve and R.V. Wattiaux. *Ann. Rev. Physiol.* 1966, **28**, 435.

25. C. de Duve. In 'Lysosomes in Biology and Pathology' J.T. Dingle and H.B. Fell eds. Amsterdam: North-Holland Publishing Company, 1969, vol. 3, p. 3.

26. B.F. Trump and A.U. Arstilla. In 'Pathobiology of Cell Membranes' B.F. Trump and A.U. Arstilla eds. New York: Academic Press. 1975, vol 1, p 1.

27. A.B. Kane, R.P. Stanton, E.G. Raymond, M.E. Dobson, M.E. Knafelc and J.L. Faber. *J. Cell Biol.* 1980, **87**, 643.

28. U.T. Brunk and J.L.E. Ericsson. *Histochem. J.* 1972, **4**, 479.

29. P. Nicotera. *J. Biol.Chem.* 1994, **269**, 30553.

30. E. Nilsson, R. Ghassemifar and U.T. Brunk. *Histochem. J.* 1997, **29**, 1.

31. L.P. Deiss, H. Galinka, H. Berissi, O. Cohen and A. Kimchi. *EMBO J.* 1996, **15**, 3861.

Oxidative Stress and Antioxidants in Type II Diabetes

MOLECULAR MECHANISMS OF CELLULAR LIPID PEROXIDATION IN DIABETES

Sushil K. Jain.
Department of Pediatrics, Louisiana State University Medical Center, Shreveport, LA 71130, USA.

1 INTRODUCTION

Diabetes mellitus is characterized by hyperglycemia caused by impaired secretion or action of insulin. Eventually, many diabetic patients develop microvascular and macrovascular complications such as thickening of capillary basement membranes, retinopathy, nephropathy, neuropathy, accelerated atherosclerosis and increased tendency to infection.

Diabetics vary in their predisposition to the development of diabetic complications. There are two schools of thought regarding the development of diabetic complications. The genetic hypothesis suggests that complications of diabetes are genetically predetermined as part of the diabetic syndrome, whereas the metabolic hypothesis suggests that cellular and vascular damage complications are the effects of long-term hyperglycemia. In other words, if the hyperglycemia of diabetes could be prevented or controlled, then diabetic complications would not develop. Many studies have suggested that hyperglycemia can cause oxidative cellular damage in diabetes. This review outlines the biochemical mechanisms for the elevated oxidative cellular damage in the diabetes and its potential role in the development of diabetic complications.

2 COMPLICATIONS OF DIABETES

Many epidemiological studies provide evidence that poor glycemic control in diabetic patients and elevated glucose levels in nondiabetic persons significantly increase the risk of coronary heart disease. Risk factors such as oxidation of lipoproteins, protein glycation, platelet hyperaggregability, and hypercoagulability of blood are direct consequences of hyperglycemia and contribute in varying degrees to the development of cardiovascular disease in diabetes mellitus.[1]

Diabetes is also associated with dysfunction of the connective tissues, collagen, and cellular immune and defense mechanisms.[2] Platelets from diabetic subjects exhibit enhanced adhesiveness, increased aggregability to various agonists, decreased survival, and increased generation of thromboxane.[3] Thromboxane is known to induce not only aggregation, but also vasoconstrictor activity.[3] Similarly, erythrocytes of diabetic patients are known to have several abnormalities, such as, excessive aggregation, reduced deformability, hyperviscosity, glycosylation of proteins, sorbitol accumulation, oxidative damage, phosphatidylserine externalization in the outer membrane bilayer and increased adhesivity to endothelial cells.[4] It has been suggested that both platelets and erythrocytes are involved in endothelial alteration, platelet deposition, and atherosclerotic processes and the impairment of diabetic microvascular flow and complications.[5]

Long term epidemiological studies show that as many as 60 %-75 % of patients develop retinopathy and that less than 33 % develop proliferative eye disease after 40 years of diabetes.[6] Different surveys suggest that nephropathy occurs 17-19 years after the onset of diabetes in nearly 40 %-59 % of all diabetics patients but that less than 10 % of all diabetics develop renal failure. These studies suggest that 25 %-50 % of diabetic patients do not develop serious small vessel complications despite the existence of disease even though presumably different levels of glycemic control and some degree of hyperglycemia existed throughout the many years of the disease in these patients.[6]

Young and his coworkers found a relationship between deterioration of motor, sensory and autonomic nerve function and glycemic control in diabetic patients.[7] Further, other investigators found that improved glycemic control increases nerve conduction and velocity as well as peripheral nerve function and vibratory sensation threshold. Recent clinical trials have demonstrated that improved glycemic control achieved by as little as 6 weeks of intensive insulin therapy resulted in improved motor nerve conduction velocity and faster nerve conduction velocities in the peroneal, tibial, and sural nerves after 36 months of intensive insulin treatment.[8]

3 HYPERGLYCEMIA AND OXYGEN RADICALS GENERATION

3.1. Glycation of antioxidative enzymes:

One possible mechanism by which hyperglycemia can cause oxidative damage is by decreasing the activity of antioxidative enzymes that normally protect against oxidative damage, such as glutathione peroxidase, superoxide dismutase (SOD) and catalase.[9] The decreased activity of antioxidative enzymes appears to be caused by enzyme glycation. For example, the SOD activity in erythrocytes of diabetic patients is extremely decreased compared with non-diabetics, but there was no difference in quantity using the ELISA method with a monoclonal antibody to SOD.[9] Oberly[10] has reviewed the effect of diabetes on the reduced activity of SOD, catalase, and peroxidase in both animals and humans.

3.2. Oxygen radical generation by glucose autoxidation:

Glucose can autoxidize in a cell-free system under physiological conditions via enediol tautomer formation which generates hydroxyl radicals and produces hydrogen peroxide and ketoaldehydes.[11] Mashino and Fridovich[12] found that alpha-ketoaldehyde, a product of enediol autoxidation rather than of enediol itself, can lead to the generation of superoxide radicals. Gillery et al.[13] proposed that oxygen free radicals may form in increasing amounts during the course of diabetes mellitus by an electron exchange occurring between the sugar moiety of glycated proteins and molecular oxygen.

3.3. Oxygen radical generation by glycated proteins and glycoxidation products:

Mullarkey et al.[14] have demonstrated that non-enzymatic glycation of reactive amino groups in model proteins increases the rate of free radical production at physiological pH to nearly

50-fold over non-glycated protein. Superoxide generation was confirmed by electron paramagnetic resonance measurements using spin-trap phenyl-t-butyl-nitrone. Both Schiff base and Amadori glycation products were found to generate free radicals in a ratio of 1:1.5. Free radicals generated by glycated protein increased peroxidation of membranes of linoleic/arachidonic acid vesicles nearly 2-fold over controls, suggesting that the increased glycation of proteins in diabetes may accelerate vascular wall lipid oxidative modification. Similarly, Sakurai et al.[15] have shown that glycated polylysine, a model compound of glycated protein, produces superoxide radicals. These investigators suggest that the oxidation of polylysine requires iron and can, therefore, be blocked by using deferrioxamine. The iodine uptake confirmed that the oxidation involves an enediol structure in glycated polylysine. The exposure of unsaturated phospholipid liposomes to a glycated-Fe^{3+}-ADP system caused the production of thiobarbituric acid-reactive substances, which were completely inhibited by vitamin E and SOD. However, Sakurai et al.[15] did not document generation of hydroxyl radicals by the autoxidation of glycated polylysine.

In contrast, other investigators[16] suggest that N epsilon-(carboxymethyl)lysine, N epsilon-(carboxymethyl) hydroxylysine and the fluorescent cross-link pentosidine are formed by sequential glycation and oxidation reactions between reducing sugars and proteins.[16] These compounds, termed glycoxidation products, accumulate naturally in tissue collagen with age and at an accelerated rate with diabetes. Studies on glycation and oxidation using model proteins *in vitro* suggest that these products are biomarkers of more extensive underlying glycative and oxidative damage to the protein. Structural characterization of the cross-links and other products accumulating in collagen may help elucidate the relationship between oxygen radicals and collagen damage associated with diabetes.

3.4. Oxygen radical generation by the glucose metabolite (NADPH):

Our study[17] has documented increased membrane lipid peroxidation and increased fragility of erythrocytes treated with high levels of glucose. These studies have reported that glucose-induced lipid peroxidation in erythrocytes was blocked with fluoride, an inhibitor of glucose metabolism; with vitamin E, an antioxidant; with parachloromercuriobenzoate and metyrapone, inhibitors of the cytochrome P-450 system; and with dimethylfurane, diphenylamine, and thiourea, scavengers of oxygen radicals. These studies[17] suggest that increased glucose oxidation leads to increased levels of glucose metabolites such as NADPH, which stimulates the cytochrome P-450 like activity of hemoglobin in erythrocytes or of microsomes in various tissues, resulting in increased production of oxygen radicals leading to cellular lipid peroxidation. An increase in the activity of cytochrome P-450 has been found in certain tissues of diabetic patients and animals.[18] However, other investigators suggest[19] that hyperglycemia activates aldose reductase, which in turn converts glucose to sorbitol in the presence of NADPH; because the glutathione reductase system that detoxifies intracellular peroxide also requires NADPH, its diversion would lead to membrane oxidation.

In conclusion, glucose can react with proteins *in vivo* to form stable covalent adducts (glycation). The aldehyde group in the glucose is condensed with amino groups on proteins via Schiff's base linkage and this aldimine product could be re-arranged to the corresponding ketoamine and amadori products, which can generate oxygen radicals during autoxidation. However, generation of oxygen radicals may also be independent of glycation. For example, oxygen radicals may be generated from the autoxidation of glucose or ketoaldehydes, or from the increased activity of cytochrome P-450 system, such as that caused by an excess formation of NADPH from the increased glucose metabolism. Generation of different kinds of active oxygen radicals can oxidize membrane lipids or proteins and inactivate enzymes, which can impair cellular function and integrity and lead to cell injury.

4 EVIDENCE FOR CELLULAR LIPID PEROXIDATION IN DIABETES

The evidence for oxidative damage in diabetes has been reported as far back as 1979 by Sato et al.[20] These authors reported that the average level of lipid peroxide in plasma is higher in diabetic patients than in normal people and that diabetic patients with angiopathy had higher lipid peroxide levels than other diabetic patients. Since an atheroma contains high levels of lipid peroxide, Sato et al.[20] further suggested that high levels of lipid peroxide in plasma may cause an increase in lipid peroxide levels in the intima of the blood vessel, which may then initiate the formation of atherosclerosis. Subsequent studies from other laboratories have confirmed that the plasma of human diabetics has elevated levels of lipid peroxidation products by using different methods of assessment, such as, thiobarbituric acid-reactive substances, fluorescent substances like lipofuscin, conjugated diene products, and chemiluminescence.[21-26] Levels of lipid peroxidation products in plasma were correlated with the severity of diabetic complications such as retinopathy and nephropathy.[27] The increase in lipid peroxidation products in plasma has also been documented in rats with alloxan- or streptozotocin-induced diabetes.[22] However, this increase disappeared when their hyperglycemia in rats was controlled with insulin.[22] Recent studies have found increased and similar *in vitro* oxidizability of low density lipoprotein (LDL) fractions of plasma from diabetic patients and identified autoantibodies against oxidatively modified LDL in Type II diabetic patients again suggesting that LDL oxidation occurs *in vivo* in diabetes.[28] Morel and Chisholm[29] detected increased lipid peroxidation products in a lipoprotein fraction containing very low density lipoprotein (VLDL) and LDL obtained from rats with streptozotocin-induced diabetes. The enhanced oxidation in the diabetic VLDL+LDL fractions correlated with the *in vitro* toxicity of lipoprotein fractions to proliferating fibroblasts. In contrast, high density lipoprotein was not cytotoxic. The fact that insulin treatment of diabetic animals inhibited both oxidation and cytotoxicity of VLDL+LDL shows that the increased oxidation and development of cytotoxic activity in the diabetic VLDL+LDL fraction was related to the hyperglycemia.[29] The increased oxidizability of LDL isolated from diabetic patients could be reduced to control levels by a 6-week standard treatment with probucol. Evidence for elevated oxidative stress in diabetes have been reported in erythrocytes, liver, brain, heart, retina, lense, kidney and nerve. Additionally, Gallaher et al.[30] have recently shown a 5-fold increased excretion of urinary MDA in streptozotocin-induced diabetic rats over that in nondiabetic rats.

In summary, the human body is continuously generating reactive oxygen species, collectively called oxidative stress. These reactive oxygen species, if not detoxified, can result in oxidative cellular damage and impaired tissue functions. For example, increased oxidative stress can cause oxidative modification of plasma lipoproteins, associated with an increased risk of cardiovascular disease and atherosclerosis. Under normal conditions, the body has the ability to detoxify reactive oxygen species through antioxidative defence mechanisms. However, it seems that in diabetes, an imbalance exists between oxidant stress and antioxidative defence mechanisms favoring the former, which can result in cellular damage and lead to diabetic complications.

References

1. N.B. Ruderman, J.R. Williamson and M. Brownlee. *FASEB J.* 1992, **6**, 2905.
2. J. F. Tarsio, L. A. Reger and L.T. Furcht. *Diabetes.* 1988, **37**, 532.
3. S.K. Jain, K.S. Krueger, R. McVie, J.J. Jaramillo, M. Palmer and T. Smith. *Diabetes.* 1998, **21**, 1511.
4. R. L. Jones and C.M. Peterson. *Amer J. Med.* 1981, **70**, 339.
5. L.O. Limpson. *Nephron.* 1985, **39**, 344.
6. S. Strowig and P. Raskin. *Diabetes Care.*1992, **15**, 1126.
7. R.J. Young, C.C.A. Macintyre, C.N. Martyn, R.J. Prescott and D.J. Ewing. *Diabetologia.* 1986, **29**, 156.
8. P.Reichard, A.Britz, P. Carlsson, I. Cars, L.Lindblad, B.Y. Nillson and U.Rosenqvist. *J. Intern. Med.* 1990, **280**, 511.
9. K. Arai, S. Maguchi, S. Fujii, H. Ishibashi, K. Oikawa and N. Taniguchi. *J. Biol. Chem.* 1987, **262**, 16969.
10. L. W. Oberly. *Free Rad. Biol. Med.* 1988, **5**, 113.
11. S.P. Wolff, Z.Y. Jiang and J.V. Hunt. *Free Rad. Biol. Med.* 1981, **10**, 339.
12. T. Mashino and I. Fridovich. *Arch. Bioch. Biophys.* 1987, **252**, 163.
13. P. Gillery, J.C. Monboisse, F.X. Maquart and J.P. Borel. *Med. Hypoth.* 1989, **29**, 47.
14. C.J. Mullarkey, D. Edelstein, and M. Brownlee. *Biochem. Biophys. Res. Commun.* 1990, **173**, 932.
15. T. Sakurai, K. Sugioka and M. Nakano. *Biochim. Biophys. Acta.* 1990, **1043**, 27.
16. J.W. Baynes. *Diabetes.* 1991, **40**, 405.
17. S.K. Jain. *J. Biol. Chem.* 1989, **264**, 21340.
18. L.V. Favreau and J.B. Schenkman. *Diabetes.* 1988, **37**, 577.
19. M. Inouye, A.Tsutou, T. Mio and K. Sumino. *Horm. Metab.Res.* 1994, **26**, 353.
20. Y. Sato, N. Hotta, N. Sakamoto, S. Matsuoka, N. Ohishi and K. Yagi. *Biochem. Med.* 1979, 21, 104.
21. S. K. Jain, R. McVie, J. Duett and J.J. Herbst. *Diabetes.* 1989, **38**, 1539.
22. S. K. Jain, S.N. Levine, J. Duett and Hollier B. *Metabolism.* 1990, **39**, 971.
23. P. Rajeswari, R. Natarajan, J.L. Nadler and D. Kumar. *J.Cell Physiol.* 1991,**149**, 100.
24. S.K. Jain and R. McVie. *Metabolism.* 1994, **43**, 306.
25. S.K. Jain and S.N. Levine. *Free Rad. Biol. Med.* 1995, **18**, 337.
26. S.K. Jain, S.N. Levine, J.J. Duett and B. Hollier. *Diabetes.* 1991, **40**, 1241.
27. M. Hayakawa and F. Kuzuya. *Nippon Ronen Igakkai Zasshi* 1990, **27**, 149.
28. S.K Jain, R. McVie, J.J. Jaramillo, Y. Chen. *Free Rad. Biol. Med.* 1998, **24**, 175.
29. D.W. Morel and G.M. Chisolm. *J. Lipid Res.* 1989,**3 0**, 1827.
30. D.D. Gallaher, A.S. Csallany, D.W. Shoeman and J.M. Olson. *Lipids.* 1993, **28**, 663.

THE BIOCHEMICAL BASIS OF DIABETIC COMPLICATIONS - A ROLE OF OXIDATIVE STRESS ?

Jan Aaseth and Ole Wilhelm Bøe.

Department of Clinical Chemistry, Hedmark County Hospitals, N-2400 Elverum, Norway.

1 INTRODUCTION

Insufficiently or suboptimally controlled diabetes is characterized by unphysiologically high extracellular levels of glucose. The excesses of glucose move easily intracellularly into some cells, e.g. vascular endothelial cells and periphereal nerves, whereas the influx into other cells, such as muscle cells, is insulin-dependent, and thus restricted. It has been shown that a good metabolic control, as reflected by physiological values of blood glucose or near-physiological levels of glycosylated hemoglobin (HbA1c), is of importance to avoid complications.[1] However, the biochemical basis of diabetic complications is largely unknown.

2 GLYCATION AND OXIDATIVE STRESS

All kinds of proteins, including enzymes, transport molecules and structural components, will provide target sites for glycation. Recently Schleicher et al.[2] characterized a group of advanced glycated end products (AGE), denoted carboxymethyl-lysine derivatives (CML), that could be detected in heart, kidneys and particularly in the arteries. These products are formed after oxidative modification of the primary glycated species, and they appeared to serve as endogenous markers for oxidative damage.[2]

In healthy individuals, the intracellular glucose levels, e.g. in endothelial cells, are kept low, owing to its rapid intracellular break-down. The abundant influx of glucose in uncontrolled diabetic hyperglycemia is accompanied by activation of several cellular functions and associated with superoxide production. Generation of oxygen free radicals in the diabetic state has been reported by several authors.[3,4] The radicals may be formed either by activation of the membrane-bound NADPH oxidase; or in mitochondria; or by copper- or iron-catalyzed autoxidation with an intermediate step of cupric or ferric ion reduction analogues to chemical reduction induced by glucose in the classic Benedict reaction *in vitro*. In the presence of oxygen, the reduced metal ions can lead to formation of toxic superoxide anions, extra- or intracellularly.[5]

Usually, superoxide is detoxified rapidly by the enzyme superoxide dismutase (SOD), leading to formation of hydrogen peroxide, that is subsequently transformed to water in the cytosolic compartment of the cells, through the action of the enzyme glutathione peroxidase (GSH-Px) using glutathione (GSH) as an essential cofactor.

3 GLUTATHIONE DEPLETION IN DIABETES

Poorly controlled diabetes is characterized by derangement of cellular defence mechanisms against oxygen radicals. Most important in the failure to handle oxygen radicals is a dramatic reduction in the cellular concentration of the reducing substance glutathione (GSH), in endothelial cells,[6,7] retinal cells [8] and other tissues. Thus, in preliminary studies we have found red cell GSH levels reduced, in average to 50 % of the mean reference value, 2.6 mmol/L,[9] in diabetics with blood glucose in the range 16-20 mmol/L (n = 5). Several mechanisms are supposed to contribute to the GSH deprivation: Firstly, the abundance of glucose involves activation of the polyol pathway, leading to transformation of glucose to fructose, a pathway that is dependent upon available NADPH. This leads to overconsumption of the coenzyme NADPH, that is essential for the enzymatic regeneration of oxidized GSH.[10] Secondly, the only cellular mechanism for generation of this coenzyme, NADPH, is by the metabolic break-down of glucose via the pentose-phosphate shunt, which is inhibited significantly - to about 50 % of the regular rate when glucose is increased from about 5 to about 30 mmol/L, as assessed from studies on endothelial cells in culture.[6] Thirdly, it should be taken into account that available reduced GSH can interact directly with the aldehyde glucose; as the CO group of glucose react directly with the SH group of GSH leading to formation of a *hemithioacetal,* that interacts further with the amino group of GSH, leading to the formation of a cyclic compound, a *thiazolidine.*[11]

The reduced cellular GSH in diabetic states is associated with reduced GSH-Px activity, explaining the increased lipid peroxidation observed e.g. in the sciatic nerve in experimental diabetic neuropathy.[12] The increased generation of superoxide radicals characterizing diabetic hyperglycemia, combined with increased tissue levels of hydrogen peroxide owing to reduced GSH-Px activity, can lead to generation of the highly reactive hydroxyl radical, particularly in the presence of free iron or copper ions.

It is well known that patients suffering from increased retention of *iron*, as seen e.g. in hemochromatosis, has increased risk of developing diabetes and diabetic complications. Glycation of the antioxidative metalloproteins transferrin or SOD gives rise to liberation of prooxidative metal ions.

4 ANTIOXIDANT TREATMENT

Lipoic acid supplementation (20, 50 or 100 mg/kg) prevented the GSH depletion in the sciatic nerve in experimental diabetes, and prevented the electrophysiological changes.[12] Lipoic acid - in reduced form - can interact with socalled mixed disulphide bindings between protein thiol groups and GSH, thereby releasing GSH into the intracellular cytosolic space, thus activating the GSH-Px enzyme system and restoring the enzymatic scavenging activity.[11] Furthermore, lipoic acid apparently also possesses an inherent scavenging activity itself.[13] It has been administered orally to human patients (dose: 600 mg/day) with positive results, including reduced albuminuria.[13] Another thiol-containing drug that presumably acts by similar mechanisms to normalize endothelial functions and albuminuria, is captopril.[14,15]

Vitamin E administered in high doses (900 mg/day for 4 months) is reported to improve insulin action in non-insulin-dependent diabetic patients, and it is supposed to reduce oxidative stress.[13] Thus, tocopherol depleted cells are more vulnerable for diabetic damage than tocopherol adequate cells.[12] According to recent studies by Schleicher et al.[2] several antioxidants can inhibit the oxidative formation of CML-derivatives *in vitro*: In decreasing order, desferrioxamine, an iron chelator vitamin E, catalase, superoxide dismutase, aminoguanidine, and lipoic acid, prevented against such AGE formation. Previously, Jeejeebhoy et al.[16] have reported that diabetic neuropathy developed after chromium-depleted parenteral nutrition could be reversed by chromium supplementation. While previous authors have postulated that this insulin-mimetic activity is mediated by a specific chromium-containing «glucose tolerance factor», it is now tempting to suggest that the effect is caused by an «antioxidant-mimetic activity» of trivalent chromium.[5,17] Thus, addition of chromium(III)chloride could block the oxygen radical-mediated hemolysis induced by copper salts in an *in vitro* model system.[5] Zinc salts appeared to be inefficient in these studies, whereas scavenging thiols such as mercaptodextran and thioctic acid did protect the red cells.[18]

Another antioxidant investigated in diabetics *in vivo* as well as in high glucose conditions *in vitro*, is vitamin C (ascorbic acid).[19] Most of the studies report reduced ascorbic acid levels extra- and intracellularly in advanced diabetic cases. Chemically, ascorbic acid has a structure that resembles that of glucose. Since vitamin C and glucose share the same carrier in noninsulin dependent tissues, the elevated blood glucose in diabetes competitively inhibits the cell entry of vitamin C, thus resulting in localized vitamin C defiency, e.g. in vascular endothelial cells.[13] The possibility that ascorbic acid in high concentrations protects intracellular proteins against glycation can not be rejected. A protein protecting potential of ascorbic acid given in high doses was postulated several years ago by Linus Pauling, but his hypothesis has not been fully investigated.

Another approach to raise the depressed GSH levels is by inhibition of the polyol pathway using aldose reductase inhibitors.[10] However, an optimized glycemic control is still the superior method to restore the free radical scavenging activity of vulnerable tissues.

5 CONCLUSION

The development of diabetic complications, including micro- and macrovascular diseases, as well as retinopathy, neuropathy and nephropathy, appears to be accelerated by increased generation of free oxygen radicals in the cells and tissues. GSH, that usually acts as a protector against oxygen radicals, is depleted from vulnerable tissues in cases of uncontrolled or advanced diabetes. Localized vitamin C defiency in the same tissues results from the fact that glucose and vitamin C share the same carrier in noninsulin dependent cells. The activities of cooperative factors to GSH, such as the metalloenzymes GSH-Px and SOD are also reduced in several tissues of poorly treated cases of diabetes, probably owing to the continous protein glycation. The levels of vitamin E are reported to be low, as well.

A new hypothesis providing a biochemical basis for diabetic complications can be induced from the above observations, including both the traditional belief in a crucial role of glycation of important proteins, and the more recent discovery of the profound toxic influences of free oxygen radicals on human tissue constituents. It is reasonable to suggest that diabetic complications can be prevented or retarded by administration of appropriate antioxidants, in addition to traditional therapeutic principles.

References

1. The DCCT Research Group. *N. Engl. J. Med.* 1993, **329**, 977.
2. E. D. Schleicher, E. Wagner and A.G. Nerlich. *J. Clin. Invest.* 1997, **99**, 457.
3. C. J. Mullarkey, D. Edelstein and M. Brownlee. *Biochem. Biophys. Res. Commun.* 1990, **173**, 932.
4. J. W. Baynes. *Diabetes.* 1991, **40**, 405.
5. J. Aaseth, L.G. Korkina, I.B. Afanas'ev. *Acta Pharmacol. Sin.* 1998, **19**, 203.
6. A. Kashiwagi, T. Asahina, Y. Nishio, et al. *Diabetes.* 1996, **45**, 884.
7. N. E. Cameron and M. A. Cotter. *Diabetes.* 1997, **46** (Suppl. 2), S31.
8. C.-D. Agardh, E. Agardh, Y. Qian and B. Hultberg. *Metabolism.* 1998, **47**, 269.
9. J. Aaseth. In *Trace Elements in Health and Disease.* Almquist & Wiksell, Stockholm. 1985, 134.
10. M. C. Bravi, P. Pietrangeli, O. Laurenti, et al. *Metabolism.* 1997, **46**, 1194.
11. P. C. Jocelyn. *Biochemistry of the SH Group.* Academic Press, London 1972.
12. P. A. Low, K. K. Nickander and H. J. Tritschler. *Diabetes.* 1997, **46** (Suppl. 2), S38.
13. L. Packer, H. J. Witt and H. J. Tritschler. *Free Radical Biol. Med.* 1995, **19**, 227.
14. M. Chopra, H. Beswick, M. Clapperton, et al. *J. Cardiovasc. Pharmacol.* 1992, **19**, 330.
15. A. Gazis, S. R. Page and J. R. Cockcroft. *Diabetologia.* 1998, **41**, 595.
16. K. N. Jeejeebhoy, et al. *Am. J. Clin. Nutr.* 1977, **27**, 505.
17. M. Yonoha, Y. Ohbayashi, N. Noto, et al. *Chem. Pharm. Bull.* 1980, **28**, 893.
18. J. Aaseth, L. Benov and S. Ribarov. *Acta Pharmacol. Sin.* 1990, **11**, 363.
19. S. R. J. Maxwell, H. Thomason, D. Sandler, et al. *Ann. Clin. Biochem.* 1997, **34**, 63.

ANTIOXIDANTS AND DIABETES MELLITUS

P. Knekt and A. Reunanen.

National Public Health Institute, Mannerheimintie 166, 00300 Helsinki, Finland.

1 INTRODUCTION

Autoimmune destruction of pancreatic beta cells leads to insulin dependent diabetes mellitus (IDDM). Oxygen-derived free radicals play an important role in the destruction process.[1] Nitroso compounds have been shown to be toxic to beta cells in animals.[2] Accordingly, vitamin E, which is a free radical scavenger and blocker of nitrosamine formation, may provide protection against IDDM. Pharmacological doses of vitamin E have been shown to reduce insulin resistance,[3] which plays a leading role in the pathogenesis of non-insulin dependent diabetes mellitus (NIDDM). The hypothesis that vitamin E status predicts the occurrence of IDDM and NIDDM was studied in a prospective population study.

2 POPULATIONS AND METHODS

Both the IDDM study [4] and the NIDDM study[5] were carried out as case-control studies nested within a cohort of individuals with no previous history of diabetes mellitus at the baseline examination. Under the Sickness Insurance Act, all diabetics needing drug therapy are entitled to reimbursement of drug costs, eligibility for which requires a detailed medical certificate from an attending physician. A central register of all patients receiving drug reimbursement is kept by the Social Insurance Institution. Individuals in the study populations of the present study were linked to that register using the unique social security code assigned to each Finnish citizen.

The IDDM study was based on a cohort of 7526 men aged 15-99 years and free from diabetes. During follow-up, from the baseline examination in 1973-1976 to the end of 1994, 19 diabetes cases occurred. The mean age of the diabetes patients was 26 years (range 21-46). Three controls per case were selected by individual matching for sex, age and municipality; thus a total of 57 controls were chosen.

The NIDDM study was based on 1427 men and women from 25 communities, 15-99 years old and free from diabetes. The population was monitored for occurrence of NIDDM from the baseline in 1968-1972 to the end of 1986. During this period NIDDM occurred in 55 men and 51 women. For each diabetes patient two controls were individually matched for sex, age, and municipality. A total of 201 controls were chosen.

Serum levels of selenium, α-tocopherol and ß-carotene were determined for the cases and controls based on serum samples stored at -20 °C at the baseline examination. Serum levels of α-tocopherol and ß-carotene were determined using high performance liquid chromatography, and of serum selenium using the direct graphite furnace atomic absorption spectrometric method.

The statistical analyses were based on the conditional logistic model.

3 RESULTS

High levels of serum α-tocopherol were associated with decreased risk of insulin dependent diabetes mellitus. The mean level in cases was 5.3 mg/l and in controls 6.4 mg/l (p for difference=0.07). The cholesterol-adjusted relative risk of the disease between the highest and lowest tertiles of the α-tocopherol distribution was 0.12 (95 % confidence interval (CI) =0.02-0.85). No association was observed for serum selenium status; the mean levels in cases and controls were 55.4 µg/l and 55.8 µg/l, respectively.

An inverse association was observed between serum α-tocopherol and serum ß-carotene and incidence of non-insulin dependent diabetes mellitus. The relative risks between the highest and lowest tertiles were 0.61 (CI = 0.32-1.15) and 0.45 (CI= 0.22-0.92), respectively. Adjustment for serum cholesterol, body mass index, and smoking abolished the associations, however. The adjusted relative risks were 1.24 (CI=0.54-2.83) for serum α-tocopherol and 0.98 (CI=0.41-2.36) for β-carotene. Serum selenium was not associated with occurrence of NIDDM; the mean level for cases was 64.8 µg/l and for controls 63.6 µg/l.

4 CONCLUSIONS

The findings corroborated the hypothesis that high vitamin E status provides protection against IDDM. On the basis of this small observational study it is, however, impossible to say whether the association is causal or related to other as yet undefined factors, and thus further epidemiologic studies on this topic are suggested.

Our finding of a lower risk of NIDDM in individuals with higher levels of serum antioxidants, particularly those with higher levels of serum alpha-tocopherol, is in agreement with an earlier finding.[6] However, the significance of the reduction disappeared after adjustment for cardiovascular risk factors. Our results therefore do not support the hypothesis that low levels of dietary antioxidants cause NIDDM, but rather suggest that the association is mediated through factors more closely connected with the pathogenesis of NIDDM.

References
1. L. W. Oberley, *Free Radi. Biol. Med.* 1988, **5**, 113.
2. E. Wilander and R. Gunnarsson, *Acta Path. Microbiol. Scand. Sect. A*. 1975, **83**, 206.
3. G. Paolisso, A. D'Amore, D. Giugliano, A. Ceriello, M. Varricchio and F. D'Onofrio, *Am. J. Clin. Nutr.* 1993, **57**, 650.
4. P. Knekt, A. Reunanen, J. Marniemi, A. Leino and A. Aromaa. *J. Int. Med.* 1998, **244**, (in press).
5. A. Reunanen, P. Knekt, R-K. Aaran and A. Aromaa, *Eur. J. Clin. Nutr.* 1998, **52**, 89.
6. J. T. Salonen, K. Nyyssönen, T-P. Tuomainen, P. H. Mäenpää, H. Korpela, G. A. Kaplan, J. L. Lynch, S. P. Helmrich and R. Salonen, *Br. J. Med.* 1995, **311**, 1124.

BODY IRON STORES, VITAMIN E AND THE RISK OF TYPE 2 DIABETES

Jukka T. Salonen[1], Tomi-Pekka Tuomainen [1], Kristiina Nyyssönen[1], Kari Punnonen[2].

[1]Research Institute of Public Health and [2]Department of Chemistry, University of Kuopio, P.O. Box 1627, SF 70211 Kuopio, Finland.

1 OXIDATIVE STRESS AND TYPE 2 DIABETES

A role has been suggested for free radical stress and lipid peroxidation in the etiology of non-insulin-dependent diabetes mellitus (NIDDM).[1,2] In cross-sectional studies, patients with NIDDM have had increased production of reactive oxygen species in mononuclear cells, enhanced oxidative damage to DNA and elevated levels of indicators of lipid peroxidation.[3-6] In addition, measures of lipid peroxidation have been correlated with serum insulin levels.[4]

There is some evidence from *in vitro* studies of the mechanisms by which free radical stress could induce insulin resistance and thus cause NIDDM. Rudich and coworkers observed that oxidative stress reduces the insulin responsiveness of adipocytes.[7] Recently, they also showed that a prolonged oxidative stress attenuates the activation of glucose transporters (GLUT1 in myocytes and GLUT4 in adipocytes) by insulin, impairing glucose transport and inducing insulin resistance.[8,9] Olsen and coworkers reported that physiologic concentrations of hydrogen peroxide can inhibit insulin receptor and reduce insulin signalling in adipocytes.[10]

2 HAEMOCHROMATOSIS, IRON ACCUMULATION AND TYPE 2 DIABETES

Type 2 or non-insulin dependent diabetes mellitus (NIDDM) is a common complication of iron overload diseases such as haemochromatosis; 50-80 % of patients with hereditary haemochromatosis (HH) develop diabetes.[11-12] HH is the most common autosomal recessive genetic disorder in Northern European populations. We found earlier in over 1000 middle-aged men elevated fasting serum insulin and blood glucose levels in persons with high serum ferritin.[13] Iron is a catalyst of free radical stress, and oxidative stress may impair insulin action.

We tested the hypothesis that iron accumulation in the body predicts the development of NIDDM in a prospective 4-year follow-up study of 985 randomly sampled non-diabetic men aged 42-60 from East Finland, area with high NIDDM incidence.[14] We estimated body iron stores by the ratio of levels of transferrin receptor (TfR) and ferritin in frozen serum samples drawn at baseline examinations for 41 men who later developed NIDDM during the 4-year follow-up and for 82 controls who were matched for age, four measurements of obesity, baseline glucose and insulin and eight other strongest risk factors.

In a logistic model, men with high iron stores (lowest quartile of TfR/ferritin <9.4 µg/µg) had a 2.4-fold (95 % confidence interval (CI) 1.03 to 5.5, p=0.043) risk of NIDDM, compared with men with lower iron stores. In a step-up model adjusting for the strongest other predictors, baseline serum triglycerides and glycosylated proteins, the respective odds ratio was 2.5 (95 % CI 1.1 to 6.0, p=0.039).[14]

This is the first study showing an association between body iron stores and the incidence of diabetes. Our data provide support to the theory that elevated body iron stores, even in the range not considered haemochromatotic, contribute to the development of type 2 diabetes. As the incidence of NIDDM is very high and rising in populations of Northern European origin, the screening for mutations associated with haemochromatosis by genotypic analysis, monitoring of iron status and iron-depleting treatment could potentially constitute new important measures in the primary prevention of type 2 diabetes.

3 VITAMIN E AND THE INCIDENCE OF TYPE 2 DIABETES

As there were no previous epidemiologic studies of the role of antioxidants with regard to the incidence of diabetes, we studied whether low vitamin E status is a risk factor for incident type 2 diabetes in a population-based follow-up study.[15] Diabetes assessed at baseline and at the four-year follow-up in 944 men aged 42-60, who had no diabetes at the baseline examination.

Oral glucose tolerance test was performed at the four-year follow-up. A subject was defined diabetic if he had either (a) fasting blood glucose ≥6.7 mmol/l, or (b) 2-h post glucose-load blood glucose ≥10.0 mmol/l, or (c) clinical diagnosis of diabetes with either dietary, oral or insulin treatment.

Forty-five men developed diabetes during the follow-up. In a multivariate logistic regression model including the strongest predictors of diabetes, a low lipid-standardised plasma vitamin E (below median) was associated with a 3.9-fold (95 % confidence interval 1.8 to 8.6, p=0.0008) risk of incident diabetes. A decrement of 1 &mol/l of uncategorised unstandardised vitamin E concentration associated with an increment of 22 % in the risk of diabetes, when allowing for the strongest other risk factors as well as serum low density lipoprotein cholesterol and triglyceride concentrations (p=0.0004).

Our findings were subsequently confirmed in another population-based follow-up study in Finland.[16] These findings showed a strong independent association between low vitamin E status before follow-up and an excess risk of type 2 diabetes and provide some support for the theory that free radical stress has a role in the causation of type 2 diabetes.

3 CONCLUSION

Data from both basic research and epidemiologic studies concerning the role of oxidative stress, catalysts of oxidative stress and antioxidants in the etiology of type 2 diabetes are very limited. Current findings are very provocative and certainly warrant further both prospective cohort studies as well as antioxidant supplementation and iron depletion trials in persons with antioxidant deficiencies and elevated iron stores, respectively.

References

1.   L. Oberley, *Free Radical Biol. Med. 1988*, **5**, 113.
2.   J.M. Gutteridge, Fr*ee Radic. Res. Commun. 1993*, **19**, 141.
3.   J.T. Salonen, K. Nyyssönen, T.-P. Tuomainen, P.H. Mäenpää, H. Korpela, G.A. Kaplan, J. Lynch, S.P. Helmrich, Salonen R, *Brit. Med. J.* 1995, **311**, 1124.
4.   J.T. Salonen, T.-P. Tuomainen, K. Nyyssönen, H.-M. Lakka and K. Punnonen, *Brit. Med. J.* 1998, **317**, 727.
3.   N.K. Gopaul, E.E. Anggard, A.I. Mallet, D.J. Betteridge, S.P. Wolff and J. Nurooz-Zadeh, *FEBS Lett.* 1995, **368**, 225.
4.   L.K. Niskanen, J.T. Salonen, K. Nyyssönen and M.I. Uusitupa, *Diabet. Med.* 1995, **12**, 802.
5.   P. Dandona, K. Thusu, S. Cook, B. Snyder, J. Makowski, D. Amstrong and T. Nicotera, *Lancet.* 1996, **347**, 444.
6.   G. Bellomo, E. Maggi, M. Poli, F.G. Agosta, P. Bollati, and G. Finardi, *Diabetes.* 1995, **44**, 60.
7.   A. Rudich, N. Kozlovsky, R. Potashnik and N. Bashan, *Am. J. Physiol.* 1997, **272**, E 935.
8.   A. Rudich, A. Tirosh, R. Potashnik, R. Hemi, H. Kanety and N. Bashan, *Diabetes.* 1998, **47**, 1562.
9.   A. Rudich, A. Tirosh, M. Khamaisi, D. Pessler, R. Potashnik and N. Bashan, *Diabetologia.* 1998, **41**(Suppl 1), A 34.
10. G.S. Olsen, L.L. Hansen and L. Mosthaf, *Diabetologia.* 1998, **41**(Suppl 1), A25.
11. D.L. Witte, W.H. Crosby, C.Q.. Edwards, V.F. Fairbanks and F.A. Mitros, *Clin. Chim. Acta.* 1996, **245**, 139.
12. Yaouanq JM, Diabetes and haemochromatosis: current concepts, management and prevention. *Diabete Metab.* 1995, **21**, 319.
13. T.-P. Tuomainen, K. Nyyssönen, R. Salonen, A. Tervahauta, H. Korpela, T. Lakka, G.A. Kaplan and J.T. Salonen JT, *Diabetes Care.* 1997, **20**, 426.
14. J.T. Salonen, T.-P. Tuomainen, K. Nyyssönen, H.-M. Lakka and K. Punnonen, *Brit. Med. J.* 1998, **317**, 727.
15. J.T. Salonen, K. Nyyssönen, T.-P. Tuomainen, P.H. Mäenpää, H. Korpela, G.A. Kaplan, J. Lynch, S.P. Helmrich and R. Salonen R, *Brit. Med. J.* 1995, **311**, 1124.
16. A. Reunanen, P. Knekt, R.K. Aaran, A. Aromaa, *Eur. J. Clin. Nutr.* 1998, **52**, 89.

VITAMIN C STATUS AND OXIDATIVE STRESS IN DIABETICS

D. Naidoo, O. Lux and M. Phung.

Department of Clinical Chemistry, The Prince of Wales Hospital, High Street, Randwick, NSW 2031, Australia.

1 INTRODUCTION

Oxidative stress is postulated to be increased in diabetic patients.[1] Accumulating evidence suggests that oxidative cell injury caused by free radicals contributes to the development of diabetic complications. On the other side, a decreased efficiency of antioxidant defences seems to correlate with pathological tissue changes. In this study we investigated the susceptibility of erythrocytes from diabetic patients to *in vitro* oxidation in comparison to non-diabetics. The relationship of this parameter to metabolic control and plasma levels of vitamin C was determined.

2 METHODS

Sixty-six diabetic patients and forty-eight non-diabetic subjects were investigated. Fasting bloods were collected into heparinised tubes. A 20 % v/v erythrocyte suspension was incubated at 37 °C with the free radical generator 2,2'-azobis (2-amidinopropane) dihydrochloride (AAPH). The degree of haemolysis was measured by absorbance at 405 nm and a haemolytic index (T50) was calculated from a graph of haemolysis versus time.[2] Diabetic control was determined by the levels of haemoglobin A_1C and fructosamine. Vitamin C levels were measured using an HPLC method with electrochemical detection.

3 RESULTS

We observed a statistically significant reduction in the haemolytic index in the diabetic group (Mann-Whitney $p<0.005$) compared to non-diabetics (Table 1, Fig. 1).

Table 1
Vitamin C, Fructosamine and Haemolytic Index in Diabetics and controls.

	Vit Cμmol/L	Fructosamine μmol/L	haemolytic index (min)
Diabetic (n=66)	42 ± 3	318 ± 10	64 ± 2
Non-Diabetic (n=48)	66 ± 6	222 ± 2,5	79 ± 2

All data expressed as mean ± SEM.

The correlation between the haemolytic index and fructosamine or haemoglobin A_1C levels in the diabetic group was not significant. Plasma levels of vitamin C were significantly lower in diabetic patients compared to non-diabetics (Mann-Whitney p<0.005) (Table 1), but no correlation was found with fructosamine or haemoglobin A_1C levels.

Figure 1

A Typical Haemolysis Versus Time Curve.

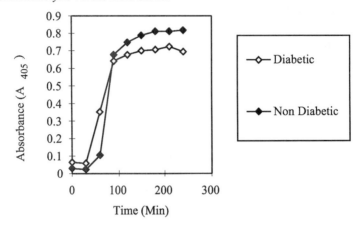

4 DISCUSSION

In this study we investigated plasma vitamin C status in diabetic individuals compared with non-diabetics and the susceptibility of erythrocytes to oxidation *in vitro* and related these parameters to glycaemic control. Our study confirmed that diabetic patients have significantly lower levels of plasma vitamin C. Previous studies have shown that low ascorbate levels in diabetes appears to be a consequence of the disease itself and not due to inadequate dietary intake of vitamin C.[3] Altered metabolism of vitamin C or increased consumption because of increased free radical production are possible mechanisms.

A significant increase in erythrocyte susceptibility to oxidation was observed as reflected by shorter T50 values. The damage induced in erythrocytes by the radical initiating reaction with AAPH can be considered a good model of antioxidant defence status in biomembranes. Previous studies suggest that AAPH damage is produced at the membrane surface, so diminished antioxidants at the membrane level, or even intracellularly, may contribute to lowered membrane ability to withstand peroxidative damage. In a subgroup of diabetics in this study (n=11) we measured red cell ascorbic acid levels and found they were significantly lower than in non-diabetics (p<0.005). Also, hyperglycaemia can cause membrane lipid peroxidation and osmotic fragility in human red blood cells via non enzymatic glycation and glucose autooxidation. This is probably another factor contributing to the observed increased haemolysis, even though there was no significant correlation between the levels of fructosamine or glycated haemoglobin to the haemolytic index.

In conclusion, there are different biochemical pathways in chronic hyperglycaemia, which can result in an additional load of oxygen species in diabetes. This study demonstrated lower levels of vitamin C in plasma and red cells of diabetic patients compared to non-diabetics, and increased susceptibility to oxidative stress *in vitro*.

References
1. S.P. Wolf, Z.Y. Jiang, J.V. Hunt.*Free Rad. Biol. Med.* 1990, **10**, 339-352.
2. A.J. Sinclair, P.B. Taylor, J. Lunec, A.J. Girling and A.H. Barnett.*Diab. Med.* 1994, **11**, 893-898.
3. C. Regnault, E.R.R. Postaire, G.J.P. Rousset, M. Bejot and G.F. Hazebroueq. *Ann.Pharmacother.* 1993, **27**, 1349-50.

ANTIOXIDANT STATUS OF INSULIN DEPENDENT DIABETICS

M. Agnes Cser, I. Sziklai-László[1], Maurice F. Laryea[2], Ingrid Lombeck[2].

[1]Bethesda Children's Hospital of the Hungarian Protestant Church, KFKI Atomic Energy Research Institute, Budapest, Hungary. [2]Children's Hospital, University of Düsseldorf, Federal Republic of Germany.

Keywords: diabetes, child, selenium, glutathione peroxidase, vitamin A, vitamin E, antioxidants.

1 INTRODUCTION

Diabetes mellitus could be best characterized by insufficient insulin synthesis leading to hyperglycaemia and glycosuria. However it is a heterogenous disease, there is not a clear separation between type I and type II diabetes.

Free radicals may play a role in the cause of the disease.[23] Hyperglycaemia can lead to the glycation of tissue proteins.[16] Glycation and glucose autooxidation[17] is generating hydrogen peroxides, hydroxyl radicals and protein-reactive ketoaldehydes. Hyperglycaemia can also lead to increased lipid peroxidation, superoxide production,[1] glycation of the lipoproteins[18] oxidative DNA damage,[8] increased platelet aggregation induced by increased thromboxane synthesis[30] and decreased production of the prostacyclin activators.[3,14] Defense against free radical damage include tocopherol (vitamin E), ascorbic acid, beta carotene-retinol (vitamin A), several metalloenzymes including GSH-Px and Se.[19] Nutritional factors have great influence in the management of diabetes mellitus and in early prevention of complications. Se being an integral, part of the GSH-Px has protecting role against oxidative tissue damage.[11] The biologically important hydrogen donors include thiol systems such as glutathione.[22] In addition to functioning as radical savangers, tocopherols (vitamin E) and beta-carotene, the major carotenoid precursor of vitamin A are known to react with singlet molecular oxygen mainly in cellular membranes.[27] The major nutritional sources of food Se for the human are some forms of proteins and plant-derived foods. Grain crops, cereals high in carbohydrates have nearly 80 % Se absorption rates.[32,36] We questioned the role of non-sugar carbohydrates derived Se intake, their effects on blood Se status and urinary Se excretion in insulin dependent juvenile diabetics.

2 MATERIALS AND METHODS

Two consecutive studies were carried out. Thirtyfive diabetic and 127 healthy, 6-17 years old children were investigated in order to measure Se and GSH-Px status. Eight diabetic and eight healthy children were selected to record total daily diet and collect urine over 24 h in three consecutive days.   All children were within normal body weight and height range, none of them was obese. 4 ml EDTA blood was taken in fasting state. Informed consent was obtained from one or both parents and of each child. Protocols were approved by the local Ethical Committee on Research. Blood and urine Se by AAS and GSH-Px (EX.1.11.1.9.) activities by coupled test systems were measured as before.[6] The coefficients of variance were 3.6 % within day and 4.6 % between test days. For GSH-Px two different plasma and erythrocyte pools were used. The intra-assay and inter-assay coefficients of variation were 3.3 % and 5.4 % in measuring plasma and 3.4 % and 5.7 % in measuring erythrocyte GSH-Px activity. Plasma glucose by glucose oxidase, plasma triglycerides, cholesterols, HDL-cholesterols by kits (Boehringer, Mannheim, Germany), urine creatinine by Meckotest (Merck, Darmstadt, Germany), retinol, tocopherols by HPLC (Waters model 501 with U6K injector) were determined. Food samples were analyzed by instrumental neutron activation analysis (INAA).

3 RESULTS

Age, gender ratio, body weight, height and haematological parameters were similar in diabetics and healthy children. Plasma cholesterols (4.5±0.9 v. 4.2±0.7 mmol/L) were similar, but HDL-cholesterols were lower (1.8±0.4 v. 2.0±0.3 mmol/L) and triglycerides (0.62±0.04 v. 0.39±0.11 mmol/L) were higher in diabetics. Retinols in plasma were lower in diabetics (1.66±0.48 v. 1.87±0.51), but γ tocopherols (1.22±0.51 v. 1.07±0.34 μol/L, p<0.05), tocopherols (18.01±4.71 v. 12.76±2.59 μol/L, p<0.001) were higher in diabetics. Significant, positive correlations were found between plasma Se and α-tocopherol concentrations both in healthy children (r=0.83, n=25, p<0.001) and diabetics (r=0.65, n=35, p<0.01). Total lipids correlated with total tocopherols only in the diabetic patients (r=0.61, n=35, p<0.01), but not in the healthy group. Se content of erythrocytes (1.4±0.2 v. 1.1±0.2 μol/L), of blood (1.1±0.1 v. 0.8±0.1 μol/L), and of plasma (0.9±0.1 μol/L v. 0.7±0.1 μol/L), were significantly higher in diabetics. The GSH-Px activity in the plasma was 93±12 U/L in healthy children and 118±17 U/L, significantly higher in diabetics. Plasma GSH-Px activity increased with enhanced blood glucose levels in diabetics (r=0.56, n=35, p<0.001). The erythrocyte GSH-Px activity was 6.4±0.8 U/gHb in the healthy children and 6.9±1.3 U/gHb in diabetics. Total calory intakes were not different, but diabetics consumed significantly more proteins (62±3 v. 36±4 g/d), more colloid carbohydrates (370±10 v. 280±8 g/d) and less fat (90±3 v. 135±2 g/d) in absolute amount or calculated for kg body weight. The Se intake from carbohydrate sources was significantly higher in diabetics (33±7 μg/d) compared to healthy children (17±8 μg/d). The Se content of food samples produced in Hungary and ready for consumption, such as breads, rice, noodles, spaghetti sorts, potato are separately described.[31] The urinary Se output was similar in diabetics (15.8±1.5 μgSe/g creatinin) in healthy (10.7±1.7 μgSe/g creatinin) children.

4 DISCUSSION

When selenium intake is low because of low Se content in food ingredients resulted by low soil Se concentrations, a well-balanced diet of diabetic patients could allow them to consume more Se than that of the healthy population. The higher Se intake with carbohydrates could explain our observations, since grain products are the primary sources for selenium in the infant and childhood diet[25] and diabetics consumed 20 % more carbohydrates than the healthy children. Fat has negligible, but proteins have high Se content.[25] Therefore the absolute total daily Se intake would also be higher in the diabetics than in the healthy children.

Our present results on blood Se contents are in agreement with previous observations on Hungarians, Germans[5] and with findings of Swedish[12,13] and Finnish authors.[33] Finnish diabetics had higher blood Se status than healthy when Se supply was low, and the difference disappeared after Se was supplemented by soil fertilization[33] resulted in much higher blood Se status in the population. In diabetics with retinopathy or impaired kidney functions blood Se status was not different than that in the healthy.[15,25] Our results demonstrated that diabetics had more Se both in cellular and extracellular compartment of blood, opposite of the findings of authors reporting results in adult diabetics with decreased erythrocyte GSH-Px activities.[15] Erythrocyte Se content of healthy were lower than that of diabetics but lower than that of healthy German children.[5] Plasma GSH-Px activity has been proven to be an appropriate marker of short-term changes of Se intake.[31] Both plasma and erythrocyte GSH-Px activities were dependent of plasma or erythrocyte Se concentrations in healthy children,[7] indicating a low Se state observed in Hungary similar to observations in New Zealand.[20] No such correlations were detected in diabetics, suggesting that they their higher Se status was induced by different diet regimes and higher Se intake.

Blood glucose levels and glucosuria seem to alter the GSH-Px activities in plasma of the diabetic children. Whether the hyperglycaemia could induce an increased plasma GSH-Px activity is an open question, no data of literature support or oppose our findings. We believe that good glycaemic control support the optimal use of the present micronutrient status. All-trans-retinol and α-tocopherol levels depend on the intestinal absorption.[10] *In vivo* studies revealed that Se was bound to high density lipoproteins shortly after ingestion in the form of dimethyl selenide and lipoprotein apoproteins used selenomethionine.[2] The distinct differences in associations between Se status and lipids and lipoproteins in diabetics and healthy children can only be explained by the difference between their general metabolic states. GSH-Px activity has a protective role against tissue damage caused by peroxides produced from lipid metabolism.[28]

Further biochemical details are still missing and waiting for explanations. Defences against free radical damage include tocopherol and retinol concentrations. Natural alpha-tocopherols are potentially beneficial antioxidants of the prevention of oxidative damages. Our results on vitamin E are in accordance to American[9] investigators who have found that diabetics had higher plasma tocopherol levels than healthy normal subjects. It is likely that diabetics may attempt to compensate for the increased adhesiveness of diabetic platelets. The status of vitamin A in diabetic patients is not well studied. At low concentrations vitamin A stimulates insulin release while at high concentrations it has an inhibitory effect which may be mediated by imparment of intracellular glucose oxidation.[4] At present, there is no evidence to suggest that diabetic patients may have vitamin A deficiency.[21] Urinary Se output was not different in diabetics and this can be explained by the facts that they did not have proteinuria and creatinine clearance were also normal. We conclude that the activity of selenium-dependent antioxidative systems of insulin dependent young diabetic patients in good glycaemic state was higher that that of healthy children.

Reference

1. Baynes J.W. *Diabetes*. 1991, **40**, 405-412.
2. Burk R.F. *Biochimica Biophysica Acta*. 1974, **372**, 255- 265.
3. Butkus A., Skrinska V.A., Schumacher O.P. *Thromb. Res*. 1980, **19**, 211-216.
4. Chertow B.S., Buschmann R.JH., Kaplan R.L. *Diabetes*. 1979, **28**, 754-761.
5. Cser A., László-Sziklai I., Menzel H., Zaβ R., Lombeck I. In: Mengen- und Spurenelemente 11. Arbeitstagung, 1991, Gebrüder Mugler GmbH., Oberlungwitz, Sachsen 1991, pp. 1-8.
6. Cser A., Sziklai-László I., Menzel H., Lombeck I. *J. Trace Elem. Electrolytes Health Dis*. 1993, **7**, 205-210.
7. Cser M. A., Sziklai-László I., Menzel H., I. Lombeck. *J.Trace Elements Med. Biol*. 1996, **10**, 167-173.
8. Dandona P., Thusu K., Cook S., Snyder B., Makowski J., Armstrong D., Nicotera T. *The Lancet*. 1996, **347**, 444-445.
9. Darby W.J., Ferguson M.E., Furman R.H., Lemley J.M., Ball C.T., Meneely G.R. *Ann. NY. Acad. Sci*. 1994, **52**, 328-333.
10. Favier A. In: Selenium in Medicine and Biology. Néve J.A. Favier eds., Walter de Gruyter, *Berlin*. 1989, pp. 29-50.
11. Ganther H.E., Hafeman D.G., Lawrence R. A., Serfass R. E., Hoekstra W. G.In: Trace elements in human health and disease. Vol.II.eds. Prasad A.S., Oberlas D. Acad. Press New York, 1985, pp.165-234.
12. Gebre-Medhin M., Ewald U., Platin L.O., Tuvemo T. *Acta Paediatr. Scand*. 1984, **73**, 109-114.
13. Gebre-Medhin M., Ewald U., Tuvemo T. *Uppsala J. Med. Sci*. 1988, **93**, 57-62.
14. Gerarad J.M., Stuart M.J., Rao G.H.R., Steffes M.W., Mauer S.M., Brown D.M., White J.C. *J. Lab. Clin. Med*. 1980, **95**, 950-954.
15. Holler C.F., Ulberth W., Osterode K., Irsigler K. *Bioavailability, Ettlingen*. 1993, May 9-12. , pp. 235-238.
16. Hunt J.V., Wolff S.P. *Free Rad. Res.Comms*. 1991, **12-13**, 115-123.
17. Hunt J.V., Dean R.T., Wolff S.P. *Biochem. J*. 1988, **256**, 205-212.
18. Koschinsky T. In:Abnormalities in Subclinical Diabetic Angiopathy. Eds.:Weber B., Burger W., Danne T., *Pediatr. Adolesc. Endocrinol. Basel, Karger*. 1992, **22**, 32-43.
19. Machlin L.J., Bendich A. *FASEB*. 1987, **1**, 441-445.
20. McKenzie R.L., Rea H.M., Thomson C.D., Robinson M.F. *Am J. Clin. Nutr*. 1978, **31**, 1423-1418.
21. Moradian A.D., Moirley J.E. *Am. J. Clin. Nutr*. 1987, **45**, 877-895.
22. Niki E. *Chem. Phys. Lipids*. 1987, **44**, 227-253.
23. Oberley L. W. *Free Rad. Biol. Med*. 1988, **5**, 113-124.
24. Pennington J.A., Hendricks T.C., Douglass J.S., Petersen B., Kidwell *J. Food Addit. Conram*. 1995, **12**, 809-820.
25. Pennington J.A.T., Schoen S. A. *Internat. J. Vit. Nutr. Res*. 1996, **66**, 350-362.
26. Rückgauer M., Zeyfang A., Krusae-Jarres J.D. Clinische Chemie, 1997, **12**, 805-809.
27. Sies H., Murphy M.E., Di Mascio P., Stahl W. In:Lipid-soluble antioxidants. *Biochemistry and Clinical Applications*. A.S.H.Ong & L.Parker 1992 Birkhäuser

Verlag, Basel, Switzerland pp. 160-165.

28.   Sundee R.A., Hoekstra W.G. *Nutr. Rev.* 1998, **38**, 265-273.
29.   Sziklai László I., Cser M.A.: Evaluation of selenium content of food, infant formulas and human milk samples from Hungary by INAA. In press 1999.
30.   Thomas G., Skrinska V., Lucas F.V., Schumacher P. *Diabetes.* 1985, **34**, 951-954.
31.   Thomson C.D., Rea H.M., Doesburg V.M., Robinson M.F. *Br. J. Nutr.* 1977, **37**, 457-460.
32.   Thomson C. D., Robinson M.F. *Am. J. Clin. Nutr.* 1980, **33**, 303-323.
33.   Wang W.C., Mäkelä A.L., Näntö V., Mäkelä P. Biol. *Trace Element. Res.* 1995, **47**, 355-364. In: Methods of Enzymology. Vol.77, ed. Jacoby W.B., Academic Press Inc. London, pp. 325-333.
34.   World Health Orgaization Geneva: Trace elements in human nutrition and health. Macmillan-Ceuterick, India-Belgium, 1996.

Dietary Intake, Bioavailability and Antioxidative Effects of Flavonoids and Phenolics

DESIRABLE VERSUS HARMFUL LEVELS OF INTAKE OF FLAVONOIDS AND PHENOLIC ACIDS

Vibeke Breinholt.

The Danish Veterinary and Food Administration, Institute of Food Safety and Technology, Division of Biochemistry and Molecular Toxiclogy, Mørkhøj Bygade 19, 2860 Søborg, Denmark.

1 INTRODUCTION

Flavonoids and phenolic acids are plant components ubiquitously present in fruits and vegetables commonly consumed by humans. The two groups of compounds constitute a major portion of the daily intake of non-nutrients, and have for several years been regarded as potential human anticarcinogens, and as protective factors against cardiovascular diseases.

Epidemiological studies, however, are not conclusive regarding the potentially protective role of the flavonoids against human cancer, whereas some epidemiological data point towards a protective function of the flavonoids against human coronary heart disease. The actions of phenolic acids have not been the topic of major epidemiological studies despite their high dietary levels compared to other phytoprotectants, and despite the existence of some documentation of anticarcinogenic actions in experimental animals and their potentially cardioprotective activities. However, it is foreseen that this group of compounds will receive increasing attention in the following years with regard to their potentially health-promoting effects in humans.

The current manuscript will provide a short review of the present knowledge regarding beneficial and adverse effects imposed by dietary flavonoids and phenolic acids in experimental animals and humans, and will present some of the proposed mechanisms involved. The observed effects will be discussed with respect to the dietary levels required to elicit a protective or a harmful action, and the approximate dose-levels of the compounds found in an average human diet.

2 ANTIOXIDANTS AND THEIR POTENTIAL ROLE IN CANCER AND CARDIOVASCULAR DISEASE

Oxidative processes and the generation of reactive oxygen species (ROS) are involved at various steps in the cancer process[1,2] and presumably also play a role in the development of cardiovascular diseases. ROS have the ability to induce alterations of the genetic material by oxidizing specific DNA-bases or indirectly by oxidizing proteins or lipids important in maintaining integrity and homeostasis of the individual cell. The produced functional damage can trigger events such as mutagenesis, carcinogenesis and aging, events that antioxidants might counteract. Increasing evidence furthermore suggest that oxidation of low-density-lipoproteins (LDL) in human beings is atherogenic. Avoidance and delay of LDL oxidation by dietary antioxidants might thus be beneficial in the prevention of atherosclerosis and consequently coronary heart disease.

Endogenous antioxidant defenses, such as catalase and superoxide dismutase and metal binding proteins are inadequate to prevent oxidative damage completely, and thus diet-derived antioxidants are important in maintaining health. In recent years much attention has focused on the potentially protective role of antioxidants in the prevention of cancer and heart disease in humans, and in particular the preventive role of antioxidants present in commonly consumed foods.

The flavonoids and phenolic acids have received major attention as potentially protective factors against cancer and heart diseases in part by virtue of their potent antioxidative properties, and their ubiquitous presence in a wide range of commonly consumed foods.

3 DIETARY INTAKE OF FLAVONOIDS AND PHENOLIC ACIDS

Current flavonoid intake levels in northern Europe have been estimated at around 50-150 mg/d for total flavonoids, including flavanols (catechins), flavanones, flavones, flavonols, and anthocyanins.[3] Similar exposure levels to flavonoids are found in other European contries.[4] Intake data from the United States, however, indicate that the total flavonoid intake might be in the order of 1 g/d.[5] In populations that consume high amounts of tea or red wine, which are particularly rich in catechins, the total intake of flavonoids may indeed approach the estimated U.S. level.

Intake levels of phenolic acids are expected to be somewhat higher, than what is evident for the flavonoids, since their concentrations are very high in several important fruits and vegetables, e.g. apples, potatoes, cabbage, tomatoes and in beverages like wine and coffee (Table 1). The intake of phenolic acids are thus expected to exceed several grams per day, however, more exact information on the human exposure to this group of compounds is warranted. Despite the fact that the simple phenolic acids is one of the most abundant groups of potentially human dietary anticarcinogens very little is known about the actual human exposure to this group of compounds. Sporadic reports on the content of phenolic acids in selected food items are found in the literature (Table 1), however more detailed surveys on the daily intake of this groups of compounds still need to be conducted.

Table 1

Content of phenolic acids in various foods.

Phenolic Acid	Dietary Source	Content of important food source
Caffeic acid	apples, potatoes, coffee beans, white wine, olives, spinach, cabbage, asparagus, lettuce	fresh apples (100-130 mg/100 g)
Chlorogenic acid	coffee beans, apples, pears, cherries, plums, peaches, blueberries, tomatoes	fresh apples (100 mg/100 g) coffee (8% of the fresh bean) soy beans (1%)
Ferulic acid	wheat, rice, corn, tomatoes, spinach, cabbage, asparagus, soy, beans	chia seeds (200 mg/100 g)
Coumaric acid	white wine, tomatoes, spinach, cabbage, citrus fruits (peel), strawberries, parsley	
Gallic acid	cherries, black and green, grapes	red wine (95 mg/L)

4 ABSORPTION AND METABOLISM OF DIETARY FLAVONOIDS AND PHENOLIC ACIDS

With the exception of some of the isoflavonoids and the catechins, which are extensively absorbed in humans and experimental animals, flavonoids are in general poorly absorbed. Generally less than 1 % of the total administered dose reaches the systemic circulation, and absorption most often ranges from 0.2 to 0.5 %. The non-absorbed residue in the gastrointestinal tract, however, undergoes extensive metabolism, mediated by microbial enzymes in the colon, leading to the formation of a wide range of simple phenolic acids[6,7,8] (Fig. 1). In contrast to the parent compound the simple phenolic degradation products are readily absorbed, with only a small portion of the administered flavonoid leaving the gastrointestinal with an intact flavan nucleus.[6] As the flavonoid degradation products, produced in the colon, exhibit distinct biochemical activities, of which antioxidation is just one, the effect of flavonoid administration to experimental animals or flavonoid consumption in humans, might thus rather be a combined effect of the parent compound and the simple phenolic degradation products, than a result of the flavonoid *per se*.

After systemic absorption, the simple phenolic degradation products can undergo additional modifications (Fig. 1). *In vitro*[9] and unpublished *in vivo* data from our laboratory furthermore suggest that flavonoids via hepatic enzymes, primarily the cytochrome P450 1A isozymes,[9] can be converted by addition of hydroxyl groups or removal of methoxy groups. As most, if not all, of the biochemical and biological activities associated with the flavonoids, are highly dependent on the number and position of substituents on the 3-ring flavan nucleus, it is evident that hepatic or microsomal biotransformation of flavonoids might dramatically change the inherent properties of the parent compound (Fig. 2).

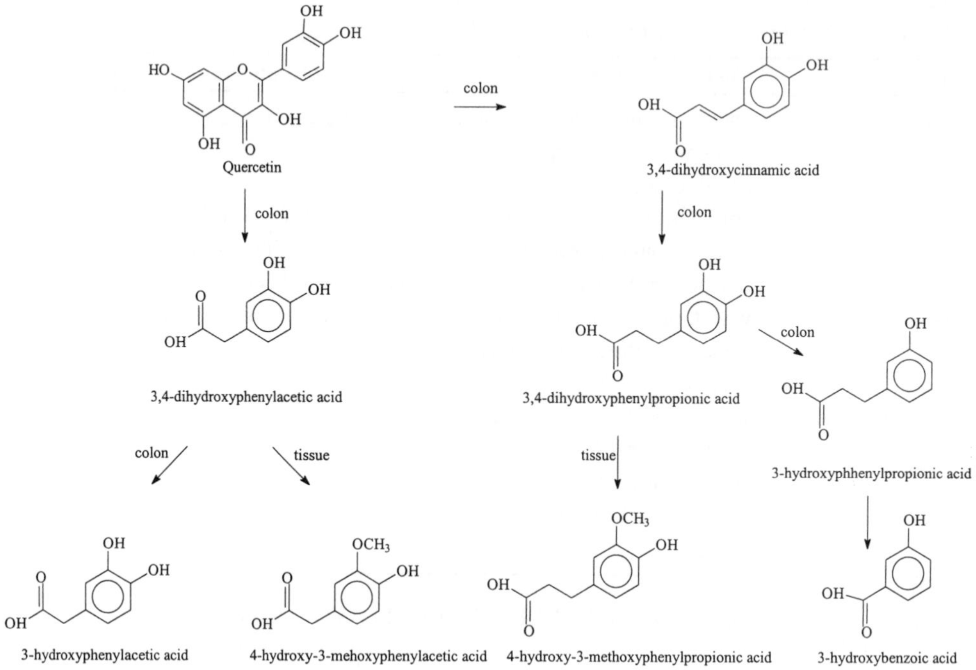

Figure 1

Representative metabolic conversion of quercetin in colon and tissue.

Hormonal activity is, for instance, lost when the estrogenic flavanol naringenin is converted into eriodictyol,[9,10] and chrysin, which exhibits weakly cytochrome P4501A inducing capacity, is hydroxylated to form apigenin which does not induce this enzyme system (11, unpublished data). Chrysin is furthermore far less cytotoxic than its dihydroxylated metabolite luteolin,[12] which on the other hand is a much better antioxidant than chrysin.[13] The *in vivo* formation of these flavonoid metabolites, however, is only minor compared to the total amount of flavonoid administered. The biological relevance of these metabolites is presently not known. Metabolites formed in the are also quantitatively more important.

Based solely on the flavonoid plasma concentrations, given in Table 2 for a small number of flavonoid aglycons and glycosides, it is thus questionable whether humans do ingest enough flavonoids to sustain a sufficiently high blood concentration of the parent compound to influence parameters such as tissue xenobiotic enzymes or blood antioxidant enzymes. As mentioned above the flavonoid degradation products presumable play an important role in the overall protective actions following flavonoid exposure, and it is thus

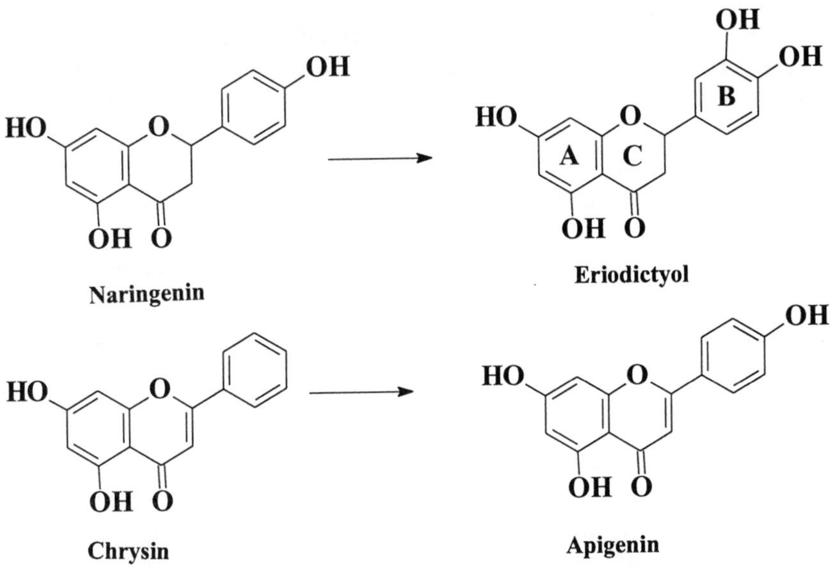

Figure 2
Interconversion between flavonoid subgroups.

conceivable that the intact flavonoid molecule in concert with its degradation products exert the protective functions observed in experimental animals and in humans. The blood level of intact flavonoids as well as of their simple phenolic degradation products need to be further investigated in order to determine the optimal blood levels of flavonoids required to elicit an effect in humans. This information will also allow us to evaluate more critically the human relevance of flavonoid-induced actions in experimental animals, with regard to both adverse and beneficial actions.

Table 2
Peak plasma concentrations of flavonoids in humans after oral administration.

Flavonoid	Dose (mg/kg b.w)	Plasma C_{max} (μg/ml)	Reference
Quercetin	65	< 0.1	54
(+)-Catechin	32	0.064	55
3-methoxycatechin	30	11	56
Diosmin	10	0.4	57
Quercetin-3-rutinoside	1.4	0.09	58

Like the flavonoids, phenolic acids are also universally present in all plant-derived foods and thus in most diets. Phenolic acids (i.e. hydroxycarboxylic acids with phenolic hydroxyl groups) are widely distributed in nature in the forms of their esters, ethers or in their free forms. Some of the hydroxycinnamic acids occur most frequently as simple esters with carboxylic acid or glucose, while hydroxybenzoic acids are present mainly in the form of glucosides. Furthermore phenolic acids often occur in the plant tissue as conjugates with other natural compounds such as flavonoids, hydroxy fatty acids, and sterols, or bound to cell wall polymers. Like the phenolic degradation products of the flavonoids the plant-derived phenolic monomers are also substantially absorbed from the gastrointestinal tract.[7,8,14] The metabolic fate of the phenolic acids in man as well as experimental animals has only been sparsely investigated, however, microbial metabolism of the phenolic acids in the gut, seem to account for the major portion of metabolites formed.[7,8] Phenolic acid degradation is thus characterized by reduction of double bonds, and decarboxylation and demethylation reactions.

5 BENEFICIAL AND HARMFUL EFFECTS OF FLAVONOIDS AND PHENOLIC ACIDS ON THE DEVELOPMENT OF CANCER IN EXPERIMENTAL ANIMALS AND HUMANS

The role of flavonoids in carcinogenesis has been, and still is very controversial. Studies aimed specifically at investigating the potentially modulatory role of flavonoids in experimental animals have been inconclusive due to conflicting results where either significant protective effects, no protective effect or even an enhanced cancer response have been observed, although cancer protection is the most frequent observation. As shown in Table 3 several of the members of the flavonoid family have been found to exert chemopreventive activities in experimental animals against a wide range of cancer types induced by a wide range of carcinogens. The mechanisms of protection are thought to be a combination of several distinct actions. In experimental animals several of the most commonly consumed flavonoids have been found to exhibit a wide range of biochemical and biological activities of potential importance in human cancer protection. The diverse properties of the flavonoids include antioxidant action,[15-17] anti-promoting activity,[18-20] and the ability to inhibit certain enzymes involved in carcinogen activation.[21-24] Furthermore, several of the flavonoids have recently been found to elicit an estrogenic response by interacting with the estrogen-receptors,[10, 25] in addition to affecting multiple enzyme activities involved in cell proliferation and signal transduction, such as kinases, phospholipases and phosphodiesterases (for a review see 26). Provided that the flavonoids reach the target tissue and at a sufficiently high concentration, the described chemopreventive actions might be relevant in humans.

Some of the flavonoids, however, are genotoxic[27-30] and exhibit pro-oxidative activities *in vitro*[16,31] and might thus be harmful to humans and experimental animals. For instance the flavonoid, quercetin, has been found to produce tumors in the rat,[32,33] act as a co-carcinogen of estradiol-induced kidney tumorigenesis in the hamster[34] and of azoxymethanol-induced colon cancer in rats.[35] More recently the isoflavonoid genistein was also found to promote azoxymethanol-induced colon carcinogenesis in the rat model.[36] Although, some of the cancer data on quercetin, have been criticised as over-interpreted, the recent and very elaborate studies on quercetin and genistein,[35,36] provide evidence that some flavonoids do have the potential to act as carcinogens in experimental animals. However, whether the cancer-inducing mechanisms apply to humans, is not yet certain. Some of the arguments against the possible role of quercetin as a human carcinogen is that there is no evidence for carcinogenicity in close to 20 studies employing quercetin doses between 0.25 and 10 % (for review see 37), the level of quercetin in the diet is relatively low, only a small fraction of the ingested quercetin reaches the systemic circulation, and epidemiological studies do not confirm any risk of neoplastic disease in subjects with a normal intake of flavonoids.

From Table 3 it is also evident that the doses of flavonoids at which a harmful or a beneficial effect on cancer is observed by far exceed what humans can normally ingest from dietary sources. It is thus questionable that the observed adverse as well as beneficial effects of flavonoids in experimental animals at all apply to humans. In several epidemiological studies investigating the relationship between diet and cancer, a protective effect of the consumption of fruits and vegetables on various types of cancer has been found,[38-40] however, the role of flavonoids in human cancer prevention is far from conclusive.

Table 3
Effect of flavonoids on tumorigenesis in experimental animals.

Flavonoid	Dose (mg/kg/day)	Carcinogen (target organ)	Decrease (%) in tumor incidence	Reference
Biochanin A	50 (18 wks.)	MNU (breast)	60	59
Myricetin	17 (1 d)	B[a]P (lung)	28	60
Catechin	125 (68 weeks)	Spontaneous (breast)	33	61
Luteolin	250 (15 wks.)	20-MC (fibrosarcoma)	40	62
Quercetin	250 (15 wks.)	20-MC (fibrosarcoma)	48	62
Catechin	250 (15 wks.)	20-MC (fibrosarcoma)	0	62
Diosmin	1000 (22 wks.)	4-NQO (tongue)	68	63
Hesperetin	1000 (22 wks.)	4-NQO (tongue)	78	63
Rutin	1000 (50 wks.)	AOM (colon)	0	64
Quercetin	500 (50 wks.)	AOM (colon)	77	64
Genistein	1000 (3 ds)	DMBA (breast)	60	65
Genistein	250 (55 wks.)	AOM (colon)	-40	66
Apigenin	5,10 μM (11 wks.)	UVA/B (skin)	25-47	67

Abbreviations: MNU= N-nitroso-N-methylurea; B(a)P= benzo(a)pyrene; 20-MC= 20-methyl-cholanthrene; 4-NQO= 4-nitroquinoline-1-oxide, AOM= azoxymethanol, 7,12-dimethyl-benzanthracene, UVA/B= ultraviolet light A and B.

Four epidemiological studies have so far investigated the association between flavonoid intake and cancer risk. Three of the studies did not provide evidence for a protective role of flavonoids on cancers of the alimentary or respiratory tract, stomach, colon or lung[41-43] at flavonoid intake levels ranging from 12-42 mg/d. What is interesting to note is that in all of these studies a significant decrease in the incidence of heart disease was observed. This observation could be taken to indicate that the blood or tissue levels of flavonoids required to elicit an effect on heart disease is somewhat lower than what is required to modulate the cancer process, that flavonoid access to the target tissues differs between the two ailments, or that the major protective mechanisms involved are different. In a study by Knecht et al.[44] an inverse association between the intake of flavonoids and the incidence at all sites of cancer was observed. This association was mainly due to lung cancer, which presented a relative risk of 0.54. In this study the comsumption of apples showed an inverse association with lung cancer after adjustment for the intake of other fruits and vegetables.

Like the flavonoids phenolic acids also exhibit the ability to inhibit various cancer types initiated by various carcinogens (Table 4). The doses required to elicit an effect are still high, but, the high intake of phenolic acids taken into consideration, the experimentally employed doses are somewhat closer to the human exposure level, than is evident for the flavonoids.

Table 4

Effect of phenolic acids on tumorigenesis in experimental animals.

Phenolic acid	Dose (mg/kg/d)	Carcinogen (target organ)	Decrease (%) in tumor incidence	Reference
Caffeic acid	11000 (50 wks.)	B(a)P (forestomach)	40	68
Caffeic acid	5000 (30 wks.)	spontaneous tumors	17	69
Chlorogenic acid	70 (7 wks.)	AOM (colon, ACF)	0	70
Gallic acid	20 μM	TPA (skin)	43	71
Propyl gallate	20 μM	TPA (skin)	76	71
Caffeic acid	5000 (35 wks.)	MMNG (forestomach)	-75	72
Chlorogenic acid	250 (6,12 wks.)	AOM (colon, regression of ACF)	86,72	73
Caffeic acid	500 (7 wks.)	4-NQO (tongue)	27	74
Ferulic acid	500 (7 wks.)	4-NQO (tongue)	39	74
Chlorogenic acid	500 (7 wks)	4-NQO (tongue)	33	74

Abbreviations: B[a]P, benzo(a)pyrene; AOM, azoxymethanol; ACF, aberrant crypt foci; TPA, 12-tetradecanoylphorbol-13-acetate; MMNG, N-methyl-N-nitrosoguanidine; 4-NQO, 4-nitro-quinoline-1-oxide, 7,12-dimethylbenzanthracene, UVA/B, ultraviolet light A and B.

At present no epidemiological studies have been conducted to investigate the potential role of phenolic acids on human cancer. The wide-spread occurrence of phenolic acids in foods of plant origin and in particular the fact that important foods sources, such as wine, tea and apples, which are rich in flavonoids also contain phenolic acids in abundance, severely complicate the interpretation of the role of flavonoids in epidemiological studies, as the observed effect might just as well be mediated by the phenolic acids. If no biological marker, such as serum flavonoid level is employed or the intake of phenolic acids is concurrently assessed it is difficult, if not impossible, to determine the individual or the combined role of the flavonoids and phenolic acids in ameliorating human diseases.

5 BENEFICIAL AND HARMFUL EFFECTS OF FLAVONOIDS AND PHENOLIC ACIDS ON THE DEVELOPMENT OF CARDIOVASCULAR DISEASE IN EXPERIMENTAL ANIMALS AND HUMANS

Several *in vitro* and animal studies have provided promising data on the potentially protective role of flavonoids towards heart disease. In addition to their ability to prevent or delay LDL oxidation, flavonoids are capable of altering a wide range of additional processes which might also play a role in their overall cardioprotective actions. Some flavonoids have for instance been found to inhibit platelet adhesion and aggregation, to act as inhibitors of lipoxygenases and cyclooxygenases, hereby depressing the synthesis of thromboxane A2, to promote fecal excretion of bile acids, which will have a lowering effect on the serum triglyceride level, and to disaggregate preformed thrombi (for review see 26). Some of the commonly occurring flavonoids, such as kaempferol, apigenin and naringenin, furthermore exhibit estrogenic activities,[10] which might also be protective against human heart disease.

In contrast to the weak association between flavonoid intake and cancer protection in humans, an inverse association between coronary heart disease and flavonoid intake has been observed in four out of six epidemiological studies conducted (Table 5). In the first study conducted by Hertog and coworkers [45] the risk for mortality of coronary heart disease was 65 % lower in the highest tertile of flavonoid intake (42 mg/d) when compared to the lowest (12 mg/d). In three other studies the same inverse relationship between flavonoid intake and coronary heart disease was observed.[4,46,47] In two of these studies [4,45] tea was found to be the major source of flavonols, which was the only flavonoid subgroup investigated, and it was found to be inversely correlated with both coronary heart disease and stroke. In a study conducted in Wales [48] where tea drinking is extensive no protective effect of flavonol or tea consumption could be found. In contrast, increased mortality of ischaemic heart disease and total mortality was observed in the tertile of high flavonol intake (43 mg/d) when compared to the lowest tertile (14 mg/d). The most plausible explanation to the adverse effects of flavonoid intake is that high tea consumption in Wales tend to be correlated with a less healthy lifestyle, such as excessive smoking and a high fat intake.[48] The largest prospective cohort study concerning flavonoid intake and coronary heart disease was conducted in the U.S. and employed 34.000 males.[49] In this study no correlation between flavonoid intake and heart disease was observed, although a weak but non-significant inverse correlation was found for flavonoid intake and coronary mortality in subjects who had previously been diagnosed with coronary heart disease. In the Finnish cohort study [47] only a weak association between dietary flavonoid intake and coronary heart disease was observed. In this study, however, only the flavonol intake was measured. Inclusion of the flavanol intake, which seems to dominate the overall flavonoid intake in Finland,[50] might have resulted in different conclusions from this study. All in all, epidemiological evidence points to a possible protective role of flavonoids on cardiovascular diseases, however, the present data are far from conclusive.

Table 5

Epidemiological studies on flavonoid intake and coronary heart disease and stroke.

Population	Age (follow-up/years)	Protection	Flavonoid intake (mg/d)
The Netherlands, Zutphen, 805 men, CHD	65-84 (5)	+	12-42
Finland, 5130 men and women, CHD	30-69 (20)	+ +	> 5.0
The Netherlands, Zutphen, 552 men, stroke	50-69 (15)	+	> 30
U.S.A. Health Professionals, 34 789 men,CHD	40-75 (6)	-	7-40
U.K., Caerphilly, 1900 men, CHD	49-59 (14)	-	14-43

From *in vitro* studies it is evident that the phenolic acids share with the flavonoids the ability to modify a wide range of processes involved in the development of atherosclerosis, as mentioned previously. Simple phenolics such as caffeic, coumaric and protocatechuic acids are capable of preventing LDL oxidation.[51] Evidence also exists that caffeic and coumaric acids act synergistically with α-tocopherol, extending the antioxidant capacity of LDL by recycling α-tocopherol from the α-tocopherol radical,[52] and chlorogenic acid, caffeic acid and ferulic acid all exhibit inhibitory activity toward lipoxygenases and cyclooxygenases.[53] Despite the *in vitro* evidence that simple phenolic acids might be beneficial in the etiology of human heart diseases, no epidemiological studies have so far been conducted to investigate the possible protective role of simple phenolics on cardiovascular disease.

6 CONCLUSION

A large number of epidemiological studies have emphasized the importance of fruits and vegetables in the prevention of various pathologies, such as cancers and cardiovascular diseases. Contradictory *in vitro* and animal experimental data exist concerning the role of flavonoids and phenolic acids in ameliorating human diseases. More *in vivo* studies are needed to determine conclusively whether flavonoids and phenolic acids inhibit the growth of atherosclerotic plaques, thus reducing the risk of artherosclerosis and whether the compounds are capable of exerting a beneficial effect towards human cancer.

Despite the promising data concerning the health-promoting properties of flavonoids and phenolic acids, their administration to humans in a pure state is not forseen. Under certain conditions, such as during disease or, in general, when the intake of fruits and vegetables is inadequate, it could be desirable to increase the intake of antioxidants by increasing the consumption of foods rich in these food constituents.

References
1. Cerutti P., *Lancet*. 1994, **344**, 862.
2. Witz G. *Soc. Exp. Biol. Med.* 1991, 675.
3. Justesen U., Knuthsen P. and Leth T. *J. Chrom. A.*, In press, 1998.
4. Hertog M.G.L, Kromhout D., Aravanis C., Blackburn H., Buzina R., Fidanza F., Giampaoli S., Jansen A., Menotti A., Nedeljkovic S., Pekkarinen M., Simic B.S., Toshima H., Feskens E.J.M., Hollman P.C.H. and Katan M.B. *Arch. Int. Med.* 1995, **155**, 381.
5. Kuhnau J. *World. Rev. Nutr. Diet.* 1976, **24**, 117.
6. Gross M., Pfeiffer M., Martini M., Campbell D., Slavin J. and Potter J. *Cancer Epidemiology, Biomarkers & Prevention.* 1996, **5**, 711-720.
7. Scheline R.R. *Acta. Pharmacol. Toxicol.* 1966, **24**, 275-285.
8. Scheline R.R. *Acta. Pharmacol. Toxicol.* 1968, **26**, 189-205.
9. Nielsen S.E., Breinholt V., Justesen U., Cornett C. and Dragsted L.O. *Xenobiotica*. 1998, **28**(4), 389-401.
10. Breinholt V. and Larsen J.C. *Chem. Res. Tox.* 1998, **11**(6), 622.
11. Canivenc-Lavier M.-C., Vernevaut M.-F., Totis M., Siess M.-H., Magdalou J. and SuschetetM. *Toxicology.* 1996, **114**, 19.
12. Breinholt V. and Dragsted L. *In vitro Mol. Toxicol.* 1998, **11**(2), 193.
13. Rice-Evans C.A., Miller N.J. and Paganga G. *Free Rad. Biol. Med.* 1996, **20**(7), 933.
14. Teuchy H. and Van Sumere C.F. *Arch. Int. Physiol. Biochem.* 1971, **79**, 589-618.
15. Ratty A.K. and Das N.P. *Med. Metabol. Biol.* 1988, **39**, 69-79.
16. Laughton M.J., Halliwell B., Evanes P. and Hoult J.R.S. *Biochem. Pharmacol.* 1989, **38**, 2859.
17. Cholbi M.R., Paya M. and Alcaraz M.J. *Experimentia.* 1991, **47**, 195.
18. Birt D.F., Walker D., Tibbels M.B., and Bresnick E. *Carcinogenesis.* 1986, **7**, 959.
19. Kato R. Nakadate T., Yamamoto S. and Sugimura T. *Carcinogenesis.* 1983, **4**, 1301.
20. Wei H., Tye L., Bresnick E. and Birt D.F. *Cancer Res.* 1990, **50**, 499-502.
21. Vernet A. and Siess M.H. *Fd. Chem. Toxicol.* 1986, **24**, 857.
22. Siess M.H., Leclerc J., Canivenc-Lavier M.C., Rat P. and Suschetet M. *Toxicol. Appl. Pharm.* 1995, **130**, 73.
23. Beyeler S., Testa B. and Perrissoud D. *Biochem. Pharmacol.* 1998, **37**, 1971.
24. Guengerich F.P. and Kim D-H. *Carcinogenesis.* 1990, **11**, 2275-2279.
25. Gustavson J.Å. Abstract presented at Symposium on Phytoestrogen Research Methods. September 21-24, 1997, Tucson, Arizona, U.S.A., 1998.
26. Manach C, Regerat F., Texier O., Agullo G., Demigne C. and Remesy C. Bioavailability, metabolism and physiological impact of 4-oxo-flavonoids. *Nutrition Res.* 1996, **16**(3), 517-544.
27. Czeczot H., Tudek B., Kusztelak J., Szymczyk T., Dobrowolska B., Glinkowska G., Malinkowski J. and Strzeecka H. *Mutat. Res.* 1990, **240**, 209.
28. MacGregor J.T. and Jur, L. *Mutat. Res.* 1979, **54**, 297.
29. Brow, J.P. *Mutat. Res.* 1980, **75**, 243-277.

30. Das A., Wang J.H. and Lien E.J. *Progress Drug Res.* 1994, **42**, 133.
31. Arouma O.I. *Fd. Chem. Toxic.* 1994, **32**(7), 671.
32. NTP. NIH Publication No. 91-3140, 1991.
33. Dunnick J.K. and Hailey J.R. *Fundam. Appl. Toxicol.* 1992, **19**, 423-431.
34. Zhu B.T. and Liehr J.G. *Toxicol. Appl. Pharmacol.* 1994, **125**, 149.
35. Pereira M.A., GrubbsC.J., Barnes L.H., Olson G.R., EtoI., Juliana M., Whitaker L.M., Kelloff G.J., Steele V.E. and Lubet R.A. *Carcinogenesis.* 1996, **17(6)**, 1305.
36. Rao C.V., Wang C-X., Simi B., Lubet R., Kelloff G., Steele V. and Reddy B.R. *Cancer Res.* 1997, **57,** 3717.
37. Formica J.V. and Regelson W. *Fd. Chem. Toxic.* **33(12)**, 1061.
38. Hartman P.E., Shankel D.M. *Environ. Mol. Mutagen.*1990, **15**, 145.
39. Dragsted L.O., Strube M. and Larsen J.C. *Pharmacol. Toxicol.* 1993, Suppl.1, 121.
40. Wattenberg. *Cancer.* 1992, **52**, 2085s.
41. Hertog M-G.L. *Proc. Nutr. Soc.* 1996, **55**, 385.
42. Hertog M-G-L, Feskens E-J.M., Hollman P.C.H., Katan M.B. and Kromhout D. *Nutr. Cancer.* 1994, **22**, 175.
43. Goldbohm R.A., van den Brandt P.A., Hertog M.G.L., Brants H.A.M. and van Poppel G. *Am. J. Epidemiol.* s 141, s 61,1995.
44. Knecht P., Järvinen R., Seppänen R., Heliovaara Markku, Teppo L., Pukkala E and Aromaa A. *Am. J. Epidemiol.* 1997, **146**, 223.
45. Hertog M.G.L., Feskens E.J.M., Hollman P.C.H., Katan M.B. and Kromhout D. *Lancet.* 1993, **342**, 1007.
46. Keli S.O., Hertog M.G.L., Feskens E.J.M. and Kromhout D. *Arch. Int. Med.* 1996, **156**, 637.
47. Knect, P., Jarvinen, R. Reunanen, A. and Maatela, J. BMJ, 1996, **312**, 478-481.
48. Hertog M.G., Sweetnam P.M., Fehily A.M., Elwood P.C. and Kromhout D. *Am. J. Clin. Nutr.* 1997, **65**, 1489.
49. Rimm E.B., Katan M.B., Ascherio A., Stampfer M.J. and Willet W.C. *Ann. Int. Med.* 1996, **125**, 384-389.
50. Kumpulainen. Abstract presented at the Second International Conference on Natural Antioxidants and Anticarcinogens in Nutrition, Health and Disease. June 24-27, 1998, Helsinki, Finland.
51. Abu-Amsha R., Croft K.D., Puddey I.B., Proudfoot J.M. and Beilin L.J. *Clin. Sci. Colch.* 1996, **91(4)**, 449-458.
52. Laranjinha J., Viera O., Madeira V. and Almeida L. *Arch. Biochem. Biophys.* 1995, **323(2)**, 373-381.
53. Huang, M.T., Lysz, T., Ferraro, T., Abidi, T.F., Laskin, J.D. and Conney, A.H.Cancer Res. 1991, **51(3)**, 813-819. 1.
54. Gugler R., Leschik M., Dengler H.J. *Eur. J. Clin. Pharmacol.* 1975, **2**, 229.
55. Lee M.J., Wang Z.Y., Li H. *Biomarkers. Prev.* 1995;4:393-399.
56. Hackett A.M., Gruffiths L.A., Vermeille *M. Xenobiotica.* 1985, **15**, 907.
57. Cova D., De Angelis L., Giavarini F., Palladini G., Perego R. *Int. J. Pharmacol. Ther.*

Toxicol. 1992, **30**, 29.

58. Hollman P.C.H., van Trijp J.M.P., Buysman M.N.C.P., van der Gaag M.S., Mengelers M.J.B., de Vries J.H.M. and Katan M.B. Thesis, 1997, Chapter 6.
59. Gotoh T., Yamada K., Yin H., Ito A., Kataoka T. and Dohi K. *Jpn. Cancer Res.* 1998, **89**, 137.
60. Chang R.L., Huang M.-T., Wood A.W., Wong C.Q., Newmark H.L., Yagi H., Sayer J.M., Jerina D.M. and Conney A.H. *Carcinogenesis.* **6(8)**, 1127.
61. Bhide S.V., Azuine M.A., Moushumi L. and Telang N.T. *Breast cancer Res. Treat.*, 30, 233.
62. Elangovan V., Sekar N. and Govindasamy S. *Cancer Lett.* 1994, **87**, 107.
63. Tanaka T., Makita H., Ohnishi M., Mori H., Satoh K., Hara A., Sumida T., Fukutani K., Tanaka T. and Ogawa H. *Cancer Res.* **57(2)**, 246.
64. Descher E.E., Ruperto J., Wong G. and Newmark H.L. *Carcinogenesis.* 1991, **12(7)**, 1193.
65. Lamartiniere C.A., Moore J.B., Brown N.M., Thompson R., Hardin M.J. and Barnes S. *Carcinogenesis.* **16**(11), 2833.
66. Rao C.V., Wang C.Y., Simi B., Lubet R., Kelloff G., Steele G., and Reddy B.S. *Cancer Res.* **57**, 3717.
67. Birt D.F., Mitchell D., Gold B., Pour P., and Pinch H.C. *Anticancer Res.* 1997, **17**, 85.
68. Wattenberg L. W, Coccia J.B., and Lam L.K.T. *Cancer Res.* 1980, **40**, 2820.
69. Hirono I., Ueno I., Hosaka S., Takanashi H., Matsushima T. and Natori S., 1980,00
70. Exon J.H., Magnuson B.A., South E.H. and Hendrix K. *J. Toxicol. Environ. Health.* 1998, **53**(5), 375.
71. Gali H.U., Perchellet E.M., Klish D.S., Johnson J.M. and Percehllet J.P. *Int. J. Cancer.* 1992, **51**(3), 425 1998
72. Hirose M., Mutai M., Takahashi S., Yamada M., Fukushima S. and Ito N. *Cancer Res.* 1991, **51**, 824.
73. Morishita Y., Yoshimi N., Kawabata K., Matsunaga K., Sugie S., Tanaka T. and Mori. H., *Jpn. J. Cancer Res.* 1997, **88**(9), 815.
74. Tanaka T., Kojima T., Kawamori T., Wang A., Suzui M., Okamoto K. and Mori H. *Carcinogenesis.* 1993, **14**(7), 1321.

ANTIATHEROGENICITY AND ANTIOXIDATIVE PROPERTIES OF POLYPHENOLIC FLAVONOIDS AGAINST LDL OXIDATION

Michael Aviram, Rohtem Aviram, and Bianca Fuhrman.

Lipid Research Laboratory, Technion Faculty of Medicine, The Rappaport Family Institute for Research in the Medical Sciences and Rambam Medical Center, Haifa, Israel.

1 INTRODUCTION

Oxidative modification of low density lipoprotein (LDL) is thought to play a key role during early atherogenesis. Oxidized-LDL (Ox-LDL) is taken up by macrophages at enhanced rate via the scavenger receptors,[1-3] leading to the formation of lipid-laden foam cells, the hallmark of early atherosclerosis.[4]

Aggregation of LDL represents another lipoprotein modification with atherogenic properties, as aggregated LDL (Agg-LDL) was found to be taken up by macrophages at increased rate, leading to foam cell formation.[5]

Both Ox-LDL and Agg-LDL are found in atherosclerotic lesions,[6] and several lines of evidence imply that both modifications occur *in vivo* in the arterial wall. Recently, we have shown that extensive oxidation of LDL leads to its aggregation.[7] The resistance of LDL to oxidative modification depends on extrinsic, as well as on intrinsic factors.[8] Increased resistance of LDL to oxidation was observed after treatment with various synthetic pharmaceutical agents.[9,10] An effort is being made however to identify natural food and food products which can offer antioxidant defense against LDL oxidation. We have previously shown that dietary supplementation with nutrients rich in polyphenolic flavonoids, such as olive oil,[11] red wine,[12] or licorice root extract,[13] increases the resistance of LDL to oxidative modification.

2 POLYPHENOLIC FLAVONOIDS AND LDL OXIDATION

Dietary consumption of nutrients rich in polyphenolic flavonoids was shown to be inversely associated with morbidity and mortality from coronary heart disease.[14] The French paradox, i.e. low incidence of cardiovascular events in spite of a diet high in saturated fat, was attributed to regular drinking of red wine, which contains a substantial amount of potent antioxidant polyphenols. The polyphenolic flavonoids are a family of related compounds that include flavanols, flavonols, flavonones, anthocianidins, flavones, flavans, and the phenylpropanoids or hydroxycinnamic acid derivatives. Polyphenolic flavonoids are components of a wide variety of edible plants, fruits, vegetables, and of beverages such as tea, coffee, beer and red wine. Some polyphenols appear to possess antioxidative characteristics and their antioxidant activity is related to their chemical structure, their rate of reaction with the relevant free radicals, and their partition coefficient between aqueous and lipid environment.[15] Most flavonoids were shown to be water soluble chain-breaking agents of the peroxyl and the alkoxyl radicals,[16] and they are also chelators of transition metal ions such as iron ions. We have studied several nutrients that contain diverse polyphenolic flavonoids for their antioxidative capacity against LDL oxidation *in vitro* and *in vivo*.

2.1. *In vitro* studies

Red wine contains a substantial amount of polyphenols in comparison to white wine since it is prepared with the skin of the grapes that contain the flavonoids. The most important water soluble polyphenols in red wine are the flavonols quercetin and myricetin, and the 3-flavanols catechin and epi (gallo) catechin. Another nutrient rich in polyphenols is the root of the plant licorice (*Glycyrrhiza glabra*). Licorice root contains flavonoids from the flavan and chalcone subclasses, which have lipophilic characteristics and exert potent antioxidative activity.

We have also analyzed recently extracts from orange and grapefruit peels known to contain the flavonones hesperidin and naringin, respectively. Ginger root extract is an additional source of a different class of polyphenols- the gingerols.

We have studied the *in vitro* antioxidative potency of the above nutrients (Fig. 1A), along with the purified polyphenol flavonoids (Fig. 1B) to inhibit copper ions-induced oxidation of LDL. LDL (100 µg of protein/ml) was incubated at 37 °C for 4 h with 10 µM $CuSO_4$ in the absence (Control), or in the presence of increasing concentrations of red wine, licorice root extract, orange peel extract, grapefruit peel extract and ginger root extract. All of these nutrients inhibited copper-ion induced LDL oxidation in a dose- dependent manner. Fig. 1A demonstrates the IC_{50} (the concentration needed to inhibit $CuSO_4$-induced thiobarbituric acid reactive substances (TBARS) formation in LDL by 50 %) of each nutrient examined. Figure 1B demonstrates the antioxidative capacities of the pure polyphenols derived from the above nutrients. The flavonoids gingerol from the ginger root, glabridin from the licorice root, and catechin and quercetin from the red wine, were most potent antioxidants against LDL oxidation.

We have also studied juices from fruits and vegetables for their antioxidative capabilities against LDL oxidation. We found that juices (used at 5 % volume concentration) from lemon, fennel, onion, tomato, pomegranate and, pepper (which contain polyphenolic flavonoids), were most potent antioxidants against copper ion-induced LDL oxidation in the order presented above (Fig. 2).

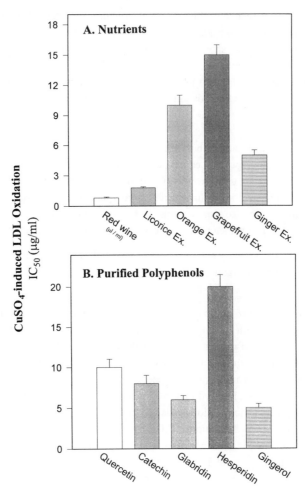

Figure 1
Antioxidative effect of polyphenols-rich nutrients against LDL oxidation.

2.2. Human studies

We have studied the *in vivo* effect of polyphenols-rich nutrients on LDL oxidation by a dietary supplementation to human volunteers of red wine or licorice extract.

The effect of red wine consumption with meals on the susceptibility of LDL to oxidation was studied in 17 healthy males who consumed 400 ml of red wine per day for a period of 2 weeks. Red wine, unlike white wine consumption resulted in a significantly reduced susceptibility of the volunteers' isolated LDL to copper ions-induced oxidation as determined by a 46 %, 72 % and 54 % reduction in the formation of TBARS, lipid peroxides and conjugated dienes, respectively. This antioxidant effect could be related to the elevation in LDL-associated polyphenolic flavonoids.[12]

The protective effect of licorice extract against LDL oxidation *in vivo* was studied after an oral administration of licorice root extract (encapsulated in a softgel) at a concentration of 100 mg/d for a period of 2 weeks. Oral supplementation with licorice resulted in a time-dependent gradual reduction in the susceptibility of LDL to copper-ions induced oxidation. This was evident by a 110 minutes prolongation of the lag phase required for the initiation of LDL oxidation, and by a 44 % reduction in lipid peroxides formation.[13]

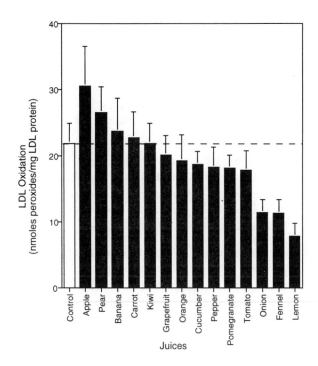

Figure 2
The effect of fruit and vegetable juices on copper ion-induced LDL oxidation.

3 POLYPHENOLIC FLAVONOIDS AND LDL AGGREGATION

Aggregation of LDL represents another atherogenic lipoprotein modification. Aggregated LDL (Agg-LDL) is taken up by macrophages at enhanced rate leading to foam cell formation.[5] Several lines of evidence imply that LDL aggregation occurs in the arterial wall,[6,17] but little is known about the mechanism responsible for this modification *in vivo*. We have recently demonstrated that extensive oxidation of LDL leads to its aggregation.[7] Thus, we have studied the effect of polyphenols (which were previously characterized as potent antioxidants against LDL oxidation) on LDL aggregation.

In vitro incubation of LDL (100 μg of protein/ml) for 2 h at 37 °C with red wine (10 %, v:v), catechin or quercetin (50 μM), resulted in a 30 %, 70 % or 50 % reduced LDL aggregation (induced by vortexing), respectively. These results suggest that polyphenols directly affect LDL aggregation, probably due to their binding to the lipoprotein with a subsequent effect on the lipoprotein hydrophobic domains.

4 POLYPHENOLIC FLAVONOIDS AND ATHEROSCLEROSIS

In order to study the effect of nutrients rich in polyphenolic flavonoids, along with purified polyphenols, on atherosclerosis development in relation to the susceptibility of LDL to oxidation and to aggregation, we have used the atherogenic apolipoprotein E deficient (E°) mice model, as these mice develop rapid and severe atherosclerosis, and since their LDL is highly susceptible to oxidation and to aggregation.[6,18]

E° mice were supplemented for up to 6 weeks in their drinking water with placebo (1.1 % alcohol), red wine (0.5 ml/day/mouse), or with the main polyphenols in red wine, catechin or quercetin at a concentration of 50 μg/day/mouse.[19] These atherosclerotic mice were also supplemented with licorice extract (200 μg/day/mouse) or with its major polyphenol glabridin (20 μg/day/mouse).

Analyses of the mice aortic arch lesions after 6 weeks of polyphenol consumption revealed that the atherosclerotic lesion area in mice that consumed red wine, quercetin, catechin, licorice extract, or glabridin, was significantly reduced by 48 %, 46 %, 39 %, 30 % and 50 %, respectively, in comparison to the atherosclerotic lesions area in mice that received placebo[13] (Fig. 3A).

Since foam cells in early atherosclerotic lesion result from increased uptake of LDL by macrophages, we next investigated the cellular uptake rate of LDL derived from E° mice that consumed polyphenols. Incubation of J-774 A.1 macrophages for 5 h at 37 °C with LDL (100 μg of protein/ml) derived from E° mice that consumed catechin, quercetin or red wine, resulted in 31 %, 40 % and 52 % reduced LDL-induced cellular cholesterol esterification (implies reduced cellular lipoprotein uptake) respectively, in comparison to the effect induced by LDL derived from the placebo receiving mice. These results suggest that the reduced atherosclerotic lesion formation in E° mice that consumed polyphenols may be attributed to reduced uptake of their LDL by macrophages.

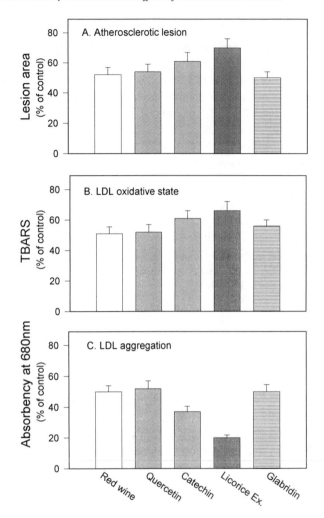

Figure 3

Dietary consumption by E° mice of polyphenolic flavonoids: antiatherogenic effect.

Next, we have studied the relationship between the reduced development of atherosclerotic lesions in mice, following polyphenols consumption, and LDL modifications. We have demonstrated that LDL derived from E° mice that consumed red wine, quercetin, catechin, licorice, or glabridin for 2 weeks, was less oxidized, even in its basal state (not induced by copper ion or other oxidants), by 49 %, 48 %, 39 %, 34 % or 44 % respectively, in comparison to LDL isolated from mice that consumed placebo (Fig. 3B). Furthermore, consumption of red wine, quercetin, catechin, licorice or glabridin by the E° mice also resulted in reduced susceptibility of their LDL to aggregation induced by vortex (Fig. 3C).

5 SUMMARY

This review article provides evidence for the antiatherogenic effects of dietary consumption of nutrients rich in polyphenols. We have demonstrated that dietary consumption by the atherosclerotic E° mice of red wine or licorice extract, as well as of their purified potent polyphenolic flavonoids, inhibited the progression of aortic atherosclerosis in these mice. This effect was associated with reduced LDL oxidative state, as well as reduced propensity of their LDL to oxidation and also to aggregation. The polyphenols' antioxidative and antiaggregative effects on LDL, could be shown to be associated with binding of these polyphenols to the LDL particle via the formation of ether bonds.[13]

References
1. Steinberg D., Parthasarathy S., Carew TE, Khoo JC, Witztum JL. Beyond cholesterol: modifications of low-density lipoprotein that increase its atherogenicity. *New Engl. J. Med.* 1989, **320**, 915-924.
2. Aviram M. Oxidative modification of low density lipoprotein and atherosclerosis. *Isr. J. Med. Sci.* 1995, **31**, 241-249.
3. Witztum J.L. and D. Steinberg. Role of oxidized low density lipoprotein in atherogenesis. *J.Clin. Invest.* 1991, **88**, 1785-1792.
4. Brown M.S. and J.L. Goldstein. Lipoprotein metabolism in the macrophage: implications for cholesterol deposition in atherosclerosis. *Annu. Rev. Biochem.* 1983, **52**, 223-261.
5. Suits AG, Chait A., Aviram M., Heinecke JW. Phagocytosis of aggregated lipoprotein by macrophages: low density lipoprotein receptor dependent foam cell formation. *Proc. Natl. Acad. Sci. U.S.A.* 1989, **86**, 2713-2717.
6. Aviram M., Maor I., Keidar S., Hayek T., Oiknine J., Bar-El Y., Adler Z., Kertzman V., Milo S. Lesioned low density lipoprotein in atherosclerotic apolipoprotein E-deficient transgenic mice and in humans is oxidized and aggregated. *Biochem. Biophys. Res Commun.* 1995, **216**, 501-513.
7. Maor I., Hayek T., Coleman R. and Aviram M. Plasma LDL oxidation leads to its aggregation in the atherosclerotic apolipoprotein E deficient mice. *Arterioscler. Thromb. Vasc. Dis.* 1997, **17**, 2995-3005.
8. Frei B. and J.M. Gaziano. Content of antioxidants, pre-formed lipid hydroperoxides, and cholesterol as predictors of the susceptibility of human LDL to metal ion-dependent and -independent oxidation. *J. Lipid Res.* 1993, **34**, 2135-2145.
9. Parthasarathy S.,S.G. Yang, J.L. Witztum, R.C. Pittman and D. Steinberg. Probucol inhibits oxidative modification of low density lipoprotein. *J. Clin. Invest.* 1986, **77**, 641-644.
10. Pentikainen M.O., K.A. Lindstedt and P.T. Kovanen. Inhibition of the oxidative modification of LDL by Nitecapone. *Arterioscler. Thromb. Vasc. Biol.* 1995, **15**, 740-747.

11.   Aviram M. and E. Kassem. Dietary olive oil reduces the susceptibility of low density lipoprotein to lipid peroxidation and inhibits lipoprotein uptake by macrophages. *Ann. Nutr. Metab.* 1993, **37,** 75-84.

12.   Fuhrman B., A. Lavy and M. Aviram. Consumption of red wine with meals reduces the susceptibility of human plasma and low density lipoprotein to undergo lipid peroxidation. *Am. J. Clin. Nutr.* 1995, **61,** 549-554.

13.   Fuhrman B., Buch S., Vaya J., Belinky P.A., Coleman R., Hayek T. and Aviram M. Licorice etanolic extract and its major polyphenol glabridin protect LDL against lipid peroxidation: *in vitro* and ex-vivo studies in humans and in the atheroscleroteic apolipoprotein E deficient mice. *Am. J. Clin. Nutr.* 1997, **66,** 267-275.

14.   Hertog M.G.L., Feskens E.J.M., Hollman P.C.H., Katan M.B., Kornhout D. Dietary antioxidant flavonoids and risk of coronary heart disease; the Zutphen Elderly Study. *Lancet.* 1993, **342,** 1007-1011.

15.   Rice-Evans C.A., N.J. Miller and G. Paganga. Structure-antioxidant activity relationships of flavonoids and phenolic acids. *Free Rad. Biol. Med.* 1996, **20(7),** 33-956.

16.   Afansiev I.B., Dorozhko A.J., Brodski A.V., Kostyuk A.V., Potapovitch A.I. Chelating and free radical scavenging mechanisms of inhibitory action of rutin and quercetin in lipid peroxidation. *Biochem. Pharmacol.* 1989, **38,** 1763-1769.

17.   Hoff H.F., O'Neil J. Lesion-derived low density lipoprotein and oxidized low density lipoprotein share a liability for aggregation, leading to enhanced macrophage degradation. *Arterioscler. Thromb. Vasc. Dis.* 1991, **11,** 1209-1222.

18.   Hayek T., Oiknine J., Brook J.G. and Aviram M. Increased plasma lipoprotein lipid peroxidation in apo E-deficient mice. *Biochem. Biophys. Res. Commun.* 1994, **201,** 1567-1574.

19.   Hayek T., Fuhrman B., Vaya J., Rosenblat M., Belinky P., Coleman R., Avishay E. and Aviram M. Reduced progression of atherosclerosis in apolipoprotein E-deficient mice following consumption of red wine, or polyphenols quercetin or catechin, is associated with reduced susceptibility of LDL to oxidation and aggregation. *Arterioscler. Thromb. and Vasc. Biol.* 1997, **17,** 2744-2752.

ANTIOXIDANT POTENCY AND MODE OF ACTION OF FLAVONOIDS AND PHENOLIC COMPOUNDS

Joseph Kanner.

Department of Food Science, ARO, The Volcani Center, Bet Dagan, Israel.

1 INTRODUCTION

Flavonoids and many phenolic compounds are secondary metabolites synthesized by plants, mostly following various stresses. These include mechanical, chemical, UV radiation, environmental and microbiological stresses. Most of these have been reported to induce generation of active oxygen species in plants. The active oxygen species appear to induce a cascade of reactions ending in the production of a series of polyphenolic compounds, which protect the plant from cell injury.[1,2]

Currently, there is considerable interest in elucidating the role of dietary antioxidants in human nutrition and medicine. Flavonoids and phenolic compounds have gained much interest as natural antioxidants.[3-6] They represent important groups of bioactive compounds in foods which may prevent the development of many diseases including atherosclerosis and cancer.[7,8]

The mode of action of many antioxidants depends on their capability of acting as electron donors, electron acceptors, decomposers of peroxide and hydroperoxides, metal activators and deactivators, and UV absorbers. Polyphenolic compounds have been identified as fulfilling most of the criteria described above.

Some flavonoids have potential SOD like activity. They are effective scavengers of hydroxyl radicals, inhibit peroxyl radical mediated and heme catalyzed lipid peroxidation, scavenge HOCl and $NO_2^{\cdot}$ and may enhance production of prostaglandins and leukotriens by inhibiting self destruction of cyclooxygenase and lipoxygenase. However, at high concentrations they inhibit the end products generated from these enzymes.

2 ELECTRONIC STRUCTURE AND ANTIOXIDANT ACTIVITY

Numerous authors have investigated the antioxidative activity of flavonoids and polyphenols. Several attempts have been made to elucidate the relationships between chemical structure and activity. Phenols contain a hydroxyl group attached to a benzene ring. Interaction between the mobile electrons of the ring and the 2 pairs of non-bonded oxygen electrons, can result in the transfer of one of these electrons to carbon orbitals of the benzene ring, forming a mesomeric form.[9,10] The mesomeric form causes increased attraction of the oxygen to the benzene ring and a decreased attraction of oxygen to hydrogen, thus facilitating rupture of this bond. This effect makes the hydrogen atom labile, acting as an electron and hydrogen donor, or as an excellent antioxidant.

Several factors affect the electron or hydrogen displacement from the phenolic hydroxyl group, such as additional OH or OR groups in the ortho or para position. This increases the antioxidant effect of polyphenolic compounds. The use of polyphenolics and flavonoids such as quercetin, epicatechin, epicatechin gallate, epigallocatechin gallate, caffeic acid and tannins as antioxidants in foods, commenced at the beginning of this century and became more intensive in the forties.[11] Letan[12] was one of the first to search the relationship between structure and antioxidant activity of quercetin and some of its derivatives. Most recently several attempts have been made to elucidate structure-activity relationships.[13-17]

The capability of polyphenolic compounds to donate electrons was demonstrated by their reaction with cytochrome-C, metmyoglobin or ferric iron (Kanner, results not published).

Table 1

Structural aspects and antioxidant activity of flavonoids.

Flavonoid Class	Substitution OH	Redox Potential E/Vvs.NHE	LPO I_{50} μM	Ferryl reduction K_1/mol^{-1}s^{-1}	TEAC mM
Flavanone					
Naringenin	5,7,4'	0.76	1137	4	1.5
Eriodictyol	5,7,3'4'	0.36	26	25	---
Taxifolin	3,5,7,3'4'	0.37	22	18	1.9
Flavone					
Apigenin	5,7,4'	0.71	2754	31	1.4
Luteolin	5,7,3'4'	0.41	17	63	2.1
Flavonol					
Galandin	3,5,7		0.56	6	----
Kaempferol	3,5,7,4'	0.39	6	115	1.3
Quercetin	3,5,7,3'4'	0.29	6	279	4.7
Myricetin	3,5,7,3'4'5'	0.20	1057	2	3.1

Rice-Evans & Miller Biochem. Soc. Trans. 1996, 24, 790.
Jorgensen & Skibsted Free Red. Res. 1998, 28, 335.

The radical scavenging capacity of flavonoids was demonstrated with superoxide, hydroxyl and peroxyl radicals.[18-20] The rate constant reactivity with hydroxyl radical is high, but this result is not surprising due to the generally high reactivity with aromatic compounds. However, for O_2·⁻, the rate constant is relatively low. Most recently it was demonstrated that cyanidin, an anthocyanin, has the capability of scavenging O_2·⁻ at a rate of $2.2 \cdot 10^5 M^{-1}S^{-1}$ at pH 7.0. That of the aglycone (anthocyanidin) is 10 fold higher ($1.9 \cdot 10^6$ $M^{-1}S^{-1}$).[21] Of great interest is the reactivity of flavonoids and phenolic compounds with peroxyl radical. This rate was found to be in the range of 10^8 to$10^7 M^{-1}S^{-1}$.[18,19] The reactivity of the flavonoids with free radicals was found to depend on the structure and several groups which are allocated around the heterocyclic and aromatic B ring.[22-24] On the basis of the results collected from several studies which are presented in table 1, it is possible to summarize that flavonoids such as the flavanones (eriodictyol, taxifolin) the flavone (luteolin) and flavonol (quercetin and myricetin) containing an ortho dihydroxy structure like catechol are highly active against membrane lipid peroxidation, or as electron donors to a ferryl system.[13,14,16] These results emphasize the fact that hydroxyl groups around the B ring increases antioxidative activity. The 2,3-double bond which is necessary for maintaining a conjugated double bond along the entire molecule appears to be of great importance. Galandin, a flavonol (OH substitution at 3,5,7, of heterocyclic and A ring position) which does not contain hydroxyls around the B ring, was found to work as a good antioxidant against enzymatic and non enzymatic lipid peroxidation.[13]

The 2,3 double bond in conjugation with 4 carbonyl and 3 hydroxyl on the heterocyclic ring greatly increases antioxidant activity of the flavonoid. The presence of the carbonyl with heterocyclic ring and the 2-3 double bond could participate in radical stabilization that also increased antioxidant activity. This structure increases the capability of the molecule to donate electrons and to reduce ferryl and to work as a scavenger of free radicals.[13,14,18,19]

Data from some studies have shown that it is possible to predict the antixodative effect by the redox potential of the flavonoid.[13,15,16] However, the solubility and steric effects may be affected by the structure of the molecule (glucosides or other adducts) which could increase or decrease its activity. Addition of a 3-hydroxy group to luteolin to yield quercetin increases the reactivity for reduction of ferryl myoglobin. However, glycoside formation at this position like rutin, dramatically lowers the reactivity.[16] In contrast with these results our results have shown that malvidin 3-glucoside works almost 3 fold better than malvidin in inhibiting membranal lipid peroxidation catalyzed by ferryl mygolobin.[24] The effectiveness of an antioxidant could be affected by the substrate in which this antioxidant is found such as molecules in emulsion, membranes, organelles, low density lipoprotein etc, and of course by the catalyst which catalyzes the reaction such as: free metals, hemeproteins, H_2O_2-activated hemeprotein, $HO^{\cdot}$, NO_x, HOCl organic hydroperoxide and others.[24]

The antioxidant activity of anthocyanins and other polyphenolics is strongly affected by the system and catalyzers in use. In a membrane system catalyzed by H_2O_2-activated myoglobin, malvidin 3-glucoside was found to be the most effective antioxidant and resveratrol the least, whilst in the system catalyzed by iron redox cycle, the most effective antioxidant was resveratrol > malvidin 3-glucoside = malvidin > catechin.[24] Our results and those published most recently by Satue'-Gracia et al.[25] have demonstrated that the use of stable radicals or compounds which break down to free radicals seem not be relevant as oxidizable substrates or catalyzers.[26,27] They may not provide data which are relevant to the mechanism of action of those antioxidants in biological systems.

Reaction of free radicals with flavonoids or other phenolic compounds leads to the generation of a semiquinone radical by the following reaction:

$$LOO^{\bullet} + PhOH \rightarrow LOOH + PhO^{\bullet} \tag{1}$$

The stability of the phenoxyl radical is very important.[18,28] This could greatly affect its capability of interacting with a second free radical by the following termination rection

$$LOO^{\bullet} + PhO^{\bullet} \rightarrow LOOPhO \tag{2}$$

Indeed, an effective antioxidant has been shown to react in a 1:2 stoichiometry: one antioxidant molecule interacts with two free radicals.[18,28-30]

Rate constants for hydrogen abstraction by the phenoxyl radical have been measured in non aqueous media.[31] The results have demonstrated that the reaction between a phenoxyl radical with other phenolic compounds is ~100-300 fold higher than the reaction of a peroxyl radical with phenolic compounds.

$$PhO^{\bullet} + ArOH \rightarrow PhOH + ArO^{\bullet} \tag{3}$$

$$LOO^{\bullet} + ArOH \rightarrow LOOH + ArO^{\bullet} \tag{4}$$

This could explain the possible reduction of active oxidized antioxidants in a mixture, an effect which produces synergism. It is also known that phenoxyl radical partially interacts with a second phenoxyl radical producing a dimer, similar to a termination reaction.

$$PhO^{\bullet} + PhO^{\bullet} \rightarrow PhOOPh \tag{5}$$

If the reduction potential of the phenoxyl radical is high enough it could propagate rather than interrupt chain reaction. A series of articles by Stocker group has shown that increasing the amount of α-tocopherol in LDL incresed the rate of LDL oxidation by several catalyzers including H_2O_2-activated peroxidase.[32] It was also reported that coumaric acid could enhance oxidation of α-tocopherol.[33] We have also demonstrated in a microemulsion of linoleate peroxidized by myoglobin, that coumaric acid enchances lipid peroxiation probably by the following reaction:

$$PhO^{\bullet} + LH \rightarrow PhOH + L^{\bullet} \tag{6}$$

A part of the semiquinone radicals, especially those formed from dihydroxy phenols like catechol, could interact with oxygen producing quinones and superoxide radicals.[34] Such a reaction could decrease the antioxidant effects of the phenolic compounds, or produce a prooxidant effect.

$$^{\bullet}O\text{-}PhOH + O_2 \rightarrow O=Ph=O + O_2^{\bullet -} \tag{7}$$

Due to this possible effect it is recommended to enchance the reactivity of phenolic antioxidants by the addition of SOD and catalase into the system.

3 METAL CHELATION BY FLAVONOIDS AND PHENOLICS

Considerable use has been made of the ability of phenolic compounds to form complexes with metals. Some of these comlexes are natural pigments of plants. Metal chelation could lead to: **a)** activation of the metal as a catalyst of oxidation, **b)** the chelation could deactivate the metal to oxidation and **c)** the chelation could lead to a complex which works as a hydroperoxide decomposer.[34,35,37]

Ferrous ion is known to be activated by complexation with EDTA leading to increased reduction of oxygen to $O_2^{\cdot-}$. This interaction could lead to a prooxidant effect, deleterious to biological systems. Ferrozine complexes ferrous ion in a very stable form which prevents the reduction of oxygen and production of $O_2^{\cdot-}$. Several flavonoids such as catechin, quercetin and myricetin are known to chelate iron and to produce a chromophor which can be determined spectrophotometrically. Myricetin and quercetin in the presence of iron at pH 7.0 were found to lead to autoxidation and generation of hydroxyl radicals. This was not detected with catechin at the same pH (Kanner, results not shown).

The redox chemisty of iron is itself directly affected by coordination ligand. The effects of these ligands on the electronic structure and the reactivity of iron is critical for understanding the possible antioxidant role of polyphenols in biological systems.[34] Chelation to some flavonoids or phenolics compounds could affect the redox potential of iron and its solubility in water/lipid phases.

The hydroxyl and carbonyl groups around the heterocyclic and B ring can chelate metals, such as Fe, Cu, Mn, Ca Mg or Al.[9,38] The groups which contain oxygen with 2 pairs of non-bonded electrons are important for complexation of the metals. Several flavonoids were found to produce relatively stable complexes with iron and copper, among the flavonols: fisetin, quercetin, kaempferol and rutin, among the flavanones: taxifolin, naringenin and the flavanol, chatechin. For chelation of iron, the hydroxyls at position 3'4' are important and also, 3-OH appear to be more important than 5-OH. The structural feature of a carbonyl at the 4 position is necessary for the metal chelation of many flavonoids. Several authors have shown that the 4 carbonyl and 3 or 5 hydroxyl groups act cooperatively to chelate metals ions.[38-40]

Iron mobilization from hepataytes by flavonoids was determined by Morel[40] which found that catechin is a better chelator than quercetin or diosmetin.

Most recently it was found by Gorelik[41] et al. and Ferrali et al.[42] that flavonoids such as catechin and quercetin can inhibit iron dependent membrane lipid peroxidation and oxidation of oxymyoglobin or oxyhemoglobin to met forms.

One could assume that the antioxidant effect of flavonoids depends not only on their radical scavenger capability but also on their metal chelation effect. This effect could delocalize the metal from membranes or other targets for oxidation, sterically preventing the interaction between the metal with the biological target or prevent interaction with oxygen or hydroperoxides or work as a hydroperoxide decomposer. Metal ions chelated by phenolic compounds could form compounds of iron in its reduced form which catalytically breaks down hydroperoxides and by a site specific mechanism scavenge the free radicals to an alcohol or a non-radical product.[43,57]

Of great importance is the reaction between phenolic compounds and metal enzymes such as peroxidases, cyclooxygenases and lipoxygenases. In the presence of cyclooxygenase, phenol or phenolic compounds stimulate enzyme reaction and prostaglandin generation at relatively small concentrations. However, at high concentration they inhibit prostaglandin production.[44-46]

The stimulation of cosubstrates by such phenolic compounds or flavonoids, could be explained by the reduction of the peroxyl radical formed during the reaction of cyclooxygenase and by the protection of the enzyme from self inactivation by the peroxyl radical or by ferryl [34,47] as shown by the following reactions:

$$P\text{-}Fe^{+3} + O_2 \text{ AAOOH} \rightarrow \text{Comp I} + O_2 \text{ AAOH(PGH}_2) \qquad [8]$$

$$\text{Comp I} + \text{AAH} \rightarrow \text{Comp II} + \text{AA}^\bullet \qquad [9]$$

$$\text{AA}^\bullet + 2O_2 \rightarrow O_2\text{AAOO}^\bullet \text{ (PGG}_2{}^\bullet) \qquad [10]$$

$$\text{PGG}_2{}^\bullet + \text{PhOH} \rightarrow \text{PGG}_2 + \text{PhO}^\bullet \qquad [11]$$

$$\text{Comp II} + \text{PhOH} \rightarrow P\text{-}Fe^{+3} + \text{PhO}^\bullet \qquad [12]$$

$$\text{net AAH} + 2\text{PhOH} + 2O_2 \rightarrow \text{PGH}_2 + 2\text{PhO}^\bullet \qquad [13]$$

If the phenolic compound concentration is high it could interact with Comp I and Compound II and inhibit propagation by reaction [9] and [12].

Such inhibition has been demonstrated in the past and most recentlly by Bakovic and Dunford[47] and Jang et al.[48]

Lipoxygenases in animals utilize arachidonic acid, the same substrate such as cyclooxygenase, to generate a single stereospecific hydroperoxide as the primary product. In plants lipoxygenase could oxidize an unsaturated fatty acid with a 1.4 pentadiene structure. Studies on purified soybean enzyme indicate that the native inactive form contains a high spin ferrous state in the active center. Binding and reaction with a lipid hydroperoxides results in oxidation of the iron to a high spin ferric state:

$$\text{ROOH} + Fe^{+2} \rightarrow \text{RO}^\bullet + \text{HO}^- + Fe^{+3} \qquad [14]$$

Increased production of leukotrienes, has been implicated in several human diseases. There has been considerable interest in the development of lipoxygenase inhibitors for therapeutic use.[49,50] Lipoxygenase and 15-lipoxygenase mRNA have been detected in human atherosclerotic lesion[51], raising the possibility that the enzyme promotes LDL oxidation *in vivo*.[52] Many phenolic compounds especially flavonids found in plants were found to inhibit lipoxygenases.[53-56] Recently it was found by us that similar to cyclooxygenase, lipoxygenase reactivity is stimulated by low concentration of phenolics and inhibited at high concentration. Small concentration of wine phenolics were found to inhibit co-oxidation of ß-carotene in the presence of linoleic acid, however, not the production of conjugated dienes.[57] German[58] showed that wine phenolics inhibit both plant and animal lipoxygenases at high concentration. However, at low concentration they increase lipoxygenase end products. The increase in end products formation after a long incubation is consistent with the hypothesis that the enzyme is stabilized against self inactivation by leak of free radicals which can oxidize the protein. We believe that a high concentrataion of flavonoids or wine phenolics inhibits lipoxygenase mostly by scavenging the enzyme free radical complex or by reducing the active ferric form of the enzyme to the native ferrous inactive form.

4 COMPLEXATION BY PROTEINS AND SOLUBILITY

Polyphenol-protein complexation depends on the characteristics of the protein, pH and the structure of the phenolic compound such as molecular size, conformational structure and water solubility.

The interaction of polyphenols and protein is thought to take place in two successive phases, one by hydrogen bonding and the other by hydrophobic interaction between protein sites and aromatic rings.[59,60] Phenolic compounds could interact with serum albumin.[61,62] This complexation should be taken into consideration when polyphenolic compound are extracted from blood during evaluation of the bioavailability of those compounds. Furthermore, in this specific interaction of orientated phenolic compounds with proteins these substances could be specific targets such enzymes, endothelial cells, apoproteins, LDL, receptors, cytokines, etc. Most recently the effect of bovine serum albumin on the antioxidant activity of several phenolic compounds, were determined by Heinonen et al.[63] Vinson et al.[64] have used the LDL to bond polyphenolic antioxidant, as a method which may be used to predict a good antioxidant activity *in vivo*.

The antioxidative effect of many polyphenols are not only dependent on their specific structure, and binding capability to macromolecules but also by their solubility. This effect was recently determined by several authors[65-67] and was found to depend on an interfacial phenomena. Flavonoids are also known to interact specifically with phospholipids producing reversible complexes.[68,69]

The many factors which affect the capability of phenolic compounds and flavonoids to act as antioxidants, could easily explain the great difference in their activity found by many researchers. Thus, it must be emphasized that it is necessary to determine as much as possible the antioxidant activity in a system which could simulate the *in vivo* conditions or to act with these compounds in *in vivo*.

More importantly, a good dietary antioxidant is not only one of low redox potential or high free radical scavenging capability, but one of high bioavailability which could reach the target of oxidation *in vivo* and at a critical concentration in order to provide protection.

References

1.   J.J. Macheix, A. Fleuriet and J. Billot. Fruit Phenolics. CRC Press. Boca Raton, Florida, 1990.
2.   R.M.M. Crawford, G.A.F. Hendry and B.A. Goodman. Oxygen and Environmental Stress in Plants. *The Royal Society of Edinburgh, Edinburgh Proceeding (B)* **102**, 1994.

3.   E.N. Frankel, J. Kanner, J.B. German, E. Parks and J.E. Kinsella. *Lancet*, 1993, **341**, 454.

4.   M.G.L. Hertog, E.J.M. Feskens, P.C.H. Hollman, M.B. Katan and D. Kromhout. *Lancet*. 1993, **352**, 1007.

5.   J.E. Kinsella, E.N. Frankel, B.I. German and J. Kanner. *Food Techol.* 1993, **47**, 85.

6.   J. Kanner, E.N. Frankel, B.J. German and J.E. Kinsella. *J. Agric. Food Chem.* 1994, **42**, 64.

7.   J.V. Formica and W. Regelson. *Food Chem. Toxicol.* 1995, **33**, 1061.

8.   K.T. Kumpulainen and J.T. Salonen. Natural Antioxidanta and Food Quality in Atherosclerosis and Cancer Prevention. *The Royal Soc. of Chem. Cambridge.* 1996.

9.   P. Ribe reau-Gayon. Plant Phenolics. *Oliver and Boyd, Edinburgh,* 1972.

10.   R.J. Fessenden and J.S. Fessenden. Organic Chemistry Willard Grant Press, Boston, 1982.

11.   J.R. Chipault, G.R. Mizuno and W.L. Lundberg. *Food Technol.* 1956, **10**, 209.

12.   A. Letan. *J.Food Sci.* 1966, **31**, 518.

13.   S.A.B.E. van Acker, D.J. van der Berg, M.N.J.L. Tromp, D.H. Griffioen, W.P. van Bennekom, W.J.F van der Vijgh and A. Bast. *Free Radic. Biol. Med.* 1996, **20**, 331.

14.   C. Rice-Evans, N.J. Miller and G. Paganga. *Free Radic. Biol. Med.* 1996, **20**, 933.

15.   G. Cao, E. Sofic and R.L. Prior. *Free Radic. Biol. Med.* 1997, **22**, 746.

16.   L.V. Jorgensen and L.H. Skeibsted. *Free Radic. Res.* 1998, **28**, 335.

17.   A Arora, M.G. Nair and G.M. Strasburg. *Free Radic. Biol. Med.* 1998, **24**, 1355.

18.   W. Bors, W. Heller, C. Michel and M. Saran. *Methods Enzymol.* 1990, **186**, 343.

19.   W. Bors, C. Michel and M. Saran. *Methods Enzymol.* 1994, **234**, 420.

20.   S.V. Jovanovic, S. Steenken, M. Tosic, B. Marjanovic and M.G. Simic. *J. Am. Chem. Soc.* 1994, **116**, 4846.

21.   H.Yamasaki, H. Uefuji and Y. Sakihama. *Arch. Biochem. Biophys.* 1996, **332**, 183.

22.   W.A. Pryor, J.A. Carnicelli, L.T. Devall. B. Tait, B.K. Trivedi, D.T. Witiak and M. Wu. *J. Org. Chem.* 1993, **58**, 3521.

23.   T. Tsuda, K. Shiga, K. Ohshima, S. Kawakish and T. Osawa. *Biochem. Pharmacol.* 1996, **52**, 1033.

24.   T. Lapidot, S. Harel, R. Granit and J. Kanner. *J. Agric. Food Chem.* 1998 (in press).

25.   M.T. Satue'-Gracia, M. Heinonen and E.N. Frankel. *J. Agric. Food Chem.* 1997, 45, 3362.

26.   C. Rice-Evans and N.J. Miller. *Methods of Enzymol.* 1994, **234**, 279.

27.   H. Wang, G. Cao and R.L. Prior. *J. Agric. Food Chem.* 1997, **45**, 304.

28.   J. Ruiz, A. Perez and R. Pouplena Quant. *Struct. Act. Relet.* 1996, **15**, 219.

29.   C.E. Boozer, G.S. Hammond C.E. Hamilton and J.N. Sen. *J. Am. Chem. Soc.* 1955, 77, 3233.

30.   J. Winterle, D. Dulin and T. Mill. *J. Org. Chem.* 1984, **49**, 491.

31.   M. Foti, K.U. Ingold and J. Lusztyk. *J. Am. Chem. Soc.* 1994, **116**, 9440.

32.   P.K. Witting, J.M. Upston and R. Stocker. *Biochemistry.* 1997, **36**, 1251.

33.   J. Laranjinha, O. Vieira, V. Madeira and L. Almeida. *Arch. Biochem. Biophys.* 1995, **323**, 373.
34.   J. Kanner, B. German and J.E. Kinsella. *CRC Critical Reviews in Food and Nutrition.* 1987, **25**, 317.
35.   Y. Osawa. in: Anthocyanins as Food Colors, Markakis P. ed. Academic Press, New York, 1982, 41.
36.   D.M. Miller, G.R. Buettner and S.D. Aust. *Free Radic. Biol. Med.* 1990, **8**, 95.
37.   S. Harel. *J. Agric. Food Chem.* 1995.
38.   M. Thompson and C.R. Williams. *Anal. Chim. Acta.* 1976, **85**, 375.
39.   A. Puppo. Phytochemistry. 1992, **31**, 85.
40.   L. Morel, G. Lescot, P. Cogrel, O. Sergent, N. Pasdelup, P. Brissot, P. Cillard and J. Cillard. *Biochem. Pharmacol.* 1993, **45**, 13.
41.   S. Gorelik. The effect of natural antioxidants on lipid and myoglobin stability in muscle food. *Thesis, University of Jerusalem.* 1996.
42.   M. Ferrali, C. Signorini, B. Caciotti, L. Sugherini, L. Ciccoli, D. Giachetti and M. Comporti. *FEBS Letters* 1997, **416**, 123.
43.   P. Spiteller and G. Spiteller. *Biochem. Biphys. Acta.* 1998, **1392**, 23.
44.   W.L. Smith. WEWLands *J. Biol. Chem.* 1971, **24**, 271.
45.   F.E. Dewhirst. *Prostaglandins.* 1980, **251**, 7329.
46.   W.E.W. Lands Prostaglandins. 1982, **24**, 271.
47.   M. Bakovic and H.B. Dunford. *Biochemistry.* 1994, **33**. 6475.
48.   M. Jang, L. Cai, G.O. Udeani, K.V. Slowing, C.F. Thomas, C.W.W. Beecher, H.H.S. Fong, N.R. Farnsworth, A.D. Kinghorn, R.G. Menta, R.C. Moon and J.M. Pezzuto. *Science* 1997, **275**, 218.
49.   R.A. Lewis, K.F. Austen and R.J. Soberman. N. Engl. *J. Med.* 1990, **323**, 645.
50.   J.R. Cashman. *Pharm. Res.* 1985, **253**, 261.
51.   S. Yla-Herttuala, M.E. Rosenfeld and S. Parthasarath, E. Sigal, T. Sarkioia, J.L. Witztum and D. Steinberg. *Proc. Natl. Acad. Sci. USA.* 1987, **87**, 6959.
52.   J.W. Heinecke. *Current Opinion in Lipidology.* 1997, 8, 268.
53.   M.A. Moroney, M.J. Acaraz, R.A. Forder, F. Carey and J.R.S Coult. *J. Pharm. Pharmacol.* 1988, **40**, 782.
54.   J. Baumann, F. von Bruchhausen and G. Wurm. *Prostaglandins.* 1980, **20**, 627.
55.   M.J. Langhton, P.J. Evans, M.A. Moroney, J.R.S Hoult and B. Halliwell. *Biochem. Pharmacol.* 1991, **42**, 1673.
56.   E. Nagalabu and N. Lakshmaiah. *J. Clin. Biochem. Nutr.* 1996, **21**, 123.
57.   J. Kanner. Natural antioxidants and anticarcinogens in nutrition, health and disease. *Helsinki.* 24-27 June, 1998.
58.   J.B. German. *Wine and human health.* Udine 9-11 October, 1996.
59.   C.M. Spencer, Y. Cai, R. Martin, S.H. Gaffney, P.N. Goulding, D. Magnolato, T.H. Lilley and E. Haslam. *Phytochemistry.* 1988, **27**, 2397.
60.   E. Haslam and T.H. Lilley. *CRC. Crit. Rev. Food Sci. Nutr.* 1988, **27**, 1.

61. B. Bartholome, I. Estrella and T. Hernandez. *Polyphenolic Commun.* 1996. Julay 15-18, 439.
62. B.K. Maralidhara and V. Prakash. Int. *J. Pept. Protein Res.* 1995, **46**, 1.
63. M. Heinonen, R. Dietrich, M.T. Satue'-Gracia, S.W. Huang, J.B. German and E.N. Frankel. *J. Agric. Food Chem.* 1998, **46**, 917.
64. V.A. Vinson, J. Jang, Y.A. Dabbagh, M. Serry and S. Cai. *J. Agric. Food Chem.* 1995, **43**, 2798.
65. K. Schwarz, E.N. Frankel and J.B. German. *Fett/Lipid.* 1996, **98**, 115.
66. E.N. Frankel, S.W. Huang, J. Kanner and J.B. German. *J. Agric. Food Chem.* 1994, **42**, 1054.
67. I.A. Hopia, S.H. Huang, K. Schwarz, J.B. German and E.N. Frankel. *J. Agric. Food Chem.* 1997, **45**, 3033.
68. S.W. Huang. E.N. Frankel, J.B. German and R. Aeschbach. *J. Agric. Food Chem.* 1997, **45**, 1992.
69. E. Bombardelli and M. Spetta. Cosmet. *Toileties.* 1991, **106**, 69.
70. A. Saija, M. Scalese, M. Lanza, D. Marzullo, F. Bonina and F. Castelli. *Free Radic. Biol. Med.* 1995, **19**, 481.

BIOMARKERS OF FLAVONOID INTAKE IN HUMANS

Anna Ferro-Luzzi and Giuseppe Maiani.

Unit of Human Nutrition, National Institute of Nutrition, Rome, Italy.

1 INTRODUCTION

A recurrent theme throughout the presentations on flavonoids at this meeting, has been the limitation of the current understanding of their role in humans and their relevance to health under real life conditions. Apparently the amount and chemical form of dietary flavonoids and, most importantly, degree of their absorption from the human gut as well as kinetics of absorption, distribution and excretion, remain unclear. For example, Kumpulainen and colleagues have shown that flavonoids are present in the Finnish diet in much larger amounts than previously thought.[1]

This paper will introduce the concept of biomarkers and review the current state of the knowledge on biomarkers for flavonoids, focussing on quercetin as the paradigm. Quercetin was chosen, as it has been investigated more extensively than any other flavonoid, has a strong antioxidant capacity and is present in appreciable amounts in most human diets.

2 BIOMARKERS

Biomarkers are important tools for the study of the effects on health of dietary exposure. A biomarker has been usefully defined, in the context of diet and health as any biochemical, molecular, genetic, immunological or functional measure of events occurring in a biological system following exposure to dietary components thought relevant to health. This definition derives from a combination of several others developed over the years, mostly in the context of toxicological evaluations (Van Poppel, Verhagen, et al. 1997 ID: 8678). Biomarkers have been usefully classified as: biomarkers of exposure, namely measures that indicate whether exposure to a given agent has taken place; biomarkers of effect, that provide an indication of biological events such as presence of specific metabolites and/or adducts, and of organic/tissutal/functional changes; biomarkers of susceptibility, which identify specific individuals possibly at greater risk to the effects of exposure for reasons such as genetics, behaviour, occupation, etc. (Fennel, 1990 ID: 8677). Appropriate biomarkers can be developed when there is a good understanding of the mechanisms of the action and metabolic pathways of a given compound. In the case of flavonoids, the information needed concerns the proportion of dietary flavonoids passing across the intestinal wall in humans, whether they are absorbed as parent compounds or metabolites, metabolic pathways of absorbed compounds, turnover rates, capacity to build up effective body concentrations and be stored in tissues as such or as active metabolites, and the changes they induce during their passage through the body. Ideally, a biomarker should be: specific, reliable, sensitive, reproducible, quantifiable, quantitatively relative to exposure (linear dose response), detectable in trace quantities, inexpensive to assay, available by non- or modestly invasive techniques, applicable in epidemiological surveys and clinical trials (large numbers, logistic considerations, etc.).[3]

So far, none of the flavonoids identified as potentially relevant for human health have biomarkers of exposure that fulfill all the above prerequisites. This is not surprising given that the health-promoting attributes of flavonoid have gained scientific prominence only in recent years.

2.1. Biomarkers of exposure to dietary flavonoids

Biomarkers of dietary exposure would predict the oral load of individual flavonoids. However, development of such biomarkers is seriously hampered by the remarkable variability of flavonoid contents in the various food items in the diet.

The uncertainties raised by the large variability of type and amount of flavonoids present in the diet is further compounded by the lack of data sets allowing the calculation of their contents in the diet, difficulty of analysing dietary duplicates and lack of suitable methods for measuring these substances. Thus, all the well-known limitations relative to the dietary methods for assessment of macro and micronutrients apply in the case of flavonoids.[8]

2.2. Biomarkers of internal exposure

Given that the direct estimate of dietary flavonoids is of uncertain value, it becomes crucially important to rely on alternative methods for investigating their impact on health, such as plasma levels or urinary concentrations of the parent compound or its metabolites. However, development of such biomarkers of internal exposure requires good insight into the kinetics of absorption, distribution and disposition of the compound of interest. These parameters are still being investigated, and the current knowledge on the bioavailability of the various forms of flavonoids in humans is incomplete.

The field has been dominated by the following five assumptions:
1. the human ileum lacks the enzymes for splitting beta glycosydic bonds
2. such a splitting can be performed by the colon flora, thus liberating the aglycones only in the colon
3. absorption of glycosydes can take place only after the splitting the glycosidic bond has liberated aglycone
4. the colon can absorb aglycones
5. aglycones are better absorbed than glycosides.

Research conducted over the last four-to-five years has demonstrated that most of the above assumptions are flawed.

As concerns human colon microflora, there is little doubt that it contains enzymes capable of splitting the beta-glycosidic bond.[9] As hydroxylation of this bond enhances the susceptibility of the molecule to undergo ring cleavage, the destiny of the flavonoid aglycones in the colon would be that of further degradation, which would make dietary flavonoids unavailable for absorption. However, it appears that the cleaving of the ring might proceed at a slower rate than glycosidase activity, thus intact flavonoid aglycones would be present in the colon.[10] The conceptual problem is that, in principle, the colon has very little absorptive capacity, except for water and electrolytes. Thus, it is doubtful that the colon can be an active site for absorption of flavonoid aglycones.

However, while there are scanty resident flora in the upper ileum, the terminal portion of the ileum is not sterile and it appears that there is an appreciable number of bacteria.[11] Furthermore, exposure of the intestinal microflora to specific substrates (e.g. beta glucosidase) appears capable of inducing the activity (300 fold) within 24 h.[12] Therefore, it is plausible that some flavonoid aglycons could be liberated in the lower ileum, a site enabling their absorption .

Animal studies are of limited value when studying human digestion, because there is a marked species-specificity in the kinetics of absorption and physiology of the gastrointestinal system. For example, the stomach of the rat and the mouse, being practically layered by bacteria and yeast, is a totally unsuitable model for human studies.[11] Very few controlled trials have been conducted in man so far; earlier studies have a limited value, as megadoses were usually employed and also because the analytical methods available in those years lacked the sensitivity and specificity necessary to detect the micromolar concentrations of plasma and urine. Thus, the first study where an oral load with 4 g quercetin was given,[13] found no detectable presence of quercetin in plasma or in urine.

As late as 1995, controlled trials and experiments employing appropriate protocols provided reliable evidence on the kinetics of absorption, disposal and metabolism of flavonoids, mostly quercetin and some isoflavonols.

The Manach group, using the rat model, was among the first to show that both quercetin aglycone and its glycoside (rutin) are absorbed, producing similar plasma levels of their metabolites.[14] Their data suggest that most absorption would take place in the colon, once the aglycone has been liberated. However, the mechanism of transport of the aglycone across the intestinal wall still remained undefined, although the authors thought a passive transport, similar to that of bile acids, was plausible, justified by the high gradient of concentration of flavonoids (a factor of 14) existing between the lumen of the cecum (about 1400 µmol/L) and plasma (about 100 µmol/L). On the other hand, studies conducted on Caco2 cells incubated with quercetin as well as with quercetin glycosides by Noteborn and colleagues found that quercetin as well as all glycosides tested, except for Quercetin 4' glucoside, act as glucose-transport carrier inhibitors.[15] They also found that after a 2-h incubation, no quercetin aglycone was detectable in the cells.

These results differ from preliminary data obtained by our laboratory, where quercetin aglycone was found in Caco2 cells incubated with quercetin glucoside, reaching a peak concentration at 24 h and declining thereafter until 72 h. These data suggest these cells not only have the capacity of splitting the glycosidic bond but also to further metabolise aglycone (Fig 1.).[16]

This conflicting evidence does not allow firm conclusions to be drawn and more research is needed to clarify the mechanism by which flavonoids pass through the gut wall into the circulation.

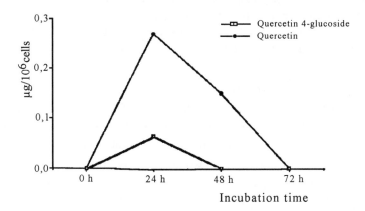

Figure 1
Presence of quercetin aglycone in Caco2 cells incubated in a medium containing quercetin-glucoside (70 μmol/L).[16]

3 HUMAN STUDIES OF THE BIOAVAILABILITY OF FLAVONOIDS

In 1995 new evidence was obtained by Hollman and colleagues in The Netherlands, who carried out an absorption study on nine ileostomy patients, using a simplified model where the events taking place in the colon that confound the results of balance studies are eliminated.[17] Quercetin was administered orally in various forms; onion quercetin glucoside, quercetin aglycone and pure quercetin rutinoside. Absorption was defined as the difference between the administered dose and the amount recovered in the ileostomy effluent over the next 13 h. This study established, for the first time, that intact glycosides can be absorbed by the human ileum (Table 1).

Table 1
Mean cumulative excretion over 13 h in ileostomy effluent after an oral load with three different forms of quercetin(Qu) (modified from 17).

Supplement	Intake (as aglycone) mg	Excretion in ileostomy effluent mg	Absorption %
Onion	89	37	52
Qu-3-rutinoside	100	72	17
Qu aglycone	100	66	24

Moreover, aglycone bioavailability quite unexpectedly appeared to be lower than that of betaglucoside, with only 24 % of the quercetin aglycone oral load absorbed compared to 52 % for the quercetin glucoside in the onion preparation. The study also provided evidence that the sugar linked to the aglycone might play an important role in determining the bioavailability of the phenolic compound, as only 17 % of quercetin rutinoside was absorbed as compared to the higher proportion of the onion quercetin glucoside. This finding led the authors to speculate that flavonoid glycosides might be actively transported through the gut wall, with sugar carriers defining the amount and speed of transport.[17]

Table 2

Concentrations of flavonols in plasma and in urine after daily ingestion of 750 ml of red wine, 50 g of fried onions or three cups of black tea for four days.[18]

	Red Wine	**Fried Onions**	**Black Tea**
PLASMA *			
Quercetin	8 ± 3	16 ± 5	8 ± 4
Kaempferol	1 ± 1	1 ± 0	4 ± 2
Isorhamnetin	3 ± 2	2 ± 1	1 ± 1
Urinary excretion			
Quercetin	112 ± 33	153 ± 66	76 ± 45
Kaempferol	72 ± 25	95 ± 139	202 ± 116
Isorhamnetin	106 ± 58	55 ± 26	29 ± 27

* $x \pm SD$, average of two blood samples per subject.

Variability of the bioavailability of food quercetin was further quantified in a study conducted by on 12 subjects with intact colon (Table 2).[18] Blood was collected on the 4th day. Quercetin absorption was found to be higher from onions, and much lower when ingested from wine or black tea. Subjects receiving onion achieved plasma quercetin values that were double those of subjects consuming red wine or black tea, despite the presence of similar amounts of quercetin in the portions consumed (14 to 16 mg/d for 4 days), 6 glasses of wine, 3 cups of tea, and 50 g of onions.

A similarly wide variation in bioavailability has been reported both for catechins and isoflavones. Maiani and colleagues investigated absorption of catechins and found ample interindividual difference in plasma epigallocathechin gallate after green tea drinking,[19] and epicatechin after red wine drinking (Maiani, et al. in preparation). The absorption kinetics of the two classes of catechins differed, with epigallocathechin gallate (EGCG) peaking 30 minutes after ingestion, while epicatechin from red wine peaked at 50 minutes and did not return to basal values after 120 minutes (Fig. 2).

Absorption of green tea catechins-epigallocathechin gallate, epigallocathechin and epicatechin-was investigated also by Lee et al.[20] who measured the form in which these circulate in blood. This study confirmed the wide interindividual variability of plasma levels, as well as diverse peak time for the three catechins. It also showed that -at 1 h after the challenge -most of the circulating catechins were conjugated, either in the glucuronide or sulfate form.

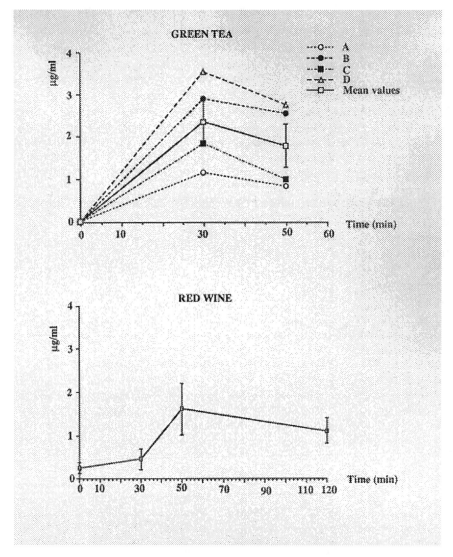

Figure 2

Plasma concentration (μg/ml) of epigallocathechin gallate (EGCG) before and after green tea ingestion,[19] and of epicatechin (EC) before and after red wine drinking (Maiani et al. In preparation) respectively.

Isoflavonoids bioavailability in man has been investigated to a certain extent, as great importance is attached to their capacity of binding to estrogens receptors and thus to behave as estrogens agonists/antagonists. However, the metabolism and the effective dose are still incompletely understood. Isoflavonoids have been reported in plasma and urine of subjects habitually consuming high quantities of soy product, such as the Japanese who excrete 20-30 times more isoflavones in their urine than Westerners. Plasma concentrations of daidzein and genistein, the two main isoflavones of soy, have been investigated under experimental conditions. Plasma levels of daidzein and genistein in 12 subjects significantly increased within 6 h from an oral challenge of soy bean milk as source of isoflavones, then decreasing to basal values within 24 h.[21] An almost linear dose-response was obtained by the provision of three increasing dosages of isoflavones, respectively 0.7, 1.3 and 2.0 mg of total isoflavones/kg body weight. These amounts contain almost identical quantities of daidzein and genistein (44 % genistein and 56 % daidzein) and correspond to 25, 46 and 1 mg/d of daidzein and 19, 36 and 56 mg/d of genistein. Plasma concentration reached the same levels for the two compounds, peaking at the same time, after 6.5 h from ingestion, and reverting to basal values after 24 h. As the urinary excretion of daidzein was higher 21 %, on all three doses, than that for genistein (9 %), it appears that the bioavailability of daidzein might be slightly higher than that of genistein.[22]

Another dietary trial conducted by Finnish and Canadian scientists, using a soy product as source of isoflavones for 28 days, confirmed the expected increase in plasma concentrations of daidzein and genistein.[23] However, the prolonged supplementation resulted in much higher plasma values than those of the study reported above[21] (Fig. 4). Although the kinetics parameters of isoflavonols absorption and excretion are not available, it seems reasonable to postulate that the excretion half-life allowed a build-up of the plasma levels.[24-27]

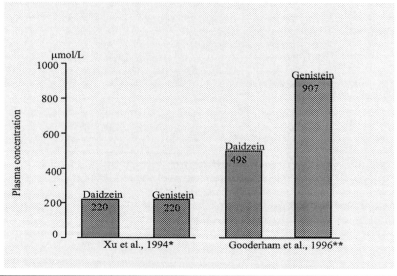

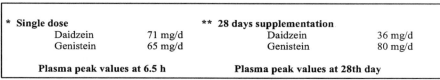

* Single dose		** 28 days supplementation	
Daidzein	71 mg/d	Daidzein	36 mg/d
Genistein	65 mg/d	Genistein	80 mg/d
Plasma peak values at 6.5 h		**Plasma peak values at 28th day**	

Figure 4

Concentration of isoflavones in human plasma after acute ingestion[21] or long term supplementation[23] with soy product.

A linear dose-response of plasma levels following changes in intake represents a critical feature of a putative biomarker of exposure. Several human studies have demonstrated that such a linear dose-response between dietary intake and plasma levels indeed exists for quercetin and kaempferol[18] and possibly for some of the investigated isoflavonoids.[21] Fig. 5 illustrates the results obtained for quercetin and kaempferol, in healthy volunteers consuming increasing amounts of the specified flavonoid with different foods.

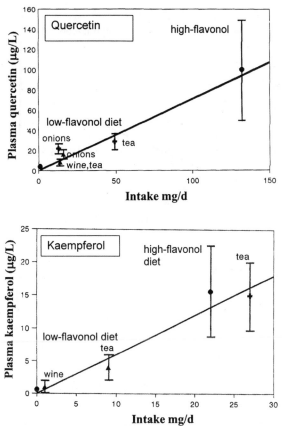

Figure 5

Dose-response to dietary challenges in humans: Plasma levels of quercetin and of kaempferol.[18]

It thus appears that, at least for the investigated compounds, namely quercetin and kaempferol, possibly also daidzein and genistein, plasma levels could serve as useful biomarkers of their intake. This conclusion might not necessarily extend to conditions and compounds different from those tested, as many parameters can affect the bioavailability in humans of the same flavonoid.

Moreover, urinary excretion of some flavonoids has been shown to be in good agreement with dietary intakes. For quercetin and kaempferol, a good linear correlation was obtained between ingestion and renal excretion of subjects at various levels of intake, as shown in Fig. 6.[18] Urinary excretion had an intraindividual coefficient of variation similar to that of plasma levels, i.e. about 22 %, but its interindividual coefficient of variation was found to be larger (36 %) than that of plasma (23 %). The correlation between urine and plasma quercetin concentrations was r=0.46, and the intraindividual correlation of the same individual on different diets was r=0.77.[18]

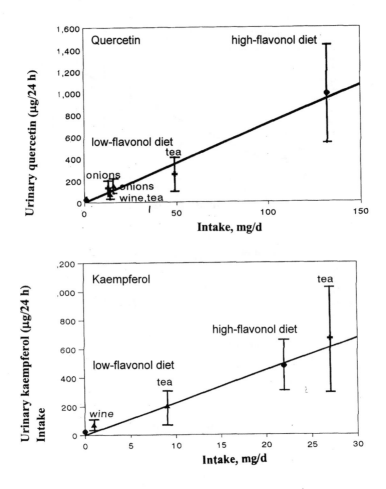

Figure 6

Dose-response to dietary challenges in humans: Urinary levels of quercetin and of kaempferol in volunteers on different diets.[18]

A similar dose-response was obtained for daidzein and genistein, although less well documented, and apparently not perfectly linear at lower levels of ingestion.

An insight of the kinetics of absorption, distribution and disposition of flavonoids is essential for the understanding of their health impact. In the case of quercetin, a first investigation conducted on two subjects with intact colon established that quercetin glucoside from onions is quickly absorbed, with peak plasma levels reached in about 3 h, and a half-life of absorption of about 1 h.[28] The distribution time has a half-life of 3.8 h and an elimination half-time of about 17 h. The study was later repeated on a further 9 subjects,[29,30] who received either onion (quercetin glucoside), or apples (various quercetin glucosides) or pure quercetin rutinoside, in quantities that provided approximately the same amount of quercetin aglycone.

Preliminary data showed that the kinetics of absorption of the glycosides differed among the various foods, as apple quercetin glycosides peaked in plasma at 2.5 h, and the rutinoside supplement took 7 h to peak in plasma. On the other hand, the disposition kinetics appeared to be more uniform, with an elimination half-life of about 25 h for both onions and apple quercetin. In terms of the onion quercetin glycosides, this study gave kinetic parameters that differed from those of the previous study, peak plasma levels reached much earlier, at 0.7 h, and a much longer elimination half-life, 25 h (Table 3).

Table 3

Kinetic parameters of absorption and disposition of quercetin.[17,28-30]

	1st Study 1995	**2nd Study 1997**
Peak level	2.9 h	0.7 h
Absorption half-life	0.9 h	n.a.
Distribution half-life	3.8 h	n.a.
Elimination half-life	16.8 h	25 h

The above kinetic parameters have led to two conclusions. The first was that, under conditions of constant intake, a steady state plasma level would be reached in about 3-4 days; this implies that plasma quercetin is a short-term biomarker, allowing the estimate of the dietary exposure over the previous 4 days.[18]

The second conclusion is drawn from the 25 h half-life of the elimination kinetics, the implication being that the level of quercetin in the human body is likely to progressively accumulate on a diet providing a constant supply of this flavonoid. That such a build-up occurs is confirmed by the study conducted by Conquer and colleagues, who supplemented 27 healthy adults in a double-blind feeding trial, with 1.0 g pure quercetin aglycone (a value about 50 fold greater than the amount present in western diets) plus 1.0 g of other unspecified flavonoids, 200 mg rutin, 200 mg bromelain, for 28 days. Exceptionally high plasma quercetin values were found, 427 mg/ml, when blood was sampled after 28 days of supplementation.[31]

This finding raises the question whether excess quercetin can be stored in body tissues. Tissue levels of flavonoids are of great interest because they would signify bodily ability to build up stores of potentially useful antioxidants, to be mobilised when and where they are needed, rather than relying on a constant intake. Unfortunately, there are no data in the literature as yet on tissue levels, but such a mechanism is considered unlikely because of the active and fast conjugation process that circulating flavonoids appear to undergo. Thus, plasma might be the main storage organ.

Another important parameter that needs to be addressed is the form in which flavonoids are present in the blood and in tissues–if they exist. Two studies have been conducted so far to establish the form in which quercetin circulates in the blood, and they have given conflicting results. The first study was conducted in rats, and found that flavonoid metabolites circulate in blood tightly bound to plasma albumin.[14] The authors established that administration of quercetin aglycone or rutin led to the formation of the same quercetin-albumin complex. Albumin-quercetin complex was found equally efficient as unbound quercetin in preventing oxidation of human LDL.

The other study was conducted by Paganga et al.[32] on normal free-living subjects. Employing an analytical method that allowed to detect and simultaneously quantify several families of flavonoids, these authors identified the presence in plasma of several glycosylated forms of quercetin, anthocyanins and phloridzin, thus suggesting that the human intestine is capable of absorbing unmodified flavonoids and also that the parent compounds then circulate unchanged in the blood.

4 CONCLUSIONS

The state of the art on biomarkers of flavonoids in the context of human health has been reviewed. It has been noted that, while the field is still very young, the first, important steps have been taken to develop and validate exposure biomarkers. An attempt has been made to calculate the specificity, sensitivity and reproducibility of putative biomarkers, range and confidence intervals of the estimate of dietary exposure, when one single measure is carried out.[18] The results appear promising, but preliminary as yet. More research is needed to confirm the findings as they were obtained mostly on small groups of selected individuals, which may not claim to represent other vulnerable or susceptible or otherwise diverse groups. Also, other flavonoids important to human health need to be investigated, such as the isoflavonoids, resveratrol among them, and various chemical forms in which they are present in food as we consume it, the synergism and the competitive inhibition, etc. Both plasma and urinary levels appear to be sensitive to changes in the diet and could be developed into useful biomarkers.

Questions that remain open relative to biomarkers of flavonoid exposure are:
- Quantification of the relationship between diet and plasma or urine under diverse conditions;
- Establish if entero-hepatic cycling of flavonoids or their metabolites occur;
- Standardisation of study protocols and analytical techniques
- Clarification of mechanism of uptake of aglycones as well as of glycosides
- Estimation of the error of measurement.

Acknowledgements

The authors wish to warmly acknowledge the assistance of Ms. Giovina Catasta for technical assistance in the preparation of the manuscript.

References

1.   Kumpulainen J. T. Flavonoids and phenolic acids in foods and diets. NAHD'98, June 24-27, Helsinki, Finland. 1998.

2.   Cook N.C. and Samman S. Flavonoids -Chemistry, metabolism, cardioprotective effects, and dietary sources. *Nutritional Biochemistry.* 1996, **7**, 66-76.

3.   Henderson R.F., Bechtold W.E., Bond J.A., Sun J.D. The use of biological markers in toxicology. *Critical Reviews in Toxicology.* 1989, **20**, 65-68.

4.   Maiani G., Salucci M., and et al. Antioxidant preservation in fresh vegetables stored under MAP conditions. INN. 1998.

5.   Soleas G. J., Diamandis E., and Goldberg D. M. Resveratrol: A molecule whose time has come? And gone? *Clinical Biochemistry.* 1997, **30**(2), 91-113.

6.   Sato M., Suzuki Y., Okuda T., Yokotsuka K. Contents of Resveratrol, Piceid, and Their Isomers in Commercially Available Wines Made from Grapes Cultivated in *Japan. Biosci.Biotech.Biochem.* 1997, **61**, 1800-1805.

7.   Celotti E., Ferrarini R., Zironi R., Conte L.S. Resveratrol content of some wines obtained from dried Valpolicella grapes: Recioto and Amarone. *J.Chromatogr.* 1997, 47-52.

8.   Bingham S.A. Biomarkers used to validate dietary assessments in human population studies. In: Crews HM, Hanley AB, eds. Biomarkers in food. Chemical risk assessment. The Royal Society of Chemistry ed. Cambridge, UK: Thomas Graham House, Science Park, Milton Road. 1995, 20-26.

9.   Durand M., Bernalier A., and Dore J. Biochemistry of fermentation. Conference Proceeding,---,1994. Barcelona, Spain, ILSI Europe. Workshop on Colonic Microflora: *Nutrition and Health.* 14-16 September.

10.   Formica J. V. and Regelson W. Review of the biology of quercetin and related bioflavonoids. *Food and Chemical Toxicology.* 1995, **33**(12), 1061-1080.

11.   Wilkins T. D. The 25 Trillion Bacteria Within All Of Us. ILSI Europe. Workshop of Colonic Microflora: Nutrition and Health, 14-16 September, Barcelona, ---. 1994.

12.   Rumney C. J. *in vivo* and *in vitro* models of the human colonic flora. ILSI Europe. Workshop on colonic microflora: Nutrition and Health, ---. 1994, 14-16 September, Barcelona, Spain.

13.   Gugler R., Leschik M., Dengler H.J. Disposition of quercetin in man after single oral and intravenous doses. *Eur.J.Clin.Pharmacol.* 1975, **9**, 229-234.

14.   Manach C., Morand C., Texier O., et al. Quercetin metabolites in plasma of rats fed diets containing rutin or quercetin. *The Journal of Nutrition.* 1995, **125**, 1911-1922.

15.   Noteborn H. P. J. M., Jansen E., Benito S., and Mengelers M. J. B. Oral absorption and metabolism of quercetin and sugar-conjugated derivatives in specific transport systems. *Cancer Letters.* 1997, **114**, 175-177.

16.   Salucci M. The effect of selected flavonoids on cell cycle progression of Caco2. 1999. Institute of Physiopathology, University of Pavia, Italy.

17.   Hollman P. C. H. and et al. Absorption of dietary quercetin glycosides and quercetin in healthy ileostomy volunteers. *American Journal of Clinical Nutrition.* 1995, **62**(1276), 1282.

18.   De Vries J. H. M. Assessment of flavonoid and fatty acid intake by chemical analysis of biomarkers and of duplicate diets. 1-127. 1997. Wageningen Agricultural University, Division of Human Nutrition and Epidemiology, The Netherlands.

19.   Maiani G., Serafini M., Salucci M.., Azzini E., Ferro-Luzzi A. Application of a new high-performance liquid chromatographic method for measuring selected poliphenols in human plasma. *J.Chromatogr.* 1997, 311-317.

20.   Lee M-J., Wang Z.Y., Li H., et al. Analysis of plasma and urinary tea polyphenols in human subjects. *Cancer Epidemiol.Biomark.Prev.* 1995, **4**, 393-399.

21.   Xu X., Wang H-J., Murphy P.A., Cook L., Hendrich S. Daidzein is a more bio-available soymilk isoflavone than is genistein in adult women. *J. Nutr.* 1994, **124**, 825-832.

22.   Kelly G. E., Joannou G. E., Reeder A. Y., Nelson C., and Waring M. A. The variable metabolic response to dietary isoflavones in humans (43829). *Proceedings of the Society for Experimental Biology and Medicine.* 1995, **208**, 40-43.

23.   Gooderham M. J., Adlercreutz H., Ojala S. T., Wahala K., and Holub B. J. A soy protein isolate rich in genistein and daidzein and its effects on plasma isoflavone concentrations, platelet aggregation, blood lipids and fatty acid composition of plasma phospholipid in normal men. *Journal of Nutrition.* 1996, **126**, 2000-2006.

24.   Adlercreutz H., Markkanen H., and Watanabe S. Plasma concentrations of phyto-estrogens in Japanese men. *Lancet.* 1993, **342**, 1209-1210.

25.   Herman C., Adlercreutz T., Goldin B.R., et al. Soybean phytoestrogen intake and cancer risk. *J. Nutr.* 1995, **125**, 757S-770 S.

26.   Kurzer M.S., Lampe J.W., Martini M.C., and Adlercreutz H. Fecal lignan and isoflavonoid excretion in premenopausal women consuming flaxseed powder. *Cancer Epidemiology, Biomarkers & Prevention.* 1995, **4**(4), 353-358.

27.   Musey P. I., Adlercreutz H., Gould K. G., Collins D. C., Fotsis T., Banwart C., Mäkelä T., Wähälä K., Brunow G., and Hase T. Effect of diet on lignans and isoflavonoid phytoestrogens in chimpanzees. *Life Sciences.* 1995, **57**(7), 655-664.

28.   Hollman P. C. H., Gaag M. V. D., Mengelers M. J. B, van Trijp J. M. P., De Vries J. H.M., and Katan M.B. Absorption and disposition kinetics of the dietary antioxidant quercetin in man. *Free Radical Biology & Medicine.* 1996, **21**(5), 703-706.

29.   Hollman P.C.H. Bioavailability of flavonoids. *Eur.J.Clin.Nutr.* 1997, **51**, S 66-S 69.

30.   Hollman P.C.H., van Trijp J.M.P., Mengelers M.J. B., De Vries J.H.M., and Katan M.B. Bioavalability of the dietary antioxidant flavonol quercetin in man. *Cancer Letters.* 1997, **114**, 139-140.

31.   Conquer J. A., Maiani G., Azzini E., Raguzzini A., and Holub B. J. Supplementation with quercetin markedly increases plasma quercetin concentrations without effect on selected risk factors for heart disease in healthy subjects. *Journal of Nutrition* 1997, **128**, 593-597.

32.   Paganga G. and Rice-Evans C. The identification of flavonoids as glycosides in human plasma. *FEBS Letters.* 1997, **401**, 78-82.

DIETARY FLAVONOIDS AND ANTIOXIDANT PROTECTION

Piergiorgio Pietta.

ITBA-CNR, Via F.ll Cervi, 93, 20090 Segrate (Milano), Italy.

1 INTRODUCTION

Flavonoids are natural polyphenolic compounds that are present in a large variety of vegetables, fruits and beverages, and may be considered an important source of antioxidants.[1] Flavonols, flavones, anthocyanins, catechins and proanthocyanidins are the most common classes, and constitute the main part of total polyphenol dietary intake.[2] This can range between 50 and 800 mg/day, depending on the consumption of vegetables and fruit (providing mainly flavonols, flavones and their dehydroderivatives), and of specific beverages, such as red wine, tea, and unfiltered beer. These last contain anthocyanins, catechins and phenolic acids, and may account for up to 80 % of the total polyphenol supply. Indeed, a glass of red wine or a cup of tea (particularly, green tea) can afford approximately 200 mg of total flavonoids, that represents a large amount when compared with those achievable from vegetables (onions, 40 mg/100 g; green salad, 1 mg/100g) or fruits (6-10 mg/apple; 1-2 mg/peach; 10 mg/orange).[3]

2 FLAVONOID PROPERTIES

A multitude of studies have suggested that flavonoids exert a large array of activities, including interaction with enzymes involved in cell division and proliferation, platelet aggregation and detoxification.[4] However, most attention has been devoted to the antioxidant activity of flavonoids, caused by their ability to reduce free radical production either by inhibiting enzymes or chelating transition metal ions responsible for free radical generation.[5] In addition, flavonoids for their lower one-electron redox potential, are able to reduce the highly oxidizing reactive oxygen species (ROS).[6] The capacity of most common dietary flavonoids to behave as *in vitro* antioxidants has been largely investigated, and a list of their TEAC (Trolox Equivalent Antioxidant Capacity) is available.[7] Unfortunately, little is known about the efficacy of dietary flavonoids *in vivo*, and this is due to the scarce knowledge on their bioavailability and uptake in humans.

3 FLAVONOID ABSORPTION

Only recently, it has been proved that flavonoids from dietary sources are absorbed at an extent that may promote an antioxidant effect. According to most authors, flavonol, flavone and isoflavone glycosides are firstly hydrolysed to aglycones, and these are then absorbed undergoing glucuronidation and sulfation when crossing the intestinal membrane.[8,9] However, it has been reported that quercetin-glycosides from fried onions are better absorbed than quercetin itself, even though the analytical evidence supporting the presence of quercetin-glycosides in blood is lacking.[10] Nevertheless, the assumption that also glycosides are absorbable is correct, as recently proved by the LC-MS detection of quercetin-3-rutinoside in blood of volunteers fed olive oil supplemented tomato puree.[11] As far as catechins are concerned, epigallocatechingallate and epicatechingallate have been detected in human blood after intake of green tea infusions or decaffeinated green tea extracts.[12]

In conclusion, the existing data suggest that dietary flavonoids are absorbed preferentially in the aglycone form, although also glycosides may be absorbed. Likely, flavonoid glycosides for a significant absorption require to be splitted in the small and large intestine, and this is in agreement with the time (about 6 h) needed to reach peak plasma levels of the resulting modified aglycones. On the other hand, free aglycones may be absorbed in the first part of the intestine, and this is consistent with the earlier rise (1-2 h after ingestion) in their plasma concentration.[13] However, more work with advanced and validated procedures is needed to further prove this assumption.

Concerning the extent of absorption, there is a large consensus that dietary flavonoids are uptaken at very low extent. The percentage of absorption normally does not exceed few percents of the ingested dose, as by measuring the blood levels of intact flavonoids and their conjugates. Food composition may represent an important factor that affects bioavailability. Thus, proteins may bind to polyphenols reducing their availability; alcohol may improve it, as evidenced by the increased uptake of red wine phenolics as compared with that achieved after alcohol-free red wine.[14] In addition, recent data support for an improved absorption of specific flavonoids in presence of fats. Thus, catechins from green tea, oligomeric proanthocyanidins from grape seeds, silibinin from milk thistle are absorbed at higher extent when administered as phospholipid complexes (Phytosomes) rather than free compounds.[15] Similarly, quercetin was detected in blood after consumption of onions fried with margarine, and rutin after intake of tomato puree mixed with olive oil. These findings are of great interest, and they should be carefully considered when explaining the mechanism of flavonoid absorption.

4 FLAVONOID METABOLISM

During absorption across the intestinal membrane, flavonols, flavones, isoflavones and catechins are partly transformed to their glucuronides and sulfates.[8,9] Subsequently, this small fraction of the absorbed flavonoids is metabolized by the liver enzymes, resulting in more polar conjugates to be excreted into urine or returned to the duodenum via gall-bladder. However, the major part of ingested flavonoid is not absorbed and is largely degraded by the intestinal microflora. The bacterial enzymes catalyse several reactions, including hydrolysis, cleavage of the heterocyclic oxygen containing ring, dehydroxylation and decarboxylation. A variety of phenolic acids are produced, depending on the structure of the flavonoid involved.[16] These phenolic acids can be reabsorbed, subjected to conjugation and O-methylation in the liver, and then may enter into circulation. This aspect is relevant for the antioxidant protection, mainly for two reasons. The first one is that phenolic acids may account for a large fraction of the ingested flavonoids (30-60 %),[12] and the second is that some of these acids, for their catechol structure, possess a radical-scavenging ability comparable to that of their intact precursors.[17] This suggests that these metabolites may take part in the antioxidant protection.

5 *IN VIVO* FLAVONOID ANTIOXIDANT POTENTIAL

Based on the available data, it may be assumed that dietary flavonoids may display their first antioxidant defence in the digestive tract, by limiting ROS formation and scavenging them. Once absorbed, either as aglycones and glycosides or, at larger extent, as phenolic acids, they continue to exert an antioxidant effect, although other systemic activities based on mechanisms different form red-ox are possible. The major question is how the antioxidant protection takes place *in vivo*. In other words, is the antioxidant effect due only to the increased plasma levels of flavonoids and their metabolites or are other concerted mechanisms also possible? Recently, it has been proved that a single dose of green tea catechins either free (Greenselect) or as phospholipid complexes (Greenselect Phytosomes) produces a transient decrease of plasma ascorbate and total glutathione. This decrease is well correlated with the time course of plasma catechin concentration, and it is accompanied by a rise of the plasma antioxidant capacity.[15] No modification of vitamin E or ß-carotene was observed. On the other hand, after long-term consumption of green tea, the levels of vitamin E in RBC membranes and LDL, ß-carotene in plasma and polyunsaturated fatty acids in RBC membranes were improved, while hydrophilic antioxidants remained practically unchanged.[18] Likely, long-term intake of green tea guarantees a base-line plasma concentration of catechins and their metabolites, which is able to induce an improvement of lipophilic vitamin levels. This modification may be explained by assuming that the antioxidant protection can be exerted through a cascade involving endogenous antioxidants, which react differently according to their polarity and redox potential.[19] More specifically, flavonoids and their metabolites are capable to reduce the highly oxidising ROS, becoming less aggressive aroxyl radicals. In turn, some of these free radicals for their hydrophilic character may oxidise ascorbate, which may be regenerated by glutathione. This may explain the transient decrease of plasma ascorbate and total glutathione, following an ingestion of green tea. On the contrary, the oxidation of lipophilic vitamins by aroxyl radicals is unlikely. As a result, vitamin E and ß-carotene are spared from the attack of aroxyl radicals at expenses of polar antioxidants, and may defend RBC PUFA against oxidative modifications. However, this sparing effect requires a long-term consumption of green tea, and it is accompanied by homeostatic recovery of the hydrophilic antioxidants.

6 CONCLUSIONS

Dietary flavonoids represent an important source of antioxidants, since their intake may reach 800 mg/day. Recent years, many papers have been published on the *in vitro* antioxidant activity of flavonoids. However, their antioxidant efficacy *in vivo* has been less documented, possibly due to the limited knowledge on their biokinetics. Only recently, it has been proved that a small fraction of the ingested dietary flavonoids is absorbed either in the aglycone or glycoside form, while the major part is extensively degraded to different phenolic acids. Both the absorbed flavonoids and their metabolites may display an *in vivo* antioxidant activity, which seems to be exerted through a cascade involving differently the physiologic antioxidants.

References
1. P.G. Pietta, P.L., Mauri, P. Simonetti and G. Testolin, *Fresenius J. Anal. Chem.*, 1995, **352**, 788.
2. K. Herrmann, *Z. Lebensm Unters Forsch*, 1988, **186**, 1.
3. C.T. Ho, C.Y. Lee and M.T. Huang, "Phenolic compounds in food and their effect on health." American Chemical Society, Washington, D.C., 1992, Vol.2., p.54.
4. N.E. Cook and S. Samman, *J. Nutr. Biochem.* 1996, **7**, 67.
5. W. Bors, W. Heller, C. Michel and M. Saran, *Methods Enzymol.*, 1990, **186**, 343.
6. S.V. Jovanovich, S. Steenken, Y.Hara and M.G. Simic, *J. Am. Chem. Soc.* 1996, **2**, 2497.
7. C. Rice-Evans, N.J. Miller, P.G. Bolwell, P.M. Bramley and J.B. Pridham, *Free Rad. Res.*, 1995, **22**, 375.
8. C. Manach, F. Regerat, O. Texier, G.Agullo, C. Demigné and C. Remesy, *Nutr. Res.*, 1996, **16**, 517.
9. T. Ishikawa, M. Suzukawa, T. Ito, H. Yoshida, M. Ayaori, M., Nishikawi, A.Yonemura, Y. Hara, H. Nakamura, *Am. J. Clin. Nutr.*, 1997, **66**, 261.
10. P.C. Hollman, M.V. Gaag, M.J. Mengelers, J.M. Van Tryp, J.M. De Vries and M.B. Katan, *Free Rad. Biol. Med.*, 1996, **21**, 703.
11. C. Gardana, P. Rossetti, P. Simonetti, M. Porrini and P.G. Pietta, *J. Chromatogr.*, in press.
12. P.G. Pietta, P. Simonetti, C. Gardana, A. Brusamolino, P. Morazzoni and E. Bombardelli, *Biofactors*, 1998, **8**, 111.
13. R.A. King and D.B. Bursill, *Am. J. Clin. Nutr.*, 1998, **67**, 867.
14. P. Simonetti, F. Brighenti, and P.G. Pietta, Polyphenol Communications 98, in press
15. P.G. Pietta, P. Simonetti, C. Gradana, A. Brusamolino, P. Morazzoni and E. Bombardelli, Biochem. Mol. Biol. Int., in press.
16. P.G. Pietta, C. Gardana, P.L. Mauri, *J. Chromatogr.*, 1997, **693**, 249.
17. I. Merfart, J. Heilmann, M. Weiss, P.G. Pietta and C. Gardana, *Planta Med.*, 1996, **62**, 289.
18. P.G. Pietta, P. Simonetti, C. Roggi, A. Brusamolino, N. Pellegrini, L. Maccarini and G. Testolin, "Natural Antioxidants and Food Quality in Atherosclerosis and Cancer Prevention", eds N. Kumpulainen, J.T. Salonen, Royal Society of Chemistry, Cambridge, 1996, p. 249.
19. P.G. Pietta, P. Simonetti, *Biochem. Mol. Biol. Int.*, 1998, **44**, 1069.

TROLOX EQUIVALENT ANTIOXIDANT CAPACITY OF AVERAGE FLAVONOIDS INTAKE IN FINLAND

J.T. Kumpulainen, M. Lehtonen and P. Mattila.

Agricultural Research Centre of Finland, Food Chemistry Research, L-Building, 31600 Jokioinen, Finland.

1 INTRODUCTION

Flavonoids, a group of aromatic secondary plant metabolites belonging to a class of phenolic substances have recently attracted much research interest among food chemists, nutritionists as well as clinical researchers. This is because these compounds have antibacterial, antiviral, antioxidative, antiatherogenic and anticarcinogenic properties in living organisms.

Flavonoids are classified according to degree of unsaturation, and degree of oxidation of the flavone skeleton. Accordingly, the major classes are flavones, flavonols, flavanones, isoflavones, anthocyanidins and cathecins. Within the above classes further chemical differentiation is possible based on the number and nature of substituent groups, such as hydroxyl groups, attached to the rings.

Further structural complexity is due to a common occurrence of the compounds as *O*- or *C*- glycosides in which one or more of the flavonoid hydroxyl groups are bound to a sugar or sugars by an acid labile hemiacetal bond. Thus, a common flavonoid, such as kaempferol occurs in nature in any one of the 214 different forms. Altogether, over 5000 different flavonoids have been identified. The analytical chemistry of flavonoids is very difficult due to the fact that the glycosides must be hydrolyzed to aglycones in order to be able to separate and identify the compounds. However, the optimization of hydro-lysis is difficult as it depends on stability of the flavonoid species and the matrix in question.

Hertog et al.[1] have analyzed vegetables and fruits for contents of 5 commonly existing flavonoids in the Netherlands and reported high levels of quercetin in onions, kale and broccoli. Kaempferol was found in high concentrations in kale, endive and leek.

High concentrations have been reported in tea, particularly in green tea, red wine, onions, citrus fruits and apples.[1-5]

Previously reported total dietary intake estimates of flavonoids in various populations vary from 2.6 mg/d to several hundred mg/d, depending on the food consumption pattern.[2-5]

The antioxidative potencies of different flavonoids vary strongly depending on their chemical structure. Various *in vitro* tests have been established in order to compare the antioxidative potencies of flavonoids. [6-7]

These tests demonstrate that important structural requirements for the antioxidative capacity are as follows: **1)** The O-dihydroxystructure of the B-ring, **2)** the 2,3 double bond in conjugation with a 4-oxo function and, **3)** the presence of both 3- (a) and 5-(b) hydroxy groups for maximal radical scavenging capacity. Thus, over four-fold differences exist in the antioxidative capacity among various flavonoids. Furthermore, the most potent flavonoid possesses a five-fold antioxidative capacity compared to vitamin C or Trolox.[6] This demonstrates that flavonoids are very potent antioxidants and, considering that many of them have been shown to be bioavailable,[8-12] one may postulate that their adequate intake may be important for the prevention of age- and oxidative stress related chronic diseases, which is supported by recent studies.[13-14]

2 EXPERIMENTAL

2.1. Sampling and sample preparation

Samples of fruits, beverages, vegetables and berries most commonly consumed in Finland were collected from Finnish whole sale companies and supermarkets. Care was taken to include the most important varieties and brands. Samples were collected in such a way that, based on the average food consumption data in Finland in 1997, the most important plant foods were included in the samples. Furthermore, the major collection harvesting areas of wild and cultivated berries were represented. In addition, certain berries known to contain very high amounts of flavonoids were included although their average consumption figures were low. Altogether 77 samples of fruits, 89 samples of vegetables, 151 samples of berries and 60 samples of beverages including tea and wines were collected. Altogether 377 samples were collected. Subsamples collected were pooled and homogenized in a blender to make 92 final samples. Pooled samples were frozen and freeze-dried afterwhich they were stored frozen at - 20 °C before analysis.

2.2. Analytical methods

Analytical method employed in this study was a modified method of Hertog et al.[15]

2.2.1. Extraction and hydrolyzation of flavonoids. Freeze-dried samples of 0.5 g were treated by adding 40 ml of 62.5 % aqueous methanol containing 2 g/l of 2,3 tert-butyl-4-hydroxyanisol (BHA). The extract was ultrasonicated for 5 min, then 10 ml of 6 M HCl was added. The extract was nitrogenated for 1 min, heated up to 90 °C and kept in a shaking water bath for two hours. The extract was then cooled to room temperature, diluted to 100 ml with methanol and ultrasonicated for 5 min resulting in an extract ready for HPLC analysis.

Beverages (juices, teas and wines) were prepared by adding 25 ml of 62.5 % aqueous methanol as above to a 15 ml sample followed by ultrasonification for 5 min and addition of 10 ml of 6 M HCl. The rest of the sample handling was the same as that for the dried samples, except that hydrolysis time was 4 h at 90°C for all beverages.

The catechins in beverages were extracted by adding 3 x 3 ml of ethylacetate into a

5 ml sample. The extraction was performed in a test-tube fitted with a corkscrew cap by manually shaking the sample for 1 min. Ethylacetate layer was separated from the water layer and evaporated to dryness with nitrogen stream. Residue obtained was solubilized into methanol and diluted to 20 ml with methanol, resulting in a sample ready for the HPLC analysis. Approximately 2 ml of the final sample extact was filtered using 0.2 μm filter (Acrodisk 13 CR PTFE Gelman Sciences) before HPLC analysis. Injection volume was 10 μl and all samples were analysed in duplicate.

    2.2.2. Standards. Flavonoid standards were purchased from Roth Ltd., Aldrich Ltd., and Fluka Ltd. All solutions used in the method were of HPLC purity (higher than 99.5 % purity). Stock standards (C= 0.1 mg/ ml) were prepared into either methanol (Mallinckrodt) or N,N-dimethylformamide (Mallinckrodt)-methanol mixture. Stock standards were maintained at -18°C protected from light. Working standards were made by diluting stock standards to yield 2-4 μg/ml. The standards, their purities and analytical wavelengths used in the quantification by diode array detector (DAD) are presented in table 1.

Table 1
Flavonoid standards, their purities and DAD analytical wavelengths used.

Standard	Producer	Purity	Wavelength (nm)
(+)-catechin	ROTH	> 98 %	280
(-)-epicatechin	ALDRICH	97%	280
(-)-epicatechin gallate	ROTH	> 98 %	280
epigallocatechin	ROTH	> 98 %	270
epigallocatechin gallate	ROTH	> 98 %	270
eriocitrin	ROTH	HPLC	280
narirutin	ROTH	HPLC	280
rutin	ROTH	HPLC	254
naringin	ROTH	HPLC	280
myricitrin	ROTH	HPLC	254
hesperidin	ROTH	97%	280
myricetin	ROTH	HPLC	370
eriodictyol	ROTH	HPLC	280
fisetin	ALDRICH	HPLC	370
quercetin	ROTH	HPLC	370
naringenin	ROTH	HPLC	280
luteolin	ROTH	HPLC	329
hesperetin	ROTH	HPLC	280
kaempferol	FLUKA	> 96 %	370
isorhamnetin	ROTH	HPLC	370
apigenin	ROTH	HPLC	329
rhamnetin	ROTH	HPLC	370
galangin	ROTH	HPLC	270
quercitrin	ROTH	HPLC	254

2.2.3. HPLC Analysis. The analyses were carried out using Hewlet Packard HPLC 1090 Series II High Performance Liquid Chromotograph (HPLC) equipped with HP1090 Series II DAD and an eight channel electrochemical Coulometric Array Detector (CAD; Esa, Inc. USA). HPLC and DAD intrumentation was controlled by the HP 3D Chem Station computer program. CAD was controlled using Esa CoulArray Version 1.001 computer program. Flavonoid separation was done by an Inertsil (GL Sciences, Inc. Japan) ODS-3 (4.0 x 150 mm, 3 µm) column. Mobile phases used were: 0.025 M NaH_2PO_4 (pH 3.5) containing 1% methanol and (solution A) and, 0.025 M NaH_2PO_4 (pH 3.5) containing 80 % methanol (solution B). Gradient program used was 5-95 % B over 0-60 min. and 95 % B over 60-63 min. using a stream velocity of 0.9 ml /min. Column and CAD electrode temperatures were 35 °C.

UV-spectra was followed over the range of 190-500 nm with a resolution of 2 nm. Five different wavelengths were used for flavonoid quantification (254, 270, 280, 329, 370 nm). CAD was operated using potentials 0-700 mV (100 mV intervals).

2.2.4. Quantification. Quantification by DAD was done using the external standards method. Flavonoids were quantified by the wavelength showing the highest UV response for the flavonoid in question (see table 1). Calibration curves were made over a range of 1-8 µg/ml (4 standards). Detector response was linear over the concentration range used. For all standards r^2 was higher than 0.998. Quantification was by peak area measurement. Compounds were identified using UV spectras over the range of 190-500 nm.

Quantification using CAD was done by comparing the detector responses among the known and unknown compounds. Quantification was by peak height. CAD was calibrated over the concentration range of 2-4 µg/ml. Identification by CAD was done by comparing retention times and oxidation tendency at various potentials of the known and unknown peaks.

2.3. Calculation of average dietary intake and TEAC of flavonoids in Finland

The average dietary intake of fruits, beverages, vegetables and berries was obtained from a recent dietary survey of the Finnish National Public Health Institute. This survey was based on a food consumption of several thousand Finns living in both cities and countryside in 1997.[16] For the calculation of TEAC, following conversion factors were used:[6] epicatechingallate 4.9; epigallocatechingallate 4.8; quercetin 4.7; eriodictyol 4.7; epigallocatechin 3.8; myricetin 3.1; epicatechin 2.5; catechin 2.4; luteolin 2.1; naringenin 1.5; apigenin 1.4; hesperitin 1.4; kaempferol 1.3 and isorhamnetin 1.1.

3 RESULTS

3.1. Average dietary intake of flavonoids in Finland

The total average intake of flavonoids in Finland based on 1997 food consumption data was 55.2 mg/d (Fig. 1).

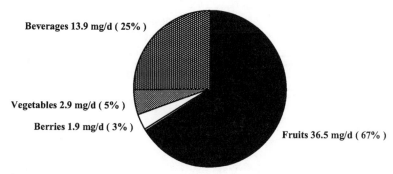

Figure 1

Contribution of food groups to the total average intake of flavonoids in Finland.

Fruits contributed 36.5 mg/d (67 %) followed by tea, wine and other beverages (altogether 13.9 mg/d = 25 %), vegetables (2.9 mg/d = 5 %) and berries (2.0 mg/d = 3 %) to the total intake. Among fruits, high contents of flavonoids were found in oranges (409 mg hesperetin/kg and 118 mg naringenin/kg). Oranges were by far the best source of average flavonoid intake in Finland representing 28.5 mg/d followed by other citrus fruits and apples (Fig. 2).

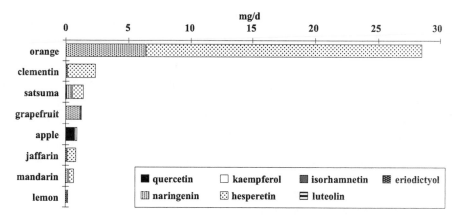

Figure 2

Contribution of various fruits to the average intake of flavonoids in Finland.

Tea beverage, particularly green tea, dispayed the highest flavonoid content (286 mg epigallocatechingallate/kg and 70 mg epicatechingallate/kg) among beverages. Tea, besides containing the most potent flavonoids in terms of TEAC, contained also the greatest variety of flavonoids. Tea alone represented almost 12 mg/d average intake followed by orange juice, 1.8 mg/d (Fig 3.).

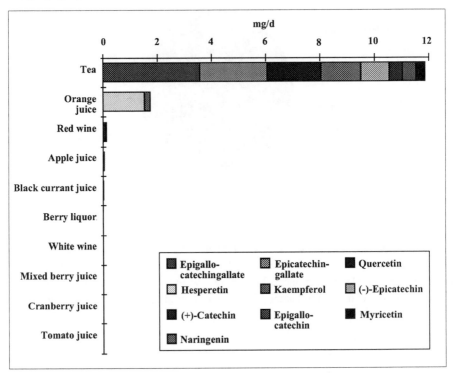

Figure 3
Contribution of various beverages to the average intake of flavonoids in Finland.

Red wine, while a good source of flavonoids, is very little consumed in Finland and contributes negligible to the total flavonoid intake.

Among vegetables onions, particularly red onions (307 mg quercetin/kg), had the highest contents, then asparagus (60 mg kaempferol/kg) and cabbages (47 mg kaempferol/kg). Onion group contributed the most flavonoid intake among vegetables (1.7 mg/d) followed by the cruciferous group (0.45 mg/d), spinach (0.4 mg/d) and tomatoes (0.35 mg/d) (Fig. 4).

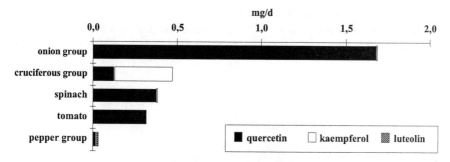

Figure 4

Contribution of the most important groups of vegetables to the average intake of flavonoids in Finland.

Among berries cranberries (104 mg quercetin/kg and 69 mg myricetin/kg), lingonberries (100 mg quercetin/kg), black currants (41 mg quercetin/kg and 53 mg myricetin/kg), rowanberries (106 mg quercetin/kg and 10 mg myricetin/kg), and blueberries (37 mg both quercetin and myricetin/kg) had the highest contents. Lingonberries contributed most (0.75 mg/d) to the total flavonoid intake followed by black currants (0.44 mg/kg) and blueberries (0.35 mg/d) (Fig. 5).

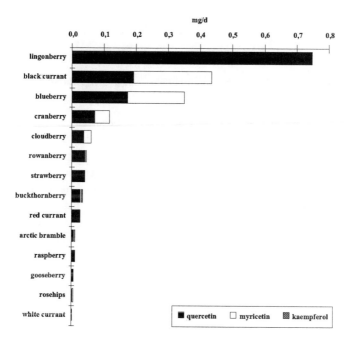

Figure 5

Contribution of berries to the average intake of flavonoids in Finland.

Among the most important flavonoids citrus flavonoids, particularly hesperetin, contributed most, 28.3 mg (51 %) to the total average intake (Fig. 6). Other major flavonoids were naringenin 8.3 mg (16 %), then quercetin 7.0 mg (13 %) and epigallocatechingallate 3.6 mg (6 %).

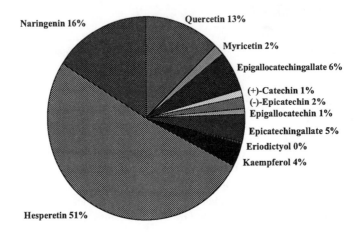

Figure 6
Contribution of the most important flavonoids to the total average intake of flavonoids in Finland.

3.2. Trolox equivalent antioxidant capacity (TEAC) of average flavonoids intake in Finland

TEAC of average flavonoids intake in Finland was 118.4 equivalents (Fig. 7). The greatest contribution was provided by beverages (43 %) followed by fruits (40 %), vegetables (10 %) and berries (7 %). Here one can see the effect of strong antioxidants present in tea, vegetables (onions) and berries, whose TEAC value is almost double compared to the intake value. Naturally, the bioavailabilities and *in vivo* antioxidative effects of the flavonoids in question still need to be elucidated. It is known, however, that quercetin is bioavailable not only as an aglycone but also as a monoglycoside.[10] Furthermore, it has been shown that many other compounds present in foods affect the bioavailabilty of flavonoids, such as alcohol or vegetable oils.[17]

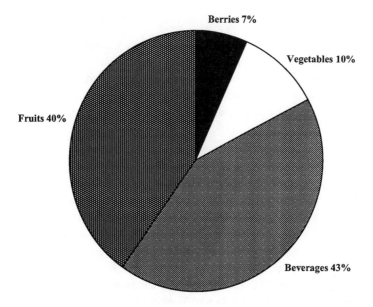

Figure 7
Contribution of food groups to the average TEAC intake of flavonoids in Finland. Total TEAC = 118,4 equivalents.

4 DISCUSSION

Anthocyanidins were not included in the present flavonoid intake estimate. Their contribution to the total intake and TEAC of flavonoids may not, however, be very significant due to the low average consumption of berries and red wines in Finland which are most likely the best sources of anthocyanidins in the Finnish diet. They would, however, change the TEAC more favorably for the berries and red wine due to their high antioxidative capacity.[6]

The present data show that the average flavonoid intake in Finland is almost 20 times higher than that previously estimated on the basis of Hertog's analytical data which included only five flavonoids.[18] Leth et al[3] have reported the flavonoid intake in Denmark to be 28 mg/d including the citrus flavonoids. They found the major sources to be onions, tea, oranges, and orange juice. The most prevalent flavonoid aglycones were found to be quercetin, kaempferol, hesperitin, and naringenin. However, their sampling included neither as many food items nor flavonoids as the present study. It is clear that great individual and population group differences must exist in flavonoids intake depending on food consumption habits, such as drinking of tea, red wine and consumption of citrus fruits and berries. It remains to be clarified what the effects of such differences in flavonoids intake might be on the morbidity or mortality of such populations.

Acknowledgements

The contributions of Mr. Pasi Tapanainen and Ms. Tarja Vikman for their skilful help in analysing the samples, Mr. Tauno Koivisto for the data processing and Ms. Sari Aalto for the technical help in preparing the manuscript are gratefully acknowledged. I am also grateful to Dr. Marja-Leena Ovaskainen of the National Public Health Institute and Ms. Pirkko Uutela of our research unit for their help in providing the average food consumption data of the Finns for 1997.

References

1. M.G.L. Hertog, P.C.H. Hollman, M.B. Katan, *J. Agric. Food. Chem.* 1992, **40**, 2397.
2. M.G.L. Hertog, D. Kromhout, C. Aravanis. et al., *Arch. Intern. Med.* 1995, **155**, 381.
3. T. Leth and U. Justesen, in "Polyphenols in food", Cost 916, Bioactive plant cell wall components in nutrition and health, First workshop, Aberdeen, Scotland, 16-19 April 1997. Ed. R. Armadò, H. Andersson, S. Bardócz, and F. Serra. EUR. 18169, p. 39 .
4. K. Herrmann, Z. Lebensm. *Unters. Forsch.* 1988, **186**, 1.
5. C.T. Ho, C.Y. Lee and M.T. Huang, "Phenolic compounds in food and their effect on health." American Chemical Society, Washington, D.C. 1992, Vol.2., p.54.
6. C. Rice-Evans, N. Miller, P.G. Bolwell, Bramley J.B. Pridham, *Free. Rad. Res.* 1995, **22**, 375.
7. G. Cao, E. Sofic, R.L.Prior, *Free. Rad. Biol. Med.* 1997, **22**, 749.
8. C. Manach, F. Regerat, O. Texier, G.Agullo, C. Demigné and C. Remesy, *Nutr. Res.* 1996, **16**, 517.
9. T. Ishikawa, M. Suzukawa, T. Ito, H. Yoshida, M. Ayaori, M., Nishikawi, A.Yonemura, Y. Hara, H. Nakamura, *Am. J. Clin. Nutr.* 1997, **66**, 261.
10. P.C. Hollman, M.V. Gaag, M.J. Mengelers, J.M. Van Tryp, J.M. De Vries and M.B. Katan, *Free Rad. Biol. Med.* 1996, **21**, 703.
11. C. Gardana, P. Rossetti, P. Simonetti, M. Porrini and P.G. Pietta, J.Chromatogr. in press.
12. P.G. Pietta, P. Simonetti, C. Gardana, A. Brusamolino, P. Morazzoni and E. Bombardelli, *Biofactors.* 1998, **8**, 111.
13. M.G.L. Hertog, E.J.M. Feskens, P.C.H. Hollman, M.B. Katan and D. Kromhout, *Lancet.* 1993, **342**, 1007.
14. S.O. Keli, M.G.L. Hertog, E.J.M. Feskens, D. Kromhout, *Arch. Intern. Med.* 1996, **156**, 637.
15. M.G.L. Hertog, P.C.H. Hollman, D.P. Venema, *J. Agric. Food. Chem.* 1992, **40**, 1591.
16. Anon, The 1997 Dietary Survey of Finnish Adults, Publications of National Public Health Institute, BB/1998, Helsinki, Finland.
17. P. Simonetti, F. Brighenti, and P.G. Pietta, Polyphenol Communications 98, in press.
18. P. Knekt, R. Järvinen, A. Reunanen, J. Maatela, *B.M.J.* 1996, **312**, 478.

ANTHOCYANINS IN RED WINES: ANTIOXIDANT ACTIVITY AND BIOAVAILABILITY IN HUMAN

T. Lapidot, S. Harel, R. Granit and J. Kanner.

Agricultural Research Organization, Volcani Center, Department of Food Science, P.O.Box 6, Bet Dagan 50250, Israel.

1 INTRODUCTION

Anthocyanins are natural plant pigments belonging to the flavonoid family and represent a substantial constituent of the human diet. Many foods and especially red grapes and wines contain large amounts of flavonoids which are mostly anthocyanins.

The anthocyanins differ from other natural flavonoids in the range of colors that can be derived from them and by their ability to form resonance structures by changes of the pH.[1] The reported[2-4] beneficial effects of red wine were suggested to derive from phenolic compounds, mostly flavonoids, which demonstrated powerful antioxidant properties against low density lipoprotein oxidation and the general antioxidant effect of red wines correlated well with their total phenolic content.[4,5] Anthocyanins are the main class of flavonoids in red wines, and they appear to significantly contribute powerful antioxidant properties to them. The 3-glucosides, dephinidin, cyanidin, petudin, and malvidin are present in red wines, but malvidin 3-glucoside, malvidin 3-glucoside acetate and malvidin 3-glucoside coumarate are the most abundant pigments.[6]

Anthocyanins exist in an aqueous phase in a mixture of four molecular species, and their relative concentrations depend on pH.[1] At pH 1-3 the flavylium cation is red colored, at pH 5 the colorless carbinol pseudo-base (pb) is generated and at pH 7-8 the blue-purple quinoidal-base (qb) is formed which could turn to a chalcone.

In grapes and wines the anthocyanins are in the flavylium form. However, during digestion they may reach higher pH's forming the carbinol pseudo-base, quinoidal-base or the chalcone and these compounds appear to be absorbed from the gut into the blood system.

Recently, much attention has been paid to the antioxidant properties of flavonoids, which seem to protect tissues against oxygen free radicals and lipid peroxidation. Circulating low-density lipoproteins (LDL) are one of the fundamental targets of deleterious oxidation, resulting in the accumulation of atherogenic lipoprotein in human *in vivo*.

The socalled "French paradox" is characterized by the fact that the inhabitants of some regions of France have a lower than average rate of coronary heart disease (CHD) despite their consumption of high levels of saturated fatty acids and cholesterol, which have been positively correlated with increased risk of CHD. This paradox has been attributed by us to the non-alcoholic compounds, the phenolic antioxidants, which are very abundant in red wines. These phenolic antioxidants act protectively via prevention of the oxidation of low-density-lipoproteins (LDL) and the inhibition of platelet aggregation.[2-4] The presence of significant amounts of anthocyanins in red wine contributes to its powerful antioxidative activity. Indeed, anthocyanins have been found to be strong antioxidants, in varied *in vitro* studies.[7-10] However, in humans, the bioavailability of the flavonoids present in foods and especially in wines, is an unclear important issue.

Most recently, Hollman et al.,[11] in a human study with ileostomy subjects, demonstrated the absorption from the intestine of onion quercetin glycosides. Several studies in rats showed that citrus flavonols or flavones and their glycosides may reach the blood stream after oral administration.[12,13] The absorption of the citrus flavanone, naringin, in humans has also been studied recently.[14,15] The absorption properties of anthocyanins have been evaluated in the rat,[16] but no such study has been performed in humans. New evidence for the possible presence of anthocyanins in human plasma has been demonstrated.[17]

The aim of the present study was to evaluate the antioxidant activity of these compounds, in several model systems where metals catalyzed generation of reactive oxygen and lipid species and to determine the potential bioavailability of several anthocyanins from red wine, in humans.

2 MATERIALS AND METHODS

Linoleate oxidizing activity was assayed spectrophotometrically.[4] The technique consists of following the increase in conjugated dienes and their absorbance at 240 nm. The test sample contained 1.5 ml of buffered linoleate at pH 7.0, 0.1-0.4 ml active fraction, and distilled water in a mixture as follow: linoleate 2 mM, linoleate hydroperoxide (2 μM); Tween 20, 0.05 %, phosphate buffer, pH 7.0; 0.05 M, DETA (diethylenetriaminepenta-acetic acid), 0.5 mM. The blank sample contained all the reagents except the catalyzers.

Isolation of the microsomal fraction from muscle tissues was done by a procedure described previously.[18] Protein determination assay were conducted by the modified Lowry procedure,[19] using BSA as standard.

Microsomes for lipid peroxidation assay were incubated in air in a shaking water bath at 37 °C. The reaction mixture contained 1 mg of proteins/ml and 4 ml of 50 mM acetate buffer, pH 7.0. The thiobarbituric acid reactive substances (TBARS) were determined by a procedure of Bidlack et al.[20]

3 SUBJECTS AND STUDY DESIGN

Six healthy fasting volunteers (3 women and 3 man) aged 25 to 45, participated in our study, which was carried out in our laboratory. On each of the experiment days, they did not consume any kind of product rich in polyphenols (vegetables, fruits, tea, etc.) On the control day, every hour for 12 h, they drank 300 ml of water and collected urine into sterile tubes, which were frozen immediately (-20 °C). Several weeks later, on the first test day, they repeated the same procedure but replaced the fourth dose of water with white wine (the fourth was taken 4 h after the beginning of the experiment and 1 h after breakfast). After a further 2 weeks, on the second test day, the same volunteers repeated this experiment with 300 ml of red wine, containing 218 mg of anthocyanins, in place of the white wine.

The concentration of each urine tube was determined with a creatinine test kit (Sigma Diagnostics, USA). The urine was concentrated by freeze-drying and the dried samples were stored at -20 °C, for HPLC determination of phenolics in the dried urine samples were prepared by dissolving in a solution of 10 % ethanol in water, at a volume which gave a creatinine concentration of 150 mg/dl (=13.26 mmol/l). The samples were then centrifuged for 3 min at 14,000 g in a refrigerated centrifuge (Sigma, 201 M). One part of the supernatant was acidified with 10 % (v/v) of concentrated HCl, to bring its pH to 1. First HPLC analysis was carried out immediately after the acidification and a second analysis was carried out following incubation of the sample with the HCl at room temperature for 24 h in the dark. The same procedure was carried out with the wine and malvidin-3-glucoside standard solution.

HPLC analysis was carried out according to Lamuela-Raventos and Waterhouse,[21] with modification as described by Lapidot at al.[22] All the samples were filtered through a 0.45 μm nylon filter (Lida, USA). The HPLC system used (LC-10A, Shimadzu, Germany) consisted of an auto injector, a photodiode array detector (SPD-M10Avp) and a software system which controlled all the equipment and carried out data integration and processing (CLASS-VP, Shimadzu). A cartridge column (Nova-Pak C18 Å 4 μm, 3.9 x 150 mm, Waters) were used. The injection volume was 20 μl and a precolumn (Nova-Pak C18 Guard-Pak, Waters). The linear gradients for the separation were: Solvent A: 0.05 M dihydrogen ammonium phosphate adjusted to pH 2.6 with orthophosphoric acid; Solvent B: 20 % A with 80 % ethanol; and Solvent C: 0.2 M orthophosphoric acid (pH 1.6). The total time for the separation was 80 min, with a flow rate of 0.7-1 ml/min. The identification of anthocyanins in wine and urine were determined using anthocyanin standards and by comparison to Lamuela-Raventos and Waterhouse[21] results. The concentration of the pigment solution was determined according to $E_{518nm} = 33,000 \ 1*mol^{-1}*cm^{-1}$.[23] The R^2 of the standard curve was 0.9934.

3 RESULTS

The structure of the three main forms of anthocyanins are presented in Fig. 1. The main anthocyanin was malvidin 3-glucoside, followed by malvidin 3-glucoside acetate and malvidin 3-coumarate. The incubation of malvidin 3-glucoside at pH 5, transforms the pigment rapidly to a colorless form, the pseudo-base. If the red pigment is brought to pH 7, a quinoidal-base, purple pigment is formed instantly. Both compounds, could form a chalcone - which is yellow during incubation at pH 7-8.

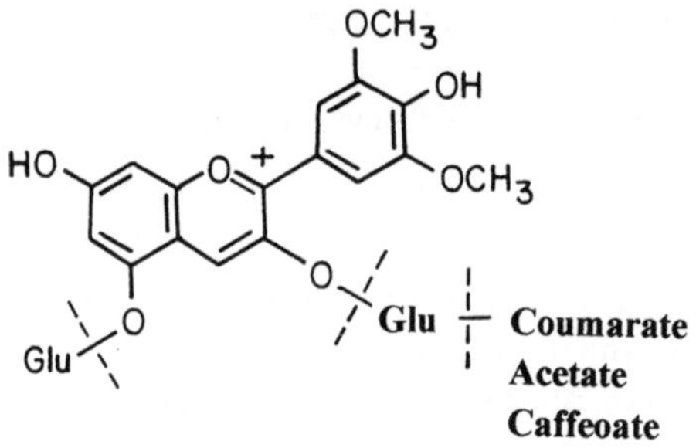

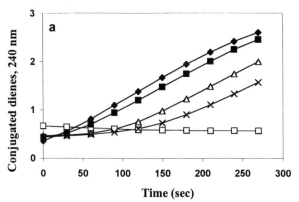

Figure 1

Structure of major anthocyanins in red grapes and wine.

The catalytic oxidation effect of myoglobin in the presence of linoleate dispersed in a microemulsion at pH 7.0 and 25 °C was determined. During this reaction myoglobin catalyzed the oxidation of linoleate via generation of free radicals which break down the heme ring, producing bleaching of the heme and a rapid decrease in the Raman spectra at 408 nm. The radicals generated in this reaction oxidized linoleate, forming conjugated dienes which increased the spectra at 234-240 nm. This reaction also generates carbonyls which had absorption peaks at 280 nm. Addition of pseudo-base malvidin-3-glucoside to this microemulsion at a concentration of 3 μM, significantly inhibited peroxidation. Fig. 2(a-b) presents data on the effectiveness of the pseudo-base and quinoidal-base malvidin 3-glucoside in inhibiting peroxidation of linoleate by myoglobin. Both compounds were found to be superior to catechin, a well known antioxidant.

Figure 2

Inhibition of lipid peroxidation (conjugated dienes) by wine anthocyanins, induced by myoglobin (2.5 μM) at pH 7.0, 23°C.

a) pseudo-base generated at pH 4.0, control, ◆; catechin 1 μM, ■ ; catechin 5 μM, ✖ ; pb malvidin-3-glucoside 1.6 μM, Δ; pb-malvidin-3-glucoside 6.5 μM, ;

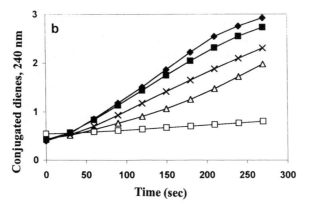

*b) quinonidal base generated at pH 7.0, control, ◆; catechin 1 µM, ■ ; catechin 4 µM, ✖ ;
q-malvidin-3-glucoside 1 µM, ∆; q-malvidin-3-glucoside 4 µM, .*

The Cabernet Sauvignon red wine chromatogram (Fig. 3) shows the presence of four main groups of anthocyanins: anthocyanidin-3-glycosides, with the aglycones identified as delphinidin, cyanidin, petunidin, peonidin and malvidin, respectively (peaks # 1-5, retention time (RT) 26.53-38.72 min); anthocyanidin-3-glycoside-acetate (peaks #6-8, RT 43.23-53.05 min). The third group (peaks #9-12, RT 61.05-64.36 min) seems to be composed of anthocyanidin-3-glycoside-coumarates, according to Lamuela-Raventos and Waterhouse.[21] The peaks with RT of 65.26 and 65.88 min seem to be anthocyanin dimers, characterized by their specific absorption at 431 nm.[6,24]

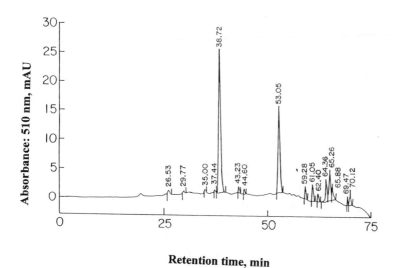

Fig. 3
HPLC chromatogram of Cabernet Sauvignon wine. Detection at 510 nm.

After ingestion of red wine, human urine does not contain any pigments, which could be identified as anthocyanins. However, concentrated urine adjusted to pH 1.0 by HCl addition, immediately developed a pink pigment, which absorbed at 500 nm (Fig. 4). The pink pigment was observed in the urine only after red wine ingestion (spectrum not shown). The pink pigment disappeared when the urine pH was adjusted back to 5.0. Such characteristic changes are very similar to those of anthocyanins, which convert from flavylium cation to a colorless pseudo-base and vice-versa, in response to pH changes.[25]

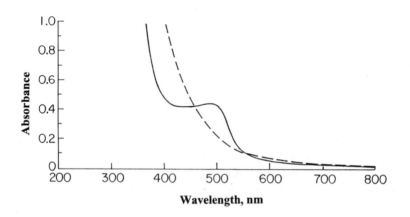

Figure 4
Spectra of human urine after red wine drinking. (---) urine; (—)urine at pH 1.0.

A typical chromatogram of urine from a volunteer after ingestion of red wine, before incubation with HCl was evaluated. The chromatogram shows two peaks, at retention times of 65.02 and 65.70 min. These peaks seem identical to those of the compounds from red wine that appeared at the retention times of 65.26 and 65.88 min, respectively. The spectra of those peaks from red wine and urine are presented in fig. 5 (a-b).

The absorption spectra of these two compounds did not show strong pH sensitivity as did the spectra of the rest of the pigments, which are described below. This fact supports the characterization of these compounds as anthocyanin dimers.[26] When the urine was incubated with 1N HCl for 24 h at room temperature, several other peaks were derived after HPLC analysis, (as presented in Table 1). The peaks that appeared at RT 33.25, 41.31 and 54.92 min showed typical spectra of anthocyanins. These peaks did not appear on the urine chromatograms after white wine ingestion. The rest of the peaks, which appeared after incubation with HCl, showed spectra similar to those of anthocyanins but with a hypochromic shift in the absorption maximum towards 487 nm. After white wine ingestion urine adjusted to pH 1.0 with 1 N HCl, did not contain any pigment, but after incubation at room temperature for 24 h it showed some of the pigments with the specific absorbance at 487 nm. After water drinking urine did not contain any pigment even after 24 h of incubation with HCl. No changes were detected in the anthocyanins profile when the red wine, or malvidin-3-glucoside standard solution were incubated for 24 h with HCl, by the same procedure as for the urine incubation. Each of the compounds derived at RT 32.58 and 39.34 min from urine after red wine drinking, and which had an absorption maximum at 487 nm, was collected with the mobile phase, into sterile tubes, after the separation by the HPLC. Those solutions having pH 1.9 were scanned by spectrophotometer and showed spectra maxima at 280 and 487 nm, by the HPLC detector. When the solutions was adjusted to pH 5.0, the absorption at 487 nm disappeared; when the pH was adjusted back to 1.9, the maximum at 487 nm rose immediately.

Fig. 5

Spectra of pigments as derived from HPLC analysis, scanned by diode array detector.

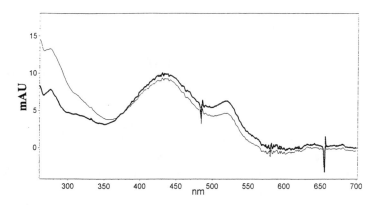

a) (—) The compound at peak RT 65.26 min, from red wine. (—)The compound at peak RT 65.02 min, excreted in urine after red wine ingestion

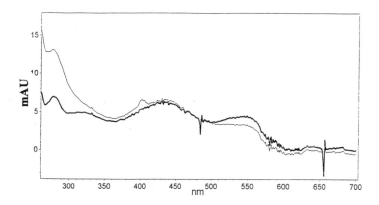

b) (—)The compound at peak RT 65.88 min, from red wine. (—) The compound at peak RT 65.70 min, excreted in urine after red wine ingestion.

Table 1

Structural aspects and antioxidant activity of flavonoids.

Flavonoid Class	Substitution OH	Redox Potential E/Vvs.NHE	LPO I_{50}	Ferryl reduction $K_1/mol^{-1}s^{-1}$	TEAC
		µM			**mM**
Flavanone					
Naringenin	5,7,4'	0.76	1137	4	1.5
Eriodictol	5,7,3'4'	0.36	26	25	---
Taxifolin	3,5,7,3',4'	0.37	22	18	1.9
Flavone					
Apigenin	5,7,4'	0.71	2754	31	1.4
Luteolin	5,7,3',4'	0.41	17	63	2.1
Flavonol					
Galandin	3,5,7	0.56	6	--	--
Kaempferol	3,5,7,4'	0.39	6	115	1.3
Quercetin	3,5,7,3',4'	0.29	6	279	4.7
Myricetin	3,5,7,3',4',5'	0.20	1057	2	3.1

VanAcker et al Free Rad. Biol. Med. 1996, 20, 331. Rice-Evans & Miller Biochem. Soc. Trans. 1996, 24, 790. Jorgensen & Skibsted Free Red. Res. 1998, 28, 335.

The anthocyanins and the "anthocyanin-like" compounds (with an optical absorption maximum at 487 nm) were absorbed from the gut into the blood system and it was possible to identify them in the urine after 1-2 h. The excretion kinetics during 12 h. of the experiment, was altered with the volunteers. Three representative diagrams are presented in. Each of the volunteers ingested 218 mg of anthocyanins, calculated on the basis of malvidin-3-glycoside molecular weight. The calculated amount of anthocyanins and anthocyanin-like compounds in the urine was between 1.0 and 6.7 % of that ingested. The amount of anthocyanins collected in the urine, as calculated from the peaks at 65.02 and 65.70 min, were 1.5 to 5.1 % of the amount of those anthocyanins ingested by the volunteers.

4 DISCUSSION

Many foods, especially red grapes and wines, contain large amounts of anthocyanins. The antioxidative effects of those compound have been described by several researchers.[7,10,27 33] However, information is lacking on the mechanism by which these compounds exert their antioxidative activity in biological systems. The pH- transformed forms of malvidin and malvidin 3-glucoside, the pseudo-base and quinoidal base remained active as very effective antioxidant in both model systems of lipid peroxidation which we determined. These results are important as most probably during ingestion of those compound they may undergo transformation at the high pH of the intestine and blood system.

The pseudo-base and quinoidal base of malvidin 3-glucoside were found to act as antioxidants better than the aglycones, in the linoleate model system activated by myoglobin. It may be that the anthocyanidin-3-glucoside is better capable of interacting with the heme which is located in a crevice of the myoglobin molecule. Data from several authors on the antioxidant activity of anthocyanins and their aglycones are conflicting. Tsuda et al.[29] found that the aglycones act better than glucosides, but Wang et al.[33] reported that some glucosides act better than the aglycones.

The antioxidant activity of anthocyanins and other polyphenolics are strongly affected by the system and catalyzers used.

Our results and those published most recently by Satue'-Gracia et al.[10] demonstrated that the use of radical assays which are not relevant to oxidizable substrates or catalyzers[33,34] may not provide data which are relevant to the mechanism of action of these antioxidants in biological systems.

It should be considered that the catalysts, the substrates to be oxidized and the dispersion media (cytosol, liposomes, membranes, LDL) all may affect the effectiveness of the antioxidants. In addition, the present results are of only very limited relevance to human nutrition if the bioavailability of each of these antioxidants is different, very low, or does not exist at all. More important, a good antioxidant is not only a compound which donates electrons, but one of high bioavailability which could reach the target of oxidation *in vivo* and at a critical concentration in order to provide protection.

In the present study the difference in the administered dose could explain the absence of malvidin-3-glucoside in the urine. The antioxidative action of the absorbed anthocyanins *in vivo* could lead to their molecular modification, and destruction of the original form.

Glucuronidation and sulfation and methylation seem to be part of the metabolism of these compounds in humans. These modifications could start in the small intestine, liver or kidney.[14,15] The flavonoid, naringin, from grapefruit was found to be excreted in human urine mainly as naringenin glucoronide.[14]

The study reported here presents data on detection of anthocyanins in human urine and follows the kinetics of their excretion after normal consumption of wine. Our study also presents, in part, data on the bioabailability in humans of several anthocyanins which are dietary antioxidants found in wines and in many fruits and vegetables.

References

1.   G. Mazza and R. Brouillerd. *Food Chem.* 1987, **25**, 207.

2.   E.N. Frankel, J. Kanner, J.B. German, E. Parks and J.E. Kinsella. *Lancet*. 1993, **341**, 454.

3.   J.E. Kinsella, E. Frankel, B. German and J. Kanner. *Food Technol*. 1993, **47**, 85.

4.   J. Kanner, E. Frankel, R. Granit, B. German and J.E. Kinsella. *J. Agric. Food Chem.* 1994, **42**, 64.

5.   A.S. Meyer, O.S. Yi, A.D. Pearson, A.L. Waterhouse and E.N. Frankel. *J. Agric. Food* Chem. 1997, **45**, 1638.

6.   G. Mazza. *Critical Reviews in Food Science and Nutrition*. 1995, **35**, 341.

7.   H. Tamura and A.Yamagami. *J. Agric. Food Chem.* 1994, **42**, 1612.

8.   T. Tsuda, M. Watanabe, K. Ohsima, S. Norinobu, S.W. Choi, S. Kaeakishi, and T. Osawa, T. *J. Agric. Food Chem.* 1994, **42**, 2407.

9.   J. Kanner "Anthocyanins as Antioxidants and Metal Catalyzed Reaction" Presented at the 213th ACS National Meeting, San Francisco. 1997.

10. T.M. Satué-Gracia, M. Heinonen and E.N. Frankel. *J. Agri. Food Chem.* 1997, **45**, 3362.

11. P.C.H. Hollman, J.H.M. de Vries, S.D. van Leeuwen, M.J.B. Mengelers and M.B. Katan. *Am. J. Clin. Nutr.* 1995, **62**, 1276.

12. M. Nieder. Munch. *Med. Wochenschr.* 1991, **133**, s 61.

13. D. Cova, L. De Angelis, F. Giavarini, G. Palladini and R. Perego. *Int. J. Clin. Pharmacol. Ther. Toxicol.* 1992, **30**, 29.

14. U. Fuhr and A.L. Kummert. *Clin. Pharmacol. Therp.* 1995, **58**, 365.

15. B. Ameer, R.A. Weintraub, J.V. Johnson, R.A. Yost and R.L. Rouseff. *Clin. Pharmacol. Ther.* 1996, **60**, 34.

16. P. Morazzoni, S. Livio, A. Scilingo and S. Malandrino. Drug Res. 1991, **41**(I), 128.

17. G. Paganga and C.A. Rice-Evans. *FEBS Lett.* 1997, **401**, 78.

18. J. Kanner and S. Harel. *Arch. Biochem. Biophys.* 1985, **237**, 314.

19. M.K. Markwell, S.S. Haas, L.L. Bieber and N.E. Tolbert. *Anal. Biochem.* 1978, **87**, 206.

20. W.R. Bidlack, R.T. Orita and P. Hochstein. *Biochem. Biophys. Res. Commun.* 1973, **53**, 459.

21. R.M. Lamuela-Raventos and A. Waterhouse. *Am. J. Enol. Vitic.* 1994, **45**, 1.

22. T. Lapidot, S. Harel, R. Granit and J. Kanner. *J. Agric. Food Chem.* 1998 (in press).

23. A.H. Moscowitz and G. Hrazdina. *Plant Physiol.* 1981, **68**, 686.

24. T.C. Somers. *Phytochemistry.* 1971, **10**, 2175.

25. R. Brouillard, B. Delaporte. *J. Am. Chem. Soc.* 1977, **99**, 8461.

26. R.B. Boulton, V.L. Singelton, L.F. Bisson and R.E. Kunkee. In Principles and Practices of Winemaking; Chapman & Hall: New York, 1996, pp 233.

27. K. Igarashi, K. Takanashi, M. Makino and T. Yose. *Nippon Sholenhin Kogyo Gakkaishi.* 1989, **36**, 852.

28. M.T. Meunier, E. Duroux and P. Bastide. *Plant Med. Phytoth.* 1989, **23**, 267.

29. T. Tsuda, K. Ohshima, J. Kawakishi and T.L.J. Osawa. *J. Agric. Food Chem.* 1994 a, **42**, 248.

30. T. Tsuda, M. Watanabe, K. Ohshima, S. Norinoba, J.W. Choi, J. Kawakishi and T. Osawa. *J. Agric. Food Chem.* 1994 b, **42**, 2407.

31. T. Tsuda, K. Shiga, K. Ohshima, J. Kawakish and T. Osawa. *Biochem. Pharmacol.* 1996, **52**, 1033.

32. H. Yamasaki, H. Uefuji and Y. Sakihama. *Arch. Biochem. Biophys.* 1996, **332**, 183.

33. H. Wang, G. Cao and R.L. Prior. *J. Agric. Food Chem.* 1997, **45**, 304.

34. C. Rice-Evans, N.J. Miller, G.P. Bolwell, P.M. Bramley and J.B. Pridham. *Free Rad. Res.* 1995, **22**, 375.

THE ISOFLAVAN GLABRIDIN INHIBITS LDL OXIDATION: STRUCTURAL AND MECHANISTIC ASPECTS

Bianca Fuhrman[1], Jacob Vaya[2], Paula Belinky[2], and Michael Aviram[1].

[1]Lipid Research Laboratory, Technion Faculty of Medicine, The Rappaport Family Institute for Research in the Medical Sciences, and Rambam Medical Center, Haifa, and [2]The Laboratory of Natural Compounds for Medical Use, Galilee Technological Center, Kiryat Shmona, Israel.

1 INTRODUCTION

Consumption of phenolic flavonoids in the diet was shown to be inversely associated with morbidity and mortality from coronary heart disease.[1] The polyphenolic flavonoids constitute a large class of compounds containing a number of phenolic hydroxyl groups, which exihibit antioxidant activity.[2] Flavonoids can exert their antioxidant activity by various mechanisms, e.g. by scavenging free radicals and lipid peroxide radicals, which initiate lipid peroxidation, by chelating metal ions, and also by inhibiting cellular enzymatic systems responsible for free radical generation.[3-5] The antioxidant activity of the diverse polyphenolic flavonoids is related to their chemical structure.[6] It was demonstrated that dietary supplementation of humans with nutrients rich in different polyphenolic flavonoids, such as olive oil (hydroxytyrosol), red wine (quercetin -flavonol, catechin -flavanol) or licorice extract (glabridin -isoflavan),[7-9] resulted in reduced susceptibility of LDL to oxidation. Furthermore, consumption of red wine along with the red wine-derived purified polyphenols quercetin or catechin, by the atherosclerotic apolipoprotein E deficient (E°) mice, resulted in reduced development of atherosclerotic lesions, and this effect was associated with reduced susceptibility of their LDL to oxidation.[10]

In the present study we have examined the antioxidative activity of a subclass of polyphenols isolated from licorice extract, the isoflavan glabridin, which was not tested before for its antioxidant activity against LDL oxidation.

2 STUDIES IN E° MICE

The E° atherosclerotic mice are a good model to study LDL oxidation, since their LDL is highly susceptible to oxidation and is even minimally oxidized in the plasma.[11] Administration of 20 g of glabridin/d/mouse to E° mice for a period of 6 weeks resulted in a significant (P<0.01) reduction in the susceptibility of their LDL to copper ions-induced oxidation, as measured by a 22 % inhibition in thiobarbituric acid reactive substances (TBARS) formation and by a prolongation of the lag phase by 35 minutes. An analysis of the mice aortic arch lesions after consumption of glabridin revealed a significant reduction (by 50 %) of the atherosclerotic lesion area in comparison to placebo treated mice.

3 MECHANISMS INVOLVED IN GLABRIDIN-INDUCED INHIBITION OF LDL OXIDATION

The efficiency of exogenous antioxidants in preventing LDL oxidation may be related at least in part to their interaction with the lipoprotein. Dietary administration of red wine or of licorice extract to humans resulted in an increment in total polyphenols bound to LDL.[8,9] In this study we have analyzed *in vitro* directly the capacity of glabridin to bind to the LDL particle. Supplementation of LDL (1 mg protein/ml) with 50 μM glabridin for 18 h at 37 °C, followed by extensive dialysis to remove unbound material, resulted in about 80 % binding of the glabridin to LDL.

Next we have examined the effect of glabridin supplementation on MPH-induced LDL oxidation, in comparison to the antioxidative effects of licorice extract (13 μg/ml) or vitamin E (40 μM). Glabridin (20 μM) enrichment of LDL inhibited the AAPH-induced formation of TBARS, lipid peroxides (PD) and cholesteryl linoleate hydroperoxides (CL-OOH) by 85 %, 66 % and 89 %, respectively, in comparison to 96 %, 65 % and 92 % inhibition by licorice extract, respectively, or 83 %, 61 % and 70 % inhibition, respectively, by vitamin E (Fig. 1). We have also determined the effect of glabridin on oxysterols formation in LDL incubated with copper ions or with AAPH. Glabridin (30 μM) inhibited the production of 7-hydroxycholesterol, 7-ketocholesterol, and 5,6α-epoxy-cholesterol after 6 h of incubation with AAPH by 55 %, 80 %, and and 52 %, respectively.

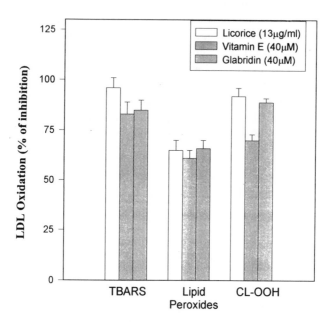

Figure 1

Glabridin supplementation of LDL inhibits AAPH-induced LDL oxidation: comparison to the effects of vitamin E and of licorice extract.

Next, we have questioned whether glabridin protects LDL from oxidation by protecting its endogenous antioxidants. Glabridin addition to LDL incubated with AAPH protected the consumption of , ß-carotene and of lycopene by 38 % and 52 %, respectively, during 30 min. of incubation, but it failed to protect the LDL-associated vitamin E from consumption.

4 STRUCTURE - ACTIVITY RELATIONSHIP OF GLABRIDIN IN RELATION TO ITS ANTIOXIDANT ACTIVITY AGAINST LDL OXIDATION

In order to investigate which part of the glabridin molecule contributes to its antioxidative activity against LDL oxidation, we have analyzed several derivatives of this isoflavan, as well as resorcinol, which is identical to the phenolic B ring in glabridin. The antioxidant activity of glabridin is expected to be related to the two hydroxyl groups in the meta position to each other at the aromatic B ring. In copper ion-induced LDL oxidation, glabridin and its derivatives which contain the two hydroxyl groups at position 2' (4'-*o*-methylglabridin, hispaglabridin A, and hispaglabridin B) substantially inhibited LDL oxidation measured as TBARS formation. On the contrary, 2'-*o*-methylglabridin, a synthesized compound whose hydroxyl at 2' position is blocked by a methyl group, inhibited LDL oxidation only to a minor extent and 2'4'-*o*-dimethylglabridin, in which both 2' and 4' positions hydroxyls are protected by methyl groups, was completely inactive. The molecular structure of glabridin and its derivatives are showed in Fig. 2A, and their antioxidant activity against LDL oxidation in demonstrated in Fig. 2B.

A

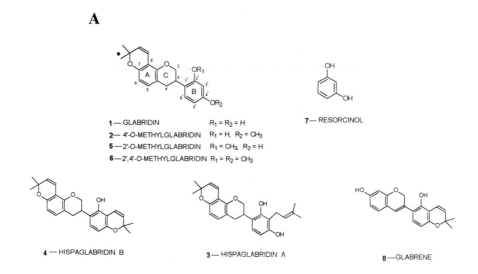

Figure 2 a

Molecular structure of glabridin and its derivatives.

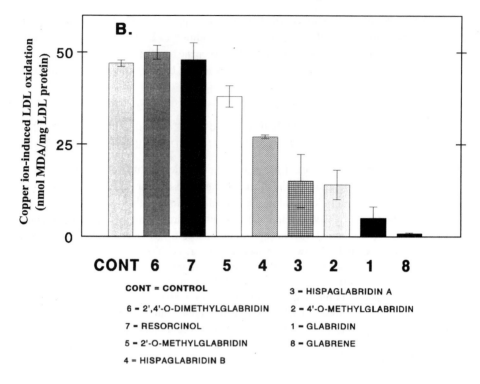

Figure 2 b

Antioxidant activity of glabridin and its metabolites against LDL oxidation.

These results indicate that the two phenolic hydroxyls are essential to obtain the antioxidant activity of glabridin. However, the hydroxyl group at the 4' position of the B ring in the glabridin molecule contributes much less than the hydroxyl group at the 2' position to the overall antioxidant activity of glabridin. Resorcinol, whose structure is identical to that of the phenol B ring of glabridin, but without its whole lipophilic moiety, exhibited low antioxidant activity against LDL oxidation, emphasizing the importance of the hydrophobic moiety of the isoflavan molecule in order to obtain efficient antioxidant activity.

References

1.   Hertog M.G.L., Feskens E.J.M., Hollman P.C.H., Katan M.B., Kornhout D. Dietary antioxidant flavonoids and risk of coronary heart disease; the Zutphen Elderly Study. *Lancet*. 1993, **342**, 1007-1011.

2.   Harborne JB. Nature, distribution and function of plant flavonoids. In: Cody, B.; Middleton E., Harborne JB., eds. Plant flavonoids in biology and medicine. New York: Alan Liss, 1986, 15-24.

3.   Afansiev I.B., Dorozhko A.J., Brodski A.V., Kostyuk A.V., Potapovitch A.l. Chelating and free radical scavenging mechanisms of inhibitory action of rutin and quercetin in lipid peroxidation. *Biochem. Pharmacol.* 1989, **38**, 1763-1769.

4.   Bors W., Heller W., Michel C., Saran M. Flavonoids as antioxidants: Determination of radical-scavenging effeciencies. *Methods Enzymol.* 1990, **186**, 343-355

5.   Laughton MJ., Evans PJ., Moroney MA., Hoult JRC., Halliwell B. Inhibition of mammalian 5-1ipoxygenase and cyclo-oxygenase by flavonoids and phenolic dietary additives: Relationship to antioxidant activity and to iron-reducing ability. *Biochem. Pharmacol.* 1991, **42**, 1673-1681.

6.   Rice-Evans C.A., N.J. Miller and G. Paganga. Structure-antioxidant activity relationships of flavonoids and phenolic acids. *Free Rad. Biol. Med.* 1996, **20**(7), 933-956.

7.   Aviram M. and E. Kassem. Dietary olive oil reduces low density lipoprotein uptake by macrophages and decreases the susceptibility of the lipoprotein to undergo lipid peroxidation. *Ann. Nutr. Metab.* 1993, **37**, 75-84.

8.   Fuhrman B., A. Lavy and M. Aviram. Consumption of red wine with meals reduces the susceptibility of human plasma and low density lipoprotein to lipid peroxidation. *Am. J. Clin. Nutr.* 1995, **61**, 549-554.

9.   Fuhnan B., Buch S., Vaya J., Belinky P.A., Coleman R., Hayek T. and Aviram M. Licorice extract and its major polyphenol glabridin protect low density lipoprotein against lipid peroxidation: *in vitro* and *ex vivo* studies in humans and in atheroscleroteic apolipoprotein E deficient mice. *Am. J. Clin. Nutr.* 1997, **66**, 267-275.

10.   Hayek T., Fuhrman B., Vaya J., Rosenblat M., Belinky P., Coleman R., Elis A. and Aviram M. Reduced progression of atherosclerosis in apolipoprotein E-deficient mice following consumption of red wine, or polyphenols quercetin or catechin, is associated with reduced susceptibility of LDL to oxidation and aggregation. *Arterioscler. Thromb. and Vasc. Biol.*1997, **17**, 2744-2752.

11.   Aviram M., Maor I., Keidar S., Hayek T., Oiknine J., Bar-EI Y., Adler Z., Kertzman V., Milo S. Lesioned low density lipoprotein in atherosclerotic apolipoprotein E-deficient transgenic mice and in humans is oxidized and aggregated. *Biochem. Biophys. Res Commun.* 1995, **216**, 501-513.

BIOAVAILABILITY AND ANTIOXIDANT PROPERTIES OF LUTEOLIN

Kayoko Shimoi[1]*, Hisae Okada[1], Junko Kaneko[1], Michiyo Furugori[1], Toshinao Goda[1], Sachiko Takase[1], Masayuki Suzuki[2], Yukihiko Hara[1] and Naohide Kinae.[1]

[1]School of Food and Nutritional Sciences, University of Shizuoka, Yada 52-1, Shizuoka 422-8526, Japan. [2]Food Research Laboratories, Mitsui Norin Co., Ltd., Miyahara 223-1, Fujieda 426-0133, Japan. * Corresponding author. Kayoko Shimoi, Ph.D.

Keywords: Luteolin; Bioavailability; Glucuronidation; Plasma; Urine; Antioxidant.

1 INTRODUCTION

Flavonoids are polyphenolic compounds occurring in a large variety of vegetables, fruits, nuts, tea and wine and have been shown to possess a wide range of biological properties such as anticarcinogenic and antioxidant activities. It has been suggested that their antioxidant activity is due to their ability to scavenge free radicals and to chelate metal ions.[1] Recent studies have focused on their chemopreventive role against free radical-associated diseases including cancer and cardiovascular disease.

In this study, we focus on luteolin. Luteolin and luteolin 7-O-β-glucoside are contained in celery, perilla leaf and seeds, green pepper, camomille tea, etc. Luteolin has been reported to exert antimutagenic,[2] and anti-inflammatory-allergic[3,4] effects, and has been recognized as a hydroxyl radical scavenger[5] and an inhibitor of protein kinase C.[6] We have demonstrated that preadministration of luteolin reduced the γ -ray (1.5 Gy)-induced frequency of micronucleated reticulocytes (MNRETs) in mouse peripheral blood[7] and suppressed lipid peroxidation in the mouse bone marrow and spleen after γ -ray irradiation and in the mouse heart and bone marrow after treatment with doxorubicin.[8,9] These results suggest that luteolin was absorbed and then acts as an antioxidant *in vivo* .

Some flavonoids have been demonstrated to be absorbed from the digestive tract, mainly the small intestine, and to circulate as metabolites in the blood of rats and humans after dosing.[10,11] However, there is little information concerning *in vivo* antioxidant activity and their metabolites. This study describes the intestinal absorption and excretion of luteolin in SD rats, and serum concentration of luteolin and its conjugates and antioxidant activity of human serum after ingestion of luteolin.

2 MATERIALS AND METHODS

2.1. Chemicals

Luteolin and chrysoeryol were isolated from pelilla seeds and purified by HPLC at the Oryza Oil & Fat Chemical Co., Ltd. (Ichinomiya, Japan). Diosmetin were obtained from Extrasynthese (Genay, France). β-glucuronidase/sulfatase and vitamin E (α-tocopherol) were purchased from Sigma (St. Louis, MO). Vitamin C (ascorbic acid) and β-glucuronidase were from Wako Pure Chem.Ind.Ltd (Osaka, Japan).

2.2. Animals and diets

Male SD rats (7-9 weeks old, 180-200 g, Japan, SLC, Inc., Hamamatsu) were housed in an air-conditioned room, fed a synthetic basic diet as described previously.[12] Three to five rats were assigned to each experimental group.

2.3. Sample preparation of blood and urine

Rat plasmas were prepared as described previously.[12] Human venous blood was collected in vacutainer tubes from a healthy female volunteer, who fasted overnight, before and 1,2,3,5 h after ingestion of luteolin (50 mg) by medical doctors.

Luteolin (50 mmol/kg in 0.5 % CMC-Na) was administered to rats, and urine samples were collected for 24 h and 48 h using metabolic cages.

2.4. HPLC analysis

HPLC analysis was performed by the method described previously[12] with some modification. The mobile phase contained the following: Solvent A, methanol with 0.03 % trifluoroacetic acid; Solvent B: distilled water with 0.03 % trifluoroacetic acid. Gradient conditions: A/B=30/70, 0-5 min; 30/70→ 65/35, 5-30 min; 65/35, 40-40 min; 65/35→ 100/0, 40-45 min. The flow rate was 0.7 ml/min. For analysis of the urine samples, Sep-Pak CN cartridges were used instead of Sep-Pak C18 cartridges. Gradient conditions: A/B=30/70, 0-5 min; 30/70→ 50/50, 5-40 min; 50/50→ 100/0, 40-45 min.

For detection of conjugates, each sample was acidified with 1M acetate buffer (pH 4.5) and was preincubated for 2 min at 37 °C. Solutions were treated with 5.4×10^2 units/ml β-glucuronidase and 0.2×10^2 units/ml sulfatase for 20 min at 37 °C, and then the same volume of 0.01M oxalic acid was added. The mixtures were centrifuged for 5min at 8,000 rpm. Supernatants were prepared in the same way as described above.

The recoveries of luteolin from plasma and urine were obtained by spiking luteolin at 1 mg/ml into each sample. Each recovery was 103.8±2.4 % (n=5), 92.07±2.4 % (n=5), respectively.

2.5. Antioxidant assay

2.5.1. Oxidation of human serum by 2,2-azobis(2-aminopropane)-dihydrochloride(AAPH). Each serum was oxidized by 30 mM AAPH for 4 h. The content of TBARS was measured by the fluorometric method of Yagi[13] with some modifications. After incubation with AAPH at 37 °C for 20 min, the content of ascorbic acid was determined by ECD-HPLC analysis of Umegaki.[14]

2.5.2. Oxidation of methyl linoleate by AAPH. The antioxidant activity of luteolin, vitamin C and vitamin E was investigated by the thiobarbituric acid (TBA) assay as described previously.[7]

3 RESULTS

As shown in Fig.1, the HPLC profiles of rat plasma after administration of luteolin by gastric intubation indicated the presence of free luteolin and glucuronide or sulfate-conjugates of unchanged luteolin and 3' or 4'-*O*-methyl luteolin (chrysoeryol or diosmetin). LC/MS analysis showed that the main metabolite (peak d) in rat plasma was a monoglucuronide (Data not shown).

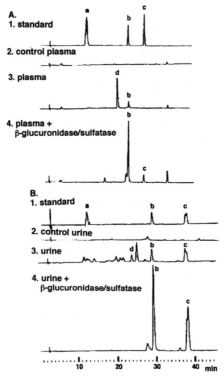

Figure 1

HPLC chromatograms of plasma extract and urine from luteolin administered rat before and after β-glucuronidase/sulfatase treatment. A: peak a (12.8 min, luteolin 7-O-β-glucoside), peak b (22.8 min, luteolin), peak c (26.9 min, diosmetin, chrysoeryol), peak d (20.0 min). B: peak a (12.0 min, luteolin 7-O-β-glucoside), peak b (28.9 min, luteolin), peak c (37.6 min, diosmetin; 37.9 min, chrysoeryol), peak d (23.61 min).

The concentration of luteolin and its conjugates in rat plasma increased to the highest level 15 min or 30 min after administration of luteolin in propyleneglycol and decreased. The conjugates were present even after 24 h after administration, although their amount was very small. The highest concentrations of luteolin and its conjugates in rat plasma were 3.08 and 14.1 nmol/ml, respectively. (Fig. 2). The intestinal absorption of luteolin in propylene glycol was faster than that in CMC-Na.

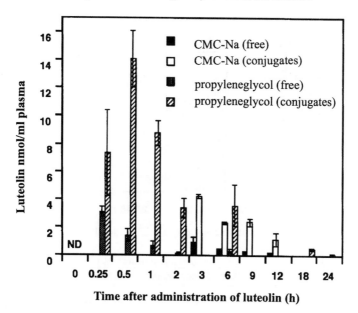

Figure 2

Concentration of luteolin and luteolin conjugates in rat plasma after oral administration of luteolin. Rats were administered luteolin in CMC-Na or propyleneglycol (50 μmol/kg). Plasma samples were treated with or without β-glucuronidase/sulfatase. Each value represents the mean ± SE (n=3).

The metabolites in urine were compared with that in plasma. Many peaks were observed in the urine extract. The retention time of the two peaks (b, d), which were also detected in the plasma, corresponded to that of luteolin monoglucuronide (24 min) and luteolin (29 min) (Fig. 1B). When rats were administered luteolin orally, almost unchanged aglycone and its glucuronide or sulfate conjugates were excreted into the urine during 24 h. The concentration of urine during 24 h and the next 24 h were 317.6±52.1 and 6.65±4.16 nmol/ml, and the excretory recoveries were 4.01±0.46 % and 0.083±0.03 % of the doses administered, respectively (n=3) (Fig.3).

Free luteolin and its monoglucuronide were also present in human serum after dosing (Data not shown). Fig. 4A shows the time course of the concentration of total luteolin in human serum treated with β-glucuronidase/sulfatase. Serum luteolin concentration at 3 h after dosing was highest (17.1 ng/ml). The serum obtained after dosing showed protective effects against AAPH-induced lipid peroxidation and degradation of ascorbic acid (Fig. 4 BC). The antioxidant activity of the serum seems to be proportional to the serum concentration of luteolin.

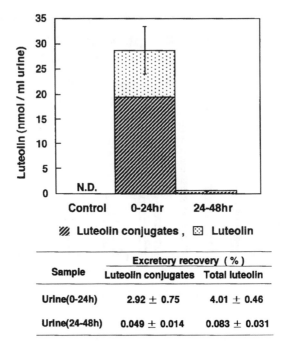

Sample	Excretory recovery (%)	
	Luteolin conjugates	Total luteolin
Urine(0-24h)	2.92 ± 0.75	4.01 ± 0.46
Urine(24-48h)	0.049 ± 0.014	0.083 ± 0.031

Figure 3
Concentration of luteolin and its conjugates in rat urine after oral administration of luteolin. Rats were administered luteolin in CMC-Na (50 μmol/kg). Urine samples were treated with or without β-glucuronidase /sulfatase. Each value represents the mean ± SE(n=3).

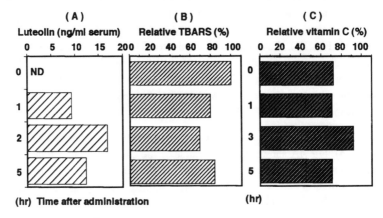

Figure 4
Concentration of luteolin and antioxidant defense in human serum. (A), Luteolin(50 mg) was ingested. Each serum was treatedtively. The conjugates remained even after 24h. The excretory recovery of luteolin and its conjugates in rats was about 4%. Luteolin and its conjugates were also detected in human serum, and the highest peak was observed 2-3h after oral administration.

Antioxidant activities of luteolin, ascorbic acid and α-tocopherol against AAPH-induced lipid peroxidation are shown in Fig. 5A. They demonstrated antioxidant activity in a dose-dependent manner. Their mixture showed a synergistic effect at the concentration where each subtance was not so effective (Fig. 5B).

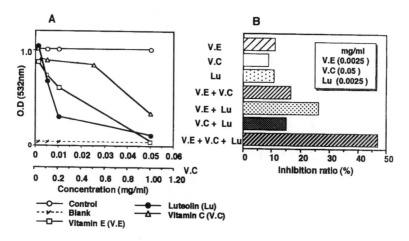

Figure 5

Antioxidant activity of vitamin E, vitamin C, luteolin and their mixture against AAPH-induced lipid peroxidation. Values are means in duplicate.

4 DISCUSSION

Recently, the bioavailability of flavonoids has been focused on in order to evaluate their *in vivo* biological activity. However, the mechanism of intestinal absorption of flavonoids in the form of both aglycone and glycoside has not yet been well elucidated. We investigated how luteolin and luteolin 7-*O*-β-glucoside are absorbed from the digestive tract using rat everted intestine and found that the main metabolite was a monoglucuronide by LC/MS analysis (Data not shown, in preparation).

Fig. 6 shows the proposed scheme of the intestinal absorption and the metabolic fate of orally administered luteolin and luteolin 7-*O*-β-glucoside. They enter the digestive tract. Luteolin 7-*O*-β-glucoside is hydrolyzed to luteolin by the intestinal microorganisms before intestinal absorption. Luteolin is absorbed most efficiently from the duodeno-jejunum. When luteolin passes through the intestinal mucosa, a rapid glucuronidation of luteolin takes place by UDP-glucuronosyltransferase. Luteolin glucuronide enters the portal vein, and the sulfation and methylation occur in the liver. Various kind of metabolites of luteolin are excreted via bile or urine. When rats were given luteolin in CMC-Na (50 μmol/kg) orally, the plasma concentration of total luteolin 3 h after dosing was 5.13±0.44 nmol/ml and the excretory recovery for 24 h as total luteolin was about 4 %. Liu et al. have reported that nearly 15 % of luteolin administered was excreted via urine (5.6 % as unmodified luteolin, 9.1 % as methylated metabolites) for 24 h.[16]

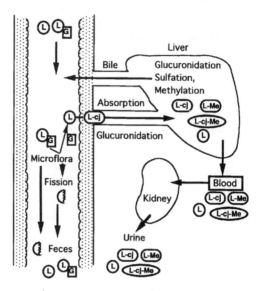

Figure 6

*Proposed scheme of metabolic fate of luteolin. L= luteolin, G= glycoside, Cj= conjugate,
Me=methylated luteolin, Cj-Me=methyl conjugate.*

The HPLC profiles of rat plasma after administration of luteolin demonstrated that
free luteolin and the glucuronide or sulfate-conjugates of luteolin and *O*-methyl luteolin
(diosmetin or chrysoeryol) circulate in the blood. Some of the luteolin can escape the
intestinal glucuronidation and hepatic sulfation/methylation, although it has been reported
that diosmetin was not detected in the free form in plasma.[15]

It is very important to know whether the absorbed flavonoids function as an
antioxidant *in vivo*. Vinson et al. have demonstrated that quercetin exerted antioxidant effect
at very low concentration (0.2 μmol/L: 0.06 μg/ml) using an *in vitro* lipoprotein oxidation
model. Rat plasma concentration shown in this study was higher than that concentration
reported by Vinson et al. Therefore, the absorbed luteolin must be a powerful antioxidant in
the rat plasma. On the other hand, the main metabolite was a monoglucuronide. However, it
is not been clarified which position of luteolin is a glucuronidation site at the present time.
Since antioxidant activity is associated in part with the number of hydroxyl groups,
antioxidant activity and metal chelating ability of this monoglucuronide might be lower than
that of luteolin. Diosmetin, 4'-*O*-methyl luteolin, exhibited hydroxyl radical scavenging
property. Therefore, even though 3'- or 4'- position of luteolin is glucuronized, those
activities do not seem to be lost completely.

Niki et al. have reported that ascorbic acid and bilirubin decreased rapidly after
AAPH-initiated oxidation of the whole blood.[17] In this study, though this is a preliminary
result, the human serum obtained after dosing showed a protective effect against lipid
peroxidation and degradation of ascorbic acid by AAPH. Free luteolin and its
monoglucuronide are present in the human serum as in the rat plasma. The antioxidant
activity of the human serum seems to be proportional to the serum concentration of free

luteolin and its conjugates. We show that the antioxidant activity of the mixture of ascorbic acid, α-tocopherol and luteolin against AAPH-induced lipid peroxidation is synergistic in this study. From these results, it is suggested that maintenance of the flavonoid level in the blood might play a role in the antioxidant status *in vivo*.

Acknowledgements

The authors thank Oryza Oil & Fat Chemical Co., Ltd. (Ichinomiya, Japan) for providing luteolin and chrysoeryol. This work was in part supported by a Grant-in-Aid for Scientific Research (C) (No.10680145) from the Ministry of Education, Science and Culture of Japan to K.S.

References

1. Cook N.C., and Samman S. *J.Nutr.Biochem.* 1996, **7**, 66-76.
2. Samejima K., Kanazawa K., Ashida H., Danno G. *J.Agric.Food Chem.* 1995, **43**, 410-14.
3. Yasukawa K.,Takido M., Takeuchi M. and Nakagawa S. *Chem.Pharm.Bull.* 1989, **37**, 1071-1073.
4. Yamamoto H., Sakakibara J., Nagatsu A. and Sekiya K. *J. Agric.Food Chem.* 1998, **46**, 862-865.
5. Husain S.R., Cillard J., Cillard P. *Phytochemistry.* 1987, **26**, 2489-2492.
6. Ferriola P.C., Cody V and Middleton E. Jr. *Biochem. Pharmacol.* 1989, **38**, 1617-1624.
7. Shimoi K., Masuda S., Furugori M., Esaki S. and Kinae N. *Carcinogenesis.* 1994, **15**, 2669-2672.
8. Shimoi K., Masuda S., Shen B., Furugori M., Kinae N. *Mutat. Res.* 1996, **350**, 153-161.
9. Sadzuka Y., Sugiyama T., Shimoi K., Kinae N. and Hirota S. *Toxicol. Lett.* 1997, **92**, 1-7.
10. Hollman P.C.H., Van der Gaag M., Mengelers M.J.B., Van Trijp J.M.P., De Vries J.H.M., and Katan M.B. *Free Rad. Biol. Med.* 1996, **21**, 703-707.
11. Manach C., Morand C., Demigne C., Texier O., Regerat F. and Remesy C. *FEBS Lett.* 1997, **409**, 12-16.
12. Shimoi K., Shen B., Toyokuni S., Mochizuki R., Furugori M. and Kinae N. *Jpn. Cancer Res.* 1997 88,453-460.
13. Yagi K. *Biochem. Med.* 1976, **15**, 212-216.
14. Umegaki K., Inoue K., Takeuchi N. and Higuchi M. *J. Nutr. Sci. Vitaminol.* 1994, **40**, 73-79.
15. Boutin J.A., Meunier F., Lambert P.H., Hennig P., Bertin D., Serkiz B. and Volland J.P. *Drug Metabolism and Disposition.* 1993, **21**, 1157-1166.
16. Liu C.S., Song Y.S., Zhang K.J., Ryu J.C., Kim M. and Zhou T.H. *J.Pharm.Biomed.Anal.* 1995, **13**, 1409-1414.
17. Niki E., Yamamoto Y., Takahashi M., Yamamoto K., Yamamoto Y., Komuro E., Miki M., Yasuda H. and Mino M. *J. Nutr. Sci. Vitaminol.* 1988, **34**, 507-512.

EFFECTS OF A FLAVONOID-RICH DIET ON PLASMA QUERCETIN AND SUSCEPTIBILITY OF LDL TO OXIDATION

Gareth McAnlis[1, 2], Jane McEneny[2], Jack Pearce[1] and Ian Young.[2]

[1]Department of Food Science, The Queens University of Belfast, Newforge Lane, Belfast, BT9 5PX. [2]Department of Clinical Biochemistry, The Queens University of Belfast, Royal Victoria Hospital, Belfast, BT12 6BJ.

1 INTRODUCTION

Flavonoids are a group of polyphenols produced naturally by plants. They occur in fruit and vegetables and are therefore an integral part of the human diet.[1] Several epidemiological studies have linked increased flavonoid consumption to lower risk of cornary heart disease (CHD) and strokes.[2,3,4]

It has been difficult to evaluate the role of dietary flavonoids as protective antioxidants in man because of uncertainties about their absorption, primarily due to the lack of sufficiently sensitive and selective analytical methods. However, recently Hollman *et al.*[5] have developed a sensitive detection technique for measuring quercetin in plasma which allows the study of their bioavailability from foods or supplements.

There have been no previous studies on the effects of consuming a varied flavonoid-rich diet on the physiological concentrations of plasma flavonoids. In view of this, the aim of the present study was to determine the *in vivo* effects of ingesting a diet rich in flavonoids.

2 MATERIALS AND METHODS

All the fruits, vegetables and beverages were purcased from a local supermarket (Sainsbury's, Forestside, Belfast). On the day of purchase the edible part of the fruits and vegetables were cleaned, chopped finely and frozen to -70 °C. Foods were then freeze dried, ground to a fine powder and analysed for flavonoids using an HPLC method developed by Hertog *et al.*[6] The daily supplementary diet contained 100 g Spanish cherry tomatoes, 30 g Feuille de Chene lettuce, 30 g of onions, 2 Braeburn apples, 500 mls of Sainsbury's standard cranberry juice and 2 tea bags for each day of the 2 week study. Tea bags (mean 3.3 g) were to be infused for 4 minutes in a standard volume of 300 mls of boiling water as suggested by the manufacturer's instructions. These foods were consumed as a component of the volunteers' normal diets.

Ten healthy volunteers, 7 males and 3 females aged between 23 and 39 years, were given a list of foods containing high concentrations of flavonoids and for two weeks prior to the beginning of the study abstained from any of the foods on the list. Fasting peripheral venous blood samples were taken at the end of the flavonoid-free diet 2 week 'washout' period (WO), after one week of the flavonoid rich diet, after two weeks of the flavonoid-rich diet and after one week of returning to normal diet (ND).

Plasma was obtained from heparinised blood samples by centrifugation at 1200 g for 10 mins at 4 °C. The quercetin present in whole plasma was quantified using HPLC according to the technique developed by Hollman *et al.*[5] The LDL was isolated and its oxidation assayed using a standard method.[7] Total antioxidant activity was measured using a COBAS FARA centrifugal analyser. [8]

3 RESULTS

Daily intake of flavonoid (quercetin, kaempferol, myricetin, apigenin) from the supplementary diet was 100.76 mg. The major sources of flavonoid intake were apples (31.8 %), cranberry juice (30.7 %), and onions (20.3 %).

The quercetin concentration in plasma after 2 weeks of washout was 15.9±37.3 ng/ml (mean of 10 observations ±SD), increased to 61.4±37.4 ng/ml after 1 week of flavonoid-rich diet, 48.4±28.9 ng/ml after 2 weeks of the flavonoid-rich diet and decreased to 16.0±21.3 ng/ml after returning to normal diet for one week (Fig. 1). Plasma quercetin concentrations were significantly ($p<0.05$) increased from the washout (WO) and the normal diet (ND) compared to both week 1 and week 2 of consuming the flavonoid-rich diet.

Despite the increase in plasma quercetin concentrations (Fig. 1) there was no significant change in the total antioxidant capacity or in the susceptibility of LDL to oxidation.

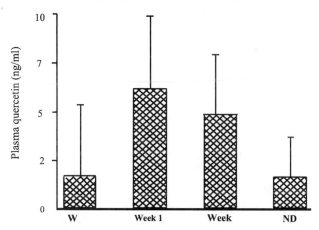

Figure 1
Baseline plasma quercetin levels; after 2 weeks washout (WO), after 1 week of flavonoid rich diet, after 2 weeks of flavonoid rich diet and finally one week after the completion of the study, volunteers having returned to their normal diet (ND). Mean of 10 observations ± SD.

4 DISCUSSION

Hertog *et al.*,[3] as part of the Seven Countries Study, found the daily flavonol and flavone intake to be approximately 13 mg/d in the USA, 6 mg/day in Finland and 33 mg/d in The Netherlands. The intake in Japan was the highest at 64 mg/d, approximately 90 % of which came from the consumption of tea. The flavonoid-rich diet used in this study contained 100.76 mg of flavonols and flavone per day, significantly greater than the average daily intake from any of the countries in the Seven Countries Study.[3]

The epidemiological and *in vitro* evidence suggest the consumption of a flavonoid-rich diet may protect against CHD.[2,3,4,9] However, there are only very weak correlations with specific foods such as apples, tea and onions.[2] Therefore, in order to assess the potential health benefits from consuming a flavonoid-rich diet it is necessary to supply the flavonoids from a variety of dietary sources.

The results from the current study clearly demonstrate that the long term consumption of a quercetin-rich diet can cause a significant increase in plasma quercetin concentrations (Fig. 1). There was no measurable rise in the total antioxidant capacity of the plasma suggesting the increase in plasma quercetin made little contribution to the endogenous antioxidants present. There was also no change in the susceptibility of LDL to oxidation probably due to the quercetin being bound to the plasma albumin.[10]

The health benefits of flavonoids have yet to be fully established although they have been shown to function in a way similar to antioxidant vitamins and protect against LDL oxidation *in vitro*. The results from this study demonstrate that the ingestion of a flavonoid-rich diet can cause a significant increase in plasma flavonoids. However, there seems to be no directly protective effect on LDL oxidation. Further research on the metabolism, distribution and excretion of flavonoids is still required to fully evaluate their potential role in human health and disease.

References
1.   J. Kuhnau. The Flavonoids: A class of semi-essential food components; their role in human nutrition. *World review of Nutrition and Diet*. 1976, **24**, 117-191.
2.   M.G.L Hertog, E.J.M. Feskens, P.C.H. Hollman, M.B. Katan, D. Kromhout, Dietary antioxidant flavonoids and risk of coronary heart disease: the Zutphen elderly study. *Lancet*. 1993, **342**, 1007-1011.
3.   M.G.L. Hertog, D. Kromhout, C. Aravanis, H. Blackburn, R. Buzina, F. Fidanza, S. Giampaoli, A. Jansen, A. Menotti, S. Nedeljkovic, M. Pekkarinen, B.S. Simic, H. Toshima, E.J.M. Feskens, P.C.H. Hollman, M.B. Katan. Flavonoid intake and long-term risk of coronary heart disease and cancer in the 7 countries study. *Archives of Internal Medicine*. 1995, **155**(4), 381-386.
4.   P. Knekt, R. Jarvinen, A. Reunanen, J. Maatela. Flavonoid intake and coronary mortality in Finland: a cohort study. *British Medical Journal*. 1996, **312**, 478-481.
5.   P.C.H. Hollman, J.M.P. Vantrijp, M.N.C.P. Buysman. Fluorescence detection of flavonols in hplc by postcolumn chelation with aluminum. *Analytical Chemistry*. 1996, **68**(19), 3511-3515.

6.   M.G.L. Hertog, P.C.H. Hollman, D.P. Venema. Optimisation of a quantitative HPLC determination of potentially anticarcinogenic flavonoids in vegetables and fruit. *Journal of Agricultural and Food Chemistry*. 1992, **40,** 1591-1598.

7.   I.F.W. McDowell, J. McEneny, E.R. Trimble. A rapid method for measurement of the susceptibility to oxidation of LDL. *Clinical Biochemistry*. 1995, **32,** 167-174.

8.   N.J. Miller, C.A. Rice-Evans. Spectrophotometric determination of antioxidant activity. *Redox Report*. 1996, **2,** 161-171.

9.   C. de-Whalley, S.M. Rankin, J.R.S. Hoult, W. Jessup, D.S. Leake. Flavonoids inhibit the oxidative modification of low density lipoproteins by macrophages. *Biochemical Pharmacology*. 1990, **39,** 1743-1750.

10. M. Ichikawa, S.C. Tsao, T.H. Lin, S. Miyauchi, Y. Sawada, T. Iga, M. Hanano, Y. Sugiyama. Albumin-mediated transport phenomenon observed for ligands with high membrane permeability. *Journal of Hepatology*. 1992, **16,** 38-49.

COMPARATIVE *IN VITRO* AND *IN VIVO* FREE RADICAL SCAVENGING ABILITIES OF A NOVEL GRAPE SEED PROANTHOCYANIDIN EXTRACT AND SELECTED ANTIOXIDANTS

D. Bagchi*, R.L. Krohn, J. Balmoori, M. Bagchi, A. Garg and S.J. Stohs.

Creighton University School of Pharmacy and Allied Health Professions, Omaha, NE 68178, USA. *Author for correspondence.

1 INTRODUCTION

Free radicals and their metabolites are increasingly recognized for their contribution to tissue injury leading to both initiation and promotion of multistage carcinogenesis.[1] Free radicals have been implicated in more than one hundred disease conditions in humans, including atherosclerosis, arthritis, ischemia and reperfusion injury of many tissues, central nervous system injury, gastritis, tumor promotion and carcinogenesis, and AIDS.[1-4] Recent studies have demonstrated that environmental pollutants, radiation, chemicals, toxins, deep-fried and spicy foods, as well as physical stress, exhibit the ability to produce enormous amounts of free radicals, resulting in oxidative deterioration of lipids, proteins and DNA, activation of procarcinogens, inhibition of cellular and antioxidant defense systems, depletion of sulfhydryls, altered calcium homeostasis, changes in gene expression and induction of abnormal proteins.[1-5]

Antioxidants function as inhibitors at both initiation and promotion/propagation/transformation stages of tumor promotion/carcinogenesis, and protect cells against oxidative damage.[3,6] The consumption of edible plants, fruits and vegetables has been demonstrated to prevent the occurrence of a number of diseases in humans and animals.[7] Vegetables, fruits and their seeds are rich sources of vitamins C and E, β-carotene, and/or protease inhibitors, compounds which might protect the organism against free radical induced injury and diseases. Grape seed proanthocyanidins have been reported to exhibit a wide range of pharmacological activity including antibacterial, antiallergic, antiviral, anti-inflammatory and vasodilatory actions.[8-10] Proanthocyanidins have been reported to inhibit lipid peroxidation, platelet aggregation, capillary permeability and fragility, and modulate the activity of enzymes including cyclo-oxygenase and lipoxygenase.[10,11]

A variety of proanthocyanidins have been shown to prevent the growth of breast cancer cells and to inhibit the enzymes involved in the replication of rhino viruses (common cold) and HIV viruses.[7] The potential of isoflavones and lignans, also known as phytoestrogens, for preventing the development of hormone-dependent cancers such as breast and prostate cancer, is attributed to their weak estrogens.[7] Proanthocyanidins may exert these effects as antioxidants, potent free radical scavengers and chelators of toxic heavy metals.[12-14]

In this study, we have assessed the comparative protective abilities of a novel IH636 grape seed proanthocyanidin extract (GSPE) with vitamin C, vitamin E succinate (VES) and ß-carotene against biochemically generated oxygen free radicals in *in vitro,* as well as against 12-*O*-tetradecanoylphorbol-13-acetate (TPA)-induced lipid peroxidation and DNA fragmentation in hepatic and brain tissues, as well as against production of oxygen free radicals in peritoneal macrophages of mice.

2 MATERIALS AND METHODS

A commercially available dried, powdered IH636 grape seed proanthocyanidin extract (batch no. AV 609016) was obtained from Inter Health Nutraceuticals Incorporated (Concord, CA). All other chemicals used in this study were obtained from Sigma Chemical Co. and were of analytical grade or the highest grade available.

2.1. Biochemical Generation of Oxygen Free Radicals

Oxygen free radicals were biochemically generated by the following procedures. Superoxide anion was produced in a total volume of 2.0 ml by incubating xanthine (100 μM) in 5 mM Tris-HCl buffer with 8 mU/ml of xanthine oxidase. The incubation mixture to generate hydroxyl radical contained 5 mM Tris-HCl, 100 μM $FeCl_3$, 100 μM ethylene diamine tetra-acetic acid (EDTA), and 100 μM xanthine in a total volume of 2.0 ml. Xanthine oxidase (8 mU/ml) was added to initiate the production of hydroxyl radical.[15]

2.2. Animals and Treatment

Female Swiss-Webster mice (20-25 gms) were obtained from Sasco (Omaha, NE). The animals were housed in a controlled environment at 25 °C with a 12 h light and 12 h dark cycle, and were acclimated for at least 3-5 days before use. All animals were allowed free access to food (Purina Rodent Lab Chow No. 5001) and tap water. VES and ß-carotene were dissolved in corn oil, while GSPE and vitamin C were dissolved in water. GSPE (25-100 mg/kg), vitamin C (100 mg/kg), VES (100 mg/kg), and ß-carotene (50 mg/kg) were orally administered to groups of animals using a feeding needle for 7 consecutive days. All groups of mice received an intraperitoneal (i.p.) injection of 1 ml of 3 % thioglycollate (DIFCO Laboratories, Detroit, MI) broth 3 days before TPA treatment to elicit peritoneal macrophages.[16] TPA was administered on the eighth day 2 h after the antioxidant treatment. Groups of mice were individually treated i.p. with 0.1 μg TPA diluted in 1 ml of sterile PBS buffer to induce an oxidative stress, and were killed 2 h post-treatment by cervcal dislocation. Control animals received the PBS buffer. The peritoneal macrophage cells were isolated, and the hepatic and brain tissues were quickly removed and the subcellular fractions were obtained as described earlier.[17] An approval (ARC#0313) from the Creighton University Animal Research Committee was obtained for this project.

2.3. Chemiluminescence Assay

Chemiluminescence, as an index of reactive oxygen species production, was measured in a Chronolog Lumivette Luminometer (Chronolog Corp., Philadelphia, PA). The assay was conducted in 3 ml glass minivials. The vials were incubated at 37 °C prior to measurement and the background chemiluminescence of each vial was checked before use. Samples were pre-incubated at 37 °C for 15 min and 4 μM luminol was added to enhance chemiluminescence. All additions to the vials, as well as chemiluminescence counting procedues, were performed under dim lighting conditions. Results were presented as counts/unit time minus background. Chemiluminescence was monitored for 6 min at continuous 30 second intervals.[17]

2.4. Cytochrome c Reduction Assay

Superoxide anion production was measured by the cytochrome c reduction assay of Babior et al.[18] The reaction mixtures were incubated with 0.05 mM cytochrome c for 15 min at 37 °C. The mixtures were centrifuged at 1,500 g for 10 min at 4 °C, and the supernatant fractions were transferred to clean tubes for subsequent spectrophotometric measurement at 550 nm. Absorbance values were converted to nanomoles of cytochrome c reduced using the extinction coefficient of $2.1 \times 10^4 \, M^{-1} cm^{-1}/15$ min.[17]

2.5. Lipid Peroxidation

Thiobarbituric acid reactive substances (TBARS) associated with hepatic mitochondria and microsomes, and brain homogenates from control and treated animals, were determined as an index of lipid peroxidation according to the method of Buege and Aust[19] and as previously published by us.[17] Malondialdehyde was used as the standard. Absorbance values were measured at 535 nm and an extinction coefficient of $1.56 \times 10^5 \, M^{-1} cm^{-1}$ was used.

2.6. DNA Fragmentation

Liver and brain samples were homogenized in lysis buffer (5 mM Tris-HCl, 20 mM EDTA, 0.5 % Triton X-100, pH 8.0). Homogenates were centrifuged at 27,000 x g for 20 min to separate intact chromatin in the pellets from fragmented DNA in the supernatant fractions. Pellets were resuspended in 0.5 N perchloric acid, and 5.5 N perchloric acid was added to supernatant samples to reach a concentration of 0.5 N. Samples were heated at 90 °C for 15 min and centrifuged at 1,500 x g for 10 min to remove protein. Resulting supernatants were reacted with diphenylamine for 16-20 h at room temperature, and absorbance was measured at 600 nm. DNA fragmentation is expressed as percent of total DNA appearing in the supernatant fractions. Treatment effects are reported as percent of control fragmentation.[20]

2.7. Statistical Analysis

Significance between pairs of mean values was determined by Student's t test. Data for each group were subjected to analysis of variance (ANOVA). Scheffe's S method was used as the post-hoc test. The data are expressed as the mean ± standard deviation of 4-6 replicates. The level of statistical significance employed in all cases was $p < 0.05$.

3 RESULTS

3.1. Comparative *in Vitro* Free Radical Scavenging Abilities of GSPE, Vitamin C, Vitamin E Succinate, a Combination of Superoxide Dismutase (SOD) plus Catalase (CAT), and Mannitol (MAN)

The abilities of various antioxidants/free radical scavengers to inhibit superoxide anion and hydroxyl radicals are presented in Table 1. Cytochrome c reduction and

Table 1

Generation of Superoxide Anion and Hydroxyl Radicals, and Inhibition by GSPE, Vitamin C (Vit C) and Vitamin E Succinate (VES).

	Superoxide Anion (SA)		Hydroxyl Radicals (OH)
Sample	Chemiluminescence Response (CPM)	Cytochrome c Reduction nmol reduced/15 min	Chemiluminescence Response (CPM)
Control	167 ± 32	0.025 ± 0.004	167 ± 32
GSPE (200 mg/l)	219 ± 48	0.18 ± 0.05	219 ± 48
Vit C (100 mg/kg)	396 ± 60	0.37 ± 0.08	396 ± 60
VES (100 mg/kg)	640 ± 97	0.70 ± 0.14	640 ± 97
SA	9704 ± 1453	2.53 ± 0.32	–
SA+GSPE (5 mg/l)	7860 ± 714*	2.10 ± 0.34	–
SA+GSPE (25 mg/l)	3687 ± 460*	1.05 ± 0.12*	–
SA+GSPE (50 mg/l)	2361 ± 345*	0.71 ± 0.16*	–
SA+GSPE (100 mg/l)	2009 ± 361*	0.47 ± 0.08*	–
SA+GSPE (200 mg/l)	1165 ± 302*	0.27 ± 0.08*	–
SA+Vit C (5 mg/l)	8557 ± 434*	2.28 ± 0.22	–
SA+Vit C (100 mg/l)	7821 ± 311*	2.15 ± 0.29	–
SA+VES (26.5 mg/l)	5855 ± 691*	1.60 ± 0.22*	–
SA+VES (75 mg/l)	5434 ± 392*	1.52 ± 0.09*	–
SA+SOD (200 µg/ml) +CAT (200 µg/ml)	1652 ± 345*	0.45 ± 0.07*	–
OH	–	–	11844 ± 1599
OH+GSPE (5 mg/l)	–	–	9741 ± 593**
OH+GSPE (25 mg/l)	–	–	4750 ± 393**
OH+GSPE (50 mg/l)	–	–	3577 ± 183**
OH+GSPE (100 mg/l)	–	–	2650 ± 395**
OH+GSPE (200 mg/l)	–	–	1196 ± 138**
OH+Vit C (5 mg/l)	–	–	10814 ± 899
OH+Vit C (100 mg/l)	–	–	10482 ± 679
OH+VES (26.5 mg/l)	–	–	7983 ± 463**
OH+VES (75 mg/l)	–	–	7640 ± 174**
OH+MAN (1.25 µM)	–	–	1552 ± 333**

*Superoxide anion (SA) and hydroxyl radicals (OH) were biochemically generated in vitro as described in the Materials and Methods Section. Data are expressed as the mean values of 4-6 experiments ± S.D. *P<0.05 with respect to the superoxide anion generating system, **P<0.05 with respect to the hydroxyl radical generating system.*

chemiluminescence assays were used to measure superoxide anion production. A combination of SOD plus CAT was used as a specific scavenger for superoxide anion. However, the reduction of cytochrome c was inhibited 82 % by using a combination of SOD plus CAT. An excellent concentration-dependent response was observed in cytochrome c reduction assays in the presence of GSPE, with 17, 59, 72, 81 and 89 % inhibitions at 5, 25, 50, 100 and 200 mg/l concentrations of GSPE, respectively (Table 1). When chemiluminescence was used to assay superoxide anion production, results similar to those observed with the cytochrome c reduction assay were obtained (Table 1). The chemiluminescence response was inhibited 83 % in the presence of SOD plus CAT.

Approximately 19, 62, 76, 79 and 88 % inhibitions in the chemiluminescence response were observed following incubation of the superoxide anion generating system with 5, 25, 50, 100 and 200 mg/l concentrations of GSPE (Table 1).

At concentrations of 5 and 100 mg/l, vitamin C inhibited the chemiluminescence response by 12 and 19 %, and cytochrome c reduction by 10 % and 15 %, respectively. When VES was added to the superoxide anion generating system at concentrations of 26.5 and 75 mg/l, the chemiluminescence response was inhibited by approximately 40 and 44 %, respectively (Table 1). Similar results were observed with respect to the cytochrome c reduction assay.

Mannitol was used as a specific scavenger of hydroxyl radical, which produced an 87 % inhibition of the chemiluminescence response at a 1.25 μM concentration. Following incubation of the hydroxyl radical generating system with 5, 25, 50, 100 and 200 mg/l concentrations of GSPE 18, 60, 70, 78 and 90 % inhibitions in chemiluminescence response were observed, demonstrating an excellent concentration dependent inhibitory effect on biochemically generated hydroxyl radicals (Table 1).

At concentrations of 5 and 100 mg/l, vitamin C inhibited hydroxyl radical generation by 9 and 12 %, respectively. VES at concentrations of 26.5 and 75 mg/l inhibited hydroxyl radicals by approximately 33 and 36 %, respectively (Table 1).

3.2. TPA-Induced Production of Free Radicals in the Peritoneal Macrophages of Mice, and Protection by GSPE and Other Antioxidants

TPA-induced enhanced production of free radicals by peritoneal exudate cells (primarily macrophages) of mice was assessed by luminol-enhanced chemiluminescence and cytochrome c reduction assays. A 6.1-fold increase in chemiluminescence response was observed following treatment of the animals with TPA. Administration of 25, 50 and 100 mg GSPE/kg for 7 consecutive days to the animals decreased the TPA-induced chemiluminescence in the peritoneal macrophages by 40 %, 55 % and 71 %, respectively. Pretreatment of animals with vitamin C (100 mg/kg), VES (100 mg/kg), a combination of vitamin C plus VES (100 mg/kg each) and GSPE (100 mg/kg) decreased the TPA-induced chemiluminescence by 16 %, 43 %, 51 % and 71 %, respectively, as compared to control samples. Administration of ß-carotene (50 mg/kg) and GSPE (50 mg/kg) for 7 consecutive days decreased the TPA-induced chemiluminescence response by 17 % and 55 %, respectively, relative to the control values (Table 2).

The effect of TPA on the production of superoxide anion by peritoneal macrophages was determined by cytochrome c reduction assay. TPA administration increased the production of superoxide anion based on cytochrome c reduction by 5.9-fold (Table 2). Administration of 25, 50 and 100 mg GSPE/kg for 7 consecutive days to the animals decreased TPA-induced cytochrome c reduction by 32 %, 53 % and 69 %, respectively. Pretreatment of animals with vitamin C (100 mg/kg), VES (100 mg/kg), a combination of vitamin C plus VES (100 mg/kg each) and GSPE (100 mg/kg) decreased TPA-induced cytochrome c reduction by approximately 15 %, 32 %, 47 %, and 69 %, respectively. Administration of ß-carotene (50 mg/kg) and GSPE (50 mg/kg) for 7 consecutive days decreased TPA-induced cytochrome c reduction by approximately 18 % and 53 %, respectively, relative to the control values (Table 2).

Table 2

Production of Reactive Oxygen Species by Peritoneal Macrophages Based on Chemiluminescence Response and Cytochrome c Reduction Following Treatment of Mice with TPA, and Protection by Antioxidants.

Sample	Chemiluminescence (CPM/3 x 10^6 cells)	Cytochrome c reduction (nmoles/15 min/3 x 10^6 cells)
Control	995 ± 156^a	4.55 ± 0.43^a
Vitamin C (100 mg/kg)	1114 ± 141^a	5.27 ± 0.50^b
VES (100 mg/kg)	1771 ± 139^b	9.10 ± 0.65^c
Vit C + VES (100 mg/kg each)	1724 ± 140^b	9.02 ± 0.58^c
ß-Carotene (50 mg/kg)	1198 ± 118^a	3.93 ± 0.70^a
GSPE (100 mg/kg)	1306 ± 94^a	5.07 ± 0.58^b
TPA	6031 ± 591^c	26.61 ± 1.60^d
TPA + Vit C (100 mg/kg)	5081 ± 335^d	22.67 ± 2.36^d
TPA + VES (100 mg/kg)	3455 ± 321^e	18.03 ± 0.83^e
TPA + Vit C + VES (100 mg/kg each)	2934 ± 132^e	14.24 ± 1.52^f
TPA + ß-Carotene (50 mg/kg)	5015 ± 199^d	1.81 ± 1.38^d
TPA + GSPE (25 mg/kg)	3592 ± 211^e	18.21 ± 1.86^e
TPA + GSPE (50 mg/kg)	2732 ± 99^e	12.65 ± 1.41^f
TPA + GSPE (100 mg/kg)	1724 ± 268^b	8.26 ± 0.84^c

Each value represents the mean ± S.D. of 4-6 mice. Values with non-identical superscripts are significantly different (p<0.05).

3.3. TPA-induced Hepatic and Brain Lipid Peroxidation and Protection by Antioxidants

Following treatment of mice with TPA increases in lipid peroxidation of 2.7-, 2.9-, and 3.1-fold were observed in hepatic mitochondria, hepatic microsomes and brain homogenates, respectively. Administration of 25, 50 and 100 mg GSPE/kg for 7 days to these animals decreased TPA-induced hepatic mitochondrial lipid peroxidation by 37 %, 41 % and 46 %, respectively, while in the hepatic microsomal fractions decreases of 47 %, 55 %, and 59 % were observed, respectively. Approximately 46 %, 53 % and 61 % decreases were demonstrated by GSPE against TPA-induced lipid peroxidation in the brain homogenates at these same concentrations (Table 3).

Administration of vitamin C (100 mg/kg), VES (100 mg/kg), a combination of vitamin C plus VES (100 mg/kg each) and GSPE (100 mg/kg) for 7 days decreased TPA-induced hepatic mitochondrial lipid peroxidation by 12 %, 36 %, 39 % and 46 %, and hepatic microsomal lipid peroxidation by 14 %, 47 %, 53 % and 59 %, respectively. Following treatment of the animals with these same antioxidants, 13 %, 45 %, 48 % and 61 % decreases, respectively, were observed against TPA-induced lipid peroxidation in brain homogenates. Administration of ß-carotene (50 mg/kg) decreased TPA-induced hepatic mitochondrial and microsomal lipid peroxidation by approximately 7 % and 12 %, respectively, while under these same conditions an 8 % decrease was observed in brain homogenate lipid peroxidation, as compared to control values (Table 3).

Table 3

TPA-Induced Lipid Peroxidation in the Hepatic and Brain tissues of mice, and the Comparative Protective Abilities of GSPE and Selected Antioxidants.

	Lipid Peroxidation (nmoles MDA/mg of protein).		
	Liver		**Brain**
Sample	Mitochondria	Microsomes	Whole homogenate
Control	2.17 ± 0.24^a	2.76 ± 0.22^a	1.62 ± 0.13^a
Vitamin C (100 mg/kg)	2.38 ± 0.19^a	2.85 ± 0.25^a	1.68 ± 0.10^a
VES (100 mg/kg)	3.05 ± 0.14^b	3.19 ± 0.29^b	1.93 ± 0.17^b
Vit C + VES (100 mg/kg each)	3.01 ± 0.17^b	3.11 ± 0.15^b	1.95 ± 0.10^b
ß-Carotene (50 mg/kg)	2.11 ± 0.12^a	2.83 ± 0.11^a	1.55 ± 0.12^a
GSPE (100 mg/kg)	2.32 ± 0.14^a	2.67 ± 0.32^a	1.71 ± 0.22^a
TPA	5.81 ± 0.34^c	8.12 ± 0.84^c	4.95 ± 0.32^c
TPA+Vit C (100 mg/kg)	5.12 ± 0.34^c	7.02 ± 0.42^c	4.32 ± 0.23^c
TPA+VES (100 mg/kg)	3.71 ± 0.39^d	4.29 ± 0.44^d	2.71 ± 0.49^d
TPA+Vit C+VES (100 mg/kg each)	$3.54 \pm 0.49^{b,d}$	$3.81 \pm 0.42^{d,e}$	2.57 ± 0.35^d
TPA+ ß-Caroten	5.41 ± 0.38^c	7.17 ± 0.46^c	4.54 ± 0.22^c
TPA+GSPE (25 mg/kg)	3.68 ± 0.39^d	4.33 ± 0.49^d	2.69 ± 0.28^d
TPA+GSPE (50 mg/kg)	3.43 ± 0.22^d	3.65 ± 0.25^e	2.33 ± 0.24^d
TPA+GSPE (100 mg/kg)	3.13 ± 0.27^b	3.31 ± 0.28^b	1.94 ± 0.40^b

Each value represents the mean $\pm$ S.D. of 4-6 mice. Values with non-identical superscripts are significantly different ($p < 0.05$).

3.4. TPA-induced DNA Fragmentation in the Hepatic and Brain Tissues, and Protection by Antioxidants

TPA-induced DNA fragmentation in hepatic and brain tissues and protection by GSPE and other antioxidants are summarized in Table 4.

TPA-induced 2.2- and 2.5-fold increases in DNA fragmentation in the hepatic and brain tissues of mice, respectively. Administration of 25, 50 and 100 mg GSPE/kg for 7 days to the animals decreased TPA-induced DNA fragmentation by 36 %, 42 % and 47 % in the hepatic tissues, and 32 %, 44 % and 50 % in the brain tissues, respectively. Pretreatment of animals with vitamin C (100 mg/kg), VES (100 mg/kg), a combination of vitamin C plus VES (100 mg/kg each) and GSPE (100 mg/kg) decreased DNA fragmentation by 10 %, 30 %, 38 % and 47 % in the hepatic tissues, and by 14 %, 31 %, 40 % and 50 % in the brain tissues, respectively. Administration of ß-carotene (50 mg/kg) for 7 days reduced TPA-induced hepatic and brain DNA fragmentation by 11 %, relative to the control values (Table 4).

Table 4

TPA-induced DNA Fragmentation in the Hepatic and Brain Tissues, and the Comparative Protective Abilities of GSPE and Selected Antioxidants.

	DNA Fragmentation	
Sample	**Liver (%)**	**Brain (%)**
Control	2.04 ± 0.30^a	1.77 ± 0.28^a
Vitamin C (100 mg/kg)	2.16 ± 0.34^a	$2.19 \pm 0.31^{a,b}$
VES (100 mg/kg)	2.63 ± 0.24^b	2.47 ± 0.36^b
Vit C+VES (100 mg/kg each)	2.54 ± 0.36^b	2.33 ± 0.31^b
ß-Carotene (50 mg/kg)	2.13 ± 0.22^a	1.97 ± 0.26^a
GSPE (100 mg/kg)	2.16 ± 0.31^a	1.89 ± 0.31^a
TPA	4.57 ± 0.51^c	4.41 ± 0.28^c
TPA+Vit C (100 mg/kg)	4.12 ± 0.31^c	3.80 ± 0.38^d
TPA+VES (100 mg/kg)	3.18 ± 0.45^d	3.03 ± 0.26^e
TPA+Vit C+VES (100 mg/kg each)	$2.83 \pm 0.23^{b,d}$	2.66 ± 0.21^b
TPA + ß-Carotene (50 mg/kg)	4.06 ± 0.29^c	3.92 ± 0.20^d
TPA + GSPE (25 mg/kg)	$2.94 \pm 0.51^{b,d}$	3.00 ± 0.16^c
TPA + GSPE (50 mg/kg)	2.67 ± 0.21^b	2.49 ± 0.24^b
TPA + GSPE (100 mg/kg)	2.43 ± 0.21^b	2.22 ± 0.19^b

Each value represents the mean ± S.D. of 4-6 mice. Values with non-identical superscripts are significantly different (p<0.05).

4 DISCUSSIONS

Proanthocyanidins are a group of polyphenolic bioflavonoids ubiquitously found in fruits, nuts, seeds and vegetables. The biological, pharmacological and medicinal properties of the proanthocyanidins have been extensively reviewed.[12-14,21] Flavonoids and other plant phenolics are reported, in addition to their free radical scavenging and antioxidant activity, to possess multiple biological activities including vasodilatory, anticarcinogenic, anti-inflammatory, antibacterial, immune-stimulating, antiallergic, antiviral and estrogenic activities, as well as being inhibitors of the enzymes phospholipase A2, cyclooxygenase and lipooxygenase.[13,14,22] The chemical properties of bioflavonoids in terms of the availability of the phenolic hydrogens as hydrogen donating radical scavengers and singlet oxygen quenchers predicts their antioxidant activity.[12-14]

In this study, the protective abilities of GSPE, a commercially available IH636 grape seed proanthocyanidin extract, vitamin C, vitamin E succinate (VES), a combination of vitamin C plus VES and ß-carotene were assessed on phorbol ester (TPA)-induced oxiative tissue and DNA damage in the hepatic and brain tissues, and activation of peritoneal macrophages. The production of reactive oxygen species by peritoneal macrophages was assessed by measuring chemiluminescence and cytochrome $\underline{c}$ reduction (Table 1). Cytochrome $\underline{c}$ reduction is a specific test for superoxide anion production,[23] while chemiluminescence is a general assay for the production of reactive oxygen species.[24] These assays clearly demonstrate the production of reactive oxygen species by peritoneal macrophages following administration of TPA, and the comparative protective abilities of GSPE, vitamin C, a combination of vitamin C plus VES, and ß-carotene. GSPE emonstrated the best protection in the chemiluminescence assay as compared to vitamin C, VES or ß-carotene at the doses which were used. The combination of vitamin C and VES demonstrated better protection as compared to either vitamin C or VES alone, which may occur as the result of regeneration of vitamin E from its oxidized form by vitamin C.[25] Similar results were obtained in the cytochrome c reduction assay (Table 1). These data indicate that GSPE, as well as other antioxidants, may be useful in preventing the *in vivo* production of reactive oxygen species.

Lipid peroxidation was assessed in the hepatic mitochondria and microsomes, and brain homogenate (Table 3), and DNA fragmentation data for hepatic and brain tissues are presented in Table 4. Lipid peroxidation and DNA fragmentation serve as indicators of oxidative tissue damage. DNA fragmentation has been demonstrated as a biochemical hallmark of apoptosis (programmed cell death) which plays a major role in developmental biology and in maintenance of homeostasis in vertebrates.[26] GSPE exhibited the best protection towards TPA-induced hepatic mitochondrial and microsomal lipid peroxidation as compared to the other antioxidants tested at the doses which were used (Table 3). A combination of vitamin C plus VES exerted a better protection as compared to vitamin C or VES alone. All antioxidants which were tested ameliorated TPA-induced increases in lipid peroxidation and DNA fragmentation in both brain and liver, with GSPE exhibiting the best protection as compared to the other antioxidants (Table 4).

The results clearly demonstrate that GSPE significantly scavenges biochemically generated superoxide anion and hydroxyl radicals, and is more potent as compared to vitamins C and E. Furthermore, GSPE can significantly attenuate TPA-induced lipid peroxidation and DNA fragmentation in the hepatic and brain tissues, and the enhanced production of oxygen free radicals in the peritoneal macrophages, significantly better than vitamin C, VES, a combination of vitamin C plus VES, and ß-carotene. These *in vitro* and *in vivo* experiments demonstrate that GSPE is a better scavenger of free radicals, highly bioavailable to the vital target organs and inhibitor of oxidative tissue damage as compared to the other antioxidants tested in this study at the doses which were used, and may therefore be useful in preventing the production of reactive oxygen species and oxidative tissue damage *in vivo*.

References
1.   H.C. Pitot and Y.P. Dragan, *FASEB J.* 1991, **5**, 2280.
2.   B.N. Ames, *J. AOAC Int.* 1992, **75**, 1.
3.   B. Halliwell, *Free Rad. Res.* 1996, **25**, 57.
4.   J.P. Kehrer, *Crit. Rev. Toxicol.* 1993, **23**, 21.
5.   S.J. Stohs and D. Bagchi, *Free Rad. Biol. Med.* 1995, **18**, 321.
6.   B. Halliwell, J.M.C. Gutteridge and C.E. Cross, *J. Lab. Clin. Med.* 1992, **119**, 598.
7.   G. Hocman, *Comp. Biochem. Physiol.* 1989, **93B**, 201.

8. I.B. Afanas'ev, A.I. Dorozhko, A.V. Brodskii, V.A. Kostyuk and A.I. Potapovitch, *Biochem. Pharmacol.* 1989, **38**, 1763.

9. M.K. Buening, R.L. Chang, M.T. Huang, J.G. Fortner, A.W. Wood and A.H. Conney, *A.H. Cancer Res.* 1981, **41**, 67.

10. H. Kolodziej, C. Haberland, H.J. Woerdenbag and A.W.T. Konings, *Phytother. Res.* 1995, **9**, 410.

11. W. Bors and M. Saran, *Free Rad. Res. Commun.* 1987, **2**, 289.

12. Z.Y. Chen, P.T. Chan, K.Y. Ho, K.P. Fung and J. Wang, *Chem.Phys.Lipids.* 1996, **79**, 157.

13. C.A. Rice-Evans, N.J. Miller and G. Paganda, *Free Rad. Biol. Med.* 1996, **20**, 933.

14. C.A. Rice-Evans and L. Packer, *Flavonoids in Health and Disease, Marcel Dekker, Inc.*, New York, 1997.

15. D. Bagchi, D.K. Das, R.M. Engelman, *M.R. Prasad and R. SubramanianEur.Heart J.* 1990, **11**, 800.

16. G. Witz and B.J. Czerniecki, *Carcinogenesis.* 1989, **10**, 807.

17. M. Bagchi and S.J. Stohs, *Free Rad. Biol. Med.* 1993, **14**, 11.

18. B.M. Babior, R.S. Kipnes and J.T. Curnutte, *J. Clin. Invest.* 1973, **52**, 741.

19. J.A. Buege and S.D. Aust, *Meth. Enzymol.* 1978, **52**, 302.

20. D. Bagchi, O.R. Carryl, M.X. Tran, R.L. Krohn, D.J. Bagchi, A. Garg, M. Bagchi, S. Mitra and S.J. Stohs, *J. Appl. Toxicol.* 1998, **18**, 3.

21. J. Masquelier, J. Michaud, J. Laparra and M.C. Dumon, *Int. J. Vitam. Nutr. Res.* 1979, **49**, 307.

22. N. Salah, N.J. Miller, G. Paganga, L. Tijburg, G.P. Bolwell and C. Rice-Evans, *Arch. Biochem. Biophys.* 1995, **322**, 339.

23. E.E. Ritchey, J.D. Wallin and S.V. Shah, *Kidney Int.* 1981, **19**, 349.

24. M.S. Fisher and M.L. Adams, *Cancer Res.* 1985, **45**, 3130.

25. G.R. Buettner, *Arch. Biochem. Biophys.* 1993, **300**, 535.

26. R.A. Schwartzman and J.A. Cidlowski, *Endocrine Rev.* 1993, **14**, 133.

EFFECT OF PARSLEY INTAKE ON URINARY APIGENIN EXCRETION, BLOOD ANTIOXIDANT ENZYMES AND BIOMARKERS FOR OXIDATIVE STRESS IN HUMANS

S.E. Nielsen, J.F. Young[1], B. Daneshvar, S.T. Lauridsen, P. Knuthsen[2], B. Sandström[1] and L.O. Dragsted.

Institute of Food Safety and Toxicology, Danish Veterinary and Food Administration, Mørkhøj Bygade 19, DK-2860 Søborg, Denmark. [1]Research Department of Human Nutrition, Royal Veterinary and Agricultural University, Rolighedsvej 25, DK-1958 Frederiksberg C, Denmark. [2] Institute of Food Chemistry and Nutrition, Danish Veterinary and Food Administration, DK-2860 Søborg, Denmark.

1 INTRODUCTION

1.1. Free radicals and antioxidants

The relationship between diet and disease in humans has been the topic of numerous investigations during recent years. Epidemiological data show a beneficial effect of a high intake of fruits and vegetables on the risk of cancer and cardiovascular diseas[1-3] The high content of antioxidants in fruits and vegetables has been related to this preventive effect due to their ability to scavenge free radicals.[4]

The human body is under constant assault by free radicals, that might cause modifications in DNA and proteins, or oxidation of low-density lipoproteins (LDL). These are reactions, that are thought to be implicated in the pathogenesis of a large number of diseases including cancer and atherosclerosis.[5,6] Free radicals are also thought to be important in the ageing process.[7]

To limit the formation of, and to scavenge the free radicals, antioxidants are needed. The body produces a range of endogenous antioxidants including the antioxidant enzymes, superoxide dismutase, glutathione peroxidase and catalase. In addition to the antioxidant enzymes, certain proteins and low molecular weight species function as endogenous antioxidants. Examples are metallothioneins, other metal ion-binding proteins and storage proteins, urate, glutathione, and ubiquinol.[8]

Although the endogenous antioxidant enzymes are highly efficient, there are some reactive oxygen species (e.g. singlet oxygen and the hydroxyl radical), where the defence is either ineffective or totally lacking. Here the body must rely on food derived antioxidants, that can scavenge these species. Besides urate and glutathione these include vitamin E and C, but also the plant derived natural antioxidants, e.g. the carotenoids and *the flavonoids*.

The flavonoids is one of the major groups of antioxidants present in fruits and vegetables. This large group of polyphenols has gained increasing interest by researchers world-wide due to experimental evidences for their ability to scavenge free radicals, their influence on several enzyme systems, anticancer-activities etc., and lately because of epidemiological evidences of an association with a lower risk of coronary heart disease 910

1.2. Human intervention with:[7] the flavone apigenin

Dietary apigenin is found mainly in parsley but also at low levels in certain vegetables, seasonings and in oranges. Apigenin possesses antioxidant activity *in vitro*.[11-14] Potent biological seasonings and in oranges. Apigenin possesses antioxidant activity *in vitro*.[11-14] Potent biological effects of this flavonoid have been described *in vitro* and *in vivo*. Apigenin has been ascribed anticarcinogenic,[15,16] anti-inflammatory[17] and other beneficial properties.

Parsley contains very large amounts of apigenin,[19] and the low concentration of other flavonoids in this plant[19] makes it suitable for an intervention study with a natural source of apigenin .

In the present study we report on the relationship between intake of daily doses of parsley and urinary excretion of apigenin. The antioxidative effect of the parsley intervention was investigated by measuring the activity of red blood cell antioxidant enzymes. In addition, a biomarker for plasma protein oxidation (protein 2-adipic semialdehyde, AAS) was measured.[20]

2 EXPERIMENTAL

2.1. Study design

Seven men and 7 women participated in a randomised cross-over study with two one-week intervention periods in succession. The subjects received a strictly controlled diet low in flavonoids during the two weeks of intervention. The energy requirement of each subject was estimated from 4 days food registration as well as body weight and degree of physical activity. All meals were prepared at the department in individual portions according to energy requirement. This diet was supplemented during one of the weeks with 20 g parsley/10 MJ providing 37.3-44.9 mg apigenin.

Blood samples were taken at start and end of each period, and the activity of red blood cell antioxidant enzymes, catalase, superoxide dismutase, glutathione reductase and glutathione peroxidase and semialdehyde (AAS) residues, a biomarker of plasma protein oxidation, were measured.[21] Urine samples were collected over 24 h from the day before the intervention started and on day 1,7,8,9,11and 14 the intervention period.[21]

2.2. Analysis of urinary apigenin

The urine samples were analysed for content of apigenin as we have previously described Briefly, aliquots of 15 ml urine to which 5 µg 5,7,8-trihydroxyflavone was added as inte standard, were enzymatically hydrolysed using β-glucuronidase and arylsulfatase to release apigenin from glucuronic- and sulphate conjugates, and subsequently solid phase extracted.

A highly sensitive HPLC method was developed for the analysis of urinary apigenin. The method implicated the use of column switching, whereby the target compounds (apigenin, an internal standard, and the potential metabolite of apigenin, acacetin (4'-OMeapigenin)) specifically were eluted from the first column, used for sample clean up, onto the second column, where a different gradient elution ensured separation from urine impurities prior to detection by ultraviolet absorbance detection.[22]

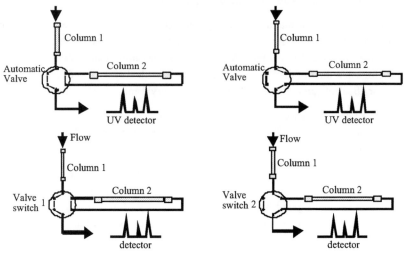

Figure 1
Schematic diagram of the column switching HPLC system.

3 RESULTS AND DISCUSSION

There has only been published one previous attempt to determine the urinary excretion of apigenin in humans after ingestion of an apigenin containing camomile extract.[23] However, due to lack of specificity and sensitivity the method failed to detect any apigenin in the urine. Janssen et al. recently investigated the plasma level of apigenin after intake of 5 g dried parsley (providing 84 mg apigenin) for seven days in a randomised cross over study.[24] However, no plasma apigenin could be detected with a limit of detection of 1.1 µmol/l and the authors concluded, that maybe apigenin was not absorbed at all.

In the present study,[21] we demonstrate that apigenin is absorbed in humans, and in average we found an excretion of 0.58 % ± 0.16 (mean ± SEM) in the 24 h urine samples during parsley intervention. The potential metabolite of apigenin, acacetin (the 4'-methoxylated derivative of apigenin) was not determined in any of the urine samples. Significant inter-individual variation in the excretion of apigenin was observed in the study, and the maximum excretion was found in a single individual with an excretion as high as 7.45 % of the dose at day seven of the parsley intervention. This apparent difference in bioavailability of apigenin could be due to variations in the intestinal uptake or in metabolism and distribution. Since identical intakes of parsley resulted in very different apigenin excretion among individuals urinary apigenin is not a valid biomarker for apigenin intake but might be a useful biomarker for apigenin absorption in human studies.

Red blood cell glutathione reductase and superoxide dismutase activity significantly increased with parsley intervention, whereas red blood cell catalase and glutathione peroxidase did not increase significantly. No significant changes were observed in plasma protein 2-adipic semialdehyde (AAS) residues, a biomarker of plasma protein oxidation. However, at the individual level, the increase in superoxide dismutase was strongly correlated with increases in catalase and a decrease in AAS. Furthermore, an overall decreasing trend in the level of all measured antioxidant enzymes was observed during the two weeks study, which may be due to the low intake of fresh fruits and green vegetables. However, the intervention with parsley seemed partly to overcome this decrease and resulted in increased or restored levels of glutathione reductase and superoxide dismutase activity.

References

1.	K.A. Steinmetz and J.D. Potter, *J. Am. Dietetic. Ass.* 1996, **96**, 1027.
2.	G. Block, B. Patterson and A. Subar, *Nutr. Cancer.* 1992, **18**, 1.
3.	G. Block and L. Langseth, *Food Tech.* 1994, **july**, 80.
4.	V.M. Sardesai, *Nutr. Clin. Pract.* 1995, **10**, 19.
5.	P. Cerutti, *Lancet.* 1994, **344**, 862.
6.	D. Steinberg, S. Parthasarathy, T.E. Carew, J.C. Khoo and J.L. Witztum, *New Eng. J. Med.* 1989, **320**, 915.
7.	H.R. Warner, *Free Rad. Biol. Med.* 1994, **17**, 249.
8.	B. Halliwell, *Free Radic. Res.* 1996, **25**, 57.
9.	M.G.L. Hertog, E.J.M. Feskens, P.C.H. Hollman, M.B. Katan and D. Kromhout, *Lancet.* 1993, **342**, 1007.
10.	M.G. Hertog, D. Kromhout, C. Aravanis, H. Blackburn, R. Buzina, F. Fidanza, S. Giampaoli, A. Jansen, A. Menotti, S. Nedeljkovic, M. Pekkarinen, B.S. Simic, H. Toshima, E.J.M. Feskens, P.C.H. Hollman and M.B. Katan, *Arch. Int. Med.* 1995, **15**, 381.
11.	J. Kühnau, *Wld. Rev. Nutr. Diet.* 1976, **24**, 1 17.
12.	B. Fernandez de Simon, J. Perez Ilzarbe, T. Hernandez, C. Gomez Cordoves and I. Estrella, *J. Agric. Food Chem.* 1992, **40**, 1531.
13.	S.A.B.E. van Acker, D.J. van den Berg, M.N.J.L.Tromp, D.H. Griffioen, W.P. van Bennekom, W.J.F. van der Vijgh, and A. Bast, *Free Radic. Biol. Med.* 1996, **20**, 331.
14.	C.G. Fraga, V.S. Martino, G.E. Ferraro, J.D. Coussio and A. Boveris, *Biochem. Pharm.* 1987, **36**, 717.
15.	D.F. Birt, D. Mitchell, B. Gold, P. Pour and H.C. Pinch, *AnticancerRes.* 1997, **17**, 85.
16.	H. Wei, L. Tye, E. Bresnick and D.F. Birt, *Carcinogenesis.* 1989, 499.
17.	S.J. Lee, K.H. Son, H.W. Chang, J.C. Do, K.Y. Jung, S.S. Kang and H.P. Kim, *Arch. Pharm. Res.* 1993, **16**, 25.
18.	M.L. Kuo, K.C. Lee and J.K. Lin, *Mutat. Res.* 1992, **270**, 87.
19.	U. Justesen, P. Knuthsen and T. Leth, *J. Chrom. A.* 1998, **799**, 101.
20.	B. Daneshvar, H. Frandsen, H. Autrup and L.O. Dragsted, *Biomarkers.* 1997, **2**, 117.
21.	S.E. Nielsen, J.F. Young, J. Haraldsdottir, B. Daneshvar, S.T. Lauridsen, P. Knuthsen, B. Sandström and L.O. Dragsted, *Br.J. Nutr.*, 1998, Accepted for publication.

22.   S.E. Nielsen and L.O. Dragsted, *J. Chrom. B*. 1998, **713**, 379.

23.   K. Tschiersch and J. Hölzl, *Pharmazie*. 1993, **48**, 554.

24.   K. Janssen, R.P. Mensink, F.J.J. Cox, J.L. Harryvan, R. Hovenier, P.C.H. Hollman and M.B. Katan, *Am. J. Clin. Nutr*. 1998, **67**, 255.

EFFECT OF FRUIT JUICE INTAKE ON URINARY QUERCETIN EXCRETION AND BIOMARKERS OF ANTIOXIDATIVE STATUS

Jette F. Young[1], Salka E. Nielsen[2], Jóhanna Haraldsdóttir[1], Bahram Daneshvar[2], Søren T. Lauridsen[2], Pia Knuthsen[3], Alan Crozier[4], Brittmarie Sandström[1] and Lars O. Dragsted.[2]

[1]Research Department of Human Nutrition, Royal Veterinary and Agricultural University, Copenhagen, Denmark. [2]Institute of Toxicology and [3]Institute of Food Chemistry and Nutrition, Danish Veterinary and Food Administration, Copenhagen, Denmark. [4]Division of Biochemistry and Molecular Biology, Institute of Biomedical and Life Sciences, University of Glasgow, Scotland.

1 INTRODUCTION

Epidemiological studies suggest that foods rich in flavonoids might reduce the risk of coronary heart diseases.[1-3] The potential beneficial mechanism could be the antioxidative properties of these compounds, but in order to exert a protective effect on coronary heart disease it is reasonable to assume that the flavonoids have to be absorbed. Few data are available on absorption of quercetin or quercetin glycosides in man. Hollman et al.[4] have carried out a single-dose study on 9 ileostomy patients which suggests that absorption of quercetin conjugates from foods may be more efficient than that of the pure aglycone. In that study the urinary excretion was 0.31%, 0.07 % and 0.12 % of the intake of conjugated quercetin from onions, rutinoside and pure quercetin aglycone, respectively. In two later studies these investigators observed an urinary excretion of 0.44 % of quercetin from apples and 1.39 % from onions.[5] The present study investigated the dose response in urinary quercetin excretion and the effect on markers of oxidative status after intake of blackcurrant and apple juice.

2 STUDY DESIGN AND ANALYSIS

The study was designed as a one week cross-over intervention where five healthy young subjects (4 women, 1 men) consumed three doses of a 1:1 mixture of blackcurrant and apple juice (750, 1000, and 1500 mL) corresponding to 4.8, 6.4, and 9.6 mg quercetin per day, respectively. The three randomised intervention weeks were separated by at least two washout weeks. Other flavonoid containing foods were excluded one week before and during each intervention period. Fasting blood samples were collected in the morning of day 1, 2, 3, 5 and 8 in each intervention period and continuous 24 h urine samples were collected at baseline and throughout all intervention periods. Plasma was analysed for trolox equivalent antioxidant capacity (TEAC), ferric reducing ability (FRAP), an unspecific lipid oxidation marker, malondialdehyde (MDA), a protein oxidation product, adipic semialdehyde substances (AAS) and ascorbate. The erythrocytes were analysed for the antioxidant enzymes superoxide dismutase (SOD), glutathione peroxidase (Gpx), glutathione reductase (GR) and catalase (CAT).

3 RESULTS

Urinary excretion of quercetin increased significantly with dose and time. The fraction of intake excreted in urine was 0.29-0.47 %, regardless of dose and without a significant variation between the five subjects, indicating that urinary quercetin excretion may be a useful biomarker of quercetin intake.

Plasma concentrations of quercetin from fasting blood samples were low, not significantly changed by dose and thus not suitable as a biomarker of quercetin intake. The juice supplied an extra ascorbic acid intake of 150, 200 and 300 mg/d during the three intervention periods, respectively, and plasma ascorbate increased during intervention. Total plasma malondialdehyde decreased with time during 1500 mL juice intervention, indicating a reduced lipid oxidation in plasma. Red blood cell catalase activity increased with dose whereas other antioxidant enzymes in red blood cells (Gpx, GR and SOD) did not change significantly. Plasma protein 2-adipic semialdehyde (AAS) residues, a biomarker of plasma protein oxidation, increased with time and dose indicating a prooxidant effect of the juice, whereas erythrocyte AAS and other unspecific markers of plasma antioxidant status (TEAC and FRAP) did not change.

4 DISCUSSION

In this study we investigated the excretion of quercetin during a low-dose intake period of several days. This is in contrast to previous studies, which have been based on administration of a single large dose, 70-100 mg.[4-6] Only a small fraction, 0.29-0.47 %, of quercetin intake was excreted in urine. This is in accordance with results reported from a single dose study with 100 mg quercetin from apples (0.44 %) but lower than the fraction of quercetin excreted from fried onions (1.39 %) during 24 h after ingestion of the dose.[5] The fraction of quercetin intake excreted into urine reached steady state after 3-4 days of intervention, indicating an elimination half-life of approximately 24 h. This is in good agreement with results from a single dose study with 100 mg quercetin glycoside from an apple preparation with an average elimination half-life from plasma of 23 h.[5]

The presence of quercetin in urine demonstrates that it has been absorbed by the gut, but the urinary content does not necessarily reflect absolute absorptive efficiency since absorbed quercetin may be metabolised, stored or excreted through other routes such as the biliary tract. However, the results from the present study suggest that urinary quercetin may be a useful biomarker of quercetin intake. Urinary quercetin was detected in all urine samples, even in the baseline samples collected at a time when the participants had avoided flavonoid containing foods for a week. The baseline level was low and could either originate from a low quercetin intake from foods other than fruit, vegetables, spices, tea and wine (which were omitted) or it could be due to a residual excretion of dietary quercetin ingested before the flavonoid-free diet was instituted.

When evaluating the effect of the present intervention study on markers of antioxidative status it should be noted that the apple- and blackcurrant juices selected for the present study contained not only the flavonoids quercetin, myricetin and kaempferol but also several other phenolic antioxidants, including simple phenolic acids, chalcons, and anthocyanins.[7,8] They also contained ascorbic acid, resulting in a significant increase in plasma ascorbate during intervention.

Our finding of a significantly increased level of Gpx with dose might indicate an effect of the juice intervention on this activity. The other antioxidant enzymes in the red blood cells did not show a consistent pattern of change. One week may seem a very short period for expression of an enzyme increase in red blood cells which lack the capability for protein synthesis, but consistent short-term effects on some of these enzymes have previously been reported, even within h.[9]

Our present finding of a decrease in total plasma malondialdehyde after an intake of 1500 mL juice per day over seven days, suggest an improvement in antioxidant status and would indicate that even low levels of flavonoids or other antioxidant constituents of the juice might decrease lipid oxidation within the plasma compartment. However, in contrast to this, the plasma protein lysine residues (AAS) were oxidised to a greater extent after the juice intervention. The prooxidant effect was immediate, and related to both dose and duration of intake. This observation suggests the presence of potentially prooxidative compounds in the juice. *In vitro* studies have demonstrated prooxidant as well as antioxidant behaviour of flavonoids.[10]

The results of the present study on AAS versus MDA illustrate that even within the blood plasma there are several subcompartments which may respond differently to a dietary challenge. These results also contradict a general pro- or anti-oxidant state in the blood compartment but indicate a differential protection or damage to specific structures, depending on their interaction with the dietary components reaching them. However, the results on enzyme activities and other markers of antioxidant status need to be confirmed and extended in long term studies.

In conclusion quercetin from apple and blackcurrant juice was absorbed and 0.29-0.47 % of intake was excreted into the urine. Juice consumption seemed to have a prooxidant effect on plasma proteins, while a putative marker of lipid oxidation, MDA in the blood plasma compartment, seemed to decrease. These effects might be related to several components of the juice and cannot be attributed solely to its quercetin content.

References

1. M. G. L. Hertog, D. Kromhout, C. Aravanis, et al. *Arch. Intern. Med.* 1995, **155**, 381.
2. P. Knekt, R. Järvinen, A. Reunanen, J. Maatela. *BMJ.* 1996, **312**, 478.
3. M. L. G. Hertog, E. J. M.Feskens, P. C. H. Hollman, M. B. Katan, D. Kromhout. *Lancet.* 1993, **342**, 1007.
4. P. C. H. Hollman, J. H. M. de Vries, S. D. van Leeuwen, M. J. B. Mengelers, M. B. Katan. *Am. J. Clin. Nutr.* 1995, **62**, 1276.

5.   P. C. H. Hollman, J. M. P. van Trijp, M. N. C. P. Buysman, M. S. van der Gaag, M. J. B. Mengelers, J. H. M. de Vries, M. B. Katan. *FEBS Letters*. 1997, **418**, 152.

6.   P. C. H. Hollman, M. S. van der Gaag, M. J. Mengelers, J. M. van Trijp, J. H. M. de Vries, M. B. Katan. *Free Radic. Biol. Med.* 1996, **21**, 703.

7.   A. B. Durkee, J. D. Jones. *Phytochem.* 1969, **8**, 909.

8.   K. B. McRae, P. D. Lidster, A. C. DeMarco, A.J. Dick. *J. Sci. Food Agric.* 1990, **50**, 329.

9.   A. N. Saghir, H. Rickards, H. S. *Pall. Exp.Neurol.* 1997, **144**, 420.

10.   G. Cao, E. Sofic, R. L. Prior. *Free Radic. Biol. Med.* 1997, **22**, 749.

ANTIOXIDANT ACTIVITY OF HYDROXYCINNAMIC ACIDS ON HUMAN LOW-DENSITY LIPOPROTEIN OXIDATION *IN VITRO*

Anne S. Meyer[1] and Mette F. Andreasen[2].

[1]Dept. of Biotechnology, Technical University of Denmark, DK-2800 Lyngby [2]Dept. of Fruit, Veg. and Food Sci., Danish Inst. of Agric. Sciences, DK-5792 Aarslev.

1 INTRODUCTION

Oxidation of low-density lipoprotein (LDL) is considered a primary event in the atherogenic process in humans.[1] Phenolic antioxidants in fruits and wine that inhibit LDL oxidation may therefore be the molecular link to the cardioprotective effect of fruit and wine consumption.[2] Consumption of whole grain products also appear to reduce development of chronic diseases, including cardiovascular diseases.[3] While much effort has recently been focused on the potential antioxidant and disease preventing activities of flavonoids, comparatively little is known about the actions of phenolic acids, including hydroxycinnamic acids.

2 HYDROXYCINNAMIC ACIDS IN CEREALS AND FRUITS

Hydroxycinnamic acids are ubiquitously present in plant foods and consist of a phenolic ring with a lateral three carbon chain[4] (C_6-C_3). In edible plants hydroxycinnamic acids are principally derived from *p*-coumaric, caffeic, and ferulic acids.[4,5] For whole grains (wheat, rye, oat, barley) the quantities reported in the literature range from apprx. 200-1000 μg/g.[6,7,8] In fruits (incl. apples, pears, plums, berries), reported values range from apprx. 1-2200 μg/g.[4,5] In both cereals and fruits the hydroxycinnamates mainly occur in various bound and esterified forms. Thus, in cereals, a considerable fraction of the ferulic and *p*-coumaric acids are covalently linked to arabinoxylans and may be linked to lignin.[9] In fruits, hydroxycinnamic acids predominantly occur as quinic acid or glucose esters with hydroxycinnamoyl-quinic acids, notably chlorogenic acid, being the most dominant form in pome and stone fruits.[4]

3 ANTIOXIDANT ACTIVITY TOWARDS HUMAN LDL OXIDATION *IN VITRO*

Several hydroxycinnamic acids and their esterified derivatives occuring in fruits and cereals have been demonstrated individually to have strong activity in reducing LDL oxidation *in vitro*.[10] When comparing the antioxidant activities of hydroxycinnamates on *in vitro* human LDL oxidation reported by different investigators[10,11,12,13] the general order of antioxidant activity appears to be caffeic > ferulic > *p*-coumaric acid (Table 1).

Table 1.

Comparison of reported antioxidant activities of hydroxycinnamic acids on human LDL oxidation in vitro.

Assay conditions	Phenolic acid concentration (µM)	Order of antioxidant activity	Ref.
LDL (200 µg protein/mL), 5 µM CuCl$_2$, 4 h, 37 °C, pH 7.4, Conjugated dienes	100 µM 5 µM	Caffeic≈ Ferulic >> p-coumaric Caffeic > Ferulic	11
LDL (30 µg protein/mL), 5 mM AAPH[a], 0-1 h, 37 °C, pH 7.4, O$_2$-consumption and decay of cis-parinaric acid[b]	0.5 - 1.5 µM	Chlorogenic > Caffeic	12
LDL (0.3 mM cholesterol), 8 µM CuCl$_2$, 4 h, 37 °C, pH 7.4, Conjugated dienes	0.1 - 10 µM	Caffeic >> p-coumaric	13
LDL (1000 µg protein/mL), 80 µM CuSO$_4$, 2 h, 37 °C, pH 7.4, Hexanal	5 µM 10 µM	Caffeic > Ferulic > p-coumaric Caffeic > Ferulic > p-coumaric	10

[a]*AAPH: 2,2'-azobis-(2-amidinopropane)hydrochloride,* [b]*Decay of cis-parinaric acid (PnA) by fluorescence.*

However, the available data are difficult to compare as different agents were used to induce LDL oxidation and the methods employed to assess antioxidant activity varied (Table 1). Despite experimental divergence, however, the differences in activity among individual hydroxycinnamates are assumingly due to variations in the hydroxylation and methylation pattern of the aromatic ring. Thus, notably the presence of the *o*-dihydroxy group in the phenolic ring (as in caffeic acid) consistently seems to enhance the antioxidant activity of hydroxycinnamic acids towards human LDL oxidation *in vitro*[10,11,13] (Table 1). This trend is also observed with other substrates than LDL.[14] Such a structure-activity relationship is analogous to other natural phenolic antioxidants and in accordance with the radical-scavenging antioxidant mechanism, where the ability to donate hydrogen depends on the number of OH-groups and where, in particular, the presence of 3´,4´ *o*-hydroxyls increase the reactivity.[15,16]

Hydroxycinnamic acids may inhibit LDL oxidation by several different mechanisms, however. These antioxidant mechanisms include free radical scavenging, metal chelation, regeneration of endogenous LDL antioxidants, and perhaps binding to apolipoprotein B in the LDL particle. The latter may be a unique antioxidant mechanism for LDL oxidation since copper-mediated oxidation of tryptophans in apolipoprotein B initiates lipid oxidation of LDL.[17] Very little is known on the effect of esterification of hydroxycinnamic acids on antioxidant activity and mechanism. Ferulic acid sugar esters from corn bran were recently shown to be better inhibitors than ferulic acid of *in vitro* LDL oxidation,[18] and we also found that esterification apparently enhanced antioxidant activity of both ferulic and *p*-coumaric acid.[10] On this basis we proposed that esterification of certain hydroxycinnamic acids to acid or sugar moieties may affect antioxidant activity by conferring differences in the ability to bind to apolipoprotein B in the LDL particle to block oxidation of tryptophan residues.[10] Furthermore, esterification may induce differences in solubility and phase distribution properties of hydroxycinnamates, which, in analogy to what has been suggested for other natural antioxidants, may be important determinants of antioxidant activity towards human LDL.[19] Further work is clearly needed to understand better the antioxidant mechanisms of hydroxycinnamic acids against human LDL oxidation.

4 PHYSIOLOGICAL SIGNIFICANCE OF HYDROXYCINNAMIC ACIDS

Some absorption of both caffeic and ferulic acid was indicated in a preliminary investigation with less than 10 volunteers.[19] However, at present there is no direct evidence that hydroxycinnamic acids exert physiologically beneficial antioxidant activity *in vivo*. More knowledge is therefore required to predict and exploit better the possible cardioprotective benefits in humans of hydroxycinnamic acid compounds.

References

1.   D. Steinberg. *Atheroscler. Rev.* 1988, **18**, 1-6.
2.   M. H. Criqui and B. L. Ringel. *Lancet.* 1994, **344**, 1719-1723.
3.   J. L. Slavin. *Crit. Rev. Food Sci. Nutr.* 1994, **34**, 427-434.
4.   K. Herrman. *Crit. Rev. Food Sci. Nutr.* 1989, **28**, 315-347.
5.   J.-J. Macheix, A. Fleuriet, J. Billot. *Fruit Phenolics*, CRC Press, Boca Raton, FL, 1990.
6.   K. Rybka, J. Sitarski, K. Raczynska-Bohanowska. *Cereal Chem.* 1993, **70**, 55-59.
7.   J. M. Zupfer, K.E. Churchill, D. C. Rasmusson, R. G. Fulcher. *J. Agric. Food Chem.* 1998, **46**, 1350-1354.
8.   Y. Xing and P. J. White. *J. Am. Oil Chem. Soc.* 1997, **74**, 303-307.
9.   T. Ishii. *Plant Sci.* 1997, **127**, 111-127.
10. A. S. Meyer, J. L.Donovan, D. A. Pearson, A. L. Waterhouse and E. N. Frankel. *J. Agric. Food Chem.* 1998, **46**, 1783-1787.
11. M. Nardini, D'Aquino, G. Tomassi, V. Gentili, M. D. Felice, C. Scaccini. *Free Rad. Biol. Med.* 1995, **19**, 541-552.
12. J. A. N. Laranjinha, L. M. Almeida, V. M. C. Madeira. *Biochem. Pharmacol.* 1994, **48**, 487-494.
13. R. Abu-Amsha, K. D. Croft, I. B. Puddey, J. M. Proudfoot, L. J. Beilin. *Clin. Science.* 1996, **91**, 449-458.
14. M.-E. Cuvelier, H. Richard, C. Berset. *Biosci. Biotech. Biochem.* 1992, **56**, 324-325.
15. B. J. F. Hudson and J. I. Lewis. *Food Chem.* 1983, **10**, 47-55.
16. C. A. Rice-Evans, N. J. Miller, G. Paganga. *Free Rad. Biol. Med.* 1996, **20**, 933-956.
17. A. Giessauf, E. Steiner, H. Esterbauer. *Biochim. Biophys. Acta* 1995, **1256**, 221-232.
18. T. Ohta, N. Semboku, A. Kuchii, Y. Egashira, H. Sanada. *J. Agric. Food Chem.* 1997, **45**, 1644-1648.
19. E. N. Frankel, S.-W. Huang, J. Kanner, B. German. *J. Agric. Food Chem.* 1994, **42**, 1054-1059.
20. E. A. Jacobson, H. Newmark, J. Baptista, W.R. Bruce. *Nutr. Rep. Int.* 1983, **28**, 1409-1417.

RADICAL SCAVENGERS AND INHIBITORS OF ENZYMATIC LIPID PEROXIDATION FROM *PLANTAGO MAJOR*, A MEDICINAL PLANT

Karl Prestkvern Skari, Karl Egil Malterud and Tuva Haugli.

School of Pharmacy, Department of Pharmacognosy, The University of Oslo, Oslo, Norway.

1 INTRODUCTION

Peroxidative and free radical mediated reactions in biological systems are subjects of intensive research, mainly due to their putative role in several serious pathological conditions, e.g. atherosclerosis, some forms of cancer, and inflammatory diseases. Many medicinal plants are rich in antioxidants and radical scavengers, and this may to some extent explain their use.

Large plantain, *Plantago major* L., is a common plant in most parts of the world. A NAPRALERT search revealed that the plant is used widely for a variety of medicinal purposes. The main use of the plant is in wound healing, but it has also been used for many other complaints, such as liver disease, inflammations, fever, cancer, renal and urinary tract disease, and asthma.

Several of these conditions may involve peroxidative/radical mediated damage. Especially noteworthy is that recent research[1,2] points towards a role for antioxidants and radical scavengers in wound healing, the most common use of *P. major*. There is, however, some evidence that other constituents of the plant, such as the polysaccharides, may be involved in this activity.[3,4]

P. major is rich in phenylethanoids[5] and flavonoids.[6] These naturally occurring phenolics have repeatedly been shown to be efficient antioxidants and radical scavengers.

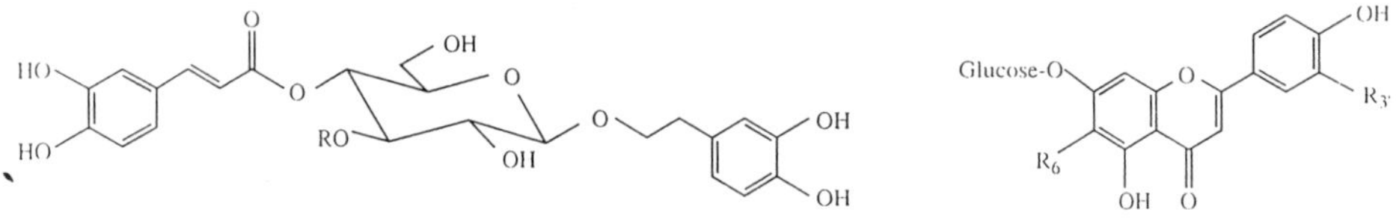

Verbascoside:	R = rhamnose	Apigenin 7-glucoside:	$R_6 = R_{3'} = H$
Plantamajoside:	R = glucose	Luteolin 7-glucoside:	$R_6 = H$, $R_{3'} = OH$
		Hispidulin 7-glucoside:	$R_6 = OCH_3$, $R_{3'} = H$
		Nepetin 7-glucoside:	$R_6 = OCH_3$, $R_{3'} = OH$

Figure 1.
Phenylethanoids (left) and some flavonoids (right) isolated from Plantago major.

Since present data on the antioxidant/radical scavenging activity of *P. major* are scant,[7,8] an investigation of the antioxidants in this plant would seem worth while. Here, we report on the radical scavenger activity of *P. major* extracts and of the radical scavenger, antioxidant and lipoxygenase inhibiting activity of the phenylethanoids verbascoside and plantamajoside.

2 EXPERIMENTAL

Plantaginis majoris herba was purchased from *Norsk Medisinaldepot*, Oslo. The plant material was extracted with 80 % aqueous ethanol, and the crude extract assayed for radical scavenging activity using the DPPH (diphenylpicrylhydrazyl) radical.[9,10] TLC was done with visualization by UV irradiation (254 nm, 366 nm) and spraying with DPPH.[11]

After removal of solvent, the crude extract was suspended in water and extracted successively with Et_2O (10-15 % of total extract), EtOAc (<5 % of total extract) and EtOAc/*n*-BuOH (10-15 % of total extract). The rest (ca 70 %) remained in the aqueous phase. All extracts were assayed for radical scavenging activity as above.

The EtOAc/*n*-BuOH extract was subjected to column chromatography over Sephadex LH20, polyamide and reverse-phase (C_{18}) Si gel, yielding spectroscopically pure (> 98 %) verbascoside and plantamajoside, ca 0.8 g of each from 10 g extract, as well as several flavonoid-containing fractions.

Radical scavenging assay on the pure substances was performed as described above, inhibition of 15-lipoxygenase catalyzed peroxidation was carried out spectrophotometrically with linoleic acid as substrate,[12] and inhibition of Fe^{2+} induced peroxidation of phospholipids by measuring formation of thiobarbituric acid-reactive substances (Method description to be published).

3 RESULTS

The crude hydroethanolic extract of *P. major* scavenged the DPPH radical (ca. 80 % scavenging during 15 minutes; c=0.25 mg/ml). The highest activity was found in the EtOAc and EtOAc/*n*-BuOH extracts (88 % and 79 % scavenged, respectively; c=0.025 mg/ml).

Column chromatography over Sephadex LH20 and NMR spectroscopy showed that the major active fractions were rich in phenylethanoids and flavonoids. Further purification yielded pure verbascoside and plantamajoside as well as a series of flavonoid-containing fractions.

In the DPPH scavenging assay, verbascoside and plantamajoside showed IC_{50} values of 13.0 μM (verbascoside) and 11.8 μM (plantamajoside) over a reaction time of 5 min. BHT (positive control) had an IC_{50} value of 302 μM. Verbascoside inhibited 15-lipoxygenase (IC_{50} 117 μM); plantamajoside was slightly more efficient (IC_{50} 96 μM), while quercetin (positive control) had an IC_{50} value of 57 μM. Iron(II) induced peroxidation of phospholipids was less influenced by verbascoside and plantamajoside (IC_{50} ca 0.7 mM for both); the positive standard BHT having an IC_{50} value of 5 μM in this system.

4 DISCUSSION

The EtOAc- and EtOAc/n-BuOH-soluble parts of a hydroethanolic extract from *P. major* show high scavenging activity towards the DPPH radical. This activity appears to be due mainly to the phenylethanoids and flavonoids in the extracts.

The phenylethanoids verbascoside and plantamajoside are good radical scavengers, possibly due to their *ortho*-dihydroxyphenyl moieties. While the radical scavenging properties of verbascoside have been reported previously,[13] plantamajoside has apparently not been investigated in this respect.

Verbascoside and plantamajoside inhibit enzymatic lipid peroxidation catalyzed by 15-lipoxygenase. To our knowledge, this effect has not been reported previously.

Likewise, they show antioxidant activity towards Fe^{2+} induced oxidation of bovine brain phospholipids, but their activity in this system is considerably less than the positive control BHT, in contrast to what is the case for DPPH scavenging. This may be due to differences in polarity; the non-polar BHT conceivably being able to enter phospholipid micelles more effectively than the polar phenylethanoids.

The flavonoid fractions from *P. major*, currently under investigation by us, are complex, and it appears that more flavonoids are present than has been reported previously.

It appears possible that the radical scavenging and antioxidant properties shown by the phenolic constituents of *P. major* may to some extent explain the medicinal properties ascribed to this plant. Further studies on this are planned.

References
1.   P. Martin. *Science.* 1997, **276**, 75.
2.   A. Shukla, A.M. Rasik and G.K. Patnaik. *Free Rad. Res.* 1997, **26**, 93.
3.   A.B. Samuelsen, B.S. Paulsen, J.K. Wold, H. Otsuka, H. Yamada and T. Espevik. *Phytother. Res.* 1995, **9**, 211.
4.   A.B. Samuelsen, B.S. Paulsen, J.K. Wold, H. Otsuka, H. Kiyohara, H. Yamada and S.H. Knutsen. *Carbohydr. Polym.* 1996, **30**, 37.
5.   H. Ravn. Caffeic acid derivatives -Chemotaxonomical, chemical, analytical and plant pathological aspects. Ph.D. thesis, Royal Danish School of Pharmacy, *Copenhagen,* 1988.
6.   S.A. Kawashty, E. Gamal-el-Din, M.F. Abdalla and N.A.M. Saleh. *Biochem. Syst. Ecol.* 1994, **22**, 729.
7.   A.M. Campos and E.A. Lissi. *Bol. Soc. Chil. Quim.* 1995, **40**, 375.
8.   S. Nishibe and M. Murai. *Foods Food Ingredients J. Jpn.* 1995, **166**, 43, *Chem. Abstr.* 1996, **124**, 82136 q.
9.   K.E. Malterud, T.L. Farbrot, A.E. Huse and R.B. Sund. *Pharmacology.* 1993, **47** (Suppl 1), 77.
10.   L. Mathiesen, K.E. Malterud and R.B. Sund. *Free Rad. Biol. Med.* 1997, **22**, 30.
11.   J. Glavind and G. Hølmer. *J. Am. Oil Chem. Soc.* 1967, **44**, 539.
12.   I.M. Lyckander and K.E. Malterud. *Acta Pharm. Nord.* 1992, **4**, 159.
13.   Q. Xiong, S. Kadota, T. Tani and T. Namba. *Biol. Pharm. Bull.* 1996, **19**, 1580.

ISOFLAVONES AND PLASMA LIPIDS IN YOUNG WOMEN - POTENTIAL EFFECTS ON HDL

Samir Samman, Philippa M. Lyons-Wall, Grace S. M. Chan and Sarah J. Smith.

Human Nutrition Unit, Department of Biochemistry, University of Sydney, NSW 2006, Australia.

1 INTRODUCTION

There is substantial structural homology between isoflavones and oestradiol: a rigid planar structure and the presence of two hydroxyl groups on the 4' carbon of ring B and carbon 7 on ring A. Some effects of oestradiol include vascular protection [1], the induction of a favourable lipoprotein profile [2] and the reduction in the propensity of low density lipoprotein (LDL) to oxidation. [3]

It has been proposed that the isoflavone components of soya beans contribute to the impact of soy protein on plasma lipids. [4] In studies with free-living subjects, isoflavone-rich soy isolate or soy flour had no effect on plasma lipids. [5,6] However, under metabolic ward conditions, pre-menopausal women who consumed an isoflavone-rich soy protein isolate had a lower plasma cholesterol concentration compared with the control period. [7] When the isoflavones were extracted from the soy protein preparation, plasma cholesterol concentrations increased relative to the control. [l8] The aim of this trial was to extend these observations by determining the effect of an isoflavone supplement, in the absence of any dietary interventions, on plasma lipids, lipoproteins and LDL oxidation in healthy premenopausal women.

2 METHODS

Healthy female volunteers aged 18-45 years were recruited to participate in the study. Subjects fulfilled the following criteria: regular menstrual cycle; no history of chronic illness; not taking any medication or oral contraceptives; not suffering from liver, bowel or gall bladder disorders; stable body weight and exercise patterns; no regular intake of soy products. The study protocol was approved by the Human Ethical Review Committee of the University of Sydney.

Subjects were randomly allocated in a single-blind manner to one of two groups, consuming either an isoflavone supplement (2 x 43 mg/d) or placebo. The active tablets contained biochanin A, 51.4 mg; formononetin, 18.6 mg; genistein, 8.6 mg; daidzein, 7.4 mg are were extracted from red clover (gift of Novogen Laboratories, Pty Ltd, North Ryde, NSW, Australia). The study commenced on day 1 of the menstrual cycle and after two complete menstrual cycles of treatment, each subject acted as their own control and crossed over to begin the second treatment for two further cycles.

Plasma total cholesterol and triacylglycerol concentrations were assayed enzymatically. Total HDL-cholesterol and HDL_3-cholesterol were assayed following precipitation[9] and HDL2-cholesterol was estimated by difference. LDL-cholesterol was determined using the Friedewald equation.[10] For the oxidation studies, LDL was isolated by ultracentrifugation[11] and subjected to oxidising conditions as described by Esterbauer and co-workers.[12] Absorbance values were converted into diene concentrations by using the extinction coefficient for conjugated dienes at 234 nm (29,500 L/mol/cm).

3 RESULTS

A carryover effect was observed for HDL_3 in the group who commenced with the isoflavones. A potential confounding factor that may have resulted in the carry-over effect of HDL_3 is the concentration of plasma oestradiol however, no significant difference was detected in plasma oestradiol at the time of sampling (130 ± 12.6 vs 125 ± 12.6 pmol/l for the placebo and treatment periods, respectively). There were no significant treatment effects on other plasma lipids, lipoproteins or oxidisability of LDL (Table 1).

Table 1.

The effect of isoflavones on plasma LDL oxidation.

	Isoflavones	Placebo
Lag phase (min)	30.4 ± 2.9	32.9 ± 3.1
Oxidation rate (nmol conjugated dienes/mg LDL protein/min)	10.2 ± 0.5	11.2 ± 1.03
Max. oxidation (nmol conjugated dienes/mg LDL protein)	361 ± 20.3	386 ± 25.4

Values are shown an mean$\pm$SEM for all parameters (n=14). Oxidation rate = change in absorbance at 234 nm/min; max. oxidation = total increase in absorbance during the assay time. Absorbance values were converted to diene concentration using the extinction coefficient 29,500 L/mol/cm.

4 DISCUSSION

There were no significant changes in plasma lipids or lipoproteins after supplementation with isoflavones over 2 menstrual cycles. However, a treatment-period interaction occurred with HDL_3 which suggests an effect of isoflavones on reverse cholesterol transport and warrants further investigation. A contributing factor in the carry-over effect of HDL_3 is the profile of plasma steroid hormones during the menstrual cycle. The possibility that HDL metabolism can be influenced by isoflavones is consistent with the known effects of oestrogen on cholesterol metabolism. Clearly, further studies are required to elucidate this potentially significant effect.

In contrast to our findings, Cassidy et al.[7] reported a 10 % reduction in plasma cholesterol concentrations when subjects consumed an isoflavone-rich diet (45 mg isoflavones/d). The 12 premenopausal women were based in a metabolic ward over a period of 4-6 months whereas subjects who participated in our study were free-living. Nevertheless, their body weights did not change and analysis of their three-day food records kept each month showed that their dietary intake remained constant during the course of the study.

Supplementation of normocholesterolaemic premenopaual women with isoflavones had no significant impact on cholesterol metabolism or LDL oxidation. Further studies are required to clarify the potential effect on HDL_3 cholesterol concentrations and the interaction with plasma steroid hormones during the menstrual cycle.

References
1.   Farhat M.Y., Lavigne M.C., Ramwell PW. *FASEB. J.* 1996, **10**, 615.
2.   Crook D., Stevenson J.C., Whitehead M.I. Estrogen replacement therapy andcardiovascular disease: effects on plasma lipid risk markers. In: Schwartz D.P., ed. Hormone Replacement Therapy. Baltimore: Williams & Wilkins. 1992, 139-170.
3.   Maziere C., Auclair M., Ronveaux M.F., Salmon S., Santus R., Maziere J.C. *Atherosclerosis*. 1991, **89**, 175.
4.   Setchell K.D.R. Naturally occurring non-steroidal estrogens of dietary origin. In McLachlan J.A. ed. Estrogens in the Environment. New York, Elsevier. 1985, 69-85.
5.   Gooderham M.J., Aldercrautz H., Ojala S.T., Wähälä K., Holub B.J. *J. Nutr*. 1996, **126**, 2000.
6.   Murkies A.L., Lombard C., Strauss B.J., Wilcox G., Burger H.G., Morton M.S. *Maturitas*. 1995, **21**, 189.
7.   Cassidy A., Bingham S., Setchell K.D.R. *Am. J. Clin. Nutr*. 1994, **60**, 333.
8.   Cassidy A., Bingham S., Setchell K. *Br. J. Nutr*. 1995, **74**, 587.
9.   Warnick G.R., Benderson J., Albers *J. Clin. Chem*. 1982, **28**, 1379.
10.   Friedewald W.T., Levy R.I., Friedrickson DS. *Clin. Chem*. 1972, **18**, 499.
11.   Gatto L.M., Samman S. *Free Rad. Biol. Med*. 1995, **19**, 517.
12.   Esterbauer H., Striegl G., Puhl H., Rotheneder M. *Free Rad. Res. Comm*.1989, **6**, 67.

FLAVONOIDS IN HERBS DETERMINED BY HPLC WITH UV-PDA AND MS DETECTION

Pia Knuthsen and Ulla Justesen.

Danish Veterinary and Food Administration, Institute of Food Chemistry and Nutrition, Mørkhøj Bygade 19, DK-2860 Søborg, Denmark.

1 INTRODUCTION

The large group of plant polyphenols is known to be natural antioxidants, and attracts particular attention because of their potential antiatherogenic and anticarcinogenic properties. The flavonoids are of major interest, as they show a relatively high prevalence in certain foods. Flavonoids are found in vegetables, fruits, herbs, and beverages as tea, wine, and fruit juices.

We have previously[1] screened plant foods on the Danish market, and combined the analytical data with intake data in order to estimate the average daily intake of flavonoids. The average intake was estimated to 28 mg per day,[2] and the major food contributors were found to be onions, tea, oranges, and orange juice. The most prevalent flavonoid aglycones were found to be quercetin, kaempferol, hesperitin, and naringenin. Quercetin is often reported as being the most abundant flavonoid in foods.[3]

Other food subjects – e.g. many herbs – have quite high flavonoid contents, but the intake is low, and their contribution to the average daily intake is therefore small. However, since they are potential sources of a high flavonoid intake, we have undertaken a study to determine the flavonol, flavone, and flavanone contents of fresh herbs available to Danish consumers.

Flavonoids occur predominantly as glycosides in nature, and are generally believed to be absorbed as aglycones by humans,[4] although recently the absorption of glycosides has been reported.[5,6] We have therefore included an investigation of glycosidic structures in this study.

2 EXPERIMENTAL

The flavonoid aglycones included in this study were the flavonols myricetin, quercetin, kaempferol, isorhamnetin, morin, and fisetin, the flavones luteolin, apigenin, chrysin, and the flavanones hesperitin, naringenin, tangeritin, and eriodictoyl.

The compounds were quantitatively determined as aglycones, obtained after hydrolysis of freeze-dried food material with hydrochloric acid in methanol according to Hertog.[7] The individual flavonoids were identified and quantified by high performance liquid chromatography (HPLC) with photodiode array (PDA) detection. The HPLC instrument consisted of a pump (Waters 616), a RP C_{18} (4.6 x 250 mm, 5 μm) column protected by a guard column, and a PDA detector (Waters 996). The mobile phase consisted of 30 % methanol and 70 % (1 % formic acid) (A), and 100 % methanol (B), with gradient elution over 50 min at a flow of 1 ml/min. UV spectra were recorded from 220-450 nm. The analysis was based on comparison of retention times and of UV spectra with commercial standards, and the limit of detection was about 0.1 mg per 100 g. We have included mass spectrometric (MS) detection in addition to the PDA detection as a verification of aglycone structures since many of the flavonoids have very similar spectra. Mass spectrometry was performed on a quadropole instrument (VG Platform II, Micromass), using atmospheric pressure chemical ionisation (APCI). The source temperature was 150 °C, the probe temperature 450 °C, the cone voltage 30 eV, and the corona discharge 1.6-1.9 kV. Negative ion mass spectra of the aglycones were aquired from 120-450 a.m.u.

Besides, LC-MS was also used to investigate the glycosidic structures. For analysis of glycosides, samples were extracted with aqueous methanol, and water (pH 2.5 with formic acid) was added to the supernatant. The mobile phase consisted of 10 % acetonitril and 90 % (1 % formic acid) (A) and 100% acetonitril (B), with gradient elution over 55 min at a flow of 1ml/min. Positive and negative ion mass spectra of the glycosides were aquired from 150-750 a.m.u.

3 RESULTS AND DISCUSSION

In Table 1 the flavonoid aglycones found in some common herbs are shown.

Table 1
Flavonoid aglycones of herbs.

Herb	Flavonoids (mg per 100g)		
Celery, leaf	Apigenin 53-114	Luteolin 9-27	
Celery, stick	Apigenin 0-2.8	Luteolin 0-0.8	
Chives	Kaempferol 2.7*	Myricetin 2.2	
Cress	Quercetin 1.8	Kaempferol 14.2	Isorhamnetin 0.6
Dill	Quercetin 7.6	Kaempferol 6.4	Isorhamnetin 2.6
Parsley	Apigenin 105-366	Kaempferol 0-2.1	
Water cress	Quercetin 0.8	Kaempferol 0.3	

One figure only indicates only one sample analysed.

The contents of apigenin was found to be very high in parsley and celery leaf, whereas the celery stick contained only low levels of apigenin. In chives, cress and dill moderate flavonoid amounts were found. It is worth noting that generally less abundant flavonoids as apigenin and luteolin are found in considerably amounts, especially since the bioavailability and biological effects of the individual flavonoids have not yet been established.

In the foods of interest, the flavonoids are mostly present as glycosides. By LC-MS determination of food extracts it is possible to investigate the composition of the herb flavonoid glycosides. E.g. it appeared that three major glycosides were present in dill as well as other glycosides in minor amounts. We suggest the three glycosides to be galacturonic or glucuronic acid of quercetin, kaempferol, and isorhamnetin, respectively.

In Fig. 1 a mass spectrum of dill extract is shown. The three molecular weights were found to be 478 for quercetin (observed as [M-H]⁻=477 in Fig. 1), 462 for kaempferol, and 492 for isorhamnetin.

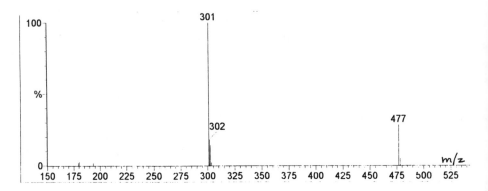

Figure 1
Mass spectrum of dill extract.

As the flavonoid amounts and composition may vary because of variety, growing conditions etc., we are at the moment analysing more samples to determine the flavonoid ranges and investigate the flavonoid glycosides of these potential high contributors of flavonoids.

References
1.   U. Justesen, P. Knuthsen and T. Leth. *J. Chromatogr.A*, 1998, **799**, 101.
2.   T. Leth and U. Justesen, in "Polyphenols in food", Cost 916, Bioactive plant cell wall components in nutrition and health, First workshop, Aberdeen, Scotland, 16-19 April 1997. Ed. R. Armadò, H. Andersson, S. Bardócz, and F. Serra. EUR 18169, p. 39 .
3.   M. Strube, L. O. Dragsted and J. C. Larsen, in "Naturally occuring antitumourigens. I. Plant phenols". Nordic Seminar- and working reports. Nordic Council of Ministers, Copenhagen. 1993, **605**. Chapter 3, p. 35.
4.   L. Griffiths, in "The flavonoids: advances in research" Ed. J. Harborne and T. Mabry. Chapman & Hall. 1982, p. 681.
5.   P. C. H. Hollman, J. H. M. de Vries, S. D. van Leeuwen, M. J. B. Mengelers and M. B. Katan. *Am. J. Clin. Nutr.* 1995, **62**. 1276.
6.   G. Papanga and C.A. Rice-Evans. *FEBS Letters.* 1997, **401**. 78.
7.   M. G. L. Hertog, P. C. H. Hollman and D. P. Venema. *J. Agric. Food Chem.* 1992, **40**, 1591.

EFFECT OF PROCESSING ON CONTENT AND ANTIOXIDANT ACTIVITY OF FLAVONOIDS IN APPLE JUICE

A.A. van der Sluis, M. Dekker and W.M.F. Jongen.

Food Science Group, Department of Food Technology and Nutritional Sciences, Wageningen Agricultural University, PO Box 8129, 6700 EV Wageningen, The Netherlands.

1 INTRODUCTION

Flavonoids are secondary plant metabolites present in fruits and vegetables. In general they occur in plants as glycosides. In epidemiological research some specific flavonoids are associated with protection against ageing diseases. This might be due to their action as antioxidant. Formation of oxygen radicals is supposed to play a key role in the development of cancer and coronary heart disease. Free radicals may attack biomolecules (lipids, proteins, DNA), which can be prevented by antioxidants. In the Dutch diet important sources of flavonoids are tea, onions and apples.[1] The most important flavonoids present in apples are flavonols (e.g. quercetin glycosides), flavanols (e.g. epicatechin) and anthocyanidins (e.g. cyanidin).[2] Both the concentration and the bioactivity of flavonoids are important in determining the health protecting capacity of a product. The effect of processing on these parameters has not yet been investigated thoroughly.

The objective of this research is to assess the effect of different processing methods on the content and antioxidant activity of health-protecting flavonoids in foods. In order to do so setting up measurement systems to quantify health claims associated with foods is required. The focus is on determining the concentration of flavonoids and antioxidant activity in apples and apple products as a function of storage and processing. Other quality factors (such as taste, colour, keeping quality etc.) will be taken into account as well. The results can be used as an input for quality modelling and optimisation of apple processing.

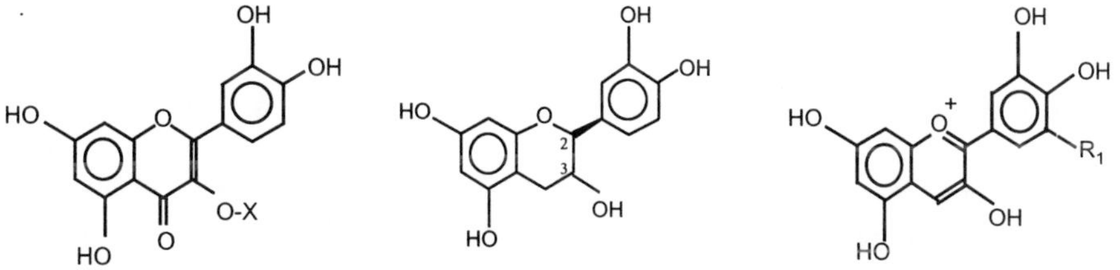

Flavonols	Flavanols	Anthocyanidins
Quercetin aglycone: X = H glycosides: X = Ar, Ga, Gl, Ru, Rh, Xy	catechin (C_3 S-configuration) epicatechin (C_3 R-configuration)	cyanidin R_1 = H delphinidin R_1 = OH

Figure 1

Basic structure of the most important flavonoids present in apples.

2 METHODS

Quercetin glycosides, catechins, chlorogenic acid and phloridzin were quantified by HPLC.[2] Antioxidant activity was assessed *in vitro*, based on a method described by Haenen.[3] This method was optimised to be able to use multititer plates and an ELISA-reader. This makes it possible to analyse large numbers of samples.[4] Lipid peroxidation was induced in male rat liver microsomes by ascorbic acid and Fe^{2+}. Inhibition of lipid peroxidation is an indication of antioxidant activity. It is described using IC_{50} (concentration at which 50 % inhibition of lipid peroxidation occurs). Three different ways to produce apple juice were compared

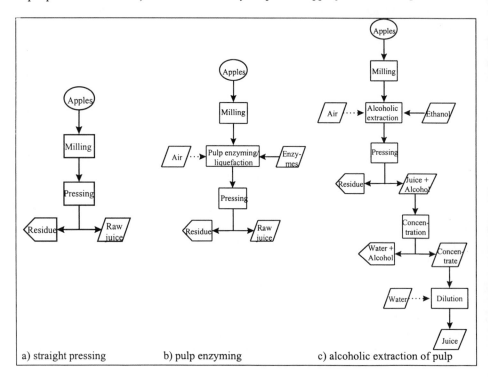

a) straight pressing   b) pulp enzyming   c) alcoholic extraction of pulp

Figure 2
Different ways to produce apple juice.

In straight pressing juice is extracted from ground apples by application of an external force. After pressing a residue remains, which is called the press cake. Often pecteolytic enzymes are used in order to obtain a higher juice yield. They are added to the ground apples (pulp) to facilitate juice extraction by degrading to some extent the pectins present in the cell walls. A specific extraction of the apple pulp is a new idea exploited in this study to obtain a juice with a higher flavonoid content. It was performed by extracting the pulp with an alcohol. The alcohol was removed from the juice by evaporation. The concentrate that is obtained during this process was diluted to 12 °Brix.

3 RESULTS

During conventional apple juice production (straight pressing or pulp enzyming) more than 80 % of the quercetin glycosides remain in the press cake, less than 10 % is found in the raw juice. For comparison a commercially available apple juice in the Netherlands (Goudappeltje) was analysed as well (Table 1). It is possible to obtain a juice with a higher amount of quercetin glycosides by extracting apple pulp with an alcohol (e.g. methanol or ethanol).

Table 1

Characteristics of differently produced apple juices.

Jonagold apples	straight pressing	pulp enzyming	Goudappeltje juice	alcoholic extraction of pulp
[Q-gly] in press cake (mg/kg)	247	222	-	94
[Q-gly] in juice (mg/kg)	2.6	<2	3.5 ± 0.2	35
IC_{50} of juice (g dry / l)	n.m.	>>12.5	8.8 ± 1.3	1.2 ± 0.1

Table 1 shows that the commercially available apple juice possesses weak antioxidant activity, the apple cultivar that was used is not known. The antioxidant activity of apple juice we produced from Jonagold apples by pulp-enzyming is lower.

Methanol-extracted apple juice produced from Jonagold apples has the highest antioxidant activity (IC_{50} is lowest). The quercetin glycosides in this juice can explain less than 10 % of the total antioxidant activity. 20 % of the total antioxidant activity can be explained if epicatechin, chlorogenic acid and phloridzin are included. The majority of the antioxidant activity is caused by other compounds like anthocyanidins and still unknown compounds.

4 CONCLUSIONS

During conventional apple juice production most of the quercetin glycosides remain in the press cake. We observed that it is possible to obtain an apple juice with an over tenfold higher amount of quercetin glycosides and an at least ten times higher antioxidant activity by extracting the apple pulp with an alcohol. This indicates the importance of the effect processing can have on the health characteristics of a product in the chain from raw material to final product.

References
1.　　M.G.L. Hertog, P.C.H. Hollman, M.B. Katan and D. Kromhout. *Nutrition and Cancer*. 1993. **20**, 21-29.
2.　　C.E. Lister, J.E. Lancaster and K.H. Sutton.. *J. Sci. Fd Agric*. 1994. **64**, 155-161.
3.　　G.R.M.M. Haenen. Thiols in oxidative stress, some implications for catecholamine toxicity. PhD Thesis, Free University, Amsterdam, 1989.
4.　　A.A. van der Sluis, M. Dekker and W. Jongen. In preparation.

PROTECTIVE EFFECT OF FLAVONOIDS ON LINOLEIC ACID HYDROPEROXIDE-INDUCED TOXICITY TO HUMAN ENDOTHELIAL CELLS

Takao Kaneko[1] and Naomichi Baba.[2]

[1]Department of Biochemistry and Isotopes, Tokyo Metropolitan Institute of Gerontology, Tokyo 173-0015, Japan. [2]Faculty of Agriculture, Okayama University, Okayama 700-0082, Japan.

1 INTRODUCTION

Lipid peroxide-induced endothelium injury has been proposed as an initiating event in the development of atherosclerosis.[1] Lipid hydroperoxides, primary products of lipid peroxidation, and their decomposed derivatives have been reported to be toxic to endothelial cells.[2] Flavonoids are polyphenol compounds found in a wide variety of plants such as vegetables, fruits, and teas, and they are expected to act as antioxidants. The protective effects of flavonoids against the cytotoxicity of linoleic acid hydroperoxide (LOOH) was examined in cultured human endothelial cells.

2 MATERIALS AND METHODS

Endothelial cells were established from human umbilical cord cords and incubated in MCDB-104 medium supplemented with fetal calf serum, growth factors, and heparin. Cells at confluency were treated for 3 h in Earle's solution containing both LOOH (50 μM) and a flavonoid (50 μM). After treatment, cell viability was determined by a coloration reaction based on the formation of a non-toxic formazan through the metabolic reduction of a tetrazolium compound by dehydrogenase activities in viable cells. Almost no cells treated with LOOH survived.

3 RESULTS AND DISCUSSION

Most flavonoids tested protected cells against LOOH-induced cytotoxicity in the case of incubation with both LOOH and flavonoids, however, their activities varied widely. Among flavones, some bearing more than three hydroxyl groups showed a protective effect against the cytotoxicity of LOOH as shown in Table 1. In particular, luteolin provided strong protection. Most flavonols with a hydroxyl group at the 3-position protected cells from injury by LOOH. Quercetin and fisetin had strong effects among flavonols. However, the substitution on rutinose with a hydroxyl group at the 3-position of quercetin, that is rutin, decreased the protective effect. Although catechins and isoflavones provided weak protection against LOOH-induced injury, flavanones had no effect. The antioxidant abilities against LOOH-induced cytotoxicity are in order of decreasing effectiveness, flavonol > flavone, isoflavone > catechin > flavanone. In contrast, when cells were incubated with flavonoids for 24 h prior to treatment with LOOH, none of the flavonoids protected cells from LOOH injury.

Table 1

Protective effect of concurrent flavonoid treatment on linoleic acid hydroperoxide-induced toxicity in human umbilical vein endothelial cells.

Class Compound	% Survival	Class Compound	% Survival'
Flavone		**Flavanone**	
Flavone	3.0 ± 2.7	Naringenin	6.8 ± 2.1
7-Hydroxyflavone	2.7 ± 0.8	Hesperetin	2.8 ± 1.4
Chrysin	3.6 ± 1.9	Hesperidin	3.5 ± 0.9
7,8-Dihydroxyflavone	29.5 ± 7.1*	Taxifolin	6.6 ± 1.1
Baicalein	2.5 ± 1.0	**Isoflavone**	
Apigenin	28.9 ± 4.5*	Daidzein	19.2 ± 5.3*
Luteolin	91.7 ± 4.7*	Biochanin A	20.4 ± 3.5*
		Catechin	
Flavonol		(+)-C	30.6 ± 13.4*
3-Hydroxyflavone	35.3 ± 4.8*	(-)-EC	8.1 ± 1.3
Galangin	61.7 ± 10.9*	(-)-EGC	1.9 ± 1.6
Kaempherol	66.2 ± 6.4*	(-)-ECG	11.4 ± 2.6
Fisetin	79.4 ± 4.5*	(-)-EGCG	5.7 ± 1.4
Morin	63.8 ± 11.0*	(-)-EGCG Glu	23.9 ± 5.3*
Quercetin	92.3 ± 5.1*	(-)-EGCG Glu$_2$	33.9 ± 6.5*
Rutin	6.3 ± 1.5	**Control**	
Quercetagetin	21.9 ± 6.5*	EtOH	2.2 ± 1.8
Myricetin	8.3 ± 2.2		

*$P<0,01$ vs. Control (EtOH)

The interaction between flavonoids and α-tocopherol was examined in this cell system. Low concentrations of flavonoids (2-10 μM) and α-tocopherol (10 μM) were used in these experiments to facilitate the observation of any interaction between flavonoids and α-tocopherol. (+)-Catechin showed a synergistic effect with α-tocopherol in protecting against LOOH-induced damage as shown in Fig. 1. Other flavonoids that protected against LOOH-induced cytotoxicity in separate treatment had an additive effect in the presence of α-tocopherol.

We previously reported that lipophilic derivatives of ascorbic acid[3] and phenolic antioxidants such as α-tocopherol and probucol[4] are effective in pretreatment, but not in concurrent treatment. In this study, flavones with more than three hydroxyl groups and flavonols were effective protectors against lipid peroxide-induced cytotoxicity in the case of incubation with both LOOH and flavonoids and had additive effects with α-tocopherol against LOOH-induced cytotoxicity. The presence of the 2,3-double bonds in conjugation with the 4-oxo function in the C ring, the *ortho*-dihydroxyl (catechol) structure in the B ring, the 3-hydroxyl group with the 4-oxo function are extremely important for the antioxidative effect on lipid hydroperoxide. These tendencies are in agreement with the results[5] of radical scavenging by polyphenol. Ingestion of vegetables and teas may be partly effective in inhibiting damage to the endothelium caused by lipid peroxides.

Flav: flavonoid, α-Toc: α-tocopherol.
*P < 0.05 vs. the sum of separate treatments with flavonoid and α-tocopherol.

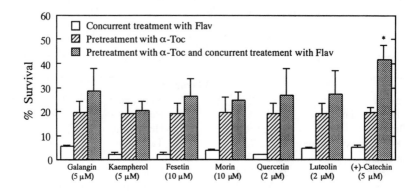

Figure 1
Effect of pretreatment with α-tocopherol (1 μM) on protection by flavonoids against linoleic acid hydroperoxide (50 μM)-induced toxicity in human umbilical vein endothelial cells.

References

1.   R. Ross and J. A. Glomset. *Nature.* 1993, **362**, 801.
2.   T. Kaneko, K. Kaji and M. Matsuo. *Chem.-Biol. Interact.* 1988, **67**, 295.
3.   T. Kaneko, K. Kaji and M. Matsuo. *Arch. Biochem. Biophys.* 1993, **304**, 176.
4.   T. Kaneko, K. Kaji and M. Matsuo. *Free Radic. Biol. Med.* 1994, **16**, 405.
5.   W. Bors, W. Heller, C. Michel and M. Saran. *Methods Enzymol.* 1990, **186**, 343.

SYNTHESIS OF ANTIOXIDANT ISOFLAVONE FATTY ACID ESTERS

Philip Lewis,[1] Kristiina Wähälä,[1] Qing-He Meng,[2] Herman Adlercreutz,[3] Matti J. Tikkanen[2].

[1]Laboratory of Organic Chemistry, Department of Chemistry, P. O. Box 55, FIN-00014, University of Helsinki, Finland. [2]Laboratory of Biochemistry, Department of Medicine, Helsinki University Hospital, Finland. [3]Institute for Preventive Medicine, Nutrition and Cancer, Folkhälsan Research Centre, Department of Clinical Chemistry, P. O. Box 60, FIN-00014 University of Helsinki, Finland.

1 INTRODUCTION

Recent publications show increasing support towards the idea that the oxidation of Low Density Lipoprotein (LDL) plays an important role in the advancement of atherosclerosis.[1-3]

It is known that circulating plasma LDL is protected against oxidation.[4] This is probably due to the high concentrations of water-soluble antioxidants present in circulating blood.[5] The oxidation of LDL, therefore, mainly occurs in the artery walls. Although LDL contains natural antioxidants such as α- and γ-tocopherol,[6] these endogenous antioxidants are thought to contribute only about 30 % to the prevention of oxidation.[7] It has hence been theoretised that additional (exogenous) lipophilic antioxidants carried in LDL could be responsible for the remaining 70 % of the antioxidant activity.

On the other hand, it has been shown that people consuming diets rich in soybean products have lower rates of cardiovascular disease.[8] This is thought to be due to soybeans being rich in phytoestrogens, especially the isoflavones genistein and daidzein. In fact *in vitro* studies have shown that both genistein and daidzein are potent antioxidants.[9-11] Unfortunately, neither genistein nor daidzein, in their aglycone form, are likely to be responsible for the missing 70 % antioxidant activity in LDL owing to their low lipophilicity. However, in view of reports suggesting that endogenous human estrogens are present in lipoproteins in the form of fatty acid esters,[12-16] it has been suggested that phytoestrogens such as genistein and daidzein could also undergo esterification *in vivo*. It is easily envisaged that such compounds could be responsible for additional antioxidant activity.

Our results show that fatty acid esters of both genistein and daidzein do show antioxidant activity and are readily incorporated into LDL particles *in vitro*.

We now present the first synthesis of isoflavone fatty acid esters. The method is both high yielding and regioselective, producing the 7-mono-, 4'-mono- and 7,4'-di-stearates and oleates of both genistein and daidzein.

2 RESULTS AND DISCUSSION

Fatty acid mono- and diesters of both genistein 1 and daidzein 2 have been synthesised selectively in high yields. Regioselectivity between the 7- and 4'- positions was achieved by utilising the difference in acidity between the 7- and 4'-hydroxyl groups. The hydroxyl group in the 7-position (ring A) is about 100 times more acidic than the B-ring 4'-hydroxyl group (CAMEO[17] calculated values).

In accordance with this difference in acidity, the selective preparation of the 7-phenolates using one equivalent of potassium t-butoxide in DMF was easily achieved. Rapid reaction with the acyl chloride (1 eq.) yielded the isoflavone fatty acid 7-mono esters in high yields, with the isoflavone 7,4'-diesters being produced as minor products. The corresponding 4'-monoesters were not produced under these conditions.

For the selective synthesis of the isoflavone 4'-monoesters, the 7,4'-diphenolates were produced by reaction of the isoflavone aglycone with 2.2-3.3 equivalents of base. The greater nucleophilicity of the 4'-phenolate compared to the 7-phenolate allowed a selective reaction with acyl chlorides (1 eq.) producing the isoflavone fatty acid 4'-monoesters. Again yields were high and the corresponding 7-mono esters were not formed under these conditions.

The use of excess base and acyl chloride afforded the isoflavone 7,4'-diesters in high yields. Strict temperature control prevented the acylation of the hydrogen-bond stabilised 5-hydroxyl group in genistein, thus allowing genistein 7,4'-diesters to be produced in preference to genistein 7,5,4'-triesters.

3 INCORPORATION INTO LDL PARTICLES

Genistein and daidzein have been calculated to have a low lipophilicity when compared to many estrogen type compounds.[18] Therefore it is reasonable to suggest that increasing the lipophilicity of the these compounds by conjugation would be beneficial to their LDL incorporation.

Our results show that the degree of incorporation of isoflavones into LDL was increased substantially upon conjugation with a fatty acid ester chain. Furthermore we have seen that both genistein and daidzein oleates were more readily incorporated than their corresponding stearate conjugates, and that dioleates are incorporated to the greatest extent. We hypothesise that the reason for the increase in oleate incorporation compared to that of the stearates is related to their 3-dimensional configuration. Initial results from molecular modelling support this theory.

1 R = H daidzein
1 R = OH genistein

References

1. Regnström J., Nilsson J., Tornvall P., Landou C., Hamsten A., *Lancet.* 1992, **339**, 1183-1186.
2. Ross R., *Nature.* 1993, **362**, 801-809.
3. Steinberg D., Parthasarthy S., Carew T. E., Khoo J.C., Witstum J.S., *N. Engl. J. Med.* 1989, **320**, 915-924.
4. Frei B., Stocker R., Ames, B.N., *Proc. Natl. Acad. Sci. USA.* 1988, **85**, 9748- 9752.
5. Frei B., *Crit. Rev. Food Sci. Nutr.* 1995, **35**, 83-98.
6. Esterbauer H., Dieber-Rothender M., Striegl G., Waeg G., *Am. J. Clin. Nutr.* 1991, **53**, 314S-321S.
7. Esterbauer H., Ramos P., *Rev. Physiol. Biochem. Pharmacol.* 1995, **127**, 31- 64.
8. Adelrcreutz H., Scand. *J. Clin. Lab. Invest.* 1990, **50**, 3-23.
9. Naim M., Gestetner B., Bondi A., Birk Y., *J. Agric. Food Chem.* 1976, **24**, 1174-1177.
10. Wei, H., Bowen, R., Cai, Q., Barnes, S., Wang, Y., *Proc. Soc. Exp. Biol.* 1995, **208**, 124-130.
11. Hodgson, J.M., Croft, K D., Puddey, I.B., Mori, T.A., Beilin, L.J., *J. Nutr. Biochem.* 1996, **7**, 664-669.
12. Janocko L., Hochberg R. B., *Science.* 1983, **222**, 1334-1336.
13. Leszczynski D.E., Schafer R.M., *Lipids.* 1990, **25**, 711-718.
14. Leszczynski D.E., Schafer R.M., *Biochem. Biophys. Acta.* 1991, **1083**, 18- 28.
15. Lavallée B., Provost P.R., Bélanger A., *Biochem. Biophys. Acta.* 1996, **1299**, 306-312.
16. Shwaery G.T., Vita, J.A., Keaney, J.F., *Circulation.* 1997, **95**, 1378-1385.
17. CAMEO (Computer Assisted Mechanistic Evaluation of Organic Reactions) by W. Jorgensen, Sterling Chemistry Laboratory, Yale University, New Haven Conneticut, 06511, USA.
18. Cunningham A.R., Klopman G., Rosenkranz H.S., *Environmental Health Perspectives.* 1997, **105**, 665-668.

Antioxidative Effects of Other Natural Antioxidants and Measurement of Oxidative Stress or Damage

CAROTENOIDS: MODES OF ACTION AND BIOAVAILABILITY OF LYCOPENE IN THE HUMAN

Helmut Sies and Wilhelm Stahl.

Institut für Physiologische Chemie I, Heinrich-Heine-Universität Düsseldorf, P.O.Box 10 10 07, D-40001 Düsseldorf, Germany.

1 INTRODUCTION

Nutrition plays an important role as a therapeutic modality in the treatment of many diseases, and an appropriate nutrient selection contributes to the prevention of disorders such as hyperlipidemia, hypertension, or vitamin deficiency. Epidemiological studies suggest a close relationship between the intake of specific dietary factors e.g. vitamins and antioxidants and the diminished risk to suffer from cancer, coronary heart diseases, or cataract.[1]

In this context carotenoids are among the compounds which attracted attention. Carotenoids are found in numerous fruits and vegetables.

Important dietary sources of carotenoids for the human are green leafy and orange to red vegetables as well as various fruits, including oranges, tangerines, or peaches.[2] ß-Carotene, α-carotene, cryptoxanthin, lutein, zeaxanthin and lycopene are the major dietary carotenoids[3] mainly provided by plant-based food. An important source for ß-carotene are carrots but spinach, broccoli or green and red peppers also contribute to human supply.

Fruits also contain considerable amounts of the provitamin A carotenoids ß-carotene and ß-cryptoxanthin; another major carotenoid which is found in fruits is lutein. Hydroxylated carotenoids in fruits and vegetables may be present either as parent carotenols or esterified with various fatty acids.[4] The carotenoid pattern of papaya and other fruits is dominated by carotenoid esters while only low amounts of free xanthophylls are found in some fruits. For instance, tangerines contain high amounts of ß-cryptoxanthin mainly present in esterified form.[5] The major carotenoid esters in tangerines were identified as ß-cryptoxanthin laurate, myristate and palmitate. Additionally, small amounts of lutein and zeaxanthin esters are detectable.

Carotenoids are also found in various kinds of seafood including lobster and salmon.[6]

2 LYCOPENE: OCCURRENCE AND STRUCTURE

While ß-carotene and lutein are found in many different kinds of fruits and vegetables, only a few products contain important other carotenoids such as lycopene. About 90 % of dietary lycopene in the US derives from tomatoes and tomato products[7] with more than 50 % from this processed food. Further sources of lycopene are watermelon, guava, rosehips and pink grapefruit.[2] Tomatoes contain about 30 mg lycopene/kg raw fruit;[8] even higher amounts are found in some of the tomato products, e.g. in tomato juice up to 150 mg lycopene/L or in tomato ketchup about 100 mg/kg. However, the lycopene content in concentrated tomato products is generally lower than expected, resulting from losses during food processing.

All-trans lycopene is the most prominent geometrical isomer in fresh tomatoes but trans-to-*cis* isomerization occurs during tomato processing and storage. In various tomato-based foods the *all-trans* isomer contributed 35-96 % to total lycopene.[9]

The chemical and biochemical activities of carotenoids are related to their unique structure, an extended system of conjugated double bonds.[10] Carotenoids are tetraterpenes formed by tail-to-tail linkage of two C-20 units, and in many carotenoids the end-groups are modified into five- or six-membered rings giving monocyclic or dicyclic compounds. Lycopene is an acyclic carotenoid which contains 11 conjugated double bonds arranged linearly in the *all-trans* form. In addition to the all-trans configuration several mono- and poly-*cis* isomeric forms may be formed. Lycopene belongs to the subgroup of carotenes consisting only of hydrogen and carbon atoms.

3 LYCOPENE: MODES OF ACTION

A number of biological effects have been attributed to carotenoids including antioxidant activity,[11] influences on the immune system,[12] control of cell growth and differentiation ,[13,14] and stimulatory effects on gap junctional communication.[15] These effects are thought to be relevant with respect to their protective properties.

Reactive oxygen species are formed in physiological processes and are capable of damaging biologically relevant molecules such as DNA, proteins, lipids or carbohydrates.[16,17] A variety of antioxidant defense systems including carotenoids are scavengers of reactive oxygen species; they are summarized under the term antioxidants.

Reactive oxygen species which are efficiently scavenged by carotenoids are $^1O_2$ and peroxyl radicals.[18] Physical quenching is a major pathway operative with respect to the deactivation of $^1O_2$.[19] Physical quenching occurs by energy transfer from the excited oxygen species to the carotenoid yielding a triplet excited carotenoid. The energy of the excited carotenoid is dissipated through vibrational interactions with the solvent to recover ground state carotenoid. The carotenoid remains intact in this process and might undergo further cycles of deactivation. The quenching of singlet oxygen by ß-carotene and other biological carotenoids occurs with rate constants approaching diffusion control; lycopene exhibited the highest rate constant.[20] Thus, lycopene is the most efficient $^1O_2$ quencher among the biologically occurring carotenoids.

Carotenoids are also able to inhibit free radical reactions.[21,22] At low concentrations and at low partial pressures of oxygen, ß-carotene was found to inhibit the oxidation of model compounds initiated by peroxyl radicals.[23] Most research has focused on the effects of β-carotene alone but also interactions with lipophilic antioxidant α-tocopherol have been reported.[24,25]

The antioxidant activity of single carotenoids and of mixtures of these compounds was studied in multilamellar liposomes.[26] Lipid peroxidation was initiated with 2,2'-azobis (2,4-dimethylvaleronitrile) (AMVN) and followed by the detection of thiobarbituric acid reactive compounds (TBARS). Among the single components lycopene and α-tocopherol most efficiently inhibited TBARS formation, with the ranking: lycopene > α-tocopherol > α-carotene > ß-cryptoxanthin > zeaxanthin = ß-carotene > lutein.

However, mixtures of carotenoids inhibited the formation of TBARS more effectively than single components when they were used at the same molar level. Such a *synergistic* antioxidant effect was most pronounced when either lycopene or lutein were present in the mixture.[26] Liposomes containing a combination of lycopene and lutein were most efficiently protected against lipid peroxidation. Suppression of lipid peroxidation by various carotenoids and α-tocopherol may be influenced by the site and the rate of radical production and/or be related to the orientation of carotenoids in lipid membranes.

Gap junctions are cell-to-cell channels providing a direct pathway for the diffusion of small molecules (<1,000 Da) between the cytoplasm of adjacent cells.[27] In additon to other functions, it has been postulated that intercellular communication *via* gap junctions plays a role in the regulation of tumor cell growth.[28]

Tumor promoters such as TPA and DDT inhibit GJC;[29] other compounds including retinoids and carotenoids[15] stimulate this communication pathway. The stimulatory effects of carotenoids on GJC were associated with their proposed cancer-preventive activities. The effects of carotenoids are not limited to pro-vitamin A compounds; retinoids which are formed from the parent compound are involved in the regulatory process.[30] In cell culture carotenoids stimulate GJC at levels of 1-10 μM; little is known about the situation *in vivo* .[31]

Depending on the dose, different effects of carotenoids on GJC *in vivo* have been reported recently.[32] Intercellular communication was measured in rat liver after application of either α-carotene, ß-carotene or lycopene in doses of 0.5, 5 and 50 mg/kg body wt/d for 5 days. No effects were observed with the lowest dose, whereas GJC was significantly increased at the 5 mg level. ß-Carotene and lycopene were more effective than α-carotene. Interestingly, GJC was inhibited when the animals received carotenoids at 50 mg/kg. In comparison to the control (100 %) GJC was diminished by ß-carotene and α-carotene to about 30-35 %; the effect was even more pronounced with lycopene (20 %). These data indicate that dose levels of carotenoids may be important and that different doses of carotenoids may have differential effects on this pathway of cellular communication. Since inhibiton of GJC has been associated with tumor promotion, high doses of carotenoids may have a negative impact on carcinogenesis, especially when initiated cells are present in an organism.

4 LYCOPENE: BIOAVAILABILITY

The uptake of carotenoids from the diet is influenced by several factors such as dietary fat, presence of fiber, or food processing.[33,34] As vitamin E, carotenoids are absorbed via the lymphatic pathway, requiring the formation of micelles from fat and bile acids. Thus, the intestinal uptake of these compounds is improved by the additional consumption of oil, margarine or butter. The particle size of uncooked food also influences carotenoid uptake. The bioavailabilty of carotenoids from pureed or finely chopped vegetables is considerably higher as compared to whole or sliced raw vegetables.

Lycopene bioavailability from a single dose of fresh tomatoes or tomato paste (23 mg lycopene), ingested together with 15 g corn oil, was compared by analyzing carotenoid concentrations in the human chylomicron fraction.[35] The lycopene isomer pattern was the same in both fresh tomatoes and tomato paste. Ingestion of tomato paste was found to yield 2.5-fold higher total and *all-trans* lycopene peak concentrations and 3.8-fold higher area under curve (AUC) responses than ingestion of fresh tomatoes. The same was calculated for lycopene *cis*-isomers, but only the AUC response for the *cis*-isomers was statistically significantly higher after and ß-carotene response. The data demonstrate that the bioavailability of lycopene is higher from tomato paste than from fresh tomatoes.

No increase in lycopene serum levels was observed after the single intake of large amounts of tomato juice. After ingestion of 700 g[36] of tomato juice corresponding to a single dose of about 80 mg of lycopene no change in serum lycopene levels was observed. In contrast, lycopene plasma levels increased significantly in human serum when processed juice was consumed. Boiling for 1h in the presence of 1% corn oil increased the bioavailability of lycopene from tomato juice significantly.[36]

Further research on the bioavailability of carotenoids will provide additional insight into the contribution of these micronutrients to the preventive effects of fruits and vegatable consumption.

Acknowledgment

Our research was supported by the Institut Danone für Ernährung (Rosenheim, Germany) and European Community (FAIR: CT 97-3100).

References

1.   W.C. Willett. *Science.* 1994, **264,** 532.
2.   A.R. Mangels, J.M. Holden, G.R. Beecher, M.R. Forman and E. Lanza. *J. Am. Diet. Assoc.* 1993, **93,** 284.
3.   W. Stahl and H. Sies. *Arch. Biochem. Biophys.* 1996, **336,** 1.
4.   F. Khachik and G.R. Beecher. *J. Chromatogr.* 1988, **449,** 119.
5.   T. Wingerath, W. Stahl, D. Kirsch, R. Kaufmann and H. Sies. *J. Agric. Food Chem.* 1996, **44,** 2006.
6.   S. Liaaen-Jensen. *New J. Chem.* 1990, **14,** 747.

7.   J.K. Chug-Ahuja, J.M. Holden, M.R. Forman, A.R. Mangels, G.R. Beecher and E. Lanza. *J. Am. Diet Assoc.* 1993, **93**, 318.
8.   D.J. Hart and J. Scott. *Food Chem.* 1995, **54**, 101.
9.   J. Schierle, W. Bretzel, I. Bühler, N. Faccin, D. Hess, K. Steiner and W. Schüep. *Food Chem.* 1997, **59**, 459.
10.   J.A. Olson and N.I. Krinsky. *FASEB J.* 1995, **9**, 1547.
11.   H. Sies and W. Stahl. *Am. J. Clin. Nutr.* 1995, **62**, 1315S.
12.   D.A. Hughes, A.J.A. Wright, P.M. Finglas, A.C.J. Peerless, A.L. Bailey, S.B. Astley, A.C. Pinder and S. Southon. *J. Lab. Clin. Med.* 1997, **129**, 309.
13.   M. Murakoshi, J. Takayasu, O. Kimura, E. Kohmura, H. Nishino, A. Iwashima, J. Okuzumi, T. Sakai, T. Sugimoto, J. Imanashi and R. Iwasaki. *J. Natl. Cancer Inst.* 1989, **81**, 1649.
14.   Sharoni Y. and Levi J. in: Natural antioxidants and food quality in atherosclerosis and cancer prevention, eds. J.T. Kumpulainen and J. Salonen (The Royal Society of Chemistry, Cambridge, 1996. p. 378.
15.   J.S. Bertram and H. Bortkiewicz. *Am. J. Clin. Nutr.* 1995, **62**, 1327S.
16.   H. Sies. *Eur. J. Biochem.* 1993, **215**, 213.
17.   B. Halliwell. *Annu. Rev. Nutr.* 1996, **16**, 33.
18.   P. Palozza and N.I. Krinsky. *Meth. Enzymol.* 1992, **213**, 403.
19.   T.G. Truscott. *J. Photochem. Photobiol. B: Biol.* 1990, **6**, 359.
20.   P. Di Mascio, S. Kaiser and H. Sies. *Arch. Biochem. Biophys.* 1989, **274**, 532.
21.   G.W. Burton and K.U. Ingold. *Science.* 1984, **224**, 569.
22.   C.A. Rice-Evans, J. Sampson, P.M. Bramley and D.E. Holloway. *Free Rad. Res.* 1997, **26**, 381.
23.   T.A. Kennedy and D.C. Liebler. *J. Biol. Chem.* 1992, **267**, 4658.
24.   P. Palozza, S. Moualla and N.I. Krinsky. *Free Radic. Biol. Med.* 1992, **13**, 127.
25.   F. Böhm, R. Edge, E.J. Land, D.J. McGarvey and T.G. Truscott. *J. Am. Chem. Soc.* 1997, **119**, 621.
26.   W. Stahl, A. Junghans, B. de Boer, E. Driomina, K. Briviba and H. Sies. *FEBS Lett.* 1998, **427**, 305.
27.   D.A. Goodenough, J.A. Goliger and D.L. Paul. *Annu. Rev. Biochem.* 1996, **65**, 475.
28.   H. Yamasaki. *Toxicol. Lett.* 1995, **77**, 55.
29.   V.A. Krutovskikh, M. Mesnil, G. Mazzoleni and H. Yamasaki. *Lab. Invest.* 1995, **72**, 571.
30.   M. Hanusch, W. Stahl, W.A. Schulz and H. Sies. *Arch. Biochem. Biophys.* 1995, **317**, 423.
31.   W. Stahl, S. Nicolai, K. Briviba, M. Hanusch, G. Broszeit, M. Peters, H.-D. Martin and H. Sies. *Carcinogenesis.* 1997, **18**, 89.
32.   V.A. Krutovskikh, M. Asamoto, N. Takasuka, M. Murakoshi, H. Nishino and H. Tsuda. *Jpn. J. Cancer Res.* 1997, **88**, 1121.
33.   J.W. Erdman, T.L. Bierer and E.T. Gugger. *Ann. NY. Acad. Sci.* 1993, **691**, 76.
34.   R.S. Parker. *FASEB J.* 1996, **10**, 542.
35.   C. Gärtner, W. Stahl and H. Sies. *Am. J. Clin. Nutr.* 1997, **66**, 116.
36.   W. Stahl and H. Sies. *J. Nutr.* 1992. **122**, 2161.

ANTIATHEROGENIC EFFECT OF LYCOPENE AND β-CAROTENE: INHIBITION OF LDL OXIDATION, AND SUPPRESSION OF CELLULAR CHOLESTEROL SYNTHESIS

Bianca Fuhrman, Avishay Elis and Michael Aviram.

Lipid Research Laboratory, Technion Faculty of Medicine, The Rappaport Family Institute for Research in the Medical Sciences and Rambam Medical Center, Haifa, Israel.

1 INTRODUCTION

Coronary heart disease develops as a result of risk factors, which promote atherogenic traits. Hypercholesterolemia is by far recognized as a major risk for atherosclerosis.[1] Oxidative modification of LDL is also believed to play a causal role in atherosclerosis.[1-4] Strategies directed to reduce plasma cholesterol levels, or to inhibit LDL oxidation, are of major clinical importance in arresting accelerated development of atherosclerosis.

Cholesterol homeostasis in cells and in plasma is maintained through feedback regulation on the cholesterol biosynthetic pathway and on the LDL-receptor activity.[5] Competitive inhibition by synthetic agents, such as statins, of the enzyme 3-hydroxy-3-methylglutaryl coenzyme A (HMGCoA) reductase, the rate limiting enzyme in the mevalonate biosynthetic pathway of cholesterol, coordinates with increased LDL-receptor activity, resulting in reduced plasma levels of cholesterol.[6] Cholesterol, which is an and product of the mevalonate biosynthetic pathway in animal cells, negatively regulates the activity of HMGCoA reductase.

ß-Carotene and its precursor lycopene, are polyisoprenoids synthesized in plants from mevalonate, and they share similar initial synthetic pathway with cholesterol. Carotenoids are transported in the human body in circulating lipoproteins, and they possess some antioxidant characteristics.[7-9] Epidemiological studies have demonstrated an association between carotenoid dietary consumption and reduced cardiovascular mortality.[10] It was proposed that carotenoids attenuate atherosclerosis development due to their protection of LDL from oxidation. However, carotenoids may inhibit atherosclerosis also via inhibition of cholesterol synthesis, since the activity of HMGCoA-reductase in animal cells was shown to be sensitive to negative regulation both by sterols and non-sterol products.

2 CAROTENOIDS AND LDL OXIDATION

Data on the ability of ß-carotene supplementation *in vitro* or *in vivo* to protect LDL from oxidation are conflicting. We[11-13] and others[14,15] have demonstrated an inhibitory effect of ß-carotene supplementation on LDL oxidation, whereas several others studies did not find such an effect.[16,17] No data are available on the ability of lycopene to directly protect LDL against oxidation.

We have examined the antioxidative effect against LDL oxidation of lycopene and of ß-carotene, in relation to LDL composition. Twelve diverse samples of LDL (100 μg of protein/ml) were supplemented *in vitro* with 3 μM of lycopene or 3 μM of ß-carotene for 30 min. at 37 °C, followed by a subsequent incubation with 5 μM $CuSO_4$.

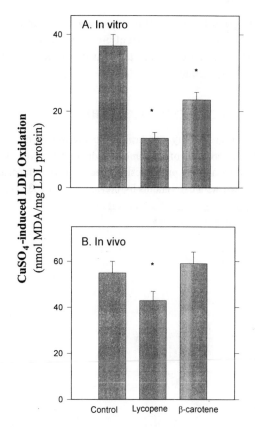

Figure 1 (A and B)
Antioxidative effect of lycopene and β-carotene against LDL oxidation.

Fig. 1A summarizes the results obtained with 7 out of 12 LDL samples, which responded to carotenoids supplementation by reduced susceptibility to copper ions-induced oxidation. Lycopene inhibited $CuSO_4$-induced LDL oxidation by 65 % in comparison to only 38 % inhibition by ß-carotene. In an attempt to find out an association between LDL composition and the response to oxidation following its enrichment with carotenoids, we have analyzed the composition of all the LDL samples. Our results demonstrated that the vitamin E content in those LDL samples, whose oxidation was inhibited by the carotenoids supplementation, was significantly ($p<0.01$) higher than that found in the "non-responder" LDLs. A direct analysis of the effect of vitamin E in combination with the carotenoids on LDL oxidizability revealed a synergistic interaction between vitamin E and lycopene and ß-carotene in the inhibition of LDL oxidation.

We have also studied the antioxidative effects of lycopene and of ß-carotene against LDL oxidation *in vivo*, in the atherogenic apolipoprotein E deficient (E° mice. Supplementation of E° mice with 50 μg of lycopene or of ß-carotene/mouse/week, for a period of 6 weeks resulted in a 22 % inhibition in $CuSO_4$-induced oxidation of LDL derived from the lycopene-supplemented mice, whereas no significant effect was observed after ß-carotene consumption (Fig. 1B).

Our results indicate a protective effect of lycopene against oxidative modification of LDL, *in vitro* as well as *in vivo*, which exceeds that of ß-carotene. However, this effect was selective to LDLs with higher vitamin E content, and was enhanced when the carotenoids were present in combination with vitamin E.

3 CAROTENOIDS AND CHOLESTEROL SYNTHESIS

Based on the concept that excess of any end product in the mevalonate pathway could regulate the production of other products of the same biosynthetic pathway, we have studied the effect of lycopene and of ß-carotene on cholesterol metabolism in macrophages. J-774 A.1 macrophages were enriched with carotenoids by pre-incubation at 37 °C with lycopene or with ß-carotene.

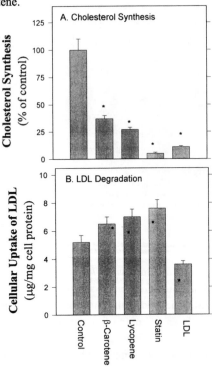

Figure 2 (A and B)
The effect of carotenoids on cellular cholesterol synthesis and on cellular uptake of LDL.

The effect of carotenoids on cellular cholesterol synthesis from ^3H-acetate was studied and compared to the inhibition of cellular cholesterol synthesis by cholesterol or by statins. ^{3}H-acetate incorporation into newly synthesized cholesterol was inhibited by 88 % and 98 % by preincubation of the cells with 60 µg of LDL-cholesterol/ml or with 1 µg of fluvastatin/ml, respectively , (Fig. 2A). Similarly, supplementation of the cells with 10 µM of lycopene or ß-carotene reduced cellular cholesterol synthesis by 73 % or by 63 %, respectively (Fig. 2A). Incubation of J-774 A.1 macrophages with 10 µM of ß-carotene or lycopene did not, however, suppressed cellular cholesterol synthesis from radio-labeled mevalonate, suggesting that these carotenoids regulate macrophage cholesterol synthesis by inhibition of the cellular HMGCoA-reductase activity, similar to the effect of statins and of cholesterol.

Since HMGCoA-reductase feedback coordinates with the LDL-receptor activity as a function of the cellular cholesterol levels, we further studied the effect of lycopene and of ß-carotene on macrophage degradation of LDL, in comparison to the effect induced by statins or by cholesterol. Fig. 2B demonstrates that macrophage enrichment with lycopene or with ß-carotene resulted in 34 % or 25 % increased cellular degradation, similarly to the effect of 1 µg/ml fluvastatin, but in contrast to a 40 % decreased cellular LDL degradation after incubation of the cells with 50 µg of LDL-cholesterol/ml.

Our results imply that cell enrichment with carotenoids can increase the removal of LDL from plasma, secondary to their suppression of cellular cholesterol synthesis. This was further evidenced by a 14 % reduction in plasma LDL cholesterol levels in 5 men that were administered 60 mg/d of tomato's lycopene for a period of 3 months.

References

1. Steinberg D., Parthasarathy S., Carew T.E., Khoo J.C. and Witztum J.L. Beyond cholesterol: modifications of low-density lipoprotein that increase its atherogenicity. *N. Engl. J. Med.* 1989, **320**, 915-924.
2. Witztum J.L. The oxidation hypothesis of atherosclerosis. *Lancet.* 1994, **344**, 793.
3. Aviram M. Oxidized low density lipoprotein (Ox-LDL) interaction with macrophages in atherosclerosis and the antiatherogenicity of antioxidants. *Europ. J. Clin. Chem. Clin. Biochem.* 1996, **45**, 135-144.
4. Aviram M. Modified forms of low density lipoprotein and atherogenesis. *Atherosclerosis.* 1993, **98**, 1-9.
5. Goldstein J.L., and Brown M.S. Regulation of the mevalonate pathway. *Nature.* 1990, **343**, 425-430.
6. Bilheimer D.W., Grundy S.M., Brown M.S., Goldstein J.L. Mevinolin and colestipol stimulate receptor-mediated clearance of low density lipoprotein from plasma in familial hypercholesterolemia heterozygotes. *Proc. Nat. Acad. Sci. USA.* 1983, **80**, 4124-4128.
7. Burton G.W. Antioxidant action of carotenoids. *J. Nutr.* 1989, **119**, 109-111.
8. Sies W., Stahl W. and Sundquist A.R. Antioxidant functions of vitamins; vitamin E and C, β-carotene, and other carotenoids. In: Beyond deficiency: new views on the function and health effects of vitamins; Savberlich HE and Machlin LY, eds; *Ann. NY. Acad Sci.* 1992, **669**, 7-20.

9.   Krinsky N.I. Antioxidant functions of carotenoids. *Free Rad. Biol. Med.* 1989, **7**, 617-635.

10. Kohlmeier L. and Hasting S.B. Epidemiologic evidence of a role of carotenoids in cardiovascular disease prevention. *Am. J. Clin. Nutr.* 1995, **62**, 137S-146S.

11. Lavy A., Ben-Amotz A. and Aviram M. Preferential inhibition of LDL oxidation by the all-trans isomer of ß-carotene in comparison to the 9- cis -ß-carotene. *Europ. J. Clin. Chem. Clin. Biochem.* 1993, **31**, 83-90.

12. ، Levy Y., Ben-Amotz A., and Aviram M. Effect of dietary supplementation of different ß-carotene isomers on lipoprotein oxidative modification. *J. Nutr. Med.* 1995, **5**, 13-22.

13. Levy Y., Kaplan M., Ben-Amotz A. and Aviram M. The effect of dietary supplementation of ß-carotene on human monocyte-macrophage-mediated oxidation of low density lipoprotein. *Israel J. Med. Sci.* 1996, **32**(6), 473- 478.

14. Jialal I., Norkus E.P., Cristor L. and Grundy SM ß-Carotene inhibits the oxidative modification of low-density lipoprotein. *Biochim. Biophys. Acta.* 1991, 134-138.

15. Jialal I. and Grundy S.M. Effect of combined supplementation with α-tocopherol, ascorbate and β-carotene on low-density lipoprotein oxidation. *Circulation.* 1993, **88**, 2780-2786.

16. Reaven P.D., Khouw A., Beltz W.F., Parthasarathy S. and Witztum J.L. Effect of dietary antioxidant combinations in humans: protection of LDL by vitamin E but not by ß-carotene. *Arteriosclerosis and Thombosis.*1993, **13**, 590-600.

17. Gaziano J.M., Hatta A., Flynn M., Johnson E.J., Krinsky N.I., Ridker P.M., Hennekes C.H., and Frei B. Supplementation with ß-carotene *in vivo* and *in vitro* does not inhibit low density lipoprotein oxidation. *Atherosclerosis.* 1995 **112**, 187-195.

β-CAROTENE NORMALIZES OXIDATIVE DAMAGE IN CAROTENOID-DEPLETED WOMEN

Betty Jane Burri,[1] Andrew Clifford,[2] and Zisca Dixon.[3]

[1]Western Human Nutrition Research Center (WHNRC), United States Department of Agriculture, San Francisco, California, USA. [2]Nutrition Department, University of California, Davis, USA. [3]Dietetics and Nutrition Department, Florida International University, Miami, FL, USA.

1 INTRODUCTION

Carotenoids are fat-soluble pigments found in fruits and vegetables. Carotenoids such as β-carotene and lycopene form part of the antioxidant defense system. The antioxidant defense system is crucial to human health, because oxidative damage has been implicated in the etiology of cancer, atherosclerosis, immunological disorders, and degenerative diseases such as cataract.[1-3] However, we do not know whether carotenoids are essential components of the antioxidant defense system. The evidence on the importance of their role in the antioxidant defense system is sparse and conflicting.

One reason for this conflicting information may be that most studies have been done by feeding carotenoid supplements to already well-fed individuals.[4-6] When these studies show no effects, it may mean that the carotenoids have no important function in oxidative damage prevention. On the other hand, it may mean that they have very useful functions that we do not observe because the baseline carotenoid concentrations are too high: the system is already saturated. Carotenoid depletion studies may provide a clearer picture of whether, and when, carotenoids are important antioxidants. There have only been a few studies of the effects of carotenoid depletion on oxidative damage. All these studies have been restricted to healthy adults. However, they consistently show deleterious increases in oxidative damage during depletion.[7-12] This report uses the information from one such carotenoid depletion study to derive preliminary estimates of the effective range of carotenoids in the antioxidant defense system of healthy adult women.

2 SUBJECTS AND METHODS

Details of the study protocol have been reported elsewhere.[8,9] Women aged 22 to 41 lived on the WHNRC metabolic unit for 120 days during the winter of 1994. They participated in a double-blind, placebo controlled trial of carotenoid depletion. They ate a six-day rotational diet of natural foods that were low in carotenoids, but provided adequate nutrients and energy. We supplemented their diet with β-carotene (Dry Carotene Beadlets, lot 011605, Roche Vitamins Inc., Nutley NJ); and with a mixed carotenoid supplement the last 20 days of the experiment (GNLD, San Jose, CA).

We measured serum retinol and carotenoid concentrations by reversed-phase liquid chromatography on a C18 column with diode array detection at 330 nm (vitamin A) and 452 nm (carotenoids). Malondialdehyde-thiobarbituric acid (MDA-TBA) concentrations were measured colorimetrically by reversed-phase liquid chromatography at 532 nm.[8] Reactive carbonyls (formaldehyde, acetaldehyde, acetone, propanol, butanone, butanal, pentanal, and hexanal) hydrazone derivatives generated by oxidatively-challenged low density lipoproteins were analyzed by reversed-phase chromatography at 360 nm.[9]

Results were analyzed by fitting a previously described random regression coefficient model[9] to the data. The saturation point, the point at which adding more β-carotene would not decrease oxidative damage, was extrapolated similarly to a low-dose extrapolation in dose-response analysis. Assumptions for calculating the saturation point were that extrapolation beyond the domain of the experimental data was valid, and that the population had a normal distribution.[9]

3 RESULTS

All serum carotenoids decreased significantly during carotenoid depletion. Serum concentrations of vitamin A did not change. Indices of oxidative damage (MDA-TBA, reactive carbonyls) increased significantly with carotenoid depletion, then normalized. MDA-TBA concentrations were maintained at baseline levels in the control group of women, who were supplemented with only 0.5 mg per day of β–carotene.[8] MDA-TBA concentrations were maintained, or normalized, in women with serum concentrations of β-carotene of 0.33 ± 0.08 μmol/L.

Reactive carbonyls increased during carotenoid depletion, then decreased during carotenoid supplementation. This decrease was measured at the end of the study, after the mixed carotenoid supplement had been provided to our volunteers. However, the major carotenoid in the supplementation mixture was β-carotene, and the concentrations of lycopene were low. We calculated that the maximal protection of LDL from oxidation occurred at serum concentrations of 2.3 ± 1.8 μmol/L of β-carotene.[9]

4 DISCUSSION

Carotenoid depletion was associated with deleterious changes in several indices of oxidative damage in platelets, red blood cells, and low density lipoproteins.[7-12] Oxidative damage to low density lipoproteins is suspected to be a major initiator of arteriosclerosis. Thus, carotenoid depletion may increase the risk of arteriosclerosis.

However, our work suggests that maximal protection from carotenoids occurs at low, physiological concentrations. We calculated that saturation and maximal protection of LDL occurred at serum concentrations of 2.3 ± 1.8 μmol/L of β-carotene, even when lycopene concentrations were low.[9] We do not know what effects higher total carotenoid or lycopene concentrations would have had on our maximal saturation calculation, but it is most likely that higher concentrations of other carotenoids would have decreased (rather than increased) this amount. We also do not know what level of long-term carotenoid intakes would result in serum concentrations of 2.3 µmol/L β-carotene. However, several studies have attained those concentrations with supplements of 15 mg/d β-carotene.[7]

Serum MDA-TBA concentrations normalized in all women with even less β-carotene, 0.5 mg per day (a total of 0.93 µmol/day β-carotene).[8] MDA-TBA concentrations were normalized in the experimental group, and maintained in the control group, by serum concentrations of 0.33±0.08 µmol/L β-carotene. Furthermore, carotenoid supplementation normalized oxidative damage in the depleted subjects, but did not improve these indices significantly over baseline values.[8,9]

Median intakes of β-carotene in the United States are 1 mg per day, and mean intakes are approximately 2 mg per day.[7,13] Total carotenoid intakes from typical Western diets would provide about double that, or 2-4 mg per day of total carotenoids.[7,13] These dietary intakes would be sufficient to maintain MDA-TBA levels in our healthy adult women. They would not provide maximal protection of oxidatively challenged LDL for all women. Nevertheless, they would be enough to provide optimal protection to some, and major protection for most of these women.

This research may explain why most epidemiological studies have shown strong correlations between β-carotene intakes and decreased mortality, but most clinical trials of β-carotene supplementation have shown no beneficial effects. Most of the supplementation studies gave β-carotene to healthy, well-fed people. They did not attempt to target people with low carotenoid intakes or status. Furthermore, many participants in the carotenoid trials attained β-carotene and total carotenoid concentrations far higher than those we calculate would provide maximal protection.[7] Trials giving smaller carotenoid supplements to targeted groups may show decreased oxidative damage consistent with decreased risks of arteriosclerosis.

References

1.   G. van Poppel and R. A. Goldbohm. *Amer. J. Clin. Nutr.* 1995, **62,** 1393 S.
2.   L. Kohlmeier and S. Hastings. *Amer. J. Clin. Nutr.* 1995, **62,** 1370 S.
3.   D. M. Snodderly. *Amer. J. Clin. Nutr.* 1995, **62,** 1448 S.
4.   α-tocopherol, β-carotene cancer preventionstudy group. *N. E. J. M.* 1994, **330,** 1029.
5.   G. S. Omenn, G. E. Goodman, M. D. Thornquist, et al. *N. E. J. M.* 1996, **334,** 1150.
6.   C. H. Hennekens, J. E. Buring, J. E. Manson, et al. *N. E. J. M.* 1996, **334,** 1145.
7.   B. J. Burri. *Nutr. Res.* 1997, **17,** 547.
8.   Z. R. Dixon, F.-S. Shie, B. Warden B et. al.. *Amer. J. Coll. Nutr.* 1998, **17,** 54.
9.   Y. Lin, B. J. Burri, T. R. Neidlinger, et. al. *Amer. J. Clin. Nutr.* 1998, **67,** 837.
10.   Z. R. Dixon, B. J. Burri, A. J. Clifford, et. al. *Free Rad. Med. Biol.* 1994, **17,** 537.
11.   S. Omaye, B. J. Burri, M. Swendseid, et. al. *J. Amer. Coll. Nutr.* 1996, **15,** 469.
12.   S. Mobarhan, P. Bowen, B. Andersen, et. al. *Nutr. Cancer.* 1990, **14,** 195.
13.   K. Alaimo, M. A. McDowell. R. R. Briefel, et al. Dietary intakes of vitamins, minerals, and fiber of persons ages 2 months and over in the United States. Advanced data from vital and health statistics, no. 258. U.S. dept. Health human services. Hyattsville, MD, 1994.

PREVENTION OF SINGLET OXYGEN DAMAGE IN 2'-DEOXYGUANOSINE BY LYCOPENE ENTRAPPED IN HUMAN ALBUMIN

Lydia F. Yamaguchi, Marisa H.G. Medeiros and Paolo Di Mascio.

Departamento de Bioquímica, Instituto de Química, Universidade de São Paulo, CP 26077, 05599-970 São Paulo, Brazil.

1 INTRODUCTION

Singlet oxygen, $^1O_2$, is of substantial importance in chemical and biological systems due to its high reactivity and involvement in physiological and pathological processes. It has been shown to be generated in biological systems and implicated in (i) defense mechanisms of living organisms such as in phagocytosis (ii) hormonal activity of prostaglandins (iii) photochemotherapy utilizing the photodynamic action of synthetic dyes (iv) clinical manifestations of toxic agents like psoralens and (v) inborn errors of metabolism exemplified by erythropoietic porphyria.[1,2]

The reactivity of $^1O_2$ with unsaturated compounds, sulfides and amino groups arises from its electrophilicity. Thus, biological targets for $^1O_2$ having the above functional groups include unsaturated fatty acids, proteins, enzymes and DNA. These lesions have been suggested to play a role in aging, mutagenesis and carcinogenesis.[2]

Lycopene is one of the major carotenoids in Mediterranean diets, found mostly in tomatoes and tomato products.[3] Numerous studies indicate that dietary carotenoids help reduce the risk of cancer, cardiovascular diseases, macular degeneration and cataracts.[4] Conjugated biomolecules such as carotenoids act as chemical and physical quenchers of 1O2 and hence provide protective mechanisms against the deleterious effects of this excited state of oxygen.[5] The quenching abilities of different carotenoids were found differ considerably.[6] Lycopene, the straight-chain isomer of ß-carotene showed the greatest quenching ability, almost two times higher than ß-carotene. Comparison of the structures of lycopene, α-carotene and ß-carotene revealed that the opening of the ß-ionone ring to a straight chain increased the quenching ability.[6]

In DNA, $^1O_2$ reacts preferentially with dGuo (2'-deoxyguanosine) residues, leading to the formation of at least four different reaction products: two 4R* and 4S* diasteromers of 4-hydroxy-8-oxo-7,8-dihydro-2'-deoxyguanosine (4-OH-8-oxodGuo), the main product, and 8-oxo-7,8-dihydro-2'-deoxyguanosine (8-oxodGuo).[7,8]

2 MATERIALS AND METHODS

2.1. Singlet oxygen generation

Two different sources of $^1O_2$ were employed: (i) photosensitization with polymer-bound Rose bengal (Sensitox®) or with Chelex® resine-associated methylene blue (MB-Chelex®) through a type II photoreaction, consisting of an energy transfer from a triplet excited state of the sensitizer (Sen) to ground state oxygen ($^3O_2$), resulting in the formation of $^1O_2$ (reaction 1);[9] (ii) thermal decomposition of the water-soluble endoperoxide 3-3'-(1,4-naphthylidene) diproprionate (NDPO$_2$) excluding the type I photoreaction (reaction 2).[10]

$$\text{Sen} + h\nu \rightarrow \text{Sen*} \xrightarrow{\; ^3O_2 \;} \text{Sen} + {}^1O_2 \qquad (1)$$

$$2\,\text{NDPO}_2 \xrightarrow{\;\Delta\;} 2\,\text{NDP} + {}^3O_2 + {}^1O_2 \qquad (2)$$

2.2. Formation and detection of 2'-deoxyguanosine and derivatives

A type II energy transfer mechanism drives the reaction of dGuo with $^1O_2$ and produces two main oxidized nucleosides, 4-OH-8-oxodGuo and 8-oxodGuo (Scheme 1).

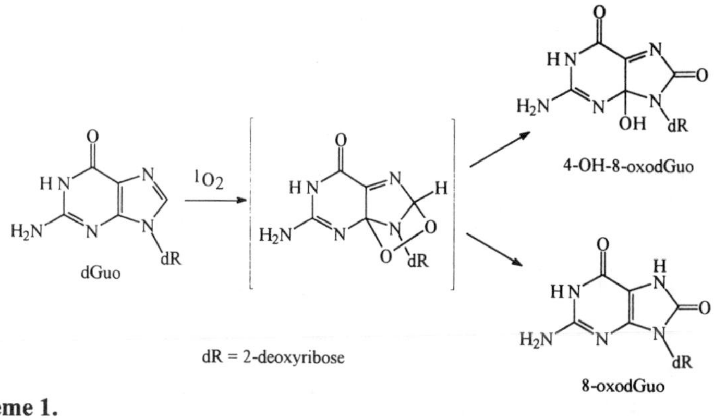

Scheme 1.

Reaction systems. System 1: Thermodecomposition of NDPO$_2$ was performed in the presence of 1 mM of dGuo and different concentrations of carotenoids in phosphate buffer 0.1 M, pH 7.4 at 37 °C. System 2: Photosensitization with 2 mg of Sensitox® or MB-Chelex® was employed to generate $^1O_2$: 1 mM of dGuo was incubated in the presence of 1 mM of FeSO$_4$ in water at 37 °C and irradiated with a 50 W tungsten lamp placed 10 cm from the solution.

Carotenoids entrapped in albumin: Carotenoids were dissolved in 100 µL of THF/0.025 % BHT. Five µL of this mixture was added to a solution of human albumin (HSA) (1 mg/mL) every 5 min for 20 min. The amount of carotenoid associated with albumin was determinated spectrophotometrically after extraction in chloroform.

The detection of 8-oxodGuo and 4-OH-8-oxodGuo: The amount of 8-oxodGuo present in the solution was measured using HPLC with UV and electrochemical detectors using a reversed phase column C-18 (Spherex, 250 x 4.6 mm, 5 µm). The mobil phase was

KH_2PO_4, 50 mM, pH 5.5 with 10 % Methanol and 2.5 mM EDTA.[7] The 4-OH-8-oxodGuo was measured by HPLC and UV detector using a normal phase amino substituted silica gel Hypersil NH_2 column (250 x 4.6 mm, 5 μm) and a mobile phase consisting of a mixture of 25 mM ammonium formate and acetonitrile (40:60).[8] Electrospray ionization mass spectrometry was also used to identify the oxidation products of dGuo after reaction with $^1O_2$.

3 RESULTS AND DISCUSSION

Results showed that lycopene was able to decrease the amount of 4-OH-8-oxodGuo (50 %) (data not shown) after 20 min irradiation with MB-Chelex®, and of 8-oxodGuo (Fig.1), using 5 mM $NDPO_2$ as a source of $^1O_2$. A control of 1 mg/ml albumin and 0.025 % BHT had no effect on production of 4-OH-8-oxodGuo or 8-oxodGuo. These data indicate that carotenoids entrapped in HSA can be an efficient quencher of $^1O_2$ and protect against the deleterious effects of this excited state molecule.

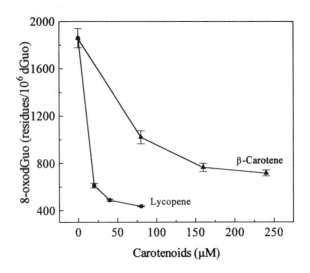

Figure 1
Effect of carotenoids on formation of 8-oxodGuo in human albumin.

In this work, we confirmed the formation of 8-oxodGuo, the oxidation product of dGuo, on reaction with $^1O_2$ (Scheme 1). Lycopene and ß-carotene were entrapped in albumin (reactions 3 and 4) to make possible the interaction of these carotenoids with $^1O_2$ in aqueous solution. Lycopene was a better quencher of $^1O_2$ than ß-carotene in protecting dGuo from $^1O_2$ generated by $NDPO_2$, a chemical generation, or MB-Chelex®, a photosensitizer.

$$^1O_2 + Carotenoids\text{-}HSA \rightarrow {^3O_2} + {^*Carotenoids\text{-}HSA} \tag{3}$$
$$^*Carotenoids\text{-}HSA \rightarrow Carotenoids\text{-}HSA + heat \tag{4}$$

The quenching efficiency of lycopene may be of interest in protecting against $^1O_2$-induced damage of biological macromolecules like DNA, as may occur in lung oxidant injuries, skin photosensitivity, erythropoietic porphyria and toxicity of certain photosensitizers used in photochemotherapy.

Acknowledgments

This work was supported by the "Fundacão de Amparo à Pesquisa do Estado de São Paulo", FAPESP (Brazil), the "Conselho Nacional para o Desenvolvimento Cientìfico e Tecnológico", CNPq (Brazil), the and "Programa de Apoio Núcleos de Excelència", PRONEX/FINEP, (Brazil), L. F. Y. is recipient of FAPESP fellowships.

References

1. Sies H. *Angew. Chem. (Inst. Ed. Engl.)* 1986, **25**, 1058-171.
2. Di Mascio P., Medeiros M.H.G., Bechara E.J.H., Catalani L.H. *Ciència é Cultura.* 1995, **47**, 297-311.
3. Gerster H. *J. Am. Coll. Nutr.* 1997, **16**, 109-126.
4. Sies H., Stahl W. *Proc. Soc. Exp. Biol. Med.* 1998, **218**, 121-124.
5. Di Mascio P., Sundquist A.R., Devasagayam T.P.A. and Sies H. In: Carotenoids, (Packer, L.), Academic Press, Orlando, *Methods. Enzymol.* 1992, **213**, 429-438.
6. Di Mascio P., Kaiser S. and Sies *H. Arch. Biochem. Biophys.* 1992, **274**, 532-538.
7. Cadet J., Berger M., Douki T., and Ravanat J.-L. *Rev. Physiol. Biochem. Pharmacol.* 1997, **131**, 1-87.
8. Ravanat J.-L.; Cadet J. *J. Chem. Res. Toxicol.* 1995, **8**(3), 379-388.
9. Greenwald R.A. Handbook of Methods for Oxygen Radical Research, CRC Press, Inc. Boca Ronton, Flórida, 1985.
10. Di Mascio P., Sies H. *J. Am. Chem. Soc.* 1989, **111**, 2909-2914.

COENZYME Q PROTECTION OF MITOCHONDRIAL ACTIVITIES IN RAT LIVER UNDER OXIDATIVE STRESS

Giorgio Lenaz, Maria Luisa Genova, Marika Cavazzoni, Marilena D'Aurelio, Bruno Nardo*, Gabriella Formiggini, Carla Bovina.

Dipartimento di Biochimica "G. Moruzzi", Università di Bologna, via Irnerio 48, 40126 Bologna, Italy, *Dipartimento di Chirurgia "A. Valsalva", Università di Bologna, via Massarenti 9, 40138 Bologna, Italy.

1 INTRODUCTION

It is generally accepted that reactive oxygen species (ROS) are deeply involved in pathological processes. Their aggressive behavior toward macromolecules causes a deterioration of cell structures which has been implicated in the onset of degenerative diseases, cancer and aging.[1] Since mitochondria are major producers of ROS through the electron transfer chain,[2] and owing to the peculiar vulnerability of the highly sophisticated machinery of oxidative phosphorylation,[3] research has largely focussed on mitochondria in order to understand ROS biochemistry and pathology. Moreover, recent developments have suggested new important roles of mitochondria in pathology: first the discovery that mitochondrial DNA (mtDNA) mutations are the basis of many diseases,[4] and secondly the importance of the mitochondria in apoptotic cell death.[5]

Here we report two studies showing the antioxidant effect of Coenzyme Q_{10} (CoQ_{10}) in rat liver. Both studies also demonstrate that exogenous CoQ_{10} is reduced in the liver cells.

The administration of this homolog to the rat, in which the major quinone is CoQ_9,[6] represents a kind of labelling allowing to distinguish endogenous from exogenous CoQ and to follow any modification of these compounds.[7]

The anticancer quinonic glycoside, adriamycin, induces oxidative stress by enhancing the production of ROS in mitochondria[8] and endoplasmic reticulum.[9]

Treatment of isolated rat hepatocytes with adriamycin induces oxidative stress[10] with production of superoxide radical and hydrogen peroxide, revealed by flow cytometric analysis with dihydroethidium and dichlorofluorescein diacetate, respectively. The mitochondrial membrane potential is collapsed by the drug treatment, as shown by flow cytometric detection of Rhodamine 123 incorporation. Concomitantly with the oxidative stress, endogenous CoQ_9 becomes largely oxidized.

On the other hand, the respiratory chain activities and ATPase of isolated mitochondria are not compromised to a major extent, suggesting that damaging ROS are largely produced in extramitochondrial compartment. The loss of mitochondrial potential is ascribed to opening of the inner membrane permeability transition pore by ROS. The basal hydrogen peroxide level and its production upon adriamycin treatment are much higher in

24-month old rats than in adult (6-month old) animals, nevertheless the mitochondrial potential is similar in both groups and is equally affected by adriamycin (unpublished).[11] Incubation of the cells with exogenous CoQ_{10} prevents ROS formation and protects both reduced CoQ and the membrane potential.[10]

It was found that the cytosolic enzyme DT-diaphorase is responsible for reduction of endogenous and exogenous CoQ, as shown by the effect of dicoumarol, an inhibitor of DT-diaphorase, preventing the protective effect of exogenous CoQ addition.[10]

During hepatic resection commonly performed before organ transplantation or removal of pathological tissues as neoplasias, the major cause of the oxidative stress is ascribable to the necessity of keeping the tissue to be transplanted under ischemic conditions to prevent haemorrhage; in fact ischemia and especially subsequent reperfusion at the end of the operation are well known causes of oxidative stress in surgery of many tissues including liver.[12] Reperfusion can lead to marked metabolic damages as it promotes the reflow of toxic metabolites in the ischemic and reperfused tissue;[12] therefore it can represent a cause of worsening ischemic damage.[13] Evidence was presented[14] that ROS, including the oxygen free radicals and strong non-radical oxidants such as hydrogen peroxide, are involved in the pathogenesis of damage by ischemia followed by reperfusion; during reperfusion ROS can be released by hepatocytes, Kupffer cells and neutrophiles.

2 MATERIALS AND METHODS

Two groups of rats, maintained in controlled conditions, were respectively i.p. injected with 1 mg CoQ_{10}/100 g body weight (0.4 ml of 0.9 % NaCl containing 1 mg CoQ_{10} solubilized with 2 % HCO-60) and with the vehicle (0.4 ml of 0.9 % NaCl containing 2 % HCO-60) once daily for 14 days. At the end of the treatment the rats were anaesthetized with Ketalar and partial hepatic ischemia was induced: 70 % of the liver (left and median lobes) was ischemized by vascular microclamping for 90 minutes and subsequently reperfused for 30 minutes. The remaining of the tissue (right and caudate lobes), utilized as control, was removed together with the ischemized part at the end of the reperfusion and all samples were stored in liquid nitrogen. Total body blood was collected at the end of the reperfusion and immediately centrifuged; the obtained plasma was stored in liquid nitrogen.

Frozen samples of liver (0.3-0.6 g) were reduced to powder, then homogenized with 1 ml water. Protein content in homogenates was evaluated by the biuret method of Gornall et al.[15]

Lipid-soluble quinones (CoQ_8, CoQ_9 and CoQ_{10} both in oxidized and reduced form) were extracted from the liver homogenates (1ml) and plasma (0,5 ml+0.5 ml 80 mM SDS) using 2 ml extraction mixture (95 % ethanol+5 % isopropanol) plus 5 ml hexane. The mixture was centrifuged and the supernatant was drawn up, evaporated to dryness and resuspended in ethanol. To prevent oxidation of reduced forms of quinones the whole procedure was performed within 20 minutes.

Antioxidant levels were determined by reverse phase HPLC using a KROMASIL 100Å-C_{18}-250x4.6 mm column provided with a precolumn; the HPLC system was equipped with a gradient controller and a photodiode array detector. The chromatographic data were processed at different wavelengths in order to analyze oxidized (275 nm) and reduced quinones.

Lipid peroxidation in liver homogenates was evaluated by determining TBARS[16] before and after addition of the free radical initiator 2-2'-azobis amidinopropane (AAPH) at various time intervals.

Succinate oxidase activity was assayed in the homogenate measuring oxygen consumption by means of a Clark's electrode according to Estabrook.[17]

3 RESULTS AND DISCUSSION

Table 1 shows the lipid-soluble antioxidant pattern in rat liver of CoQ_{10}-treated and vehicle treated rats: the major CoQ homolog is CoQ_9 both in reduced and oxidized form, relatively minor components are CoQ_8 and CoQ_{10}, both at $\leq 10\ \%$ with respect to CoQ_9 and both partly in the reduced form.

Table 1

CoQ levels in rat liver homogenate.

	CoQ_8 *red*	CoQ_9 *red*	CoQ_{10} *red*	CoQ_8 *ox*	CoQ_9 *ox*	CoQ_{10} *ox*	Total CoQ	Total **CoQ red** / Total CoQ *ox+red*
	pmol/mg protein	pmol/mg protein	pmol/mg protein	pmol/mg protein	pmol/mg protein	pmol/mg protein	pmol/mg protein	
Vehicle-treated (n=3)								
control	9 ± 10	101 ± 44	0	27 ± 3	240 ± 78	21 ± 2	398	0.28
Ischemized and reperfused	6 ± 5	73 ± 39	0	19 ± 8	n.a.	16 ± 11	n.a.	n.a.
CoQ_{10}-treated (n=4-5)								
control	7 ± 6	96 ± 23	398 ± 426	14 ± 5	158 ± 54	175 ± 62	848	0.59
Ischemized and reperfused	5 ± 5	80 ± 22	159 ± 155	13 ± 8	158 ± 9	184 ± 67	599	0.41

Values are means ±SD of double determinations; the number of experiments is shown in brackets. n.a= not assayed.

The presence of CoQ_8 in rat liver presumably reflects the intestinal absorption of this quinone synthetized by the bacterial flora.[18]

The levels of reduced CoQ_9 are lower in ischemic-reperfused lobes than in controls in both groups even if statistical significance is not reached; the high concentration of retinol esters in ischemic-reperfused lobes of vehicle treated rats prevented the detection of oxidized CoQ_9.

In the CoQ_{10}-treated animals the exogenous quinone is extensively incorporated in the liver and becomes largely reduced in both the ischemic-reperfused and control lobes, so that the total reduced quinone levels are higher in the ischemic-reperfused lobes of CoQ_{10}-

treated rats (244 pmol/mg protein) than in tha control lobes of the vehicle-treated animals (110 pmol/ mg protein. The ratio $(CoQred)/(total\ CoQox+red)$ is increased in CoQ_{10}-treated rats in comparison with vehicle-treated animals.

Administered CoQ_{10} doubles the total CoQ content in the control livers; ischemia and reperfusion in the CoQ_{10}-treated animals cause one third decrease in the total quinone level. Because of the high variability of the results it can only be speculated that CoQ is consumed or displaced during performance of its antioxidant action.

The ratio CoQ_9 red/CoQ_9 ox is slightly higher in CoQ_{10}-treated, indicating that CoQ10 could exert a sparing action towards CoQ_9 antioxidant action,[10,19] even if the two homologs have different sites of reduction owing to their different localisation in the cells.

In fact endogenous CoQ_9 is largely reduced in the mitochondrial respiratory chain, while exogenous CoQ_{10}, which does not enter mitochondria,[7] is reduced by extramitochondrial enzymes.[10,20]

In plasma a high level of reduced CoQ^{10} is found in quinone-treated rats, while CoQ_9 is only in oxidized form in both groups. CoQ_9 amount is almost double in vehicle-treated animals, though not statistically different, which might be ascribable to a solubilizing effect of the vehicle, indicating that, in spite of undesirable collateral effects, the administration via gastric gavage would be more suitable (not shown).

The potential protective action of CoQ_{10} has been tested by evaluating TBARS levels, which reflect the extent of lipid peroxidation: low basal levels of TBARS (min: 149 ± 99, max: 376 ± 360 pmol/mg protein) are found in the two groups of animals, both in control and ischemized/reperfused lobes. In both groups TBARS levels increase only 24h after the radical initiator AAPH addition (Table 2), testifying the permanently high basal level of antioxidants in liver; however, after 48h of incubation with AAPH, TBARS levels are lower in the livers of COQ_{10}-treated rats than in vehicle-treated animals.

Table 2
Protective effect of CoQ^{10} in rat liver homogenate from AAPH-induced oxidative stress.

	Control lobe		Ischemized and reperfused lobe	
	+vehicle	$+Q_{10}$	+vehicle	$+Q_{10}$
TBARS (time 0) (pmol/mg protein) (n = 3-4)	376 ± 360	161 ± 105	242 ± 73	149 ± 99
TBARS (at 24 h) (pmol/mg protein) (n = 3-4)	1070 ± 269	898 ± 122	1133 ± 372	787 ± 221
Lag period (h)	22	24	21	26
Succinate oxidase activity (nmol O_2 min^{-1} mg $protein^{-1}$) (n = 3-5)	8.57 ± 2.95	8.76 ± 5.45	$4.38\pm1.48*$	6.49 ± 4.95

Values are means $\pm SD$; the number of experiments is shown in brackets.
*Significantly different from corresponding control lobes, $p<0.05$

Accordingly, the succinate oxidase activity is significantly lower in ischemized and reperfused lobes of vehicle-treated rats (Table 2), indicating a mitochondrial damage which is overcome by CoQ_{10} administration.

In conclusion, under our experimental conditions the damage due to 90 min ischemia followed by 30 min reperfusion appears quite limited, mainly due to the high level of antioxidants in the liver. Anyway exogenous CoQ_{10} administration under strong oxidative stress conditions induces a normalising effect chiefly due to its large reduction in the body. It is likely that exogenous CoQ_{10} administration would protect liver as well as the whole body, particularly in conditions of heavy oxidative stress as reperfusion after multiple prolonged ischemias that occur during hepatic surgery. In particular our data suggest that CoQ_{10} administration effectively increases the high antioxidant power of the liver and plasma by protecting endogenous CoQ_9, mitochondrial function and particularly succinate oxidase activity and by limiting lipid peroxidation.

Acknowledgements HCO-60 was kind gift from Dr. T. Kishi, Kobe (Japan).

References
1.   A.T. Diplock. *Molec. Aspects Med.* 1994, **15**, 293.
2.   B. Chance, H. Sies and A. Boveris. *Physiol. Rev.* 1979, **59**, 527.
3.   L. Ernster. In: *Active Oxygen Lipid Peroxides and Antioxidants*, K. Yagi Ed. CRC Press, Boca Raton FL. 1993.
4.   A.H.V. Schapira. *J. Bioenerg. Biomenbr.* 1997, **29**, 105.
5.   N. Zamzami, P. Marchetti, M. Castedo, C. Zanin, J.L. Vayssière, P.X. Petit and G. Kroemer. *J. Exp. Med.* 1995, **181**, 1661.
6.   M.L. Genova, C. Castelluccio, R. Fato, G. Parenti Castelli, M. Merlo Pich, G. Formiggini, C. Bovina, M. Marchetti and G. Lenaz. *Biochem. J.* 1995, **311**, 105.
7.   M.L. Genova, C. Bovina, G. Formiggini, V. Ottani, S. Sassi and M. Marchetti. *Molec. Aspects Med.* 1994. **15**, s47.
8.   K.J. Davies and J.H. Doroshov. *J. Biol. Chem.* 1986, **261**, 3060.
9.   J. Goodman and P. Hochstein. *Biochem. Biophys. Res.Commun.* 1977, **77**, 797.
10.   R.E. Beyer, J. Segura-Aguilar, S. Di Bernardo, M. Cavazzoni, R. Fato, D. Fiorentini, M.C. Galli, M. Setti, L. Landi and G. Lenaz. *Proc.NatlAcad. Sci.USA.* 1996, **93**, 2528.
11.   P.Bernardi, K.M. Broekmeier and D.R. Pfeiffer. *J.Bioenerg.Biomembr.* 1994, **26**, 509.
12.   M. Haimovici. *Surgery.* 1979, **85**, 461.
13.   D.A. Parks and D.N. Granger. *Am. J. Physiol.* 1986, **250**, G749.
14.   H. Jaeschke. *Chem. Biol. Interact.* 1991, **79**, 115.
15.   A.G. Gornall, C.J. Bardawill and M.M. David. *J. Biol. Chem.* 1949, **177**, 752.
16.   T.F. Slater. *Methods Enzymol.* 1984, **105**, 283.
17.   R.W. Estabrook. *Methods Enzymol.* 1967, **10**, 41.
18.   T. Ramasarma. In: *Coenzyme Q*, G. Lenaz Ed., John Wiley and Sons Ltd, Chichester, 1985.
19.   V. Valls, C. Castelluccio, R. Fato, M.L. Genova, C. Bovina, G. Saez, M. Marchetti, G. Parenti Castelli and G. Lenaz. *Biochem. Mol. Biol. Int.* 1994, **33**, 633.
20.   T. Takahashi, T. Okamoto and T. Kishi. *J. Biochem.* 1996, **119**, 256.

DIFFERENT ANTIOXIDANT MECHANISMS OF α-TOCOPHEROL AND L-ARGININE RESULT IN PRESERVED ENDOTHELIAL FUNCTION IN HYPERCHOLESTEROLEMIC RABBITS

Stefanie M. Bode-Böger, Rainer H. Böger, Jürgen C. Frölich.

Institute of Clinical Pharmacology, Hannover Medical School, 30623 Hannover, Germany.

1 INTRODUCTION

Endothelium-dependent vasodilator function is impaired in atherosclerosis,[1] due to impaired biological activity of nitric oxide (NO). Rabbits fed a high-cholesterol diet show reduced NO-dependent vasodilatation.[2,3] Elevated LDL cholesterol concentrations have been shown to inhibit endothelium-dependent vasodilatation *in vitro*.[4] The fraction of LDL that plays an important role in the development of atherosclerotic lesions is oxidatively modified LDL (oxLDL).[5] Oxidation of LDL is mediated by endothelial cells, vascular smooth muscle cells, and macrophages. OxLDL is found within atheromatous lesions, but not in the normal vascular wall.[6] We have recently reported that atherosclerotic rabbit aorta releases significantly greater amounts of superoxide radicals as compared to control aortae.[7] Thus, a growing body of evidence supports the notion that oxidative stress and LDL oxidation are key events in the promotion of endothelial dysfunction and atherogenesis, which are counteracted by endothelial NO.[1]

Dietary supplementation with the precursor of endogenous NO, L-arginine, has been shown to restore the biological activity of vascular NO and to reduce the progression of atheromatous lesions in cholesterol-fed rabbits.[2,3] Lipid soluble antioxidants like vitamin E (α-tocopherol) increase the resistance of LDL to oxidation.[8] However, there are conflicting studies showing beneficial effects of dietary α-tocopherol on atheromatous lesion formation and endothelial dysfunction,[9] or no such effects,[10] in cholesterol-fed rabbits.

The present study was undertaken to compare the effects of chronic administration of L-arginine and α-tocopherol on endothelium-dependent vasodilatation, systemic NO production and vascular oxidative stress in rabbits with pre-existing hypercholesterolemia.

2 MATERIALS AND METHODS

2.1. Animals and study design

32 male New Zealand white rabbits were included in this study. The experimental protocol conformed with the Guide for the Care and Use of Laboratory Animals published by the US National Institutes of Health (NIH publication No. 85-23, revised 1985) and had been approved by the Hannover Administrative Panel on Laboratory Animal Care.

32 male New Zealand white rabbits were randomly assigned to high-cholesterol diet (1 % cholesterol chow, Altromin, Lage, Germany; N=24), or normal rabbit chow (N=8) for 4 weeks. Subsequently, cholesterol-feeding was continued with 0.5 % cholesterol for additional 12 weeks. During this second part of the study, one group of 8 cholesterol-fed rabbits received 2.0 % (wt/vol) L-arginine in drinking water (L-arginine group), and a second group of 8 rabbits received 300 mg/day of α-tocopherol by oral gavage (vitamin E group), the other 8 rabbits received no additional treatment (cholesterol group). Another group of 8 rabbits received normal rabbit chow throughout the study (control group).

24 h urines and plasma samples were collected at baseline and after 1 and and 4 months. At the end of the feeding period, the rabbits were anesthesized with sodium pentobarbital (10 mg/kg i.v.) and sacrificed, and the aorta was excised and immediately rinsed in freshly prepared, ice-cold Krebs buffer. A segment of the thoracic aorta immediately distal from the left subclavian artery was used for isometric tension recording, additional segments of thoracic aorta were used for chemiluminescence detection of superoxide anion release, and for histological examination.

The aortas were dissected free of adhering fat and connective tissue and placed into organ baths filled with oxygenated (95 % O_2, 5 % CO_2) modified Krebs solution (37 °C, pH 7.4) for isometric tension recording as described previously.[3] Cumulative concentration response curves were obtained with the endothelium-dependent vasodilator acetylcholine and the endothelium-independent vasodilator sodium nitroprusside (both 1nM to 0.1 mM) after precontraction with 1 μM noradrenaline. All drugs were purchased from Sigma (Munich, Germany).

2.2. Biochemical and physiological analyses

The production of superoxide anions was measured using the lucigenin-enhanced chemiluminescence response as described previously.[3,7] Photon emission stimulated by phorbol-12-myristate-13-acetate (PMA, 2 μM dissolved in dimethylsulfoxide) was recorded for 5 min. The specific chemiluminescence response was expressed as counts/min minus the average background activity. After the end of the measurements the rings were blotted and weighed; data were expressed as counts/min per mg of dry weight.

The procedure for preparation and oxidation of LDL was adapted from the method of Esterbauer et al.[11]

Urinary nitrate excretion was determined by GC-MS using the PFB-derivative of nitrate as described previously.[3] Quantitation was performed by selected ion monitoring at m/z 46 for endogenous nitrate and m/z 47 for (^{15}N)-nitrate used as internal standard. The detection limit of the method was 20 fmol nitrite or nitrate. Interassay variability and accuracy of the method were 3.8 % and 98 %, respectively.

Urinary creatinine was determined spectrophotometrically with the alkaline picric acid method in an automatic analyzer (Beckman, Galway, Ireland). The urinary excretion rates of NO_3^- were corrected by urinary creatinine concentration in order to limit variability due to changes in renal excretory function, as described previously.[12]

Plasma L-arginine, total cholesterol, LDL and HDL cholesterol concentrations were determined by validated laboratory methods as described previously.[13]

Segments of the proximal thoracic aorta were excised, fixed in formalin, embedded in paraffin and stained with hematoxylin/eosin for the morphometric measurement of intimal and medial cross-sectional areas by planimetry. Four sections from each animal were analyzed by observers blinded to the treatment groups, and the values were averaged.

2.3. Calculations and statistical analysis

All data are given as mean ± S.E.M. Statistical significances between groups were calculated using analysis of variance followed by Fisher`s protected least significant difference test. Treatment-induced changes within groups were tested for statistically significant differences using multiple comparison analysis of variance followed by the Scheffé f-test.

3 RESULTS

Plasma total cholesterol concentrations were significantly elevated by the cholesterol-enriched diet. Neither supplementation with L-arginine nor with vitamin E had a significant effect on plasma total cholesterol concentrations (Table 1). Dietary L-arginine induced a two-fold increase in plasma L-arginine concentrations, whereas in all other experimental groups no significant changes were observed throughout the study period (Table 1).

Table 1
Plasma L-arginine and total cholesterol concentrations.

Group	baseline	4 weeks	16 weeks
Total cholesterol [mmol/L]			
Control	1.11± 0.07	0.92 ± 0.11	0.89 ± 0.07
Cholesterol	1 .21± 0.06	49.25 ± 4.65	36.96 ± 3.05
L-Arginine	1.28 ± 0.08	41.13 ± 3.93	34.83 ± 4.24
VitaminE	1.16 ± 0.08	47.00 ± 4.71 *	36.34 ± 6.72
L-arginine [,umol/L]			
Control	137.7 ± 12.8	135.1 ±16.7	113.3 ± 20.1
Cholesterol	149.2 ± 9.5	150.9 ± 18.8	126.7 ± 116.2
L-Arginine	120.0 ± 8.3	125.6 ± 6.2	231.5 ± 10.6
Vitamin E	145.6 ± 9.9	132.0 ± 6.1	116.3 ± 11.2

*Values are mean ± S.E.M. * $p < 0.05$ vs. control.*

Endothelium-dependent relaxation of isolated aortic rings in response to acetylcholine was almost completely abolished by cholesterol feeding (Fig. 1). Treatment with L-arginine or α-tocopherol partly prevented the hypercholesterolemia-induced reduction of acetylcholine-induced vasodilatation (each $p < 0.05$ vs. cholesterol, $p < 0.05$ vs. control; Fig. 1). Endothelium-independent relaxations of isolated aortic rings induced by sodium nitroprusside showed that smooth muscle vasodilator capacity was not significantly different between any of the groups.

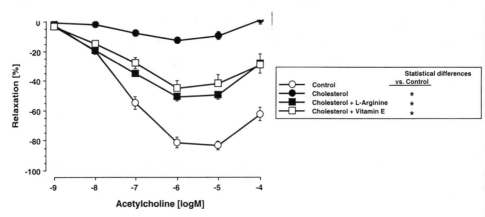

Figure 1

*Relaxation of isolated aortic rings ex vivo in response to the endothelium-dependent vasodilator acetylcholine. Aortic rings were obtained from rabbits fed 1 % cholesterol for 4 weeks followed by 0.5 % cholesterol for additional 12 weeks (Cholesterol), or the same diet plus 2 % L-arginine in drinking water (Cholesterol+ L-Arginine), or plus 300 mg/day of α-tocopherol (Cholesterol + Vitamin E). rabbits in the control group were fed normal rabbit chow during the entire study period. Data are mean ± S.E.M. * p < 0.05 vs. control.*

There was a huge increase in PMA-stimulated superoxide radical production in the cholesterol group, but not in the control group ($p < 0.05$). This increase was completely reversed in the L-arginine group ($p < 0.05$ vs. cholesterol), whereas no significant difference in PMA-stimulated superoxide production was observed between the vitamin E and cholesterol groups (Fig. 2).

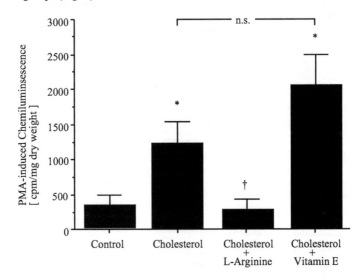

Figure 2

*Lucigenin-enhanced chemiluminescence response of isolated aortic rings after stimulation with 2 µM PMA. Data are mean ± S.E.M. * p < 0.05 vs. control. † p < 0.05 vs. cholesterol.*

The lag time of copper-mediated oxidation of isolated LDL was significantly reduced in LDL isolated from hypercholesterolemic rabbits as compared to normal controls (108 ± 17 vs. 185 ± 4 min; $p < 0.05$). Dietary supplementation with L-arginine partly restored the lag time to normal values (150 ± 14 min; $p < 0.05$ vs. cholesterol and control); however, α-tocopherol significantly prolonged the lag time to values even far beyond those in the control group (509 ± 13 min; $p < 0.01$ vs. all other groups).

Urinary nitrate excretion rates were 767 ± 28 μmol/mmol creatinine at baseline with no significant differences between the four groups. Urinary nitrate excretion decreased to 381 ± 47 μmol/mmol creatinine after 4 weeks of hypercholesterolemia ($p < 0.05$). During supplementation with L-arginine nitrate excretion rates returned to 487 ± 45 μmol/mmol creatinine ($p < 0.05$ vs. cholesterol). Supplementation with vitamin E did not affect urinary nitrate excretion as compared to cholesterol.

Intima/media ratio in histological cross sections of the thoracic aorta was increased to 4.30 ± 1.96 in the cholesterol group, whereas in the control group only the thin endothelial cell layer was visible (I/M ratio = 0). This increased intima/media ratio was mainly due to thickening of the intima. Both L-arginine and α-tocopherol administration significantly reduced intimal thickening in the aorta (to 0.97 ± 0.19 and 1.27 ± 0.07, respectively; each $p < 0.05$ vs. cholesterol; Fig. 3).

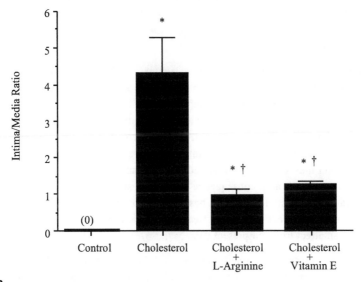

Figure 3
*Aortic cross-sectional intima / media ratio. Data are mean ± S.E.M. * p < 0.05 vs. control.
† p < 0.05 vs. cholesterol.*

4 DISCUSSION

Our study shows that chronic oral administration of L-arginine or α-tocopherol induces anti-atherosclerotic effects in cholesterol-fed rabbits. Both treatments restored endothelium-dependent vasodilator function even though treatment was initiated 4 weeks after the induction of hypercholesterolemia, and reduced the progression of atherosclerotic lesions. Concomitantly with increased urinary nitrate excretion, L-arginine inhibited the vascular release of superoxide radicals *ex vivo*, but it had only a modest inhibitory effect on copper-mediated LDL oxidation *ex vivo*. α-Tocopherol, on the other hand, strongly inhibited LDL oxidation, but it had no effect on urinary nitrate excretion or on superoxide elaboration by the vascular wall.

Two major approaches to influence the balance between NO and oxygen-derived free radicals are to increase NO production by L-arginine administration,[3] or to reduce the susceptibility of LDL to oxidation, using antioxidant drugs like vitamin E.[14] Several studies have previously shown that long-term supplementation of hypercholesterolemic rabbits with L-arginine from the very beginning of cholesterol-feeding resulted in restored endothelial function and decreased intimal plaque formation.[2,3,13] Furthermore, we have previously shown that chronic L-arginine administration reduces vascular $O_2 \cdot^-$ elaboration in isolated aortic segments *ex vivo* when it is administered starting at the beginning of the high cholesterol diet.[3,13] It has been shown that superoxide radical formation by NO synthase is reversed in the presence of excess L-arginine.[15] This concept is supported by the present data, which show that L-arginine administration reverses a pre-existing dysfunction of NO production in cholesterol-fed rabbits and reduces oxidative stress within the vascular wall.[3] Once formed, NO may react with lipid peroxyl radicals ("chain-breaking reaction"), thereby interrupting the autocatalytic reaction cascade leading to oxidative damage.[16] By this mechanism, NO may also inactivate radicals formed by other enzymes like NADPH oxidase,[17] cyclooxygenase, or lipoxygenase.[18]

Physiologically, LDL are protected against oxidation by their content of α-tocopherol and other antioxidants. Chronic oxidative stress in the hypercholesterolemic vascular wall may lead to an exhaustion of these endogenous antioxidant mechanisms, rendering the LDL more susceptible to oxidation, as indicated by the shortened lag time of the oxidation–time curve of LDL isolated from cholesterol-fed rabbits. L-Arginine had only a modest, although statistically significant effect on copper-mediated LDL oxidation *in vitro*. This effect may be explained by reduced consumption of endogenous antioxidants present within the LDL as a consequence of reduced superoxide radical formation in L-arginine-treated animals. However, it has also been reported that L-arginine in high concentrations exerts direct inhibitory effects on LDL oxidation *in vitro*,[19] and we cannot exclude such an effect from our present data.

Oxidative stress in the atherosclerotic vascular wall induces leukocyte adhesion,[20] platelet activation, and transcriptional upregulation of oxidant-responsive genes like those encoding for the adhesion molecule VCAM-1 and the chemokine MCP-1.[21] These pathophysiological events are early events in the process of atherogenesis; they are counteracted by NO. The common denominator for the regulation of cellular processes by oxidative stress and NO is activation or inactivation of NF-κB[22] and, thereby, transcriptional up- or downregulation of atherosclerosis-related genes.[23] The balance between NO and superoxide thus seems to be one major determinant for the progression or regression of atherosclerosis.[1,13]

In summary, our data suggest that both L-arginine and α-tocopherol preserve endothelium-dependent, NO-mediated vasodilation in cholesterol-fed rabbit aortic rings and inhibit aortic intimal thickening when treatment is started 4 weeks after the initiation of hypercholesterolemia. These anti-atherogenic effects seem to be due to different antioxidant mechanisms: L-arginine restores the pre-existing defect in NO formation (as assessed by urinary nitrate excretion rates) and inhibits vascular O_2·⁻ release (as assessed by vascular superoxide anion release *ex vivo*). In contrast, vitamin E decreases lipid peroxidation (as assessed by LDL oxidation *in vitro*), but it had no effect on vascular O_2·⁻ generation nor on NO formation rates. These data suggest that O_2·⁻ and oxLDL may act as sources of vascular oxidative stress independently of each other, and that both are relevant for endothelial function. Modulation of the balance between vascular oxidative stress *in vivo* and endothelial NO formation may be an effective therapeutic approach to anti-atherosclerotic therapy.

Acknowledgments

The authors gratefully appreciate the excellent technical assistance of M.-T. Suchy, F.-M. Gutzki, and K.-M. Pütz.

References

1. Böger RH, Bode-Böger SM, Frölich JC. The L-arginine–nitric oxide pathway: Role in atherosclerosis and therapeutic implications. *Atherosclerosis*. 1996, **127**, 1–11.

2. Cooke JP, Singer AH, Tsao P, Zera P, Rowan RA, Billingham ME. Antiatherosclerotic effects of L-arginine in the hypercholesterolemic rabbit. *J. Clin. Invest.* 1992, **90**, 1168–1172.

3. Böger RH, Bode-Böger SM, Mügge A, Kienke S, Brandes R, Dwenger A, Frölich JC. Supplementation of hypercholesterolaemic rabbits with L-arginine reduces the vascular release of superoxide anions and restores NO production. *Atherosclerosis*. 1995, **117**, 273–284.

4. Jacobs M, Plane F, Bruckdorfer KR. Native and oxidized low-density lipoproteins have different inhibitory effects on endothelium-derived relaxing factor in the rabbit aorta. *Br. J. Pharmacol*. 1990, **100**, 21–26.

5. Witztum JL, Steinberg D. Role of oxidized low density lipoprotein in atherogenesis. *J. Clin. Invest.* 1991, **88**, 1785–1792.

6. Ylä-Herttuala S, Palinski W, Rosenfeld M, Parthasarathy S, Carew TE, Butler S, Witztum J, Steinberg D. Evidence for the presence of oxidatively modified LDL in atherosclerotic lesions of rabbit and man. *J. Clin. Invest.* 1989, **284**, 1086–1095.

7. Mügge A, Brandes RP, Böger RH, Dwenger A, Bode-Böger SM, Kienke S, Frölich JC, Lichtlen PR. Vascular release of superoxide radicals is enhanced in hypercholesterolemic rabbits. *J. Cardiovasc. Pharmacol.* 1994, **24**, 994–998.

8. Jialal I, Grundy SM. Effect of combined supplementation with α-tocopherol, ascorbate, and β-carotene on low-density lipoprotein oxidation. *Circulation*. 1993, **88**, 2780–2786.

9.   Anderson TLG, Matz J, Ferns GAA, Anggard EE. Vitamin E reverses cholesterol-induced endothelial dysfunction in the rabbit coronary circulation. *Atherosclerosis.* 1994, **111**, 39–45.

10. Godfried SL, Combs GF, Saroka JM, Dillingham LA. Potentiation of atherosclerotic lesions in rabbits by a high dietary level of vitamin E. Br. *J.Nutr.* 1989, **61**, 607–617.

11. Esterbauer H, Striegl G, Puhl H, Rotheneder M. Continuous monitoring of *in vitro* oxidation of human low density lipoprotein. *Free Radical Res. Commun.* 1989, **6**, 67–75.

12. Böger RH, Bode-Böger SM, Gerecke U, Gutzki FM, Tsikas D, Frölich JC. Urinary NO_3^- excretion as an indicator of nitric oxide formation *in vivo* during oral administration of L-arginine or L-NAME in rats. *Clin. Exp. Pharmacol. Physiol.* 1996, **23**, 11–15.

13. Böger RH, Bode-Böger SM, Brandes RP, Phivthong-ngam L, Böhme M, Nafe R, Mügge A, Frölich JC. Dietary L-arginine reduces the progression of atherosclerosis in cholesterol-fed rabbits. Comparison with lovastatin. *Circulation.* 1997, **906**, 1282–1290.

14. Steinbrecher UP, Parthasarathy S, Leake DS, Witztum JL, Steinberg D. Modification of low density lipoprotein by endothelial cells involves lipid peroxidation and degradation of low density lipoprotein phospholipids. *Proc. Natl. Acad. Sci. USA.* 1984, **81**, 3883–3887.

15. Pritchard KA, Groszek L, Smalley DM, Sessa WC, Wu M, Villalon P, Wolin MS, Stemerman MB. Native low-density lipoprotein increases endothelial cell nitric oxide synthase generation of superoxide anion. *Circ. Res.* 1995, **77**, 510–518.

16. Hogg N, Kalyanaraman B, Joseph J, Struck A, Parthasarathy S. Inhibition of low-density lipoprotein oxidation by nitric oxide. Potential role in atherogenesis. *FEBS Lett.* 1993, **334**, 170–174.

17. Mohazzab-H KM, Kaminski PM, Wolin MS. NADH oxidoreductase is a major source of superoxide anion in bovine coronary artery endothelium. *Am. J. Physiol.* 1994, **266**, H 2568–H 2572.

18. Kukreja RC, Kontos HA, Hess ML, Ellis EF. PGH synthase and lipoxygenase generate superoxide in the presence of NADH or NADPH. *Circ. Res.* 1986, **59**, 612–619.

19. Philis-Tsimikas A, Witztum JL. L-arginine may inhibit atherosclerosis through inhibition of LDL oxidation (Abstract). *Circulation.* 1995, **92** (Suppl.), I-422.

20. Niu XF, Smith CW, Kubes P. Intracellular oxidative stress induced by nitric oxide synthesis inhibition increases endothelial cell adhesion to neutrophils. *Circ. Res.* 1994, **74**, 1133–1140.

21. Marui N, Offerman MK, Swerlick R, Kunsch C, Rosen CA, Ahmad M, Alexander RW, Medford RM. Vascular cell adhesion molecule-1 (VCAM-1) gene transcription and expression are regulated through an antioxidant-sensitive mechanism in human vascular endothelial cells. *J. Clin. Invest.* 1993, **92**, 1866–1874.

22. Peng HB, Libby P, Liao JK. Induction and stabilization of I kappa B alpha by nitric oxide mediates inhibition of NF-kappa B. J. Biol. Chem. 1995, **270**, 14214–14219.

23. Collins T. Endothelial nuclear factor-κB and the initiation of the atherosclerotic lesion. *Lab. Invest.* 1993, **5**, 499–508.

MEASUREMENT OF FREE RADICALS IN HUMANS USING ELECTRON SPIN RESONANCE SPECTROSCOPY

Anna Bini, Stefania Bergamini, Elisa Ghelfi, Anna Iannone, Marco Meli*, Maria Grazia Staffieri and Aldo Tomasi.

Department of Biomedical Sciences, University of Modena, Italy, and Department of Cardiosurgery, Hesperia Hospital, Modena, Italy.

1 INTRODUCTION

During the past 25 years, the field of free radical biology has flowered. Free Radicals are now known to play multiple roles in living systems. Concerning human pathologies, there is convincing evidence for the involvement of Reactive Free Radical Species in the initiation and development of various forms of damage and have been linked to many diseases and to the ageing process

For the most common chronic-degenerative diseases in the western world, free radicals have found an important pathogenic role. Atherosclerosis, reperfusion damage following ischemia,[1,2] neuro-degenerative diseases such as Alzheimer[3] and Parkinson,[4,5] diabetes[6-8] all find in free radical intermediates a common element. An excess of free-radical production has been linked to the ageing process. Oxidant by-products of normal metabolism can cause extensive damage to DNA, protein and lipid. Exposure to ultraviolet light, cigarette smoke and other environmental pollutants may also increase the free radical burden. The accumulation of unrepaired oxidative damage products is likely to be a major factor in cellular ageing.

A relevant number of studies have elucidated at the molecular level the formation and the free radical related damage. Mitochondrial calcium overload and permeability transition of the inner mitochondrial membrane have been shown to play an important role in production of reactive oxygen species (ROS); the expression of adhesion molecules such as lead to the production of ROS by polymorphonuclear leukocytes. The superoxide radical anion and hydrogen peroxide are known to be generated *in vivo*, and several human organs (e.g. liver and brain) are rich in iron, which appears to be mobilizable in a form that can stimulate free-radical reactions.

Tissue injury can itself cause ROS generation (e.g., by causing activation of phagocytes or releasing transition metal ions from damaged cells), which may (or may not, depending on the situation) contribute to a worsening of the injury. Attempts to prevent or limit such damage have been largely unsuccessful, principally because most of the pathways linking the formation of ROS with the end-point pathology are unknown.

The direct assessment of damaging free radical species by means of emerging technologies would not only advance our understanding of the underlying mechanisms but also facilitate prevention and intervention studies.

2 ELECTRON SPIN RESONANCE SPECTROSCOPY

Electron spin resonance (EPR) spectroscopy is considered the least ambiguous method for detecting free radicals, has a very high sensitivity, and can identify and quantify free radicals. The spectroscopic technique requires that the sample is positioned in a magnetic field and irradiated by microwaves; the technique is non-destructive, and sample can be used for further measurements. Living cells, whole tissues, and also small animals can be inserted in the cavity (equivalent to the NMR-probe) and reading performed.

2.1. Direct detection of free radicals

The application of EPR spectroscopy to biological problems can be traced back to the 50s,[9] when Commoner first applied the technique to detect free radicals in growing seeds. Unfortunately, water, the solvent where any biological reaction occurs, diminishes substantially the sensitivity of the instrument. An other hitch lays in the extremely low steady state concentration of many reactive free radicals, well under the limit of detection in water.[10]

In general, it is impossible to directly detect highly reactive free radicals formed in a complex biological model system, e.g. oxygen derived free radicals. The direct observation of more stable radicals, such as those possessing aromatic structure is quite feasible E.g. the detection of the ubiquinone radical and/or tocopheryl radical.[11] Also the ascorbyl radical is stable enough and the steady -state concentration in biological fluids allows a direct detection of the radical.

However, it has to be stressed that organic free radicals absorb in a narrow area of the EPR field; the co-presence in the same sample of different radical species will render particularly difficult the interpretation of the resulting complex spectra.

3 ASCORBYL FREE RADICAL

The possibility of an immediate detection of relatively stable free radicals makes it suitable for the application to human studies. The ascorbyl radical is the most studied species. This free radical is a resonance-stabilised tricarbonyl species that is readily formed from the one-electron oxidation of ascorbate. The radical show a characteristic doublet with hyperfine coupling value of 1.8 G and g=2.005. Because of the low reduction potential of the ascorbyl/ascorbate couple, nearly every oxidising radical that can arise in a biological system will cause the one-electron oxidation of ascorbate. The steady-state concentration of ascorbyl does not depend on ascorbate concentration, but mirrors the rate of ascorbate oxidation, thus it can be used as a measure of oxidative stress.

The typical experiment is quite simple. Blood is withdrawn, plasma separated and frozen, in alternative and for a longer preservation of the sample, plasma could be mixed with DMSO and frozen. The sample (as little as 50 μl) is thawn and immediately inserted in the spectrometer cavity. A calibrated capillary tube can be used as a container. The reading time, having the instrumental setting correctly adjusted, is about 60 seconds. In order to obtain concentration readings, a fairly accurate calibration curve can be obtained using a known concentration of a stable free radical (e.g. a nitroxide).[12]

Results have been reported in patients undergoing open heart surgery for aortic valve replacement.[13] In these patients, plasma ascorbyl radical levels were found to be significantly lower than in healthy subjects, decreased further upon ischemia, dropped to their lowest values within the first 10 min of reperfusion, and did not recover their initial values within 30 min following reflow. Ascorbyl levels at early reperfusion were slightly higher in plasma obtained from coronary sinus samples than in plasma from peripheral blood, suggesting an extra ascorbate release from the injured heart tissue.

In our experience[14] we collected blood separately from the right and left atrium during open heart surgery. While ascorbyl concentration in the blood collected from the left atrium showed the same behaviour as reported by Pietri et al.,[13] with a significant decrease observed during surgery, ascorbyl concentration in the right atrium did not change, indicating the lung as the site of ascorbate consumption.

Ascorbate is a good metal chelating agent, in particular iron and copper. Under these conditions, ascorbate initiates a Fenton-type chemistry giving rise to harmful reactive oxygen species. This was assessed in human seminal plasma. The modification in the levels of ascorbate was used to investigate non-invasively and in real time whether metal ions were catalytically active. The presence of copper (1 μM), but not iron ions caused a significant increase in ascorbyl radical, indicating that an increased copper-loading such as in some pathological conditions or in smoking subjects could cause fertility alteration in humans.[15]

Ascorbyl radical was analysed in the cerebrospinal fluid and serum of acute lymphoblastic leukaemia The radical has been investigated in samples from patients having no therapy, undergoing chemotherapy, and following therapy. A direct correlation was found between the concentration of the radical and patients undergoing chemotherapy.[16]

Hyperthermia causes ad significant increase in plasma ascorbyl radical and the authors conclude that ascorbic acid stimulates the cytotoxic action of hyperthermia, proposing the combination of hyperthermia and ascorbate treatment for tumour treatment.[17]

A recent study demonstrated that peroxynitrite, arising from the fast reaction between nitric oxide (NO) and superoxide (O_2^-) leads to generation of free radical intermediates such as the ascorbyl radical, along with other reactive free radicals such as the albumin-thiyl radical, detected by spin-trapping and uric acid-derived free radical, detected as the DMPO radical adduct in plasma whose thiol groups were previously blocked with 5,5-dithiobis-(2-nitrobenzoic acid). Peroxynitrite-mediated oxidations were also followed by oxygen consumption and ascorbate and plasma-thiol depletion. The conclusions were that peroxynitrite-mediated one-electron oxidation of biomolecules may be an important event in its cytotoxic mechanism.[18]

3.1. Ascorbyl radical and microdialysis

Ascorbyl radical can be detected using a technique, which could be applied to human studies: microdialysis. Microdialysis has been developed for pharmacological studies and applied to human studies in skeletal muscle and in the brain.[19,20] Important human studies have applied the microdialysis technique to analyse *in vivo* time-dependent changes in the levels of released neurotransmitters, related metabolites and other biologically important molecules in the extracellular space of discrete brain regions.[21,22] EPR detection of ascorbyl radical could be easily performed on the same samples obtained for studies performed for different purposes.

Animal studies have demonstrated the feasibility, however no human studies have been performed yet. Experiments has been performed to detect radicals in the frontal cortex of rats. The dialysis probe was inserted into the frontal cortex, and a doublet signal, of the ascorbyl radical detected. The ascorbyl concentration increased when the animal was challenged with cold-induced brain injury or an injection of $FeCl_2$. Using a similar approach, ascorbyl radical has been detected in the rabbit retina under ocular hyperpressure.[23]

4 SPIN TRAPPING

Many free radical species are highly reactive and short lived, and their steady state concentration is extremely low. In order to detect them it is necessary to use an indirect method: spin trapping.[24,25]

With this technique, a higher steady state concentration of free radicals is achieved, which can overcome the sensitivity problem. It has been widely applied to a multitude of different biological problems where free radicals are thought to be involved. Various spin trapping agents, both in *in vitro* and *in vivo* models, trap easily carbon- sulphur, and oxygen-centred radicals. Nitrone- or nitroso- compounds have been used as spin trapping agents. Recently a new class of phosphorylated nitrones, which gives long-lived oxygen adducts, has been introduced.[26,27]

Briefly, short-lived free radicals react with a nitrone- or nitroso-based diamagnetic compound, to form a relatively stable free radical adduct. The adduct is detected by EPR spectroscopy, giving semi-quantitative and qualitative measurements. The technique initially developed for studies *in vitro*, has been applied to *ex vivo* in animal and human studies.[25,28,29] Pitfalls and artifacts of the technique have been also carefully described.[30,31]

4.1. Spin trapping and microdialysis

An important question in free radical research concerns the site of radical production. The use of stereotaxic microdyalisis can help solving the problem. A microdialysis probe perfused with a solution containing a spin trap has been employed in animal studies to test the formation of free radical *in vivo* in selected areas.[32] Microdialysis -spin trapping can not be applied to human studies since the spin trapping agent has to be injected in the area under study, and spin trapping agents are not registered for human use. Interesting pharmacological activity of nitrone derivatives are under active study, a probable future human use is foreseen.[33,34]

4.2. Spin trapping *ex vivo*

A possible way to circumvent the impossibility of using spin trapping agents for human studies has been proposed by Tortolani et al.[35] Coronary sinus blood samples were withdrawn during open heart surgery and immediately mixed with an isosmotic spin trap. Plasma samples were extracted with toluene and analysed using EPR spectroscopy. Spectra consistent with the formation of alkoxyl and carbon-centred radical adducts were demonstrated. The species observed are free radical derived from peroxidation-linked insult and the measurement provides a quantitative index of irreversible oxidative injury *in vivo* during postischemic stress. The free radical detection may offer a powerful approach for assessing protection against reperfusion-mediated oxidative injury in the presence of anti-radical intervention, a treatment which attenuate reperfusion-mediated functional and/or tissue injury and suggests the potential for clinical application.

A second possible utilisation of the spin trapping technique has been proposed by Roselaar et al.[36] who developed an application of electron spin resonance spectroscopy, to show that a stable oxidising component or components of plasma accumulate in uraemia. The oxidant was detected by its capacity to oxidise the spin trap 3,5-dibromo-4-nitrosobenzene sulphonate, was dialyzable from plasma, and was stable over months.

5 NITRIC OXIDE (NO, NITROGEN MONOXIDE)

NO is a free radical molecule investigated in the central nervous and in the cardio-vascular system as a transmitter produced by neuronal and endothelial cells. Ischemic lesion of the brain induces nitric oxide release.[37] NO acts as a vasodilator and is a functional antagonist of the vasoconstrictor Endothelin-1.[38] For recent reviews on the diverse NO actions see.[39-41]

NO is a short-lived molecule in biological systems.. In fact, NO reacts quickly with molecular oxygen, superoxide and thiol- and iron-proteins.[42] Evaluations of NO stable by-products, like nitrate and nitrite (Griess reaction), NO-induced second messengers, like cGMP, the determination of citrulline, the haemoglobin reaction (that quantifies oxyhaemoglobin to methaemoglobin transformation by NO) have been used as indirect methods to measure NO in plasma and tissues (for review see ref. 48).

Direct NO determination can be carried out using electrochemical detection with a porphyrinic probe[43] or by EPR spectroscopy determination of various NO-iron centred compounds, including NO-haemoglobin or NO-myoglobin.[44-46] This method allow detection of NO simply withdrawing venous blood (as little as 100 µl are sufficient for the assay), freezing and introducing the sample in the EPR cavity. The method has been described in detail in.[47,48]

Nitrosyl complexes have been detected in human substantia nigra, in relation to Parkinsonís disease,[49] in tissues taken from human liver, colon and stomach tumours.[50] Blood levels of nitrosylhaemoglobin have been measured in haemodialysis patients and found significantly higher than healthy controls.[51]

A special role is played by NO in human sepsis:[41,52] on one hand NO carries out cytoprotective functions, on the other hand, overproduction appears to contribute to haemodynamic instability and tissue damage. These observations have led to the development of strategies to inhibit NO synthesis or scavenge excess NO in patients with septic shock. The capacity of red blood cells and haemoglobin to remove NO most likely accounts for the adjuvant effect of blood observed in peritonitis. Though feasible and easy to perform, the direct measurement of NO haemoglobin has never been carried out in patients with septic shock, and the possible clinical use of this approach merit a more thorough research in this field.

The role played by NO in various forms of cardiovascular shock, is less known. A recent study in the experimental animal has shown the sudden and huge increase in blood NO is concomitant to the arterial pressure decrease.[53] The association with excess nitric oxide production and the onset of this frequent and life-threatening human disease is not well characterise. The available evidence of a role for nitric oxide should stimulate the interest of investigators to explore these area more thoroughly.

References

1.   C. Scarfiotti, F. Fabris, B. Cestaro, and A. Giuliani. Free radicals, atherosclerosis, ageing, and related dysmetabolic pathologies: Pathological and clinical aspects. *Eur. Cancer Prev*. 1997. 1.

2.   S. Kuroda and B.K. Siesjo. Reperfusion damage following focal ischemia: Pathophysiology and therapeutic windows. *Clin. Neurosci*. 1997, **4**, 199.

3.   B. E. Leonard. Advances in the drug treatment of Alzheimer's disease. *Human Psychopharmacology Clinical and Experimental*. 1998, **13**, 83.

4.   A. H. V. Schapira. Oxidative stress in parkinson's disease -review. *Neuropathology and Applied Neurobiology*. 1995, **21**, 3.

5.   N.A. Simonian and J.T. Coyle. Oxidative stress in neurodegenerative diseases. *Ann. Rev. of Pharmacology and Toxicology*. 1996, **36**, 83.

6.   A. Parthiban, S. Vijayalingam, K. R. Shanmugasundaram, and R. Mohan. Oxidative stress and the development of diabetic complications -antioxidants and lipid peroxidation in erythrocytes and cell membrane. *Cell Biol Int*. 1995, **19**, 987.

7.   M. Taylor and D. Kerr. Diabetes control and complications: a coming of AGE. *Lancet*. 1996, **347**, 485.

8.   R. Smith. Experimental models suggest therapies for human diabetes. *Lancet*. 1996, **347**, 1175.

9.   B. Commoner, J. Townsend, and G. .Pake. Free radicals in biological materials. *Nature*. 1954, **174**, 689.

10.   M. C. R. Symons. Chemical and Biochemical Aspects of Electron Spin Resonance Spectroscopy. 1978.

11.   D. A. Stoyanovsky, A. N. Osipov, P. J. Quinn, and V. E. Kagan. Ubiquinone-dependent recycling of vitamin e radicals by superoxide. *Archives of Biochemistry and Biophysics*. 1995, **323**, 343.

12.   G. R. Buettner and B. A. Jurkiewicz. Ascorbate free radical as a marker of oxidative stress: an EPR study. *Free Rad. Biol. Med*. 1993, **14**, 49.

13.   S. Pietri, J. R. Seguin, P. D. d'Arbigny, and M. Culcasi. Ascorbyl free radical: a noninvasive marker of oxidative stress in human open-heart surgery. *Free Radic.*

Biol. Med. 1994, **16**, 523.

14. A. Tomasi, A. Bini, E. Ghelfi, and M. Meli. Personal observation 1998.

15. A. Menditto, D. Pietraforte, and M. Minetti. Ascorbic acid in human seminal plasma is protected from iron-mediated oxidation, but is potentially exposed to copper-induced damage. *Hum Reprod.* 1997, **12**, 1699.

16. K. Nakagawa, H. Kanno, and Y. Miura. Detection and analyses of ascorbyl radical in cerebrospinal fluid and serum of acute lymphoblastic leukemia. *Anal. Biochem.* 1997, **254**, 31.

17. K. Satoh, H. Sakagami, and K. Nakamura. Enhancement of radical intensity and cytotoxic activity of ascorbate by hyperthermia. *Anticancer Res.* 1996, **16**, 2987.

18. J. Vasquez Vivar, A. M. Santos, V. B. Junqueira, and O. Augusto. Peroxynitrite-mediated formation of free radicals in human plasma: EPR detection of ascorbyl, albumin-thiyl and uric acid-derived free radicals. *Biochem J.* 1996, **314**, 869.

19. T. Graven Nielsen, A. McArdle, J. Phoenix, L. Arendt Nielsen, T. S. Jensen, M. J. Jackson, and R. H. Edwards. *In vivo* model of muscle pain: quantification of intramuscular chemical, electrical, and pressure changes associated with saline-induced muscle pain in humans. *Pain.* 1997, **69**, 137.

20. R. D. Scheyer, M. J. During, J. M. Hochholzer, D. D. Spencer, J. A. Cramer, and R. H. Mattson. Phenytoin concentrations in the human brain: an *in vivo* microdialysis study. *Epilepsy Res.* 1994, **18**, 227.

21. U. Ungerstedt and A. Hallstr^m. *In vivo* microdialysis -A new approach to the analysis of neurotransmitters in the brain. *Life Sci.* 1987, **41**, 86.1

22. R. Kanthan, A. Shuaib, G. Goplen, and H. Miyashita. A new method of in-vivo microdialysis of the human brain. *J. Neurosci. Methods.* 1995, **60**, 151.

23. A. Muller, S. Pietri, M. Villain, C. Frejaville, C. Bonne, and M. Culcas. Free radicals in rabbit retina under ocular hyperpressure and functional consequences. *Exp. Eye Res.* 1997, **64**, 637.

24. E. G. Janzen. A critical review of spin trapping in biological systems. 115-153, 1980. Pryor W.A. *Free radicals in biology vol IV.*

25. R. P. Mason and K. T. Knecht. *In vivo* detection of radical adducts by electron spin resonance, in Oxygen Radicals in Biological Systems, Pt C, L. Packer, Editor. Academic Press Inc, San Diego. p. 112, 1994.

26. C. Frejaville, H. Karoui, B. Tuccio, F. Le Moigne, M. Culcasi, S. Pietri, R. Lauricella, and P. Tordo. 5-(Diethoxyphosphoryl)-5-methyl-1-pyrroline N-oxide: a new efficient phosphorilated nitrone for the *in vitro* and *in vivo* spin trapping of oxygen-centered radicals. *J. Med. Chem.* 1995, **38**, 258.

27. B. Tuccio, A. Zeghdaoui, J. P. Finet, V. Cerri, and P. Tordo. Use of new Beta Phosphorylated Nitrones for the Spin Trapping of Free Radicals. *Research on Chemical Intermediates.* 1996, **22**, 393.

28. G.R. Buettner and R.P. Mason. Spin-trapping methods for detecting superoxide and hydroxyl free radicals *in vitro* and *in vivo*. 1990.

29. K.J. Reszka, P. Bilski, C.F. Chignell, and J. Dillon. Free radical reactions photosensitized by the human lens component, kynurenine:an EPR and spin trapping investigation. *Free Radic. Biol. Med.* 1996, **20**, 23.

30. C. Mottley and R. P. Mason, Nitroxide radical adducts in biology: chemistry,

applications, and pitfalls, in Spin Labeling. Theory and Applications, L.J.B.a.J. Reuben, Editor. Plenum Press, New York. p. 489, 1989.

31.   A. Tomasi and A. Iannone, ESR spin-trapping artifacts in biological model systems, in Biological Magnetic Resonance. EMR of Paramagnetic Molecules, L.J.B.a.J. Reuben, Editor. Plenum Press, New York. p. 353, 1993.

32.   I. Zini, A. Tomasi, R. Grimaldi, V. Vannini, and L. F. Agnati. Detection of Free Radicals During Brain Ischemia and Reperfusion by Spin Trapping and Microdialysis. *Neuroscience Letters*. 1992, **138**, 279.

33.   E. Lancelot, M. L. Revaud, R.G. Boulu, M. Plotkine, and J. Callebert. Alpha-phenyl-N-tert-butylnitrone attenuates excitotoxicity in rat striatum by preventing hydroxyl radical accumulation. *Free Radical Biology And Medicine*. 1997, **23**, 1031.

34.   D. A. Butterfield, B. J. Howard, S. Yatin, K. L. Allen, and J. M. Carney. Free radical oxidation of brain proteins in accelerated senescence and its modulation by N-Tert butyl alpha phenylnitrone. *Proceedings of the National Academy of Sciences of the United States of America*. 1997, **94**, 674.

35.   A. J. Tortolani, S. R. Powell, V. Misik, W. B. Weglicki, G. J. Pogo, and J. H. Kramer. Detection of Alkoxyl and Carbon-Centered Free Radicals in Coronary Sinus Blood from Patients Undergoing Elective Cardioplegia. *Free Radical Biology and Medicine*. 1993, **14**, 421.

36.   S.E. Roselaar, N.B. Nazhat, P.G. Winyard, P. Jones, J. Cunningham, and D.R. Blake. Detection of oxidants in uremic plasma by electron spin resonance spectroscopy. *Kidney Int*. 1995, **48**, 199.

37.   T. Nagafuji, M. Sugiyama, T. Matsui, A. Muto, and S. Naito. Nitric oxide synthase in cerebral ischemia -possible contribution of nitric oxide synthase activation in brain microvessels to cerebral ischemic injury. *Molecular and Chemical Neuropathology*. 1995, **26**, 107.

38.   O. Durieutrautmann, C. Federici, C.Creminon, N. Foignantchaverot, F. Roux, M. Claire, A. D. Strosberg, and P. O. Couraud. Nitric oxide and endothelin secretion by brain microvessel endothelial cells -regulation by cyclic nucleotides. *Journal of Cellular Physiology*. 1993, **155**, 104.

39.   C. Szabo. Physiological and pathophysiological roles of nitric oxide in the central nervous system. *Brain Res Bull*. 1996, **41**, 131.

40.   J. Macmicking, Q. W. Xie, and C. Nathan. Nitric Oxide and Macrophage Function. *Annual Review of Immunology*. 1997, **15**, 323.

41.   M. L. Johnson and T. R. Billiar. Roles of nitric oxide in surgical infection and sepsis. *World J. Surg*. 1998, **22**, 187.

42.   J. F. J. Kerwin and M. Heller. The arginine-nitric oxide pathway: A target for new drugs, *Med. Res. Rev*. 1994, **14**, 23.

43.   C. Privat, F. Lantoine, F. Bedioui, E. Millanvoye van Brussel, J. Devynck, and M. A. Devynck. Nitric oxide production by endothelial cells: comparison of three methods of quantification. *Life Sci*. 1997, **61**, 1193.

44.   W. Chamulitrat, S.J. Jordan, and R.P. Mason. Nitric oxide production during endotoxic shock in carbon tetrachloride-treated rats. Mol Pharmacol. 1994, **46**, 391.

45.   P. Wang and J. L. Zweier. Measurement of nitric oxide and peroxynitrite generation in the postischemic heart. Evidence for peroxynitrite-mediated reperfusion injury. *J.*

Biol. Chem. 1996, **271**, 29223.

46.   A. M. Komarov, J. H. Kramer, I. T. Mak, and W. B. Weglicki. EPR detection of endogenous nitric oxide in postischemic heart using lipid and aqueous-soluble dithiocarbamate-iron complexes. *Mol .Cell Biochem.* 1997, **175**, 91.

47.   M. E. Murphy and E. Noack. Nitric oxide assay using hemoglobin method. *Methods Enzymol.* 1994, **233**, 240.

48.   A. Kozlov, A. Bini, A. Iannone, I. Zini, and A. Tomasi. Electron paramagnetic resonance characterization of rat neuronal nitric oxide production ex vivo. *Methods Enzymol.* 1996, **268**, 229.

49.   J. K. Shergill, R. Cammack, C. E. Cooper, J. M. Cooper, V. M. Mann, and A. H. Schapira. Detection of nitrosyl complexes in human substantia nigra, in relation to Parkinson's disease. *Biochem. Biophys. Res. Commun.* 1996, **228**, 298.

50.   M. C. R. Symons, I. J. Rowland, N. Deighton, K. Shorrock, and K. P. West. Electron spin resonance studies of nitrosyl haemoglobin in human liver, colon and stomach tumour tissues. *Free Radical Research.* 1994, **21**, 197.

51.   D. Roccatello, G. Mengozzi, V. Alfieri, E. Pignone, E. Menegatti, G. Cavalli, G. Cesano, D. Rossi, M. Formica, T. Inconis, G. Martina, L. Paradisi, L. M. Sena, and G. Piccoli. Early increase in blood nitric oxide, detected by electron paramagnetic resonance as nitrosylhaemoglobin, in haemodialysis. *Nephrol. Dial. Transplant.* 1997, **12**, 292.

52.   N. Fukuyama, Y. Takebayashi, M. Hida, H. Ishida, K. Ichimori, and H. Nakazawa. Clinical evidence of peroxynitrite formation in chronic renal failure patients with septic shock. *Free Radic. Biol. Med.* 1997, **22**, 771.

53.   S. Guarini, A. Bini, C. Bazzani, G. M. Ricigliano, M. M. Cainazzo, A. Tomasi, and A. Bertolini. Adrenocorticotropin normalizes the blood levels of nitric oxide in hemorrhage-shocked rats. *Eur. J. Pharmacol.* 1997, **336**, 15.

RADIOIMMUNOASSAYS OF ISOPROSTANES AND PROSTAGLANDINS AS BIOMARKERS OF OXIDATIVE STRESS AND INFLAMMATION

S. Basu.

Department of Geriatrics, Faculty of Medicine, Uppsala University , Uppsala, Sweden.

1 INTRODUCTION

Both non-enzymatic oxidation via free radicals and enzyme catalysed oxidation of unsaturated fatty acids have been implicated in a number of diseases in man including cardiovascular disorders, cancer, smoking, aging processes, Alzheimer's diseases, inflammation etc.[1-4] As consequence of the increasing interests in the basic mechanisms of these important biological processes in human diseases, there is a compelling need to develop reliable assay methods to detect the stable reaction products of these processes. Serious analytical difficulties have been encountered in the assay of various prostaglandins and free radical catalysed reaction products in the past years. The critical problems in the field of prostaglandins are mainly related to the metabolism, chemical stability, species differences and artifactual formation whereas analytical problems are more common in detecting free radical catalysed reaction products *in vivo* than *in vitro* conditions.

1.1. Isoprostanes

Lipid peroxidation through free radical oxidation of unsaturated fatty acid is presumed to be involved in oxidative injury-induced diseases. Although several methods exist to measure the reaction products of free radical induced lipid peroxidation the limitations of these methods in assessing lipid peroxidation *in vivo* are common.[5-7] Isoprostanes, a family of prostaglandin derivatives biosynthesized from arachidonic acid *in vivo* through non-enzymatic free radical catalysed mechanisms are found to be reliable parameters of oxidative injury in both human and animal models.[8-10] Recently, it has been observed that one of the major isoprostanes, (Fig. 1) which exibits potent biological activity is elevated in several syndromes that are proposed to be associated with oxidant injury.[11-13] These include smoking, high alcohol consumsion, diabetes, liver cirrhosis and vascular reperfusion etc. Levels of F_2-isoprostanes, both free in the peripheral circulation and esterified to tissue phospholipids increase highly in animal models of oxidant injury. The gas chromatography-mass spectrometry method has been used to quantify these compounds. This is a complicated analytical procedure especially considering the laborious extraction procedures and that the availability of the instrument is limited. Thus, the application of this method in large experimental and clinical studies is very difficult and practically available only to few laboratories.

1.2. Prostaglandins

Arachidonic acid oxidation through cyclooxygenases (COX) leads to the formation of a variety of prostaglandins and thromboxanes[14,15] (Fig. 1). These compounds are active biomolecules and formed constitutively in most cells throughout the body at a low concentration (pg/ml level) during basal conditions. The biosynthesis and release of prostaglandins, mainly prostaglandin $F_{2\alpha}$ ($PGF_{2\alpha}$) during various physiological (luteolysis, parturition etc.) and pathophysiological situations (inflammatory conditions) are well established.[16-18] Recent data on prostaglandins reveal that these compounds are also involved in cancer, migraine, stroke and Alzheimer's diseases among others.[19-21] However, when measuring prostaglandins *in vivo* several important factors like rapid formation, diverse occurrence and action profile, rapid metabolism, artifactual formation and species difference have to be taken into consideration. Otherwise fast and short lasting peaks (pulsatile pattern) of these compounds or their metabolites could easily be overlooked or an artifactual formation of these compounds could mislead the results.

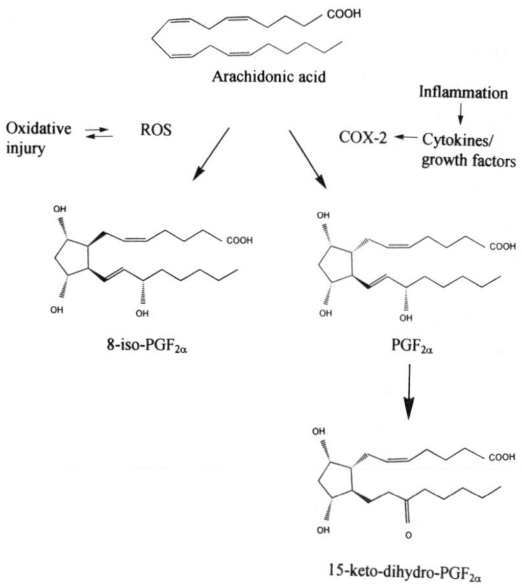

Figure 1
A schematic diagram of the formation of 8-iso-PGF$_{2\alpha}$ via non-enzymatic free radical and 15-keto-13,14-dihydro-PGF$_{2\alpha}$ via cyclooxygenase catalysed oxidation of arachidonic acid as caused by oxidative injury and inflammation, respectively.

15-Keto-13,14-dihydro-PGF$_{2\alpha}$ (15-keto-dihydro-PGF$_{2\alpha}$), a major metabolite of PGF$_{2\alpha}$ is found to be more reliable and widely applied parameter in plasma due to its longer half-life than the parent compound and does not form artifactually.[14-16] Measurement of this metabolite in frequently collected plasma samples has successfully been applied during oestrous cycle, luteolysis, parturition and inflammatory conditions in various species[16-18] (Fig. 1). However, the cross-reactivity of the earlier developed 15-keto-dihydro-PGF$_{2\alpha}$ antibodies was not studied against the recently discovered isoprostanes which is of importance due to their formation in large quantities (about 10 times higher than primary prostaglandins) during free radical catalysed oxidation of arachidonic acid (Fig. 1). The recent discovery of cyclooxygenase-2 (COX-2) as an isoform of cyclooxygenases (COX) that is shown to be inducible in macrophages, epithelial cells and fibroblast by several pro-inflammatory stimuli (cytokines, growth factors etc.) leading to COX-2 expression and release of various prostaglandins and related compounds in biological fluid which is of great interest.[22-24] Thus, the measurement of 15-keto-dihydro-$PGF_{2\alpha}$ as an indicator of *in vivo* $PGF_{2\alpha}$ biosynthesis and release in human and other species has become highlighted. Based on the earlier data on prostaglandin measurement, metabolism, artifactual formation or chemical stability profile 15-keto-dihydro-$PGF_{2\alpha}$ measurement seems to have better advantages in many respect over the measurement of PGE_2 in the assessment of inflammatory index or activation of cyclooxygenases in various cells.

1.3. Radioimmunoassay

Radioimmunoassay (RIA) has always been an important tool for the measurement of low levels of endogenous prostaglandins or their exogenous analogues.[16,17,25-28] The basic principal of a radioimmunoassay is the competition between radiolabelled and unlabelled molecules of a particular compound for the binding sites of a specific antibody raised against the same compound. As more as the particular unlabelled molecules are available in a biological sample or standard, the more radiolabelled molecule will be displaced from the binding sites of the antibody. The radioactivity of either fractions i.e. antibody bound precipitated or free supernatant fraction after separation can be quantified in order to get a standard curve and followed by the concentration of the particular compound in the samples. We have recently developed radioimmunoassays by raising antibodies specific for free 8-iso-$PGF_{2\alpha}$ and 15-keto-dihydro-$PGF_{2\alpha}$ measurement in the biological fluids. The cross-reactivity of these antibodies were well investigated against various isoprostanes and prostaglandins since the chemical structures of these substances are very close[27-28] and diverse formation of these compounds is also reported.[9,14-15] This study presents an overview of the recent development and validation of radioimmunoassays for free 8-iso-$PGF_{2\alpha}$ as an index of oxidative injury and 15-keto-dihydro-$PGF_{2\alpha}$ as an index of inflammation, and their application in experimental and clinical conditions.

2 DEVELOPMENT OF 8-ISO-$PGF_{2\alpha}$ RADIOIMMUNOASSAY

2.1. Antibody development

Having a low molecular weight 8-iso-$PGF_{2\alpha}$, like other prostaglandins, is not immunogenic itself. To obtain the molecule antigenic attachment to a larger carrier protein molecule is essential. A carboxyl group at the α-chain rendered the molecule to be coupled to an amino group of a protein by the formation of a peptide bond which is suitable for immunization purpose in animals. 8-iso-$PGF_{2\alpha}$ was coupled to bovine serum albumin (BSA) by N, N'-carbonyldiimmidazole coupling procedure with some minor changes.[27-29]

2.2. Specificity of the 8-iso-$PGF_{2\alpha}$ antibody

The specificity of the 8-iso-$PGF_{2\alpha}$ antibody is calculated by comparing the amount of substance (structurally related or endogenous, Table 1) required to displace the same amounts of radiolabelled proper compound (8-iso-$PGF_{2\alpha}$) at 50 % displacement.[27] The specificity of 8-iso-PGF$_{2\alpha}$ antibody compared with several structurally-related compounds is shown in Table 1.

Table 1

Specificity of 8-iso-PGF$_{2\alpha}$ antibody with structurally related compounds. (Reprinted from Basu[27]).

Substances		Cross-reaction %
8-iso-PGF$_{2\alpha}$		100
8-iso-15-keto-13,14-dihydro-PGF$_{2\alpha}$		1.7
8-iso-PGF$_{2\beta}$		9.8
PGF$_{2\alpha}$		1.1
15-keto-PGF$_{2\alpha}$		0.01
15-keto-13,14-dihydro-PGF$_{2\alpha}$		0.01
TXB$_2$		0.1
11β-PGF$_{2\alpha}$		0.03
9β-PGF$_{2\alpha}$		1.8
8-iso-PGF$_{3\alpha}$		0.6

3 APPLICATION OF 8-ISO-PGF$_{2\alpha}$ RADIOIMMUNOASSAY

3.1. Human

The details of the preparation of radiolabelled tracer, assay procedure and validation of 8-iso-PGF$_{2\alpha}$ assay with the above raised antibody is described elsewhere.[27]

Blood and urinary samples were collected from healthy volunteers of different ages, who had not taken any medications, to estimate the basal levels of free 8-iso-PGF$_{2\alpha}$. The peripheral plasma and urinary concentrations (representative for about eight hours urine) of 8-iso-PGF$_{2\alpha}$ from healthy volunteers were 79±9.6 pmol/l (mean ± SEM) and 3.4±0.3 nmol/l, respectively. However, to get an urinary excretion rate of 8-iso-PGF$_{2\alpha}$ the following studies were corrected with the urinary creatinine values.[18,30,32] There was no diurnal variation in the urinary levels of 8-iso-PGF$_{2\alpha}$ during the day. Neither was there any significant difference between 8-iso-PGF$_{2\alpha}$ levels in the morning urine samples compared to the 24-h urine samples.[30]

In a recent human dietary study it was shown that urinary excretion of 8-iso-PGF2α as measured by this RIA was significantly increased when the urinary concentration of nitric oxide metabolites decreased after a high linoleic acid diet for four weeks in healthy volunteers.[31] This study shows that an increased intake of polyunsaturated fatty acids presumably results in an increased oxidative stress and affects endothelial function.

4 DEVELOPMENT OF THE 15-KETO-DIHYDRO-PGF$_{2\alpha}$ RADIOIMMUNOASSAY

4.1. Antibody development

Since 15-keto-dihydro-PGF$_{2\alpha}$ is not immunogenic itself the compound was coupled to bovine serum albumin (BSA) by N, N´-carbonyldiimmidazole coupling procedure with some minor changes.[28,29]

4.2 Specificity of the 15-keto-dihydro-PGF$_{2\alpha}$ antibody

The specificity of the 15-keto-dihydro-PGF$_{2\alpha}$ antibody is calculated by comparing the amount of substance (structurally related, Table 2) required to displace the same amounts of radiolabelled proper compound (15-keto-dihydro-PGF$_{2\alpha}$) at 50 % displacement. The specificity of 15-keto-dihydro-PGF$_{2\alpha}$ antibody with several structurally related compounds is shown in Table 2.

5 APPLICATION OF THE 15-KETO-DIHYDRO-PGF$_{2\alpha}$ RADIOIMMUNOASSAY

5.1. Human

The details of the assay procedure and validation of 15-keto-dihydro-PGF$_{2\alpha}$ assay with the above raised antibody is described elsewhere.[28]

Basal levels of 15-keto-dihydro-PGF$_{2\alpha}$ in plasma and urine from healthy volunteers were determined by this radioimmunoassay. The peripheral plasma and urinary concentrations (representative for about eight hours urine) of 15-keto-dihydro-PGF$_{2\alpha}$ were 129.7±13.8 pmol/l (mean±SEM) and 0.9±0.6 nmol/l, respectively. Although measurement of 15-keto-dihydro-PGF$_{2\alpha}$ in frequently collected plasma samples is the best parameter of PGF$_{2\alpha}$ release, a large amounts of 15-keto-dihydro-PGF$_{2\alpha}$ were also seen in the urine of various species.[17,28] It may be possible that a true excretion of this metabolite still occurs in the urine which may serve as an urinary parameter of PGF$_{2\alpha}$ release which needs further investigation. In such a case a difficult problem of frequent blood sampling specially in man could be circumvented by measuring 15-keto-dihydro-PGF$_{2\alpha}$ in daily urine.

Table 2

Specificity of 15-keto-dihydro-PGF $_2$ α antibody with structurally related compounds. (Reprinted from Basu [28]).

Substances	Cross-reaction %
15-keto-13,14-dihydro-PGF$_{2\alpha}$	100
PGF$_{2\alpha}$	0.02
15-keto-PGF$_{2\alpha}$	0.43
PGE$_2$	<0.001
15-keto-13,14-dihydro-PGE$_2$	0.5
8-iso-15-keto-13,14-dihydro-PGF$_{2\alpha}$	1.7
11β-PGF$_{2\alpha}$	<0.001
9β-PGF$_{2\alpha}$	<0.001
TXB$_2$	<0.001
8-iso-PGF$_{3\alpha}$	0.01

6 ASSAY SUMMARY

The assays of both 8-iso-PGF$_{2\alpha}$ and 15-keto-13,14-dihydro-PGF$_{2\alpha}$ in the biological fluid do not need any prior extraction or purification of the biological samples thus minimizing a major origin of erroneous results. From the experience of radioimmunoassay of prostaglandins it has been observed that the advantages exceeed the disadvantages when samples were analysed unextracted than purified through various extraction and chromatographic procedures. However, the assay methods must be developed to encounter the interference problems from the samples without deteriorating the accuracy, precision and sensitivity of the assay. Further details of these radioimmunoassays are described elsewhere.[27,28]

    Taken together, both 8-iso-PGF$_{2\alpha}$ and 15-keto-dihydro-PGF$_{2\alpha}$ antibodies were found to be specific and suitable for the measurement of immunoreactive 8-iso-PGF$_{2\alpha}$ and 15-keto-dihydro-PGF$_{2\alpha}$ in the plasma and urinary samples of various species. Since PGF$_{2\alpha}$

release occurs in pulsatile fashion and metabolises rapidly to the initial metabolite 15-keto-dihydro-$PGF_{2\alpha}$ in many species, frequent sampling regime in plasma should be applied to reflect the $PGF_{2\alpha}$ release *in vivo*.

7 CONCLUSIONS

In conclusion, sensitive and specific radioimmunoassay methods for 8-iso-$PGF_{2\alpha}$, a biomarker of free radical induced oxidative injury and 15-keto-dihydro-$PGF_{2\alpha}$, a biomarker of inflammation (and other physiological and pathophysiological conditions) through the cyclooxygenase pathway have been developed and subsequently validated. The methods have been successfully applied for the bioanalysis of 8-iso-$PGF_{2\alpha}$ and 15-keto-13,14-dihydro-$PGF_{2\alpha}$ in body fluids of various species from both clinical and experimental studies. These radioimmunoassay methods, using these specific antibodies to determine the degree of oxidative modification through both non-enzymatic and enzymatic pathways, open excellent possibilities for accurate and simultaneous determination of both oxidative injury and inflammatory states in the pathogenesis of various diseases, clinical and experimental conditions. Due to rapid metabolism of these compounds sample regime is an eminent factor for the determination of these compounds in biological fluids.

Acknowledgements

This study is financially supported by the Geriatrics Research Foundation, Swedish Diabetes Foundation, Loo and Hans Ostermans Foundation for Medical Research and Swedish Council for Forestry and Agricultural Research. The author is indebted to Professor Bengt Vessby for valuable discussions

References

1. B. Halliwell and J.M.C. Gutteridge. Free radicals in biology and medicine. 2nd edition, Clarendan Press, *Oxford*. 1989.
2. M.A. Smith, S. Lawrence and G. Perry. *Alzheimer's Disease Review*, 1996, **1**, 63.
3. J.W.Baynes. *Diabetes*. 1991, **40**, 405.
4. E.R. Stadtman. *Science*. 1992, **257**, 1220.
5. J.M.C. Gutteridge and M. Grootveld. *FEBS Lett*. 1987, **213**, 9.
6. G. Cao and R.G. Cutler. Arch. Biochem. *Biophys*. 1995. 320.
7. J. M. C. Gutteridge and B. Halliwell. Trends. *Biochem. Sci*. 1990, **15**, 129.
8. J. Morrow, K. E. Hill, R. F. Burk, R.M. Nammour, K.F. Badr and L.J. Roberts II. *Proc. Natl. Acad. Sci. U.S.A.* 1990, **87**, 9383.
9. J.D. Morrow and L. J. Roberts II. *Biochem. Pharmacol*. 1996, **51**, 1.
10. J. D. Morrow, J.A. Awad, T. Kato, K. Takahashi , K. F. Badr, L. J. Roberts II and R.F. Burk. *J. Clin. Invest*. 1992, **90**, 2502.
11. J. D. Morrow, B. Frei, A. W. Longmire, S. M. Lynch, Y. Shyr, W. E. Strauss, J. A. Oates and L. J. Roberts II. *N. Engl. J. Med*. 1995, **332**, 1198.
12. A.A. Nanji, K. Samsuddin, S.R. Tahan and S.M.H. Sadrzadeh. *Pharmacol. and Exp. Therap.*1994, **269**, 1280.
13. J. D. Morrow, K. P. Moore, J. A. Awad, M. D. Ravenscraft, G. Marini, K. F.Badr, R. Williams and L.J. Roberts II. *J. Lipid Med*. 1993, **6**, 417.

14. B. Samuelsson, E. Granström, K. Greén, M. Hamberg, S. Hammarström. *Ann. Rev. Biochem.* 1975, **44**, 669.

15. B. Samuelsson, M. Goldyne, E. Granström, M. Hamberg, S. Hammarström and C. Malmsten. *Ann. Rev. Biochem.* 1978, **47**, 997.

16. S. Basu and H. Kindahl. *Vet. Med. A.* 1987, **34**, 487.

17. S. Basu, H. Kindahl, D. Harvey and K. Betteridge. *Acta vet. scand.* 1987, **28**, 409.

18. S. Basu and M. Eriksson. (Submitted) 1998.

19. G. M. Pasinetti. *J. Neurosci. Res.* 1998, **54**, 1.

20. T. A. Chan, P.J. Morin, B. Vogelstein and K. W. Kinzler. *Proc. Natl. Acad. Sci. USA.* 1998, **95**, 681.

21. K. Subbramaiah, D. Zakim, B. B. Weksler and A. J. Dannerberg. *Proc. Exp. Biol. & Med.* 199, **216**, 201.

22. J.Y. Fu, J.L. Masferrer, K. Seibert, A. Raz and P. Needleman. *J. Biol. Chem.* 1990, **265**, 16735.

23. J. A. Mitchell, P. Akarasereenont, C. Theimermann, R. J. Flower and J. Vane. *Proc. Natl. Acad. Sci. USA.* 1994, **90**, 11693.

24. J. Vane and R. A. Botting Adv. *Prost. Throm. and Leuk. Res.* 1995, **23**, 41.

25. E. Granström and H. Kindahl. *Methods Enzymol.* 1982, **86**, 320.

26. S. Basu and B. Sjöquist. *Prost. Leuk. Ess. Fatty acids.* 1996, **55**, 427.

27. S. Basu. *Prost. Leuk. Ess. Fatty acids.* 1998, **58**, 319.

28. S. Basu. *Prost. Leuk. Ess. Fatty acids.* 1998, **58**, 347.

29. U. Axen. *N,N'-Prostaglandins.* 1974, **5**, 45.

30. J. Helmersson and S. Basu. (unpublished).

31. S. Basu. *FEBS Lett.* 1998, **428**, 32.

32. A. Turpeinen, S. Basu and M. Mutanen. *Prost. Leuk. Ess.Fatty acids.* 1998 (In press)

33. J. A. Awad and J. D. Morrow. *Hepatology.* 1995, **22**, 962.

FREE RADICAL PROCESSES IN 1,2-DIMETHYLHYDRAZINE INDUCED CARCINOGENESIS IN RATS AND PROTECTIVE ROLE OF MELATONIN

A.V. Arutjunyan[1], S.O. Burmistrov[1], T.I. Oparina[1], V.M. Prokopenko[1], M.G. Stepanov[1], I.G. Popovich[2], M.A. Zabezhinsky[2], V.N. Anisimov[2].

[1]D.O. Ott Institute of Obstetrics & Gynecology RAMS. [2]N.N. Petrov Institute of Oncology, St. Petersburg, Russia.

1 INTRODUCTION

Recent study tends to emphasize the important role of free radical processes in cancerogenesis. There are many evidences that reactive oxygen species (ROS) are involved in cascade stages of procarcinogens activation being a promotors of tumor development, what itself enhances the free radicals generation and declines the efficiency of antioxidative system.[1,2] Besides many carcinogens are highly potent activators of oxygen burst and ROS generation by phagocytic cells, and this is a reason that inflammatory processes may increase incedence of cancer. There is some information about the role of free radicals in dimethylhydrazine-induced carcinogenesis.[3,4] For a variety of tumors their growth is accelerated with hormone of pinealectomized animals whereas treatment with hormone of pineal gland melatonin inhibits development of some types of tumors, mainly mammary carcinogenesis both *in vivo* and *in vitro*. Our previous investigations have shown that one of possible target for melatonin action are colon tumors.[5] Intenstines are known as one of the important sources of melatonin production in mammals.[6] It has been also shown the disturbance in circadian rhythm of melatonin secretion in colon cancer patients.[7] The aim of this study is to evaluate the effect and mechanisms of melatonin action on colon carcinogenesis induced by 1,2-dimethylhydrazine (DMH) in rats. This model is well characterized and it was shown that colon carcinomas induced by DMH are morphologically similar to human colon tumors.[8]

2 MATERIALS AND METHODS

Three month-old male Lio rats were injected 12 mg/kg of DMH in a single dose subcutaneously (experimental run I), or 5 times at weekly intervals (experimental run II) to measure free radical processes in blood serum, liver and large bowel 24 h after single exposure and 6 month after 5 exposures. The intensity of ROS generation was determined by the level of hydroxy peroxide-induced luminol dependent chemiluminescence.[9] Lipid peroxidation was assesed using determination of diene conjugates and Shiff bases,[10] carbonyl assay was used for determination of oxidatively modified proteins.[11] The measurement of riboflavin chemiluminescence was used for determination of total antioxidative activity.[12] The activity of SOD and catalase were measured by the method of Agostini et al.,[13] NADPH-diaphorase according to Kuonen et al.,[14] NO-synthase as described previously,[15] nitrites using bacterial nitrite reductase,[16] protein on Lowry et al.[17]

3 RESULTS AND DISCUSSION

The single injection of DMH induced the decrease of ROS formation in serum without changes in lipids peroxidation products (diene conjugates and Shiff bases level) and protein oxidation. Neither did total antioxidative activity (TAA) or that of Cu, Zn-superoxide dismutase (SOD) change. However, nitrite level increased while those of ceruloplasmin decreased significantly. The same changes occured in the liver, but they were accompanied by noticeable reduction of TAA and SOD activity. ROS level also dropped in the liver, following a single injection of DMH, while TAA and activity of SOD reduced. No significant changes were observed in peroxidation processes in the colon which is targeted by DMH treatment. It is possible to suggest that decrease of ROS generation which is most pronounced in serum and liver is due to enhancement of superoxide radical utilization by interaction with nitric oxide. This assumption is confirmed by data showing the increase of NO production in acute experiments with DMH (experimental run 1).

The noticeable changes in peroxidation processes have been observed in rats chronically exposed to DMH (experimental run 2), especially in their colon where its specific cancerogenic effect expressed. The data presented in the table have shown a sufficient increase of lipids and protein peroxidation resulting in stimulation of TAA and catalase activation. The activation of NO-synthase following by DMH treatment leads to enhancement of production of nitric oxide possessing a high cytostatic effect.[18]

The data obtained suggest that disbalance between level of ROS formation and antioxidative defence has occured at chronic treatment of DMH. This is a reason of hyper-sensitiviness of enterocytes to injury by ROS and peroxy radicals forming in lipid and protein peroxidation processes.

The data presented in the table clearly demonstrated that melatonin has an ability to reduce the effect of DMH on peroxidation processes in colon. It has been revealed in tendency to decrease of a slightly raised level of ROS ,and significant reduction of of lipids and protein peroxidation intensity, accompaning by inhibition of SOD and catalase activities. At the same time the supression of NO production has been observed owing to inhibition of NO-synthase and SOD activities what is prerequisite to generation of peroxynitrite, having a cytotoxic properties.[1]

The mechanism of anticarcinogenic effect of melatonin on colon carcinogenesis is unknown. It was shown that pinealectomy was followed by the increase of the crypt cell proliferation in rat colon including bowels,[20] whereas melatonin exert an inhibitory effect on cell proliferation in rodents.[21] Melatonin enhances cell-to-cell junction contacts [22] and modifies the activity of cytochromes b_5 and P_{450} [23] that may have a critical significance for its inhibitory effect on DMH-induced carcinogenesis. In the same model as we used in our experiments it has been established that the level of immunohystochemically detected melatonin in epithelium of the intestinal tract of rats with DMH-induced colon tumors was significantly reduced as compared to intact rats, whereas it was in control range in rats treated with DMH+melatonin.[5] Recently it was found that melatonin is potent scavanger of free radicals and that could be one of mechanisms of its possible antitumor effect.[24,25]

Table 1
The effect of chronic DMH and melatonin treatment on free radical oxidation andantioxidant system in rat colon.

Indexes	Control	DMH	DMH+melatonin
ROS (Rel. un./mg protein)	84.7±8.6 (n=4)	115.1±20.5 n=5)	79.5±9.0 (n=7)
Diene Conjugates (nmol/g tissue)	26,2±1.5 (n=9)	48.2±3.6 (n=6), p<0.01	15.6±2.4 (n=6), p_1<0.001
Shiff Bases (Rel.un./g tissue)	289.0±39.5 (n=9)	482.5±21.0 (n=6), p<0.001	295.4±45.7 (n=6), p<0.001
Protein oxidation (μmol CO-deriv./mg protein)	1.89±0.15 (n=5)	2.74±0.06 (n=4), p<0.05	1.92±0.12 (n=5), p_1<0.05
NO-syntase (NADPH diaphorase, inhibited by L-arginine, in nmoles of NST formasane/min/mg protein)	0.74±0.08 (n=10)	1.41±0.31 (n=6), p<0.05	0.51±0.05 (n=7), p_1<0.05
TAA (rel. un./mg protein)	1.27±0.04 (n=10)	1.64±0.09 (n=6), p<0.01	1.10±0.09 (n=7), p<0.01
SOD (rel. un./mg protein)	30.1±2.0 (n=10)	32.5±4.8 (n=6)	20.6±2.4 (n=6),p_1<0.05
Catalase (mmoles H_2O_2 min/mg protein)	4.34±0.25 (n=10)	4.94±0.30 (n=4), p<0.05	3.93±0.24 (n=7), p_1<0.05

p - DMH compared to control, p1 - DMH+melatonin compared to DMH.

The data presented in this communication have shown that antioxidative properties of melatonin may be important factor of its carcinogenic potential. They are in a good agreement with the evidence of the role of free radicals in DMH-induced carcinogenesis and its suppression by antioxidants.[3,4]

References
1.   K.Z. Guyton, T.W. Kensler. *Brit. Med. Bull.* 1993, **49**, 523-544.
2.   D. Dreher, A.F Junod. *Eur. J. Cancer.* 1996, **32A**, 30-38.
3.   A.S. Salim. *Intern. J. Cancer.* 1993, **53**, 1031-1035.
4.   S.Y. Tsai, D.M. Dunn, B.C. Pence. *FASEB J.* **6**. A1395.
5.   V.M. Anisimov, I.G. Popovich, M.A. Zabezhinski. *Carcinogenesis.* 1997, **18**, 1549-1553.
6.   G.A. Bubenik. *Hormone Res.* 1980, **12**, 313-323.
7.   I.M. Kvetnoy, I.M. Levin. *Vopr. Onkol. (Russian).* 1987, **33**(11), 29-32.
8.   K.M. Pozharisski. Pathology of Tumors in Laboratory Animals. *Vol. 1. Tumors of the Rat. IARC Sci. Publ.* 1990, **P**, 159-198.
9.   V.M. Prokopenko, A.V. Arutjunyan, T.U. Kuzminich et al. *Vopr. Med. Khimii (Russian).* 1995,**41**(3), 53-56.

10. R.S. Chio, A.L. Tappel. *Biochemistry*. 1969, **8**, 2821-2829.
11. R.L. Levine, D. Gordon, C. N. Oliver, A. Amici. *Methods of Enzymol*. 1990, **186**, 464 - 479.
12. B.L. Strongler, C.S. Soup. *Arch. Biochem. Biophys*. 1953, **47**, 8-15.
13. A. Agostini, G.G. Gerli, L. Beretta, M. Branci. *J.Clin.Chem.& Clin. Biochem*. 1980, **17**, 771-773.
14. D.R. Kuonen, M.C. Hemp, D.J. Roberts. *J. Neurochem*. 1988, **50**, 1017-1025.
15. T.I. Oparina, Ju.A. Nikolaeva, A.G. Golubev. *Bull. Exp. Biol. Med. (Russian)*. 1994, **127**(2), 216.
16. F. Madueno, M.G. Guerro. *Anal. Biochem*. 1991, **198**, 200-202.
17. O.H. Lowry, N.J. Rosenbrough, A.L. Farr, R.J. Randell. *J. Biol. Chem*. 1951, **193**, 265-275.
18. J.F. Kerwin, J.R. Lancaster, P.L.Feldman. *J. Med. Chem*. 1995, **38**, 4343-4362.
19. V. Douley-Usmar, H. Wiseman, B. Halliwell. *FEBS Lett*. 1995, **369**, 131-135.
20. B.D. Callagan. *J. Pineal. Res*. 1995, **18**, 191-196.
21. A. Lewinski, I. Rbicka, E. Wajs et al. *J. Pineal Res*. 1991, **10**, 104-108.
22. A. Ubeda, M.A. Trillo, D.H. House, C. F. Blackman. *Cancer Lett*. 1995, 241-245.
23. G. Praast, C. Bartsch, H. Bartsch et al. *Experentia*. 1995, **51**, 349-355.
24. R.J. Reiter, D. Melchorri, E. Sewerynek, B. Poeggler et al. *J. Pineal Res*. 1995, **18**, 1-11.
25. V.N. Anisimov, A.V. Arutjunyan, V.Kh. Khavinson. *Proc. Russian Acad. Sci. (Russian)*. 1996, **348**, 265-267.

PROTECTIVE EFFECTS OF "SANGRE DE DRAGO" FROM CROTON LECHLERI MUELL.-ARG. ON SPONTANEOUS LIPID PEROXIDATION

C. Desmarchelier[1], S.M Barros[2], F. Witting Schaus[3], J. Coussio[4], G. Ciccia[1].

[1]Cátedra de Microbiologia Industrial y Biotecnologia, [4]Cátedra de Farmacognosia (IQUIMEFA-CONICET). Facultad de Farmacia y Bioquímica, Universidad de Buenos Aires, Junín 956 (1113) Buenos Aires, Argentina. [2]Departamento de Análises Clínicas e Toxicológicas, Faculdade de Ciências Farmacéuticas, Universidade de São Paulo, Av. Lineu Prestes 580, (05508-900) São Paulo, Brazil. [3]PROTERRA; Madrid 166, Miraflores, Lima, Peru.

1 INTRODUCTION

"Sangre de Drago" is a red viscous latex obtained from the bark of Croton lechleri Muell.-Arg. (Euphorbiaceae) and other Croton species. Wound healing, anti-inflammatory, antiviral and antitumour properties have been claimed for "Sangre de Drago" by indigenous populations in many parts of the westernmost part of the Amazon valley, including Colombia, Ecuador, Peru and Bolivia. Several proanthocyanidins, together with catechin, epicatechin, gallocatechin and epigallocatechin, have been reported to account for up to 90 % of the dried weight of the latex.[1]

Recent studies have shown peroxyl and hydroxyl radical scavenging properties in freeze-dried "Sangre de Drago". Likewise, the latex was also highly effective in reducing *in vitro* DNA damage induced by Fe (II) salts.[2] However, a reduction in hydroperoxide-induced chemiluminescence in rat liver homogenates could not be confirmed. In order to study the capacity of "Sangre de Drago" to protect liver tissue against spontaneous lipid peroxidation, the production of TBARS in rat liver was determined both *in vitro* and *in vivo*.

2 MATERIALS AND METHODS

2.1. Plant material

"Sangre de Drago" was collected from trees growing near the Huancabamba river, district of Pozuzo, Cerro de Pasco, Peru, during August 1996. The latex was extracted by incision in the bark, filtered and freeze-dried in a Gamma A lyophilizer (Chriss, Germany). The resulting powder was then re-dissolved in distilled water according to the corresponding assay conditions, in order to assess the bioactivity. Plant material was identified by F. Witting Schaus. A voucher specimen (Witting 604) is deposited in PROTERRA herbarium (Pozuzo).

2.2. Preparation of rat liver homogenates

Adult Wistar rats of 180-200 g fed on a standard laboratory diet and water ad libitum were used. The livers was excised, perfused and homogenised with 120 mM KCl, 50 mM phosphate buffer, pH 7.4, (1:10 w/v). The samples were centrifuged at 700 x g for 10 min at 0-4 °C. The supernatant fraction was kept at -20 °C until use. Protein concentration was measured as described by Lowry et al.,[3] using bovine serum albumin as a standard.

2.3. Thiobarbituric acid-reactive substances (TBARS) assay

TBARS were determined as described in Fraga et al.[4] Rat liver homogenates, adjusted to 10 mg protein/ml in 120 mM KCl, 50 mM phosphate buffer, pH 7.4, were incubated with 1000, 100 and 10 (g dry weight/ml of "Sangre de Drago" at 37 °C tor 15 min. Sodium dodecyl sulphate (Sigma) (0.2 ml of 3 % (w/v)) was included in the reaction mixture, and after mixing, 2 ml of 0.1 N HCI, 0.3 ml of 10 % (w/v) phosphotungstic acid (Sigma) and 1 ml of 0.7% (w/v) 2-thiobarbituric acid (Sigma) were added. The mixture was heated for 60 min in boiling water, and TBARS were extracted into 5 ml n-butanol (Mallinckrodt, USA). After centrifugation, the fluorescence of the butanol layer was measured at 515 nm excitation and 555 nm emission using an Hitachi F-3010 fluorescence spectrophotometer. The values were expressed as the ratio of the amount of TBARS formed in the presence of plant extracts compared to control.

2.4. *In vivo* determinatioll of MDA

An *in vivo* method for the determination of MDA was used in order to evaluate the antioxidant capacity of the extracts to protect liver tissue against spontaneous lipid peroxidation.[5] Male Wistar rats received intraperitoneal injection of the "Sangre de Drago" solution (treated animals) or isovolumetric amounts of the solvent (control group) 4 h before being sacrificed by cervical dislocation. The livers were then removed after perfusion with 0.9 % NaCl, homogenised in three times its weight in 40 mM phosphate 140 mM NaCl pH 7.4 and centlifuged for 20 min at 1000 g at 4° C. The supernatants was then diluted in 1:5 with 0.1 M potassium phosphate buffer pH 7.0 and incubated for 2 h at 37 °C for TBARS measurement. TBARS values were expressed as nmol MDA 2 h^{-1} x mg protein^{-1}.

2.5. Statistical analysis

EC5 values were calculated using Finney probit analysis computer program, described by J. Mc Laughlin.[6]

3 RESULTS AND DISCUSSION

Peroxyl and hydroxyl radicals are important agents that mediate lipid peroxidation, thereby damaging cell membranes. The effects of "Sangre de Drago" on lipid peroxidation were determined by measuring the inhibition in the production of TBARS both *in vitro* and *in vivo*. The former assay determines the production of malondialdehyde (MDA) and related compounds which are by-products of lipid peroxidation, a free radical process that takes place in the hydrophobic core of biomembranes.[4] Different concentrations of the lyophilised latex reduced the *in vitro* production of TBARS in rat liver homogenates in a dose dependent manner, thus allowing to calculate an $IC_{50}=161$ (912-48) (g/ml). Catechin, used as a standard, was also highly effective in reducing lipid peroxidation, showing an $IC_{50} = 46$ (78-22) (g/ml).

Protective effects of the latex were also obselved *in vivo*, and a 63 % decrease in the production of MDA in livers of rats injected with "Sangre de Drago" was reached at a 200 mg/kg dose. However, an increase in the production of MDA was observed at lower doses (50 mg/kg), as compared to the control, suggesling that the latex could be prooxidant at lower doses. Data were not significant at a 100 mg/kg dose.

Acknowledgements
Research supported by the International Foundation for Science (IFS), in Stockholm, Sweden (Grant agreement No. F/2628- 1), BID-CONICET (grant PMT-SID0370) and the University of Buenos Aires (grant UBACYT FA002).

References
1.   Y. Cai, F. Evans, M. Robells, J.D. Phillipson, M. Zenk and Y. Gleba. *Phytochemisty*. 1991, **30**, 2033.
2.   C. Desmarchelier, F. Witting Schaus. J. Coussio and G. Ciccia. *J. Ethnopharmacol.* 1997, **58**, 103.
3.   O.H. Lowry, A.L. Rosebrough, A.L. Farr and R. Randall. *J. Biol.Chem.* 1951, **193**, 265.
4.   C. Fraga, B. Leibovitz and A. Tappel. *J. Free Rad. Biol. Med.* 1987, **3**, 119.
5.   V. Junqueira, K. Simuzu, L. Vidella and S. Barros. Toxicology. 1986, **41**, 193.
6.   J. McLaughlin. in: Methods in Plant Biochemistry (Harborne, J.B., ed.) pp. 1-32. Academic Press, New York, 1991.

Natural Antioxidants or Pro-oxidants in Foods and Nutrition

RELATION OF DIETARY INTAKE AND BLOOD ANTIOXIDANTS WITH ANTIOXIDANT CAPACITY IN HEALTHY NON-SMOKING MEN

A. Jeckel, H. Boeing, D.I. Thurnham[1], B. Raab[2], H.-J. Zunft.[3]

Department of Epidemiology, [2]Department of Food Chemistry and Preventive Nutrition,[3] Department of Intervention Studies, German Institute of Human Nutrition, Potsdam-Rehbruecke, Germany; [1]NI Centre for Diet and Health, University of Ulster, Coleraine, Northern Ireland.

1 INTRODUCTION

It is well known, that people eating diets high in fruits and vegetables have a lower incidence of chronic diseases. The World Cancer Research Fund/AICR report, *Food, Nutrition and the Prevention of Cancer: a global perspective*, reviewed the evidence of causal relationships with cancer and concluded that these diets reduce the risk of several cancers.[1] Ames et al.[2] pointed out, that "... antioxidants in fruits and vegetables may account for a good part of their beneficial effect as suggested by mechanistic studies." However, the evidence on the relation between dietary constituents, such as antioxidants, and the risk of cancer, should be regarded less strong than the evidence on foods and drinks.[1] This statement is based on the problem, that there is a myriad of dietary constituents, many of them being considerably correlated with each other. At present, they cannot be assessed with a sufficient degree of specificity and precision.

Measurements of antioxidants in the body may help to disentangle their role in disease. In contrast to dietary assessment methods, such biomarkers more directly reflect the activity of antioxidants in human tissues. The potential biomarker antioxidant capacity (AC) describes the concerted action of the radical scavenging properties of individual antioxidants. AC can be measured in complex mixtures such as blood plasma and is assumed to include as yet unidentified or less recognised antioxidants. In contrast to the measurement of individual antioxidant concentrations, AC specifically informs on the antiradical activity of antioxidants. Thereby it can help to investigate whether antioxidants act through this mechanism, as opposed to other mechanisms such as quenching or non-antioxidant properties, e.g. the regulation of protein kinase C activity by α-tocopherol.[3]

2 OBJECTIVE

The relation of the continuous variable AC to dietary intake and to levels of individual plasma antioxidants is described in order to better characterise this potential biomarker. In order to enhance the power to detect dietary effects, the analysis was restricted to a population being relatively homogeneous in non-nutritional determinants of AC.

3 SUBJECTS

The population of this cross-sectional study is a subsample of the EPIC-Potsdam study (EPIC= European Prospective Investigation into Cancer and Nutrition).[4-5] During recruitment of the cohort study 164 non-smoking men (age 54.5±7.4 [x±SD]) were selected to the exclusion of factors known to influence AC, e.g. certain diseases and medication.

4 METHODS

All procedures from blood draw to analyses focused on standardised procedures and minimising the depletion of antioxidants. The dependent variable of the analysis -AC- was assayed by a photochemiluminescence method.[6-7] Three different components of AC (water-soluble [ACW], water-soluble treated with uricase [ACU] and lipid-soluble [ACL]) were measured in separate plasma aliquots.

The following two groups of independent variables were collected. 14 *blood parameters* were analysed with standard methods: uric acid, vitamin C (stabilised by meta-phosphoric acid), bilirubin, α-tocopherol, cholesterol, ferritin, albumin, triglycerides, coeruloplasmin, creatinine, C-reactive protein, gamma-GT, carotenoids and retinol. A semiquantitative food frequency questionnaire (sFFQ) was applied to assess habitual *dietary intake* during the last year.[8] For analysis, the foods were allocated to 24 food groups.[9]

The statistical approach included multiple linear regression of AC allowing either the *blood parameters* or the *dietary intake* variables to enter the respective model (both in addition to the potential confounder age, professional education level and percentage of body fat). All associations given are significant on the 0.05 level.

5 RESULTS

5.1. Blood Parameter

Uric acid explained 83 % and vitamin C 4 % of the variation in ACW. Vitamin C explained 75 % and bilirubin 4 % of the variation in ACU (i.e. sum of R^2=79 %). Vitamin E and cholesterol explained 82 % of the variation in ACL.

5.2. Dietary Intake

Multiple linear regression analysis showed that dietary variables measured by the sFFQ were not able to explain variation in ACW. ACU correlated positively with fruits and cereals and inversely with alcoholic beverages (sum of R^2=16 %). ACL correlated positively with alcoholic beverages and soft drinks as well as with percentage of body fat but inversely with cheese (sum of R^2=19 %). These associations on food group level were corroborated by correlations of individual nutrients with AC.

6 CONCLUSIONS

AC reflects plasma levels of bilirubin and cholesterol in healthy non-smoking men in addition to the well known role of uric acid, vitamin C and vitamin E as antioxidants. ACU and ACL are related to dietary intake, but the effects are not strong. Nevertheless, the associations fit in with protective and adverse effects of these foods on risk of chronic diseases such as cancer (cf. ref. 1).

Most studies investigating antioxidant capacity employ assays which are dominated by uric acid. The results of the present study confirm that the biomarker ACW is not considerably confounded by dietary intake. ACU and ACL are more specific biomarkers which at the same time probably show higher associations with cancer. They are promising tools for studies which investigate disease mechanisms, but the study designs should allow for the fact that ACU and ACL may well be confounded by diet.

Next steps in the development of the biomarker AC are to investigate associations with disease incidence. The comparison of the predictive values of biomarkers, which represent distinct hypotheses of disease modulation (e.g. AC representing antiradical mechanisms), should help endeavours which aim to elucidate the importance of individual causal pathways to cancer.

References

1. World Cancer Research Fund/American Institute for Cancer Research, 'Food, nutrition and the prevention of cancer: a global perspective', American Institute for Cancer Research, Washington. 1997, Part 3, pp. 404.
2. B.N. Ames, L.S. Gold and W.C. Willett, *Proc. Natl. Acad. Sci. USA*. 1995, **92**, 5258.
3. S. Clément, A. Tasinato, D. Boscoboinik and A. Azzi, *Eur. J. Biochem*. 1997, **246**, 745.
4. E. Riboli and R. Kaaks, *Int J Epidemiol*. 1997, **26**, Suppl 1, S6.
5. S. Voß, H. Boeing, A. Jeckel, A. Korfmann, J. Wahrendorf, M. Bergmann and A. Kroke, *Ernährungs-Umschau*. 1995, **42**, 97.
6. I.N. Popov and G. Lewin, *Free Radic Biol Med*. 1994, **17**, 267.
7. I.N. Popov and G. Lewin, *J Biochem Biophys Methods*. 1996, **31**(1-2), 1.
8. S. Bohlscheid-Thomas, I. Hoting, H. Boeing and J. Wahrendorf, *Int. J. Epidemiol*. 1997, **26**, Suppl 1, S 71.
9. S. Bohlscheid-Thomas, I. Hoting, H. Boeing and J. Wahrendorf, *Int. J. Epidemiol*. 1997, **26**, Suppl 1, S 59.

DRINKING GREEN TEA LEADS TO A RAPID INCREASE IN PLASMA ANTIOXIDANT POTENTIAL

I.F.F. Benzie[1], Y.T. Szeto[1], B. Tomlinson[2] and J.J. Strain[3].

[1]Department of Nursing & Health Sciences, The Hong Kong Polytechnic University, Kowloon, Hong Kong SAR; [2]Division of Clinical Pharmacology, The Chinese University of Hong Kong, New Territories, Hong Kong SAR; [3]Northern Ireland Centre for Diet & Health, University of Ulster at Coleraine, Northern Ireland.

1 INTRODUCTION

Tea (*Camellia sinensis*), is a favoured beverage throughout the world, although the form and quantity in which it is taken varies in different geographical areas and ethnic groups.[1] Tea is rich in polyphenolic compounds with antioxidant properties, mainly quercetin and catechins, and these compounds may inhibit oxidative damage to DNA, lipid, carbohydrate and protein.[1-3] Oxidative damage is associated with various chronic diseases, including cancer, coronary heart disease (CHD), cataract, and dementia,[4,5] and it has been suggested that consumption of tea may lower the risk of chronic disease by improving oxidant/antioxidant balance.[1,6,7]

Absorption studies in human subjects are few, however, and to date have shown conflicting results.[8,9] It is not yet clear, therefore, whether the antioxidant polyphenolic compounds in tea are absorbed and can contribute to *in vivo* antioxidant defence. The aim of this study, therefore, was to measure the change in plasma antioxidant (reducing) power and to monitor the antioxidant and phenolic content of the urine following ingestion of green tea.

2 MATERIALS AND METHODS

Total antioxidant (reducing) power in plasma and urine was measured in samples collected from 10 healthy fasting subjects (5 woman, 5 men), with their informed consent, immediately prior to and at timed intervals after ingestion of 400 ml freshly prepared green tea. Subjects remained fasting, except for the ingestion of tea and sips of distilled water, for the entire period of sample collection. In addition to fasting samples, venous blood was collected into heparinised blood collection tubes at 20, 40 and 60 minutes and 2 h post-ingestion. Blood samples were kept chilled and in the dark until separation of plasma from the erythrocytes, which was within 2.5 h of blood collection. The plasma total antioxidant power was measured immediately thereafter using the FRAP assay (US patent pending) as previously described.[10] Urine samples were collected, at 30 minutes intervals, without preservative, and total antioxidant (reducing) power was measured, after appropriate dilution in distilled water, within 4 h of collection. Urinary total phenolics and creatinine were also measured.

3 RESULTS

Absorption of polyphenolic antioxidants from green tea was very fast, with the associated increase in plasma total antioxidant power peaking at 20-40 minutes post ingestion (Fig. 1). After an oral dose of tea containing >15,000 μmol antioxidant power, plasma antioxidant power increased by around 4 %; post-ingestion: mean (SEM) increase was 44 (9) μmol/l. The increase was of short duration, and plasma antioxidant power approached or reached baseline (fasting) levels by 2 h post-ingestion in most, though not all, subjects.

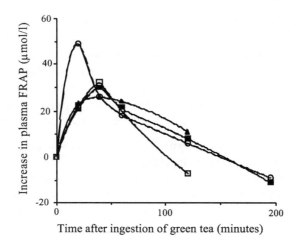

Figure 1. *Increase in plasma FRAP value (μmol/l) after ingestion of green tea; individual responses in 4 subjects.*

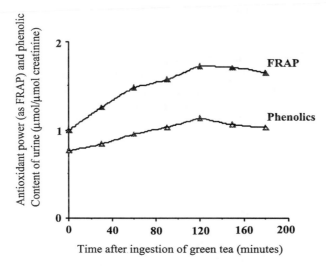

Figure 2 *Excretion of antioxidant power (as FRAP) and phenolics in urine following ingestion of tea (in one representative subject).*

Urinary excretion of absorbed polyphenolic antioxidants in tea was also fast, with peak excretion at 60-90 minutes post-ingestion. There was a significant correlation between FRAP values and total phenolics concentrations in urine ($r=0.84$; $P<0.001$) after tea was ingested, the antioxidant power and phenolic levels increasing in parallel (Fig. 2). No increase in plasma or urine FRAP values was seen in 7 subjects who repeated the study, at least 4 weeks later, drinking water in place of tea.

4 DISCUSSION

Human studies of absorption of polyphenolic antioxidants in tea have, to date, been few and their results conflicting. In this current study, which made use of the high sensitivity and reproducibility of the FRAP assay[10] for total antioxidant/reducing power, results demonstrated a clear and rapid 'spike' of antioxidant power in the plasma after ingestion of green tea. The increase, at around 4 %, was small, however, and of short duration. Excretion of absorbed antioxidant power also appeared to be rapid, mirroring excretion of phenolic compounds in the urine, and plasma antioxidant power returned to baseline within 2 h.

In summary, this study has shown that some antioxidants in polyphenolic-rich green tea are absorbed rapidly after ingestion, causing an increase in the plasma antioxidant/reducing potential over the following 2 h. Changes were relatively small and of short duration. Nonetheless regular, long-term consumption of tea may improve antioxidant defence *in vivo* and so may help lower risk of disease associated with oxidative damage.

References
1.   J.H. Weisburger. In E. Cadenas, L Packer (eds) Handbook of Antioxidants. Marcel Dekker Inc. NY, 1995, pp 469.
2.   C.A.Rice-Evans, N.J. Miller and G.Paganga. *Free Rad. Biol. Med.* 1996, **20**, 933.
3.   N.C. Cook, and S. Samman. *J. Nutr. Biochem.* 1996, **7,** 66.
4.   I. Emerit. *Free Rad. Biol. Med.* 1994, **16,** 99.
5.   B. Halliwell Oxidative stress, nutrition and health. *Free Rad. Res.* 1996, **25,** 57.
6.   P. Leanderson, A.O. Faresjo and C.Tagesson. *Free Rad. Biol. Med.* 1997, **23**, 235.
7.   L. Kohlmeier, K.G.C.Weterings, S. Steck and F.J. Kok. *Nutr. Cancer* 1997, **27**, 1.
8.   M. Serafini, A. Ghiselli and A. Ferro-Luzzi. *Eur. J. Clin. Nutr.* 1996, **50**, 28.
9.   S. Maxwell and G. Thorpe. (letter) *BMJ.* 1996, **313**, 229.
10.   I.F.F. Benzie and J.J. Strain. *Anal. Biochem.* 1996, **239**, 70.

THE ANTIOXIDANT CAPACITY OF SELECTED FOODS AND THE POTENTIAL SYNERGISMS AMONG THEIR MAIN ANTIOXIDANT CONSTITUENTS

Monica Salucci[1], Regina Lázaro[2], Giuseppe Maiani[1], Francesca Simone[1], Daymy Pineda[3], and Anna Ferro-Luzzi[1].

[1]National Institute of Nutrition, Human Nutrition Unit, Via Ardeatina 546, 00178 Rome, Italy. [2]Veterinary Faculty, Human Nutrition and Food Science Unit, University of Zaragoza, C/Miguel Servet 177. 50013-Zaragoza, Spain. [3]Instituto de Nutrición & Higiene de los Alimentos. Havana, Cuba.

1 INTRODUCTION

Oxidative stress and lipid peroxidation are the causation of a number of non communicable chronic diseases, such as cancer and atherosclerotic cardiovascular diseases.[1] Some studies have shown that the consumption of fruit and vegetables can reduce the incidence and the mortality rates of these diseases[2] and, as far as it is known, this protective effect is determined by the presence of different antioxidant compounds in these foods. An antioxidant can prevent oxidative damage by inhibiting the generation of reactive species, scavenging free radicals or raising the level of endogenous antioxidant defences.

Considering the important role of the natural antioxidant compounds content in fruit and vegetables in the prevention of human health it is interesting to measure the total antioxidant capacity of some common fruits and vegetables and the potential synergism among the main antioxidants present in these items. Previous studies using *in vitro* system or using animal system have confirmed a co-operation among antioxidants such as vitamin C, vitamin E and ß-carotene.[3] The synergistic effects among polyphenols and vitamins such as vitamin E and C have been scarcely studied and the results are not conclusive.[4,5,6]
The objective of this study is:

• to evaluate the antioxidant activity (TRAP) of selected foods (tomato, onion, lettuce, and kale) and to assess in chemical models the protective effect on the peroxidation of linoleic acid;

• to investigate potential interactions among several antioxidants present in these foods using standards such as flavonoids (rutin and quercetin), phenolic acids (caffeic acid), and vitamins (vitamin E and vitamin C).

2 MATERIALS AND METHODS

2.1. Selection of vegetables

We have selected tomato (*Solanum lycopersicum* ecotype Corbara), onion (*Allium cepa*), lettuce (*Lactuca sativa*) and kale (*Brassica oleracea*), grown in Central Italy. Fresh vegetables were purchased in a local supermarket the same day of the sample preparation.

These foods were tested as models of different kinds of antioxidants: tomato as for carotenoids (lycopene), onion as for flavonoids (quercetin), lettuce as for caffeic acid, quercetin and ascorbic acid, and kale as for phenolic acids (caffeic, ferulic and cinnamic acid).

2.2. Determination of single natural antioxidant compounds carotenoid, polyphenol and total ascorbic acid content in foods

The methods for extracting carotenoids in food and their characterisation by HPLC were described in detail by Maiani[7] with few modifications by Scott[8] and Khachik.[9] Food polyphenol extraction (flavonoids and hydroxicinnamic acids) and their characterisation by HPLC were described in detail by Hertog[10] and Maiani.[11] Finally, food ascorbic acid extraction and their characterisation by HPLC were described in detail by Margolis.[12,13] We have evaluated ascorbic acid only in tomato, lettuce and onion.

2.3. Measurement of the antioxidant potential (TRAP) of foods

Tomato, lettuce, onion and kale extracts suitable for the total antioxidant capacity analysis were obtained using the techniques described by Cao[14] and Wang.[15] The *in vitro* total antioxidant activity (TRAP) of vegetables was tested according to Ghiselli.[16]

2.4. Determination of antioxidant efficiency of foods

The principle of the method is based on the peroxidation of linoleic acid by OH˙ and the formation of malondialdehyde as final product.

Linoleic acid (10 mM) was dissolved in 1 ml of methanol, dried under nitrogen and redissolved in 2 ml of phosphate buffer according to Niki.[17] Lipid peroxidation of linoleic acid treated with lipo- or hydrosoluble food extracts (appropriately diluted) was induced by adding 10 µl of the AMVN (10 mM) for 15 min. at 37 °C. The control consisted of linoleic acid peroxidated with 10 µl of AMVN (10 mM) for 15 min at 37 °C without food extracts. The formation of MDA was detected by fluorimetric method according to Ursini.[18]

2.5. Evaluation of the synergism and/or cooperation among antioxidants

Since vitamin E is the most liposoluble component, vitamin C is the most hydrosoluble component and polyphenols are both lipo- and hydrosoluble compounds, it was needed to use two different *in vitro* models to evaluate the synergic action. TRAP assay was used to assess the hydrosoluble components while linoleic acid model has been used to assess the liposoluble ones (see point 2.4 and 2.5).

To assess the synergisms we used the solutions of both single and combined standard of pure antioxidant molecules. Concerning TRAP, the concentrations of mother solutions were: ascorbic acid 100 mM, quercetin 25 mM, rutin 25 mM, caffeic acid 25 mM, while for linoleic acid are: ascorbic acid 100 mM, α-tocopherol 50 mM, quercetin 200 mM, rutin 25 mM, caffeic acid 50 mM .

2.6. Statistical analysis

The data represent the mean ±SD of three determinations of four different food samples. The statistical analysis was conducted using parametric ('ANOVA' variance analysis) and non parametric (chi square) tests.

3 RESULTS AND DISCUSSIONS

Table 1 shows the carotenoid, polyphenol (flavonoids and hydroxycinnamic acids) and total ascorbic acid contents of selected vegetables. Results are expressed as mg/kg of food for polyphenols, as µg/100 g of food for carotenoids and as mg/100 g of food for total ascorbic acid. Tomato is rich in lycopene, ß-carotene and caffeic acid. Onion is very rich in quercetin; levels found are comparable to those reported in literature.[19] In kale, caffeic acid, cumaric acid and lutein are present in high quantity. Finally, lettuce is very rich in lutein and ß-carotene.

We can observe for each food item a high variability among all antioxidant compounds: hydroxycinnamic acids (ranging from 2 % to 20 %), flavonoids (ranging from 2 % to 34 %) and carotenoids (ranging from 13 % to 43 %). Total ascorbic acid levels in all fresh food considered, except tomato, are similar to data reported in literature.[19]

Table 1

Content of major polyphenols (mg/kg of edible portion), carotenoids (µg/100 g edible portion) and total ascorbic acid (mg/100 gr edible portion) of commercial fresh tomato, lettuce, onion and kale. Values are mean ±SD.

	Tomato	Lettuce	Onion	Kale
Flavonoids				
Quercetin	9.7 ± 3.3	24.2 ± 2.1	417.0 ± 10.4	14.2 ± 27.0
Kaempferol	18.0 ± 0.4	< 2.0°	23.9 ± 8.9	35.2 ± 7.3
Hydroxycinnamic acids				
Caffeic acid	46.7 ± 1.1	12.6 ± 1.9	33.3 ± 1.2	59.4 ± 0.2
Cumaric acid	13.0 ± 1.2	n.d.	30.0 ± 1.0	40.1± 0.9
Ferulic acid	4.0 ± 0.3	n.d.	39.0 ± 1.9	n.d
Carotenoids				
Lutein	58.7 ± 11.6	880.0 ± 321.0	-	1056.0 ± 142.0
Lycopene	2205.0 ± 997.0	n.d.	-	< 2.0°
ß-carotene	496.0 ± 100.0	556.0 ± 178.0	-	n.d.
Total ascorbic acid	10.70 ± 1.89	10.74 ± 2.14	5.15 ± 1.18	*

*n.d.= not detectable; *= not evaluated; °= below limit of detection.*

Fig. 1 shows the overall TRAP values and the separate contribution to the TRAP of the hydrosoluble and liposoluble fractions of selected plant foods. The results are expressed as µmol/g of food.

The overall TRAP of the plant foods was calculated by adding TRAP values of the hydrosoluble and liposoluble fractions. Kale has the highest TRAP value (P< 0.05) followed by lettuce, onion and tomato. Concerning the hydrosoluble fraction, kale has the highest TRAP value, too (P<0.05), followed by lettuce, tomato and onion. Also kale liposoluble fraction is the most efficient in scavenging peroxyl radical followed by onion and lettuce. Tomato, despite its high content of lycopene, has the lowest TRAP value, probably because lycopene and, carotenoids in general, are efficient quenchers of oxygen singlet rather than scavengers of peroxyl radicals.

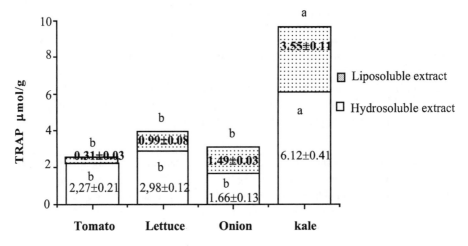

Anova : a vs b P< 0.05

Figure 1
Antioxidant activity (TRAP) of fresh selected foods and separate contribution of hydro and liposoluble extracts. Results are expressed as μmol/g of food.

Fig. 2 shows the protective effect of hydrosoluble and liposoluble fractions of selected vegetables (tomato, lettuce, onion and kale) on the peroxidation of linoleic acid.

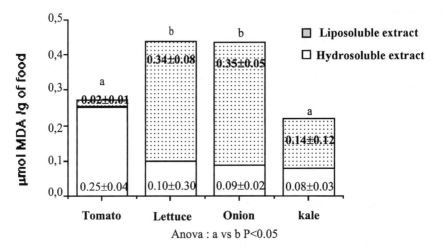

Anova : a vs b P<0.05

Figure 2
Protective effect of selected food extracts on peroxidation of linoleic acid. Results are expressed as μmol MDA/g of food.

These results are expressed as μmol MDA/g of food. Kale and tomato have the highest protective effect (0.22 and 0.27 μmol MDA/g of food, respectively, P<0.05) followed by onion and lettuce (0.44 and 0.44 μmol MDA/ g of food, respectively). The order of antioxidant effectiveness of food extracts against radicals generated in lipophilic phase is kale = tomato > lettuce = onion.

To investigate the synergisms among antioxidants we have evaluated the effect of the single molecules and their potential interactions using two different systems (hydrophilic and lipophilic). The results obtained are shown in Fig. 3, 4, 5 and they are expressed as length of the lag phase (minutes) for the hydrophilic system (TRAP) and as % of MDA produced for the lipophilic system.

To be synergistic it is necessary that the antioxidant effect of the combination of two or more compounds be more extensive than the sum of the effects caused by either compound alone. Thus, the inhibition period or lag phase, produced by a combination of two antioxidants should be longer than the sum of the inhibition period produced by either alone, and oxidation must be suppressed efficiently throughout the inhibition period.[20]

In the hydrophilic system, the antioxidant efficiency (minutes of lag phase) of individual pure molecules tested (data not shown) showed the following decreasing order: quercetin >caffeic acid >rutin = ascorbic acid (the concentration in cuvette was 4 μmol/l for ascorbic acid and 1 μmol/l for all the other compounds). Fig. 3 shows the antioxidant activity when ascorbic acid is combined with a polyphenol.

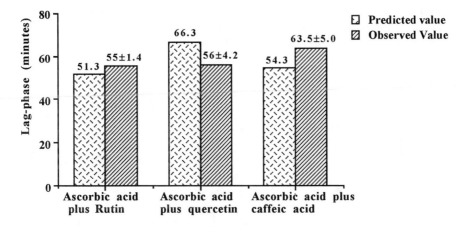

Figure 3

Effect of the combination of pure molecules on their summed antioxidant efficiency (lag phase, minutes) by monitoring the peroxidation reaction.

Predicted values are obtained by summing the individual effects, hence, these values are without SD.

We have observed a synergistic effect when combining rutin and ascorbic acid, and caffeic acid and ascorbic acid. Besides these synergistic or supra-additive effects, we have also found an antagonist effect for the combination of quercetin and ascorbic acid.

When studying these effects in a lipophilic system, we observed that among single antioxidants tested (Fig. 4) α-tocopherol has the higher protective effect followed by quercetin, ascorbic acid, caffeic acid and rutin. When combining α-tocopherol or ascorbic acid with a polyphenol (Fig. 5) α-tocopherol plus ascorbic acid and α-tocopherol plus quercetin have the strongest protective effect. These results confirm the well known cooperation between α-tocopherol and ascorbic acid and suggest a similar cooperation between tocopherol and quercetin.

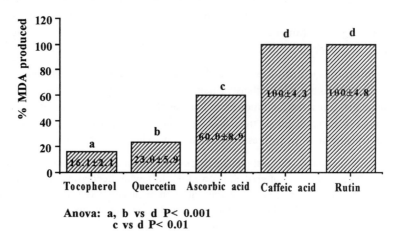

Figure 4
Measurement of protective effects on peroxidation of linoleic acid in the presence of some pure molecules. Results are expressed as % MDA produced.

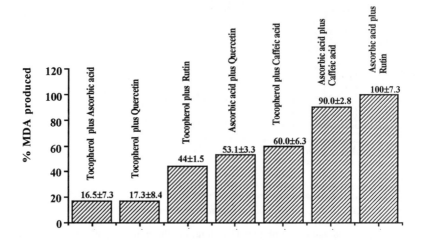

Figure 5
Measurement of protective effects on peroxidation of linoleic acid of the combined natural compounds. Results are expressed as % MDA produced.

4 CONCLUSIONS

The fresh food extracts (tomato, lettuce, onion and kale) have different total antioxidant capacity and different effect on the peroxidation of linoleic acid. Their antioxidant activity depends on the nature and on the concentration of natural antioxidants present in foods. In fact, total antioxidant capacity (TRAP) of the hydrosoluble extracts is high for all food tested, in agreement with the evidence that the TRAP is measured in an aqueous system. The order of antioxidant effectiveness of food extracts against radicals generated in the hydrophilic phase is kale >lettuce >onion = tomato.

The fresh food extracts (tomato, lettuce, onion and kale) have, also a different effect on the peroxidation of linoleic acid. Together with kale, tomato has the highest protective effect on the radicals generated in lipophilic phase as it is the richest in lipophilic antioxidants (like lycopene, β-carotene, lutein) and they are the most effective in blocking the liposoluble radicals produced. The protective effect order is kale = tomato >lettuce = onion.

Concerning the potential synergisms we observe a synergistic action in the aqueous system for rutin plus ascorbic acid, and caffeic acid plus ascorbic acid, while in the lipophilic system a simple cooperation between α-tocopherol plus ascorbic acid is observed. A similar cooperation is present between α-tocopherol plus quercetin, probably due to their lipophilic antioxidant nature.

Acknowledgement

We gratefully acknowledge financial support from Italian Ministry of Agricultural Policy. We wish to thank the European Community for giving a post-doctoral grant (FAIR CT 965080) to Dr. Lázaro. We also thank Mrs Elena Azzini and Ms Anna Raguzzini for technical advice and assistance.

References

1. Hageman J. J., Bast A., Vermeulen N. P. E. Monitoring of oxidative free radical damage *in vivo*. Analytical aspects. *Chem. Biol. Interaction.* 1992, **82**, 243-293.
2. Block G. Epidemiologic evidence regarding vitamin C and cancer. *Am. J. Clin. Nutr.* 1991, **54**, 1310S-1314S.
3. Niki E. α–tocopherol. In "Handbook of antioxidants". Ed. Enrique Cadenas and Lester Parker. *Marcel Dekker, Inc. New York, USA,* 1996, pp. 3-25.
4. Terao J.,Piskula M., Yao Q. Protective effect of epicatechin, epicatechin gallate and quercetin on lipid peroxidation in phospholipid bilayers. *Arch. Biochem. Biophys.* 1994, **308**(1), 278-284.
5. Negre-Salvayre A., Mabile L., Delchambre J. Salvayre R. α–tocopherol, ascorbic acid, and rutin inhibit synergistically the copper-promoted LDL oxidation and the cytotoxicity of oxidised LDL to cultured endothelial cells. *Biol. Trace Elem. Res.* 1995, **47**(1-3), 81-91.
6. Bors W.; Heller W.; Michel C.; Stettmaier R. Flavonoids and polyphenols. In "Handbook of antioxidants", Ed. Enrique Cadenas and Lester Parker. Marcel

Dekker, *Inc. New York, USA*. pp. 411-466 , 1996

7.   Maiani G., Pappalardo G., Ferro-Luzzi A., Raguzzini A., Azzini E., Guadalaxara A., Trifero M., Frommel T., Mobarhan S. Accumulation of ß-carotene in normal colorectal mucosa and colonic neoplastic lesions in humans. *Nutr.Cancer.* 1995, **24**, 23-31.

8.   Scott J. Observation on some of the problems associated with the analysis of carotenoids in foods by HPLC. *J. Food Chem.* 1992, **45**, 357.

9.   Khachik F., Beecher G.R. and Lusby W.R. Separation, identification and quantification of the major carotenoids in extracts of apricots, peaches, cantaloupe and pink grapefruit by liquid chromatography. *J. Agric. Food Chem.* 1989, **37**, 1465-1473.

10.   Hertog M.G.L., Hoffman P.C.H., Venema D.P. Optimization of a quantitative HPLC determination of potentially anticarcinogenic flavonoids in vegetables and fruits. *J. Agric.Food Chem.* 1992, **40**, 1591-1598.

11.   Maiani G., Serafini M., Salucci M., Azzini E., Ferro-Luzzi A. Application of a new high-performance liquid chromatographic method for measuring selected polyphenols in human plasma. *J. Chrom. B .* 1997, **692**, 311-317.

12.   Margolis S. A., Ziegler R. G., Helzllsouer K. J. Ascorbic and dehydroascorbic acid measurement in human serum and plasma. *Am. J. Clin. Nutr.* 1991, **54**, 1315 S-18 S.

13.   Margolis S. A., Duewer D. L. Measurement of ascorbic acid in human plasma and serum: stability, intralaboratory repeatability, and interlaboratory reproducibility. *Clin. Chem.* 1996, **42** (8), 1257-1262.

14.   Cao G., Sofic E., Prior R. L. Antioxidant capacity of tea and common vegetables. *J. Agric. Food Chem.* 1996, **44**, 3426-3431.

15.   Wang H., Cao G., Prior R. L. Total antioxidant capacity of fruits. *J. Agric. Food Chem.* 1996, **44**, 701-705.

16.   Ghiselli A., Serafini M., Maiani G., Azzini E., Ferro-Luzzi A. A fluorescence method for measuring total plasma antioxidant capability. *Free Rad. Biol. Med.* 1995, **18** (1), 29-36.

17.   Niki E. Free radical initiators as source of water- or lipid-soluble peroxyl radicals. *Methods in Enzymology,* 1991, **186**, 100-108.

18.   Ursini F., Maiani G., Polito A., Coassin M., Ferro-Luzzi A. TBA reactive material in human plasma and its relation to nutritional parameter. *Nutr. Rep. Intern.* 1989, **39**, 1263-1274.

19.   Anonyme. "Tabelle di composizione degli alimenti". Ed: Istituto Nazionale della Nutrizione. *Rome, Italy.* 1989.

20.   Niki E. α–tocopherol. In "Handbook of antioxidants". Ed. Enrique Cadenas and Lester Parker. *Marcel Dekker, Inc. New York.* USA, 1996, 3-25.

FLAVONOID CONTENT AND ANTIOXIDANT PROPERTIES OF BROCCOLI

Andrea Lugasi,[1] Judit Hóvári,[1] Magdolna N. Gasztonyi,[2] Ernõ Dworschák.[1]

[1]National Institute of Food-Hygiene and Nutrition, [2]Central Food Research Institute, Budapest, Hungary.

1 INTRODUCTION

Flavonoids occur naturally in plant foods and demonstrate a wide range of biochemical and pharmacological effects including antiinflammatory, antiallergic and antioxidant actions. Although broccoli (*Brassica oleracea* L. *var. italica Plenck*) is a rich source of natural antioxidants such as ß-carotene, a-tocopherol, ascorbic acid and phenolic components, its consumption in Hungary is very low. In *in vitro* studies aqueous extract from broccoli markedly suppressed the oxidation of 2'-deoxyribose investigated by the thiobarbituric acid method, inhibited the autoxidation of linoleic acid monitored by thin-layer chromatography, and scavenged hydroxyl and superoxide radicals measured by ESR method.[1]

Since free radicals are involved in the pathomechanism of different diseases it is important to know the effects of diet composition on the lipid peroxidation processes. The aim of our study was to investigate the flavonoid composition and the antioxidant efficacy of broccoli in chemical systems.

2 MATERIALS AND METHODS

Fresh broccoli was purchased from the local market. The sample was freeze dried and stored in refrigerator until analyzed. 25 mg sample was extracted twice with bidestilled water (WEB) or methanol (MEB). The total volume of the extracts was 1 ml.

2.1. Flavonoid composition

Total polyphenol content of the extracts was measured with the use of Folin-Denis reagent.[2] Total flavonoid content of the dry sample was analysed by a spectrophotometric measurement with aluminium chloride after an acidic hydrolysis.[3] The flavonol quercetin, kaempferol and myricetin and the flavone apigenin and luteolin were measured according to Hertog et al. 1992.[4]

2.2. *In vitro* measurements

Hydrogen-donating activity was measured in the presence of 1,1-diphenyl-2-pycrylhydrazyl radical (DPPH) and was expressed as the % inhibition of the color development of DPPH at 517 nm.[5] The reducing power was studied spectrophotometrically on the base of the reaction $Fe^{3+} \text{\AE} Fe^{2+}$ at 700 nm.[6] Copper-binding property was measured in hexamine buffer with tetramethylmurexide as a complexometric reagent and was expressed as the absorbance ratio at 485 vs. 530 nm of the reaction mixture.[7] The control did not contain any chelating agent and showed the highest value (3.5±0.01). Antioxidant activity of the samples was measured on the basis of its inhibitory effect on the linoleic acid autoxidation at 40 °C during a 10-days incubation period by the thiocyanate method.[8] Free radical scavenging activity of the sample was measured in $H_2O_2/\cdot OH$-luminol system with the use of a Berthold Lumat luminometer.[9] The effect of the sample on enzymatically induced lipid peroxidation was studied on rat liver microsome fraction isolated by an ultracentrifugation method.[10] Formation of the thiobarbituric acid reactive substances was determined by the method of Ottolenghi.[11] A molar absorption coefficient 1.56×10^5 (E_{532}, 1 cm) was used. The NADPH-induced lipid peroxidation depending on the incubation time was measured in a medium containing 1 mg/ml protein.

3 RESULTS

Broccoli contains a significant amount of biologically active components with antioxidant activity such as ascorbic acid (115 mg), ß-carotene (846 mg), vitamin E (621 µg) and phenolic acids (ferulic acid 1.3 mg, caffeic acid 0.8 mg, p-coumaric acid 1.3 mg) in 100 g f.w.[12] Our present investigations verified the presence of 21.5 mg flavonoids, 2.06 mg quercetin, and 4.12 mg kaempferol in 100 g fresh sample.

Aqueous and methanolic extracts from broccoli showed marked concentration-dependent hydrogen-donating activity, I_{50}-s are 0.22 and 0.47 ml, respectively (Fig. 1). Both extracts were able to reduce the ferric ions (Fig. 2), the reducing power of 1 ml aqueous and methanolic extracts were the same than that of 2.75 and 1.96 µmol ascorbic acid, respectively.

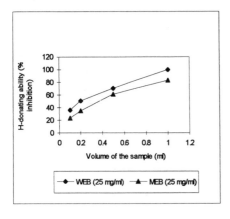 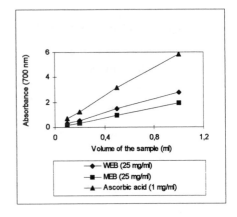

Fig. 1.
Hydrogen-donating ability of the extracts from freeze dried broccoli.

Fig. 2.
Reducing power of ascorbic acid and the extracts from freeze dried broccoli.

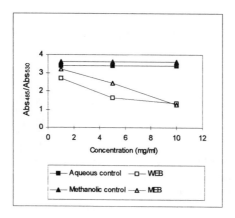

Fig. 3.
Copper-binding ability of the extracts from freeze dried broccoli.

The extracts could definitely chelate copper iones as it can be seen on Fig. 3. Aqueous and methanolic controls had higher values of Abs_{485}/Abs_{530} than the samples at different concentrations. The extracts could the autoxidation of linoleic acid at 40 °C during a 10-days incubation period prevent in concentration-depending manner (Fig. 4 and 5).

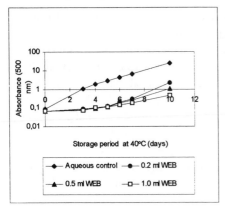

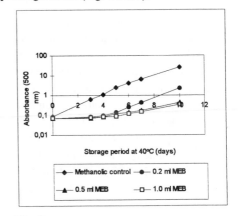

Fig. 4.
Antioxidant activity of the aqueous
extract from freeze dried broccoli.

Fig. 5.
Antioxidant activity of the methanolic
extract from freeze dried broccoli.

Aqueous extract from broccoli also inhibited the formation of the thiobarbituric acid reactive substances (MDA) in the enzymatically (NADPH) induced lipid peroxidation on rat liver microsomes (Fig. 6). Aqueous extract from broccoli showed a significant free radical scavenging activity in $H_2O_2/\cdot OH$-luminol system. The chemiluminescence light intensity was lower in the reaction mixtures containing the sample at different concentrations than the standard light (Fig. 7).

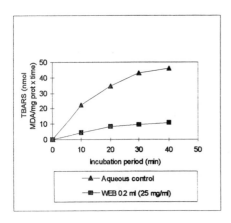

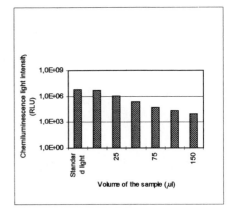

Figure 6

Effect of the aqueous extract from broccoli on NADPH-induced lipid peroxidation in rat liver microsomes.

Figure 7

Scavenging activity of the aqueous extract from broccoli (50 mg/ml) in H_2O_2/ᐧOH luminol system.

4 CONCLUSION

Broccoli is a very good source of natural antioxidants including vitamins and phenolic compounds such as phenolic acids and flavonoids. According to results of *in vitro* screening aqueous and methanolic extracts of broccoli can play an important role in prevention of lipid peroxidation as primary (chain-breaking), and/or secondary (preventive) antioxidants. Bioactive compounds in broccoli can chelate transitional metal iones that leads to the suppression of the catalytic decomposition of lipid hydroperoxides. The extracts also inhibited the autoxidation the linoleic acid and the enzymatically induced lipid peroxidation. The sample showed marked free radical scavenging activity measured by chemiluminometry. The actions mentioned above are involved the overall antioxidant activity of broccoli. Therefore frequent consumption of Brassicas may heve a role in prevention of human diseases in the patomechanism of which free radicals are involved.

Acknowledgement

This research was supported the National Scientific Research Foundation (OTKA F 017713), and the Ministry of Welfare (ETT 052/1996 02), Hungary.

References

1.   Ramarathnam, N., Ochi, H. and Takeuchi, M. In: Natural antioxidants. Chemistry, Health effects, and applications. (Ed: Shahidi, F.) *AOCS Press, Champaign, IL,* 1996, p. 76.

2.   AOAC (1990) 15th Edition, Arlington, USA, 952.03/A-C.

3.   German Pharmacopoea, No 10.

4.   Hertog, M. G. L., Hollman, P. C. H. and Katan, *M.B. J. Agric. Food Chem.* 1992, **40,**

2379.
5. Blois, *M. S. Nature.* 1958, **4617**, 1198.
6. Oyaizu, *M. Jpn. J. Nutr.* 1986, **44**, 307.
7. Shimada, K., Fujikawa, K., Yahara, K. and Nakamura, *T. J. Agric. Food. Chem.* 1992, **40**, 945.
8. Zsinka, Á. J. N., Blázovics, A. and Biacs, *P. Hung. Sci. Instr.* 1988, **64**, 11.
9. Masude, T., Isobe, J., Jitoe, A., and Nakatani, *N. Phytochemistry.* 1992, **31**, 3645.
10. Blázovics, A., Fehér, E. and Fehér, J. In: *Free radical and liver*. (Eds. Csomós, G., Fehér, J.) Springer-Verlag, *Berlin*, 1992, p. 96.
11. Ottolenghi, *A. Arch. Biochem. Biophys.* 1959, **79**, 355.
12. Food Composition and Nutrition Tables, 5th Edition (Eds. Souci, S. W., Fachmann, W. and Kraut, H.) *CRC Press, Boca Raton*, 1994.

EXAMINATION OF FLAVONOID CONTENT IN HUNGARIAN VEGETABLES

Judit Hóvári, Andrea Lugasi and Ernõ Dworschák.

National Institute of Food Hygiene and Nutrition, Budapest, Hungary.

1 INTRODUCTION

Recently several studies have proved that flavonoids have effect on the human health and thus they are important component of the human diet. The consumption of several grams of flavonoids is recommended.

The aim of the study was a qualitative and quantitative analysis of concentrations of the flavonoids in vegetables produced and consumed frequently in Hungary.

2 MATERIALS AND METHODS

31 selected vegetables were purchased from the local markets in Budapest at a period of their most frequent consumption. The edible parts of the vegetables were used to the examination. After buying the vegetables were immediately cleaned and freeze dried and stored at -18 °C until the analyses.

The flavonols (quercetin, kaempferol, myricetin) and the flanones (apigenin,luteolin) were measured according to Hertog et al.[1] Briefly, flavonoid glycosides were extracted and hydrolyzed to their aglycons with 2.0 M HCl in boiling 50 % aqueous methanol.[1] The resulting aglycons were quatified by RP-HPLC (Perkin Elmer) on a Premisphere C_{18} column (150 x 3.9 mm, 5 μm, Phenomenex, USA) using methanol/phosphate buffer (45/55 v/v, pH 2.4), as a mobile phase and UV detection (370 nm).

3 RESULTS AND DISCUSSION

Table 1 reports the flavonoid content of the fresh vegetables (mg/kg). Quercetin and kaempferol are proved to be the most widespread flavonoids in vegetables. Our present results are similar to Hertog and co-workers' observations but there are some minor differences in relation of the quality and quantity of the compounds, as well.[2] The highest quercetin concentration could be detected in the different types of onion (67.1-171.3 mg/kg) and in spinach (272.2 mg/kg). Hertog and co-workers did not measure any quercetin in spinach.[2] Significant amount of kaempferol was observed in parsnip, leek, new onion, and broccoli (66.4, 45.8, 34.3, and 30.8 mg/kg, respectively).

Table 1
Flavonoid content of vegetables (mg/kg).

Sample	Quercetin	Kaempferol	Myricetin	Luteolin	Apigenin
Purple radish	nd	10.5	nd	nd	nd
Black radish	nd	21.1	nd	nd	nd
Horse radish	5.7	25.7	nd	9.0	nd
Red beet	6.7	nd	nd	18.3	nd
Onion (old)	121.5	2.6	nd	nd	nd
Red onion	171.3	24.3	nd	nd	nd
Onion (new)	67.1	34.5	nd	nd	nd
Leek	5.0	45.8	nd	nd	nd
Cauliflower	1.5	12.5	nd	4.0	nd
Broccoli	15.4	30.8	nd	nd	nd
Kohlrabi	4.0	24.3	nd	13.0	nd
Brussels sprout	nd	12.8	nd	6.7	nd
Lettuce	16.3	nd	10.2	nd	nd
Crisped lettuce	35.0	8.4	nd	3.9	nd
Ice lettuce	13.5	nd	nd	nd	nd
Kale	nd	4.8	nd	nd	nd
Chinese cabbage	nd	7.3	nd	nd	nd
White cabbage	1.6	11.9	nd	4.2	nd
Red cabbage	9.2	nd	nd	6.3	nd
Cucumber	2.4	3.3	nd	nd	nd
Tomato	2.7	8.4	nd	nd	nd
Sweet pepper	9.4	nd	nd	10.7	nd
Californian pepper	5.1	nd	nd	11.3	nd
Carrot	3.5	nd	nd	nd	nd
Parsnip	9.9	66.4	nd	nd	nd
Swedish turnip	3.2	22.7	85.4	nd	154.0
Celery root	1.8	nd	nd	nd	24.1
Parsley leaves	nd	nd	80.8	nd	nd
Celery leaves	nd	nd	43.4	111.4	248.0
Dill	74.5	nd	7.0	nd	nd
Spinach	272.2	nd	nd	66.4	nd

nd: not detectable.

Only five of the vegetables examined in this study contained myricetin, namely swedish turnip (85.4 mg/kg), parsley leaves (80.8 mg/kg), celery leaves (43.4 mg/kg), lettuce (10.2 mg/kg), and dill (7.0 mg/kg). Regrettably the leaves of parsley, celery and dill are consumed as condiments in special Hungarian dishes therefore the participation of these vegetables in the flavonoid intake of the population is probably negligible. Based on the Hertog's observation only fresh broad been can be considered as a natural source of myricetin, but its concentration is low (26 mg/kg).

Significant amount of luteolin could be detected in celery leaves (111.4 mg/kg), spinach (66.4 mg/kg), red beet (18.3 mg/kg), kohlrabi (13.0 mg/kg), and different types of pepper (10.7-11.3 mg/kg). Only three vegetables contained apigenin, namely celery leaves (248.0 mg/kg), Swedish turnip (154.0 mg/kg), and celery root (24.1 mg/kg). Hertog et al. reported luteolin only in red bell pepper (11 mg/kg), and apigenin in celery (108 mg/kg). In a Danish survey[3] apigenin (740 mg/kg) and luteolin (200 mg/kg) were found in celery leaves, and in parsley (apigenin: 1850 mg/kg).

Opposite of our observations, Hertog and co-workers could not detect any of the flavonoids in spinach, red beet, cucumber, carrot and some of brassicas like sauerkraut, white cabbage, swedish turnip and green cabbage.[2] These discrepancies may be due to different cultivars, varietal and seasonal differences.

3 CONCLUSION

The large group of plant polyphenols attracts major interest because of their potential anticarcinogenic and other beneficial properties, presumably based on their function as natural antioxidants. Otherwise the flavonoids are compounds of particular interest because of their high prevalence in foodstuffs. Our investigation proves the presence of significant amount of different flavonoids in selected vegetables frequently consumed in Hungary. These invstigations are a part of our study on a national database of flavonoids in foddstuffs and help to discover the main sources of flavonoids and the estimation of average daily intake of quercetin, myricetin, luteolin, apigenin, and kaempferol in different groups of Hungarian population.

Acknowledgement

The research was supported by the Ministry of Welfare (ETT 052/1996 02), Hungary.

References
1.   Hertog M. G. L., Hollman P.C. H. and Venema D. P.
      J. Agric. Food Chem., 1992 a,**40,** 1591-1598.
2.   Hertog M. G. L., Hollman P.C. H. and Katan M. B.
      J. Agric. Food Chem. 1992 b, **40,** 2379-2383.
3.   Leth T. and Justesen U.
      In: *"Polyphenols in Food",* COST-916. (Eds: Amado, R., Andersson, H., Bardócz, S., Serra, F.) EC, Luxemburg 1998, 39-41.

AVENANTHRAMIDE ANTIOXIDANTS IN OATS

Lena Dimberg.

Department of Food Science, Swedish University of Agricultural Sciences, Uppsala, Sweden.

1 INTRODUCTION

Oats grains are rich in lipids and the unsaturated fatty acids, oleic and linoleic acids dominate (ca 35 % each) which is desirable from a nutritional standpoint. The palmitic acid content (ca 18 %) improves the stability of the oil, but the unsaturated fatty acids are still vulnerable to oxidation. However, oats contain various compounds with antioxidant activity, which presumably protects their own lipids from oxidation and thereby protect oat products from deterioration, but which also might give some health benefits for humans. The use of oat grains as a source of antioxidants was first proposed in the 1930s.[1,2] A specially fine-ground oat flour was marketed for antioxidant purposes and found to be effective in various food products that are sensitive to oxidation during storage.

The best known antioxidants are the tocols and oat grains contain both tocopherols and tocotrienols. Measurement of tocol concentration of hand-dissected oat groats revealed that most of the tocopherols are located in the germ while the tocotrienols are concentrated in the endosperm.[3] Investigations of minor components in oats discovered the presence of antioxidants other than tocopherols. These include the phenolic acids, ubiquitous in the plant kingdom, but also Δ^5-avenasterol, reported to have anti-polymerization activity in soybean oil at 180 °C[4], and a wide range of hydroxycinnamic esters and amides.[5-9] Some antioxidants have been identified as monoesters comprising hydroxycinnamic acids and long-chain fatty acids or alcohols[5] and others as amides, comprising hydroxycinnamic acids and anthranilic acids.[6-9] The latter group of antioxidants are trivially called avenanthramides (derived from *Avena*, the latin name for oats).

2 AVENANTHRAMIDES

2.1. Oats

The four most common cinnamic acids -*p*-coumaric acid, caffeic acid, ferulic acid or sinapic acid in combination with four different anthranilic acids (anthranilic acid, 5-hydroxyanthranilic acid, 5-hydroxy-4-methoxyanthranilic acid or 4-hydroxyanthranilic acid) give 16 combinations of avenanthramides of which 9 have so far been identified from oats. [6-9] It is possible that further combinations with other substitution patterns may exist. Oat grain contains 500-800 mg/kg avenanthramides, but both the total content and the levels of the individual compounds vary between cultivars.[8,9] However, combinations of 5-hydroxyanthranilic acid with ferulic acid (Bf), caffeic acid (Bc) or *p*-coumaric acid (Bp) tend to predominate (Fig. 1).

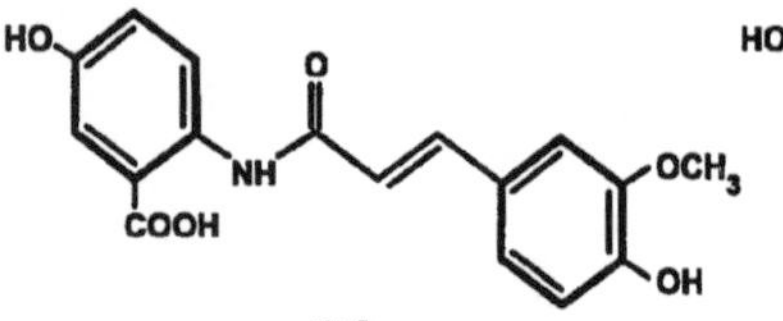

Figure 1
The three most common avenanthramides in oats.

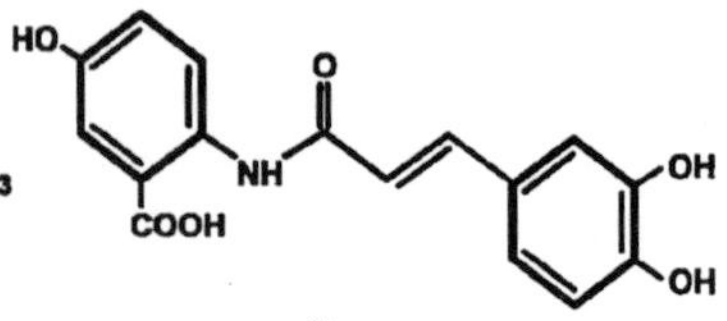

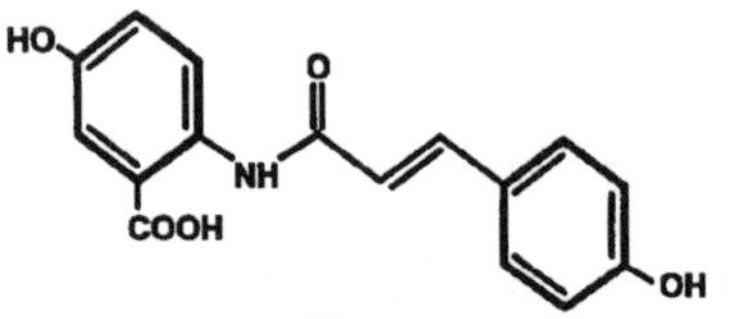

Avenanthramides have also been found in oat leaves.[10-12] There they are reported to act as phytoalexins with anti-fungal activities. Their production in the leaves is induced by fungal infection.[11-12] An enzyme, which catalyses the final step in the biosynthesis, that is the fusion of the anthranilic acid and the cinnamic acid moieties, is activated.[13] Findings by Ishihara et al.[14] indicate that Ca^{2+} is involved in the induction. All of the possible precursors can be used as substrate for the enzyme, but the highest affinities are found for 5-hydroxyanthranilic acid and for ferulic acid.[13]

In the kernel, the avenanthramides appear late in the grain maturation phase after flowering[15] and are presumably located in the outermost parts.[8] If they are produced in the kernel is presently not known.

Oat samples with a high amount of avenanthramides are evaluated as fresh tasting by a trained sensoric panel while samples with low amounts are considered as rancid.[16] Maybe this freshness is due to antioxidant activity of the avenanthramides in the oat samples.

The antioxidative activities of the avenathramides are structure related. In principal, the more methoxy or, especially, hydroxy substituents there are in the two aromatic rings, the higher antioxidative activity.[8,15] However, it should be pointed out that the test systems used in the measurement of antioxidative activity are of great importance; when a linoleic acid emulsion system is used, the activity of the avenanthramide Bf is higher than for the cinnamic acids,[8] whereas when a more hydrophilic system is used (diphenyl-picryl-hydracyl radical, DPPH) the results show the opposite, i.e. the cinnamic acids exert higher activities than Bf.[15]

To be antioxidants, oat avenanthramides must be quite stable. In fact, neither pH (acid, neutral, basic), high temperature (baking oven or boiling water bath) nor UV-light (254 nm) treatment affect them very much. This is valid both for synthetic compounds and those located within the oat tissue or oil during treatments.[8,9,15] Avenanthramides are also stable during grain storage (15 months).[8,9]

2.2. Other sources

Besides oats, avenanthramides have been found in carnation leaves (Caryophyllaceae)[17,18] and in eggs of white cabbage butterfly (Lepidoptera).[19] The avenanthramides in the carnation act, as in oat leaves, as phytoalexins[17,18] but in eggs of the butterfly as oviposition deterrents.[19] It seems that butterflies produce avenanthramides in order to prohibit other butterflies fertilizing their eggs at the same place, and thereby ensuring of their own survival. These activities are also structure related, but in this case it seems that less hydroxyl substitution results in higher activity,[19] which is contradictory to antioxidant activity.

Other biological activities reported for avenanthramides include lipoxygenase inhibition[20] and anti-histamic, anti-allergic and anti-asthmatic activities.[21]

3 CONCLUSION

As oat avenanthramides are quite stable antioxidants and probably have a high pass-through in food processing they presumably contribute to the oxidative stability of oat products and might also provide some nutritional health benefits for human beings.

References

1. L. Lowen, L. Anderson, and R.W. Harrison. *Ind. Eng.Chem*. 1937, **29**, 146.
2. F.N. Peters Jr. and S. Musher. *Indust. Eng. Chem*. 1937, **29**, 146.
3. D.M. Peterson. *Cereal Chem*. 1995, **72**, 21.
4. P.J. White and L.S. Armstrong. *J. Amv. Oil. Chem. Soc*. 1986, **63**, 525.
5. D.G.H. Daniels and H.F. Martin. *J. Sci. Fd. Agric*. 1968, **19**, 710.
6. F.W. Collins. *J. Agric. Food Chem*. 1989, **37**, 60.
7. F.W. Collins and W.J Mullin. *J. Chromatogr*. 1988, **445**, 363.
8. L.H. Dimberg O. Theander and H. Lingnert. *Cereal Chem*. 1993, **70**, 637.
9. L.H. Dimberg, L.E Molteberg, R. Solheim and W.Frølich. *J. Cereal Science*. 1996, **24**, 263.
10. L. Crombie and J. Mistry. *Tetrahedron Letters*. 1990, **31**, 2647.
11. H. Miyagawa, A. Ishihara, T. Nishimoto, T. Ueno and S. Mayama. *Biosci. Biotech. Biochem*. 1995, **59**, 2305.
12. H. Miyagawa, A. Ishihara, Y. Kuwahara, T. Ueno. and S. Mayama. *Phytochemistry*. 1996, **41**, 1473.
13. A. Ishihara, T. Matsukawa, H. Miyagawa, T. Ueno, S. Mayama and H. Iwamura. *Zeitschrift für naturforschung C-A journal of biosciences*. 1997, **52**, 756.
14. A. Ishihara, H. Miyagwa, Y. Kuwahara, T. Ueno and S. Mayama. *Plant Science*. 1996, **115**, 9.
15. L.H. Dimberg and K. Sunnerheim, unpublished.
16. E.L. Molteberg, R. Solheim, L.H. Dimberg and W. Frølich. *J. Cereal Science*. 1996, **24**, 273.
17. G.J. Niemann. *Phytochemisty,* 1993, **34**, 319.
18. M. Ponchet, J. Favre-Bonvin, M. Hauteville and P. Ricci. *Phytochemistry*. 1988, **27**,

725.

19. A. Blaakmeer, D. van der Wall, A. Stork, T.A. van Beek, A. de Groot and J.J.A. van Loon. *Journal of Natural Products*. 1994, **57**, 1145.

20. T. Wakabayashi, Y. Kumonaka, H. Ichikawa and S. Murota. *Japanese patent*. 1986, **60**, 152, 454.

21. J.P.A. Devlin and K.D Hargrave. "Pulmonary and antialle drugs" (J.P.A. Devlin, ed) John Wiley and sons: Chichester, England, 1985.

PROCESSING OF FOODS CONTAINING FLAVONOIDS AND GLUCOSINOLATES; EFFECTS ON COMPOSITION AND BIOACTIVITY

Matthijs Dekker, Addie A. van der Sluis, Ruud Verkerk and Wim M.F. Jongen.

Food Science Group, Wageningen Agricultural University, PO Box 8129, 6700 EV Wageningen, The Netherlands.

1 INTRODUCTION

The intake of various secondary plant metabolites has been associated with lower incidence in various (ageing) diseases like cancer and cardiovascular diseases. Important groups of compounds are the flavonoids (antioxidants) and glucosinolates (inducing of detoxifying enzymes). Important sources of flavonols and flavones in the Dutch diet are tea (48 %), onions (29 %) and apples (7 %).[1] The intake of these compounds with the diet has been associated with a lower incidence of cardiovascular diseases in epidemiological studies. Sources of glucosinolates are various Brassica vegetables, high intake of brassica's has been associated with higher levels of phase II enzymes, which play a role in the detoxification of various carcinogenic compounds.[2] Compositional data used in these studies is often obtained from unprocessed materials, while most foods are consumed after some sort of processing, either by the producer or by the consumer. The composition of food products can be effected by processing steps as well as the bioavailability of bio-active compounds from the final products.[3] It is therefore important to know what the effects of processing steps are on the level and activity of bioactive components such as antioxidants in foods. With this information more accurate figures can be given for epidemiological work and also product development can be directed to consumer foods with an optimal level of bio-active compounds. As examples processing effects on antioxidants in apple, tea and on glucosinolates in Brassica vegetables are presented here. The possibilities to predict the antioxidant activity of a food product from its compositional data has been tested for apple juice, black tea and green tea.

2 MEASUREMENT OF ANTIOXIDANT ACTIVITY

Lipid peroxidation was induced in male rat liver microsomes by ascorbic acid and Fe^{2+}. Peroxidation products were determined in the thiobarbituric acid assay. The inhibition of lipid peroxidation gives an indication of the antioxidant activity. For each component or food extract the inhibition of the oxidation has been determined for a range of concentrations. The activity is expressed as IC_{50}: the concentration of an individual compound or food extract at which the oxidation is inhibited for 50 %. This value can be determined from the inhibition vs. concentration curves by a fitting procedure. The assay has been optimised and made suitable for large number of samples by using microtiter plates and an ELISA reader.[4]

3 EFFECTS OF PROCESSING

3.1. Apple juice production

Flavonoid content in apple juice is only 5-10 % of the content of the apples used to produce the juice.[3] The conventional process of making juice from apples was investigated for the losses of flavonoids. It was found that the pressing process resulted in the largest losses due to binding of the flavonols to the pressing cake. Alternative processes have been shown to be possible in order to significantly improve the level of flavonoids in the final product.[4] An example is the addition of alcohol (methanol, ethanol or isopropanol) to the pulp before pressing and thereby extracting the flavonols from the insoluble particles (Fig. 1).

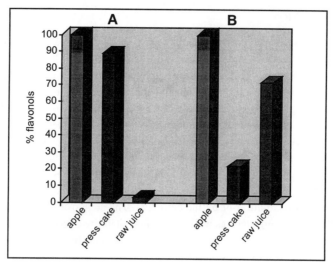

Figure 1
Effect of processing of apple juice on the partitioning of total flavonols over the juice and the press cake (A: conventional pressing; B: alcoholic extraction before pressing).[4]

The antioxidant activity of a juice produced with an alcoholic extraction process was >10 times higher than a conventional produced juice. In order to develop an apple juice with a high antioxidant activity attention has to be given to sensory properties of the juice. Too high polyphenol content is known to be responsible for astringent taste, so there will be a maximum in the level of these compounds.

3.2. Tea infusion

Flavonoid content in tea varies widely among types and brands. This has a marked effect on the antioxidant activity of tea. Green tea flavonoids consist mainly of catechins and flavonol glycosides. During the manufacturing process of black tea a part of the catechins is converted to theaflavins and thearubigens by enzymatic oxidation. The infusion process of black tea was studied.[6] The effects of infusion time and temperature on the composition and antioxidant activity of the tea extract was determined (Figure 2). The effect of time on the antioxidant activity of the tea shows that during 'normal' extraction times of 3-5 minutes, the extract contains around 50-75 % of the equilibrium activity. The infusion temperature does not have a large effect on the activity of the extract between 60 and 100 °C.

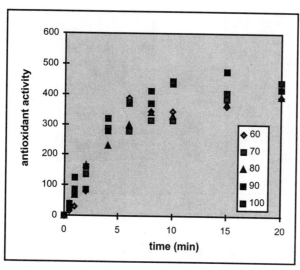

Figure 2

Effect of infusion temperature and time on the antioxidant activity of the black tea extract (antioxidant activity is expressed as the dilution factor of the tea to give a 50 % inhibition in the peroxidation assay)[3].

The compositional data show an effect of temperature and time: at higher temperature more theaflavin is found in the extract and less catechins. At short extraction times the relative contribution of the flavonol glycosides is higher due to their faster extraction rates.

3.3. Brassica processing

Brassica vegetables are well known for their high levels of bio-active glucosinolates.[7] Glucosinolates and especially their breakdown products are potential anticarcinogenic compounds due to the induction of Phase II enzymes that are involved in detoxification processes. The level of glucosinolates can be strongly affected by processing like cutting and storing in air.[7,8] This phenomenon is shown in Fig. 3 for white cabbage.

Extracts of Brassica vegetables also have an antioxidant activity. This activity is not due to glucosinolates but most likely to other compounds like flavonols, vitamins etc. Interesting results have been found for the effect of processing on antioxidant activity.[3]

After cutting white cabbage and exposing it to air an initial small decrease followed by a 50 % increase in the antioxidant activity is observed (Fig. 3). These observation has not yet been coupled to compositional changes in the cabbage.

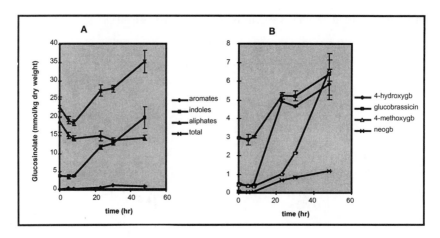

Figure 3

Levels of glucosinolates (A) in white cabbage after chopping and prolonged exposure to air at room temperature. On the right the results for the specific indolylglucosinolates are shown (B).[7]

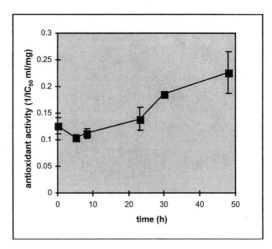

Figure 4

Effect of cutting and storing in air of white cabbage on its antioxidant activity.[3]

4 PREDICTING ANTIOXIDANT ACTIVITY FROM COMPOSITION

From the chemical analysis of food products and antioxidant activities of individual compounds it is theoretically possible to calculate the antioxidant activity of a product. This implies however that no synergistic/antagonistic or other matrix effects play a role and that all compounds with antioxidant activity are known and detectable.[4,6] For apple juice and black and green tea these calculations were performed using the antioxidant activity (expressed as $1/IC_{50}$) of the individual components or the complete mixture (equation 1).[3]

$$\frac{\sum_{i=1}^{n} C_i}{IC_{50, mixture}} = \sum_{i=1}^{n} \frac{C_i}{IC_{50, i}} = \text{dilution factor of the product to obtain 50 \% inhibition} \quad (1)$$
$$C_i = \text{concentration of componenti.}$$

In Fig. 5 the contribution of known and detected antioxidants, as calculated based upon their concentration and specific activity with equation (1), is shown as percentage of the measured total antioxidant activity.

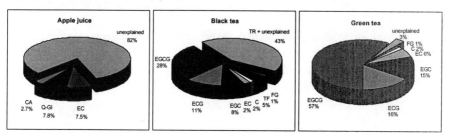

Figure 5

Contribution of known antioxidants present in apple juice, black tea and green tea (obtained by alcoholic extraction) to the total antioxidant activity of these drinks. (C: catechin, CA: cholorogenic acid, EC: epicatechin, ECG: epicatechingallate, EGC: epigallocatechin, EGCG: epigallocatechingallate, FG: flavonol glycosides, Q-Gl: quercetin glycosides, TF: theaflavins, TR: thearubigens).

The measured antioxidants in apple juice only explain some 20 % of the total activity. In black tea the majority of the measured activity can be explained by the contribution of known compounds, 43 % will be due to both thearubigens and unknown contributions. For green tea 97 % of the measured antioxidant activity can be explained by the contributions of the individual components.

5 CONCLUSIONS

It was shown that the antioxidant activity of food (extracts) can be assessed conveniently by a lipid peroxidation assay in microtiter plates. As a result of the processing of foods large effects on the antioxidant activity have been shown. Both a decrease as well as an increase in the antioxidant activity of the final product are possible. Attempts to relate the total antioxidant activity of food products to the compositional data show that sometimes matrix effects or other interactions are important. For some products (e.g. black and especially green tea) a good match between the predicted total activity and measured total activity is found while other products (apple juice) only 20 % of the activity can be explained.

References
1. M.G.L. Hertog, P.C.H. Hollman, M.B. Katan and D. Kromhout, *Nutr. Cancer*. 1993, **20**, 21.
2. R. Staack, S. Kingston, M.A. Wallig, and E.H. Jeffery, *Toxicol. Appl. Pharmacol.* **149**, 17, 1998.
3. M. Dekker, A.A. van der Sluis, R. Verkerk, A.R. Linnemann and W.M.F. Jongen, 'Agri Food Quality II' Royal Society of Chemistry, 1998, *in press*.
4. A.A. van der Sluis, M. Dekker and W.M.F. Jongen, 1998, *in preparation*.
5. A.A. van der Sluis, M. Dekker and W.M.F. Jongen, *Cancer Letters*. 1997, **114**, 107.
6. M. Dekker, A.A. van der Sluis, R. Verkerk, A.R. Linnemann and W.M.F. Jongen, 1998, *in preparation*.
7. R. Verkerk, M. Dekker and W.M.F. Jongen, 'Natural Toxicants in Food', Sheffield Academic Press, London, 1998, Chapter 3, p. 29.
8. R. Verkerk, M. Dekker and W.M.F. Jongen, 'Agri Food Quality II' Royal Society of Chemistry, 1998, *in press*.

MEASUREMENTS OF CHOLESTEROL OXIDES IN FOODS: RESULTS OF AN INTERLABORATORY COMPARISON STUDY

P. C. Dutta[4], M. F. Caboni[1], U. Diczfalusy[2], F. Dionisi[3], S. Dzeletovic[2], A. Grandgirard[5], F. Guardiola[6], J. Kumpulainen[7], V. K. Lebovics[8], J-M. Pihlava[7], M. T. Rodriguez-Estrada[1] and F. Ulberth[9].

[1]Istituto di Industrie Agrarie, Università di Bologna, Bologna, Italy. [2]Dept. of Medical Lab. Sci. & Tech., Div. of Clinical Chemistry, Karolinska Institutet, Huddinge, Sweden. [3]Department of Quality and Safety Assurance, Nestlé, Nestec Ltd., Lausanne, Switzerland. [4]Department of Food Science, Swedish University of Agric. Sci., Box 7051, 750 07 Uppsala, Sweden. [5]INRA, Unité de Nutrition Lipidique, Dijon, France. [6]Nutrition and Food Science Department, University of Barcelona-CeRTA, Barcelona, Spain. [7]Food Research Institute, Laboratory of Food Chemistry, Jokionen, Finland. [8]National Institute of Food Hygiene and Nutrition, Department of Food Chemistry, Budapest, Hungary. [9]Department of Dairy Research and Bacteriology, Agricultural University, Vienna, Austria.

1 INTRODUCTION

Cholesterol (cholest-5-en-3β-ol), a 27 carbon steroid alcohol abundant in animal tissues, is readily susceptible to oxidation.[1] Oxidation of cholesterol is enhanced by temperature, light, storage, air, pro-oxidizing agents and deficiency of antioxidants. Food ingredients and food products are frequently exposed to such factors during manufacture and handling.[2] More than 70 oxidation products of cholesterol (COPs) have been identified until now;[3-6] however, only eight of these compounds are generally reported in foods.[2,7-9] These are 5-cholesten-3β, 7α-diol (7α-OH), 5-cholesten-3β, 7β-diol (7β-OH), 5-cholesten-3β-ol-7-one (7-keto), 5-cholestan-5α, 6α-epoxy-3β-ol (5α, 6α-epoxy), 5-cholestan-5β, 6β-epoxy-3β-ol (5β,6β-epoxy), 5-cholesten-3β,20α-diol (20α-OH), 5-cholesten-3β,25-diol (25-OH), and cholestane-3β, 5α, 6β-triol (triol).[9] Some of these COPs can be produced both enzymatically and by autoxidation.[10] COPs have been shown to have a wide variety of effects both *in vitro* and *in vivo,* which may be linked to human diseases and cholesterol metabolism.[1,3,5,6,8,10-15] The only known report on the absorption of cholesterol oxides in human subjects has shown that COPs from foods are absorbed within a few hours of consumption.[16] Recently, it was reported that the lipoprotein LDL is enriched in cholesterol oxides and lipid peroxides, part of which may be of dietary origin.[17]

Literature reports on the levels of COPs in the same type of foods are extremely difficult to compare.[5,7-9,13] This inconsistency is mainly due to differences in the analytical methods used by the reporting laboratories.[5, 14,18] The importance of method validation used for the determination of COPs has been discussed some years ago,[19] and this was repeated recently.[13]

In order to harmonize the analytical methods for the determination of COPs, a Round

Robin analysis using several food samples was held during 1995,[20] where 17 leading laboratories from Europe, North America, Japan and Taiwan participated, and a part of the results of that study was reported recently.[21] The results from that study were rather similar to what is reported in the literature, i.e. a large variation in the levels of COPs in the same kind of foodstuff.[21] Besides the differences in analytical methods, sample packaging and handling may also be among the reasons of the large variations in the results.

Reviews[5,7-9,13] on COP levels in foods show that egg and milk powder are among the foods widely studied because of their frequent use as food ingredients (e.g. bakery products). The most abundant COPs in egg powders are: 7α-hydroxycholesterol, 7β-hydroxycholesterol, 5α, 6α-epoxycholestanol, 5β, 6β-epoxycholestanol, and 7-ketocholesterol.[5,7-9,13] Various drying techniques have been studied in the preparation of egg powders, as well as various drying temperatures. It has been shown that high drying temperatures increase the formation of COPs.[9,22] In milk powders, cholesterol can be oxidized during the drying process and storage. Whole milk is particularly vulnerable to cholesterol oxidation during spray drying, due to sample exposure to air and high temperatures; however, fresh milk powders are usually low in COPs.[9]

The objective of this study is to establish the critical points in the analytical methodology and select an uniform validated method for the analysis of COPs. Based on this objective, an interlaboratory study for the determination of COPs in milk powder and egg powder was conducted by a selected number of European laboratories during 1997 and the results are presented in this paper.

2 MATERIALS AND METHODS

Samples of egg powder and milk powder were packed in aluminium/plastic bags under vacuum at Nestlé, Nestec Ltd. (Lausanne, Switzerland) and shipped to participant laboratories in dry-ice by express mail. The samples were stored at -20 °C after arrival at different laboratories until further analysis. Brief descriptions of the methods used by the different laboratories are given below. Lab numbers correspond to those of the list of authors.

Many of the methods used by the participating laboratories have been published in details as follows: lab. 1 employed GC-ion trap detector mass spectrometry;[23] lab. 2 GC-MS using isotope dilution technique with added deuterium labelled COPs standards;[24] lab. 3 GC-MS in selective ion monitoring mode (SIM);[25] lab. 4 GC-MS with COPs enrichment by SPE,[26-29] lab. 5 lipid extraction by chloroform: methanol, cold saponification for 2 days with 19-OH as I.S., unsaponifiables extraction with dichloromethane, SPE enrichment of COPs (silica cartridge, 500 mg), TMS derivatisation using BSTFA/TMCS (100/1) at 60° C for 15 min, GC quantification by using a DB5 column (20m x 0.32mm x 0.25mm), COPs identification confirmed by GC-MS; lab. 6 employed GC according to Guardiola et al. [22,30] with slight modification; lab. 7 GC-MS detection and quantification,[28,31] lab. 8 employed thin layer chromatography to the separation of COPs and spectrometric quantification by an enzymatic method;[32] and finally lab. 9 used GC-MS for quantification and extraction with diethyl ether.[33] Most laboratories used 19-OH as an internal standard.

3 RESULTS AND DISCUSSION

Seven laboratories out of nine reported their results on both egg powder and milk powder samples. One laboratory reported results only on egg powder, and another laboratory reported results on milk powder only. Table 1 shows the levels of COPs found in egg powder. Some of the laboratories did not determine or find above their limit of quantification of the epoxides, triol, 7α-OH, 20-OH and 25-OH-cholesterol in this sample. Since not all the laboratories reported results of all the eight COPs, only ranges of individual COPs were taken into consideration. The range, median, standard deviation and CV% of the individual COPs in the egg powder sample are presented in Table 3, along with the corresponding values from the previous study.[20, 21] Coefficient of variation among the individual COPs are much lower compared with those obtained in the previous study. In addition, cholesterol extraction from milk powder may require particular extraction conditions due to the highly-structured nature of the milk fat globule.

Table 1

Lipid and cholesterol oxides (COPs) content in whole egg powder as determined by the participant laboratories.

COPs content (µg/gsample)

Laboratory	Lipids (g/100g)	7α-OH[a]	7β-OH	7-keto	$5\alpha,6\alpha$-epoxy	$5\beta,6\beta$-epoxy	Triol	20-OH	25-OH	Total
Lab 1	44.3	3.14	1.82	2.59	0.43	1.47	nd	nd	0.02	9.47
Lab 2	_b	3.31	1.90	2.25	0.91	3.27	0.25	_b	0.15	12.04
Lab 3	39.0	5.31	3.46	2.44	_b	_b	0.17	_b	0.32	11.70
Lab 4	40.0[c]	2.22	1.22	1.45	1.12	3.54	0.17	0.67	0.09	10.48
Lab 5	44.3	2.54	1.49	2.35	0.77	2.07	_b	_b	_b	9.22
Lab 6	_b	_b	4.36	2.15	1.58	3.13	tr	_b	0.22	11.44
Lab 7	_c	3.18	2.83	2.08	1.48	2.54	0.09	0.15	0.36	12.71
Lab 8	43.0	5.51	4.92	5.88	nd	_b	nd	nd	nd	16.31

Lab No= Corresponds to the number of the participating laboratories as given in the list of authors. [a]=*Abbreviations: 7α-OH= 7α-hydroxycholesterol; 7β-OH= 7β-hydroxycholesterol; 7-keto= 7-Ketocholesterol; $5\alpha,6\alpha$-epoxy= $5\alpha,6\alpha$-epoxycholestanol; $5\beta,6\beta$-epoxy= $5\beta,6\beta$ epoxycholestanol; Triol= Cholestanetriol; 20-OH= 20α-hydroxycholesterol; 25-OH= 25-hydroxycholesterol.* [b]=*Not determined.* [c]=*Figures provided by the manufacturer.* tr= *Traces* <0.20 ng/g. nd= *Not detected.*

The results from this study on the levels of COPs in milk powder are shown in Table 2. Similar to egg powder, some laboratories did not determine or found above the limit of quantification of the epoxides, triol, 7α-OH, 20-OH, and 25-OH-cholesterol in the milk powder sample. Simple statistical results on the individual COPs are presented in Table 4, along with the corresponding values obtained in the previous Round Robin study.[20,21] Since only a few laboratories in the previous study,[20,21] reported the values on 20-hydroxy-, and 25-hydroxy-cholesterol, these figures were not considered for both egg powder and milk powder. It is evident from Table 4 that the ranges and CV values are generally lower than those obtained in the previous study, but are still relatively high compared with egg powder. This may be attributed to the much lower levels of COPs present in this sample, which renders more difficult an accurate quantification.

Table 2

Lipid and COPs content in whole milk powder as determined by the participant laboratories.

Laboratory	Lipids (g/100g)	7α-OH[a]	7β-OH	7-keto	COPs content (μg/g sample) 5α,6α- epoxy	5β,6β- epoxy	Triol	20-OH	25-OH	Total
Lab 1	29.6	0.12	0.09	1.94	tr	0.17	nd	nd	nd	2.32
Lab 3	26.4	0.03	0.02	0.05	_b	_b	0.01	_b	0.03	0.14
Lab 4	26.0[c]	0.18	0.15	0.28	0.10	0.33	0.05	0.06	0.03	1.18
Lab 5	23.6	0.04	0.03	0.17	(0.01	0.09	_b	_b	_b	0.33
Lab 6	_b	_b	0.44	0.13	0.67	0.62	tr	_b	0.06	1.92
Lab 7	_b	0.05	0.06	0.08	0.17	0.34	0.10	nd	0.05	0.85
Lab 8	27.0	0.88	0.74	0.57	0.56	_b	1.61	nd	nd	4.36
Lab 9	_b	_b	0.04	0.08	nd	_b	0.03	_b	0.02	0.17

a, b, c and for COPs abbreviations, see Table 1. tr= Traces < 0.01 ng/g (Lab.6: < 0.05 mg/g).

General decrease in the results variability can be partly attributed to better sample handling, especially packaging and shipping. In the present study, samples were vacuum packed in high quality aluminium/plastic bags. Although the participant laboratories used their own methods, the sample work up procedures had more in common in this study compared with the previous one.[20,21] In addition, in the previous study samples for US participants were held at customs for 1 month under uncontrolled conditions.

Quantification of COPs is generally accomplished by GC, GC-MS, or by HPLC using various detection systems,[5,13,14,34,35] although HPLC has not been so widely used as GC. The reason may be the higher costs to use HPLC, detector sensitivity, and a limited availability of HPLC-MS as compared to GC-MS; this last point is a significant factor, since identification of COPs by MS is essential. Most of the laboratories used the combination of GC and GC-MS as described in the methods section.

Although in the present study variability considerably decreased, it is still high (Tables 3 and 4), which may be due to differences in some work-up steps in the methods used for COPs determination. Therefore, the critical points during the analysis steps should be considered very carefully while adopting a common method for the determination of these compounds. Some of the critical points to be mentioned here are: extraction, saponification, enrichment, use of internal standard, recovery of the COPs during work-up steps, response factors, linearity range, limit of detection and quantitation.[19,36] Other important points are the types and dimensions of the capillary columns, and the integration technique used. Even after enrichment by SPE, considerable amounts of cholesterol are present in the mixture, which makes difficult to achieve good separation between 7α-hydroxycholesterol and cholesterol. Separation between 5α, 6α-epoxycholestanol and 5β, 6β-epoxycholestanol, and between 5β, 6β-epoxycholestanol and 7β-hydroxycholesterol may also be difficult using a DB-1 and similar types of columns having a length of 30 m, i.d. of

0.25 mm, and film thickness of 0.25 mm (unpublished observation, Dutta). DB-5 and similar types of columns of similar dimensions can give better separation but still may not be sufficient. Considerable improvement in separation among the above mentioned COPs can be achieved using columns with thicker film at the expense of analysis time. It is also often observed that diffused peaks closely eluted with a COPs can produce erroneous results when quantification is done by integrating the peak areas, and need careful evaluation of such individual COP. An alternative to this is to use peak height instead of peak area which may produce better and more reproducible results (unpublished observation, Dutta).

Table 3

Simple statistical analysis of interlaboratory variation in content of COPs in egg powder - present and previous study.

COPa		No. of Lab. participated	Total Range (µg/g sample)	Median	Mean	±SD	CV (%)
7α-OH	Present study	7	2.22 - 5.51	3.18	3.60	±1.30	36
	Previous study	10	0.34 - 12.55	3.87	5.75	±4.84	84
7β-OH	Present study	8	1.22 - 4.92	2.37	2.75	±1.38	50
	Previous study	13	0.29 - 157.13	4.52	18.63	±43.13	231
7-keto	Present study	8	1.45 - 5.88	2.30	2.65	±1.35	51
	Previous study	15	0.40 - 24.56	1.57	5.98	±8.08	135
5α,6α-epoxy-	Present study	6	0.43 - 1.58	1.01	1.05	±0.44	42
	Previous study	11	0.32 - 54.44	1.24	13.70	±19.48	142
5β,6β-epoxy-	Present study	6	1.47 - 3.54	2.83	2.67	±0.79	30
	Previous study	10	0.40 - 30.63	5.73	10.85	±11.96	110
Triol	Present study	4	0.09 - 0.25	0.17	0.17	±0.07	38
	Previous study	9	0.12 - 14.54	0.66	2.65	±4.65	176

a For COPs abbreviations, see Table 1; Previous study 20, 21.

Table 4

Simple statistical analysis of interlaboratory variation in content of COPs in milk powder - present and previous study.

COPa		No. of Lab. participated	Total Range (µg/g sample)	Median	Mean	±SD	CV (%)
7α-OH	Present study	6	0.03 - 0.88	0.09	0.22	±0.33	151
	Previous study	11	0.04 - 9.73	0.13	1.19	±2.88	242
7β-OH	Present study	8	0.02 - 0.74	0.08	0.20	±0.26	132
	Previous study	14	0.03 - 2.64	0.29	0.52	±0.84	161
7-keto	Present study	8	0.05 - 1.94	0.15	0.41	±0.64	155
	Previous study	14	0.05 - 3.09	0.23	0.55	±0.80	145
5α,6α-epoxy-	Present study	4	0.10 - 0.67	0.36	0.38	±0.28	76
	Previous study	12	0.02 - 7.25	0.13	1.08	±2.18	202
5β,6β-epoxy-	Present study	5	0.09 - 0.62	0.33	0.31	±0.21	66
	Previous study	10	0.00 - 5.76	0.36	1.33	±1.99	149
Triol	Present study	5	0.01 - 1.61	0.05	0.36	±0.70	196
	Previous study	6	0.01 - 2.20	0.07	0.55	±0.88	160

a For COPs abbreviations, see Table 1; Previous study 20, 21.

Another possibility is the use of selective ion monitoring mass spectrometry, using the isotope dilution technique where deuterium labelled standard samples of individual COPs are used as internal standards. The latter technique is probably the most reliable method (lab. 2 used this technique) of quantification but very expensive indeed, and may not be afforded by many laboratories for routine analysis.

Analysis of COPs is rather complex often due to their presence at a very low level in foods and biological samples. Harmonization of methodology requires clear identification of critical steps in the procedure to reduce the sources of error. Based on the results of the present study it can be concluded that the requirements to develop an optimised procedure to better monitor these compounds, are quite high. In addition, mixed diets can contain oxidation products from cholesterol and phytosterols which make analysis far more complex and require further developments of the procedure.

References

1.   L. L. Smith, "Cholesterol Autoxidation", Plenum Press, New York, 1981.
2.   J. Sarantinos, K. O'Dea, and A. J. Sinclair, *Food Australia*. 1993, **45**, 485.
3.   L. L. Smith, *Chem. Phy. Lipids*. 1987, **44**, 87.
4.   G. Maerker, *J. Am. Oil Chem. Soc*. 1987, **64**, 388.
5.   S-K. Peng and R. J. Morin (Eds.), " Biological Effects of Cholesterol Oxides", CRC Press, Boca Raton, Florida, 1992.
6.   L. L. Smith, *Lipids*. 1996, **31**, 453.
7.   E. T. Finocchiaro, and T. Richardson, *J. Food Prot*. 1983, **46**, 917.
8.   S. Bösinger, W. Luf, and E. Brandl, *Int. Dairy J*. 1993, **3**,1.
9.   P. Paniangvait, A. J. King, A. D. Jones, and B. G. German, *J. Food Sci*. 1995, **60**, 1159.
10.   F. Guardiola, R. Codony, P. B. Addis, M. Rafecas, and J. Boatella, *Food Chem. Toxic*. 1996, **34**, 193.
11.   T. G. Toschi, and M. F. Caboni, *Ital. J. Food Sci*. 1992, **4**, 223.
12.   E. Lund, and I. Björkhem, *Acc. Chem. Res*. 1995, **28**, 241.
13.   P. B. Addis, P. W. Park, F. Guardiola, and R. Codony, "Food Lipids and Health", R. E. McDonald, and D. B. Min, D. B. (Eds.), Marcel Dekker, Inc., New York, 1996.
14.   P. C. Dutta, R. Przybylski, L-Å. Appelqvist, and N. A. M. Eskin, "Deep Frying: Chemistry, Nutrition, and Practical Applications", E. G. Perkins, and M. D. Erickson (Eds.). The AOCS press, Champaign, 1996.
15.   N. Kumar, and O. P. Singhal, *J. Sci. Food Agric*. 1991, **55**, 497.
16.   H. A. Emanuel, C. A. Hassel, P. B. Addis, S. D. Bergman,, and J. H. Zavoral, *J. Food Sci*. 1991, **56**, 843.
17.   A. Sevanian, G. Bittolo-Bon, G. Cazzolato, H. Hodis, J. Hwang, A. Zamburlini, M. Maiorino, and F. Ursini, *J. Lipid Res*. 1997, **38**, 419.
18.   S. W. Park, "Analyzing Foods for Nutrition Labeling and Hazardous Contaminants", I. J. Jeon, and W. G. Ikins, W. G. (Eds.). Marcel Dekker, Inc., New York, 1995.
19.   S. McCluskey, and R. Devery, *Trends Food Sci. & Tech*. 1993, **4**, 175.

20. L-Å. Appelqvist, Harmonization of oxysterol analysis in food and blood. Worldwide interlaboratory study with 20 leading laboratories participated. Workshop held in Lausanne, Switzerland, August 28-29, 1995.
21. L-Å. Appelqvist, *Bulletin of the International Dairy Federation.* 1996, **315**, 52.
22. F. Guardiola, R. Codony, D. Miskin, M. Rafecas, and J. Boatella, *J. Agric. Food Chem.*.1995, **43**, 1903.
23. M. T. Rodriguez-Estrada, M. F. Caboni, A. Costa, and G. Lercker, *J.High Res. Chromatogr.* 1998, in press.
24. S. Dzeletovic, O. Breuer, E. Lund, and U. Diczfalusy, *Anal. Biochem.* 1995, **225**, 73.
25. F. Dionisi, P. A. Golay, J. M. Aeschlimann, and L. B. Fay, *J. Agric. Food Chem.* 1998, **46**, 2227.
26. J. Nourooz-Zadeh, and L-Å. Appelqvist, *J. Food Sci.* 1987, **52**, 57.
27. J. Nourooz-Zadeh, and L-Å. Appelqvist, *J. Food Sci.* 1988, **53**, 74.
28. S. W. Park, and P. B. Addis, *J. Agric. Food Chem.* 1986, **34**, 653.
29. P. C. Dutta, and L-Å. Appelqvist, *J. Am.Oil Chem. Soc.* 1997, **74**, 647.
30. F. Guardiola, R. Codony, M. Rafecas, and J. Boatella, *J. Chromatogr. A.* 1995, **705**, 289.
31. K. Granelli, P. Fäldt, L-Å. Appelqvist, and B. Bergenståhl, *J. Sci. Food Agric.* 1995, **71**, 75.
32. V. K. Lebovics, M. Antal, and Ö. Gáal, *J. Sci. Food Agric.* 1996, **71**, 22.
33. C. Rose-Sallin, A. C. Huggett, J. O. Bosset, R. Tabacchi, and L. B. Fay, *J. Agric. Food Chem.* 1995, **43**, 935.
34. L. L. Smith, *J. Liq. Chrom.* 1993, **16**, 1731.
35. L. Lakritz, and K. C. Jones, *J. Am. Oil Chem. Soc.* 1997, **74**, 943.
36. F. Guardiola, A. Jordán, A. Grau, S. Garcia, J. Boatella, M. Rafecas, and R. Codony, "Recent research developments in oil chemistry", S. G. Pandalai (Ed.), Transworld Research Network, Trivandrum. 1998, **2**, p. 77.

PHYTOSTEROL OXIDES IN SOME SAMPLES OF PURE PHYTOSTEROLS MIXTURE AND IN A FEW TABLET SUPPLEMENT PREPARATIONS IN FINLAND

Paresh C. Dutta.

Department of Food Science, Swedish University of Agricultural Sciences, Box 7051, SE-750 07 Uppsala, Sweden.

1 BACKGROUND

Phytosterols are known inhibitors of cholesterol absorption in humans and this has in recent years increased the interest to fortify products with these natural compounds.[1-3] Phytosterols are structurally related to cholesterol. Unsaturated phytosterols oxidize and produce similar oxidation products as cholesterol, some of these compounds were reported in some food products recently.[4] Oxidation products of cholesterol have been received much attention in the last decades due to their undesirable health effects such as cytotoxicity, atherogenicity, cholesterol metabolism interference, mutagenicity and carcinogenicity, but studies with phytosterol oxides are very limited.[5,6] In contrast to cholesterol, phytosterol absorption in humans is much lower.[7]

2 OBJECTIVE

The objective of this study was to determine some polar oxidation products of phytosterols in raw materials (wood sterols), and in a number supplement tablet preparations based on phytosterols commercially available in Finland. In addition, a sample of pure phytosterol mixture, which was subjected to oxidation by treatment with high temperature, was analysed to compare with the unheated raw materials.

3 MATERIALS AND METHODS

Samples of pure phytosterol mixture (Kaukas Oy, Finland), and supplement tablet preparations were received from Raisio Group, Benecol Division, Raisio, Finland. After arrival, the samples were stored at -80 °C until further analyses. Since all the samples were in the free form, no saponification was needed. Enrichment of the sterol oxides by thin layer chromatography (TLC) was necessary because the samples were not soluble in the hexane:ether mixture The samples were first dissolved in a mixture of chloroform: methanol (1:1), sonicated, centrifuged, and ca. 10 mg of sample was separated on TLC silica plates by developing in the solvent system cyclohexane: diethyl ether (90:10, v/v)[4]. The polar sterol oxides area of the TLC plate was scrapped off and eluted repeatedly with chloroform: methanol (1:1, v/v). The organic solvent was evaporated under nitrogen, and subjected to silylation as described previously.[4]

3.1. Capillary column gas chromatography (GC)

In order to achieve separation of TMS-ether derivatives of phytosterol oxides, a fused silica capillary column DB-5MS (30 m x 0.25 mm x 0.50 μm; J & W Scientific, Folsom, CA, USA), fitted in a Varian GC 3700 gas chromatograph Varian, Palo Alto, CA, USA) equipped with a falling needle injector and a flame ionization detector, was used in this investigation. The GC conditions were as follows: oven temperature at 285 °C for 30 min, and then raised to 290 °C at a rate of 3 °C/min and maintained at this temperature for an additional 16 min. Helium was used as carrier gas at a pressure of 20 PSI and as a make-up gas at a flow rate of 30 ml/min. Detector temperature was at 310 °C. The peaks were computed using a HP 3396A integrator (Hewlett Packard, Avandale, PA, USA). The identification of phytosterol oxides was done by comparing the retention times of the authentic phytosterol oxides synthesized in this laboratory with the retention times and mass spectra of the test samples as described below. 5α-cholestane was used as an internal standard. All the samples were analyzed in duplicate and the mean values are presented.

3.2. Gas chromatography-mass spectrometry (GC-MS)

GC-MS analyses were performed on a GC-800 Top Series gas chromatograph (ThermoQuest, Rodano, Italy) coupled to a Voyager Mass Spectrometer with a data system MassLab 1.4 V (Manchester, UK). The TMS-ether derivatives of phytosterol oxides were separated on the same column as used for GC, the conditions were; injector temperature 230 °C and the samples were injected in a splitless mode and purge delay time was 0.8 min; a programmed oven temperature was used at 60 °C for 1 min and then raised to 260 °C at a rate of 20 °C/min and then held at this temperature for 20 min before being finally raised to 290 °C at 1°/min and kept at this temperature for an additional 15 min. The full scan mass spectra were recorded at an electron energy of 70 eV and the ion source temperature was at 200 °C.

4 RESULTS AND DISCUSSION

The content of the total polar oxidized sterols in the wood sterols and recrystalized sterols were 75 mg/100 g and 44 mg/100 g, respectively, whereas the heat treated sterols had 1380 mg/100 g. The tablet preparations; Anti K-steroli, Tri Tolosen Kasvisteroli and Kolestop (trade names for commercial phytosterol supplement products), had the total polar oxidation products at 14 mg/100 g, 26 mg/100 g and 30 mg/100 g tablets, respectively. However, only six of the polar oxidation products were identified by GC-MS by comparing the mass spectra of those with authentic samples (Table 1). Among the polar oxidized phytosterols identified, the highest amounts were observed for the epimers of epoxycampe- and sitosterol, and 7-ketocampe- and sitosterol. The total amounts of these were 338 mg/100g and 452 mg/100 g, respectively, in the heat treated sterols. Whereas, in the wood sterols, and in the recrystalized sterols, the total amounts of epoxysterols. were 33 mg/100 g and 22 mg/100 g, respectively; and the amounts of 7-ketosterols were 14 mg/100 g and 6 mg/100 g, respectively. In the tablet preparations, the amounts of epoxysterols ranged from 5 to 14 mg/100 g, and 7-ketosterols ranged from 3 to 5 mg/100 g (Table 1).

Table 1

Content of phytosterol oxides in three samples of pure phytosterol mixtures (mg/100 g) and in three tablet supplements (mg/100g tablet) manufactured in Finland.

	7α-OH	7β-OH	7-Keto	5α, 6α- Epoxy-	5β, 6β - Epoxy-	Triol	Total
Raw materials							
Wood sterol							
Sitosterol	7.3	4.3	12.3	8.0	22.6	0.4	60.0
Campesterol	0.8	0.3	1.3	0.5	2.2	nd	
Heat treated sterol							
Sitosterol	35.7	40.5	408.7	111.0	194.5	12.0	892.8
Campesterol	5.9	8.7	43.7	11.1	21.0	nd	
Recrystalized sterol							
Sitosterol	3.7	2.3	5.0	4.1	13.9	0.4	34.6
Campesterol	0.5	0.4	0.6	1.4	2.3	nd	
Tablet preparations							
Anti K-steroli							
Sitosterol	0.8	1.8	2.7	1.6	2.8	0. 2	10.7
Campesterol	0.2	0.2	0.2	tr	0.2	n	
Tri Tolosen Kasvisteroli							
Sitosterol	1.8	0.8	4.0	1.7	9.1	0.3	21.5
Campesterol	0.2	0.2	0.3	0.6	2.5	nd	
Kolestop							
Sitosterol	2.0	3.8	5.0	3.1	9.9	0.2	26.0
Campesterol	0.3	0.2	0.4	0.3	0.8	nd	

7α-OH= 7α-hydroxycampe-, and sitosterol; 7β-OH= 7β-hydroxycampe-, and sitosterol; 7-keto= 7-Ketocampe-, and sitosterol; 5α,6α-Epoxy-= 5α,6α-epoxycampe-, and sitosterol; 5β,6β-Epoxy= 5β,6β-epoxycampe-, and sitosterol; Triol= dihydroxycampe-, and sitosterol; tr= less than 0.05 mg; nd= not detected.

Absorption of phytosterol oxides in man is unknown, however, it may be assumed that the absorption of phytosterol oxides is as low as that of unoxidized phytosterols in man,[7] compared with absorption of cholesterol oxidation products (COPs).[8] Studies on daily intake of COPs by man are scarce. In a few studies in Holland, New Zealand, Australia, and in a test meal in USA,[8-11] have shown that calculated values in foods can vary widely. For example, daily consumption of ca. 2 mg of the sum of 7β-hydroxycholesterol and 5α,6α-epoxycholesterol in Holland;[9] according to the estimate by the Australian group,[11] the daily intake of cholesterol oxides may vary between 0 to 180 mg/d; whereas, an average New Zealand meal may contain 2.5 mg/meal;[10] a test meal in the USA contained 11.5 mg COPs.[8] An average intake of 6 supplement tablets from this study will contain a total amount of six polar oxidation products of phytosterols ranging from 0.07 mg to 0.15 mg, which can be considered very low.

Acknowledgment

This work was financially supported by Raisio Group, Benecol Division, Raisio, Finland.

References

1.   T. Heinemann, G. Aztmann. K. von Bergmann, *Eur. J. Clin .Invest.* 1993, **23**, 827.
2.   P.J. Jones, D.E. MacDougall, F. Ntanios, and C.A. Vanstone, *Can. J. Physiol. Pharmacol.* 1997, **75**, 217.
3.   T. A. Miettinen, P. Puska, H. Gylling, H. Vanhanen, and E. Vartiainen, *N. Engl.J. Med.* 1995, **333**, 1308.
4.   P. C. Dutta, and L-Å. Appelqvist, L-Å. *J. Am. Oil Chem. Soc.* 1997, **74** , 647.
5.   F. Guardiola, R. Codony, P.B. Addis, M. Rafecas, and J. Boatella, *Food Chem. Toxic.* 1996, **34**, 193.
6.   S. Bösinger, W. Luf, and E. Brandl, *Int. Dairy Journal.* 1993, **3**, 1.
7.   M. T. R. Subbiah, Mayo Clinic. *Proc.* 1971, **46**, 549.
8.   H.A. Emanuel, C.A. Hassel, P.B. Addis, S.D. Bergmann, and J.H. Zavoral, *J. Food Sci.* 1991, **56**, 843.
9.   P. van de Bovenkamp, T.G. Kosmeijer-Schuil, and M.B. Katan, *Lipids.* 1988, **23**, 1079.
10.   R. J. Lake, and P. Scholes, *J. Am. Oil Chem. Soc.* 1997, **74**, 1069.
11.   J. Sarantinos, K. O'Dea, and A. J. Sinclair, *Food Australia.* 1993, **45**, 485.

FORMATION OF STEROL OXIDES IN EDIBLE OILS

V.K. Lebovics[1], K. Neszlényi[1], S. Latif[1], L. Somogyi[2], J. Perédi[2], J. Farkas[2], Ö. Gaál[1].

[1]National Institute of Food Hygiene and Nutrition, Department of Food Chemistry, H-1097 Budapest, Gyáli út 3/a. [2]University of Horticulture and Food Industry, H-1118 Budapest, Ménesi út 43-45.

1 INTRODUCTION

Several oxidation products of cholesterol are considered more harmful agent in formation of atherosclerotic lesions than cholesterol itself, furthermore might have mutagenic and carcinogenic effects and influence the cholesterol biosynthesis by inhibition of HMG-CoA-reductase. These derivatives are formed by free radical reactions in the presence of oxygen.[1] More than 80 autoxidation derivatives of cholesterol have been identified. Some of these compounds are present in foodstuffs, due to the effect of heat, light, ionising radiation, processing and/or storage. There are many similarities in the structures of cholesterol and phytosterols and in those of their oxidation products. The number of publications on the occurrence, formation and possible toxic effects of analogous oxidized plant sterols and on their biological activities are limited.[2,3] Also few data has been published on the oxidative changes of plant sterols in vegetable oils.[4-5]

The purpose of this work was to study the effect of heat treatment and deep-frying on sterol oxides formation in lard and several oils.

2 MATERIALS AND METHODS

The total lipids were extracted by Folch's method and the saponification was performed with 1 M potassium hydroxide in methanol under stream of N_2 for 1 h. The nonsaponifiable part was extracted with diisopropylether and separated by thin-layer chromatography. The spots were removed from TLC plates and eluted with acetone. The phytosterol oxides were determined by means of cholesterol oxidase.[6]

The phytosterol oxides were identified by comparison of cholesterol oxidation derivatives. Since the chromatograms of cholesterol and plant sterol oxidation products are very similar, it can be supposed that the derivatives with the same R_f values have the similar structure. The phytosterol oxidation products recently are not available in chemical catalogues.

Olive, soy, maize, peanut, palm oil and palm oil-lard blend (1:1 v/v) were heated on large surface at 180 °C from 15 to 240 minutes.

Deep-frying of potatoes was carried out 10 times repeatedly without adding of fresh oil at 180 °C using sunflower, palm, sunflower-palm oil blend (30:70 v/v). 5 g of each oil were used for analysis. Determination of tocopherols were performed by HPLC.

3 RESULTS AND DISCUSSION

Analysis of sterol oxides showed that there were no detectable levels of any phytosterol oxides in control samples as well as in olive and maize oil samples heated for 15 minutes. 30 -minute- heating resulted in formation of 7 α-hydroxy-, 7ß-hydroxysterols and 7-ketosterols. The levels of total oxysterols in soy-, olive-, maize and peanut oil ranged from 14,3 to 69,4 mg kg[-1]. During 240-minute-heating the amount of total oxysterols of these oils were 91,2-196,2 mg kg[-1]. That is 3,9-6,7 % of sterols originally present in the oils were converted to oxidized products. (Table 1) Among the investigated oils palm oil seemed to be the most stable during deep-frying of potato. (Table 2) The oxidative stability of lard-palm oil blend (1:1) during 120 -minute- treatment proved to be higher than the lard itself.

Table 1

Levels of phytosterol oxides in oils heated at 180 °C*.

Oil	Phytosterol oxides	Control	15	30	60	120	180	240
				m i n	u t e s			
	7α-hydroxysterols	nd	5.63	6.97	8.69	26.76	42.24	51.69
	7ß-hydroxysterols	nd	3.66	4.34	9.14	39.84	40.76	54.39
Soy	7-ketosterols	nd	0.94	3.02	5.78	32.87	37.32	42.04
	Total oxysterols	nd	10.23	14.33	23.61	99.47	120.32	148.12
	Total sterols	3760	-	-	-	-	-	-
	7α-hydroxysterols	nd	nd	6.65	9.18	17.83	20.98	36.28
Olive	7ß-hydroxysterols	nd	nd	6.30	12.12	21.74	22.80	36.43
	7-ketosterols	nd	nd	9.71	10.75	13.52	18.35	21.44
	Total oxysterols	nd	nd	22.66	32.05	53.09	62.13	94.15
	Total sterols	1926	-	-	-	-	-	-
	7α-hydroxysterols	nd	nd	13.80	31.23	55.68	-	80.78
Maize	7ß-hydroxysterols	nd	nd	15.02	26.75	52.84	-	91.00
	7-ketosterols	nd	nd	16.92	19.07	19.76	-	24.42
	Total oxysterols	nd	nd	45.74	77.05	128.28	-	196.20
	Total sterols	4696	-	-	-	-	-	-
	7α-hydroxysterols	nd	20.58	30.74	36.70	43.66	41.29	65.50
Peanut	7ß-hydroxysterols	nd	19.40	27.12	26.94	40.89	39.19	66.40
	7-ketosterols	nd	5.17	11.49	13.19	26.17	31.61	33.09
	Total oxysterols	nd	45.15	69.35	76.83	110.72	112.09	164.99
	Total sterols	2480	-	-	-	-	-	-

*All values are given in mg kg[-1] , means of triplicate measurements area/mass ratio: 12,7 cm^{2}g[-1]. nd= not detected: < 0,5 mg kg [-1].

Table 2

Formation of oxidized sterols during potato deep-frying at 180 °C.

	Sunflower oil					Palm oil					Sunflower-palm oil blend 30:70 v/v					Lard*				
Number of frying	0	1	3	6	9	0	1	3	6	10	0	1	3	6	10	0	1	3	6	10
Sterol oxides																				
7α-hydroxysterols	nd	10.54	12.48	22.09	27.40	nd	nd	nd	nd	2.80	nd	nd	nd	1.53	3.00	nd	nd	7.10	11.09	13.56
7β-hydroxysterols	nd	10.09	14.98	17.67	27.30	nd	nd	nd	nd	2.73	nd	nd	nd	1.67	4.39	nd	nd	7.32	11.04	14.38
7-ketosterols	nd	6.94	13.74	14.98	16.61	nd	nd	nd	nd	1.42	nd	nd	nd	nd	3.00	nd	nd	nd	nd	5.68
5α,6α-epoxide	nd	nd	nd	nd	nd	nd	nd	nd	nd	0.11	nd	nd	nd	nd	nd	nd	nd	nd	nd	nd
Total oxysterols	nd	28.38	41.2	54.74	71.31	nd	nd	nd	nd	7.06	nd	nd	nd	3.20	10.39	nd	nd	14.42	22.13	33.62

*All values are given in mg kg^{-1}, means of triplicate measurements. nd= not detected: <0,5 mgkg^{-1}, area/mass ratio: 0,4 cm^2/g. *Sterol oxides in lard: 7α-hydroxycholesterol, 7β-hydroxycholesterol, 7-ketocholesterol, 5α, 6α-cholesterol-epoxide.*

The levels of α-and γ-tocopherols were substantionally decreased after 30 minute heating. In conclusion, the combination of TLC and enzymatic method can be used not only for oxycholesterols but for the determination of phytosterol oxidation products. TLC method is suitable only for the separation of those phytosterol oxides derivatives which are oxidized at different positions. During oxidation the 7-hydroxysterols are early and sensitive indicator of oxidation of phytosterols which are detectable even after short time of heating of oils.

Acknowledgement

This work was supported by Hungarian National Scientific Research Foundation T 020930.

References

1.   S. K. Peng, R. J. Morin, Biological Effects of Cholesterol Oxides, CRC Press, Boca Raton, FL, USA. 1992.
2.   E. T. Finocchiaro and T. Richardson, *J. Food Prot.* 1983, **46**, 917.
3.   S. Bösinger, W. Luf and E. Brandl, *Int. Dairy J.* 1993, **3**, 1.
4.   P. C. Dutta, JAOCS. 1997, 74, **6**, 659.
5.   J. Nourooz-Zadeh and L.A. Appelqvist, *J. Am. Oil Chem. Soc.* 1992, **69**, 288.
6.   V. K. Lebovics, M. Antal, Ö. Gaál, *J. Sci. Food Agric.* 1996, **71**, 22.

TOCOPHEROLS, CAROTENOIDS, AND CHOLESTEROL OXIDES IN PLASMA FROM WOMEN WITH VARYING SMOKING AND EATING HABITS

H. Billing[1], O. Nyrén[2], A. Wolk[2], and L.-Å. Appelqvist[1].

[1]Department of Food Science, Swedish University of Agricultural Science, P.O. Box 7051, 75007 Uppsala, Sweden, [2]Department of Medical Epidemiology, Karolinska Institutet, P.O. Box 281, 17177 Stockholm, Sweden.

1 INTRODUCTION

Plasma levels of lipid oxidation products such as cholesterol oxides have been specifically associated with certain forms of cancer.[1] Epidemiological studies have demonstrated an inverse association between intake of fruit and vegetables and risk of a wide variety of epithelial cancers. High intake of carotenoids has been suggested as one contributing factor in this connection.[2] In addition smokers are known to have a lower antioxidant status than non-smokers.[3]

Therefore we undertook an exploratory study on the effect of smoking and intake of fruit and vegetables in four combinations on levels of antioxidants and lipid oxidation products in plasma.

2 MATERIAL AND METHODS

2.1. Subjects and plasma collection

Plasma was collected from 39 women aged 39-54 years, selected from a cohort of 49 273 women in the Uppsala health care region, Sweden. In 1992-93 the women had answered a comprehensive questionnaire and based on this questionnaire, ten women from each of the following four groups were randomly chosen to participate in this study: non-smokers with fruit and vegetable intake in the highest quartile, non-smokers with fruit and vegetable intake in the lowest quartile, smokers with fruit and vegetable intake in the highest quartile, smokers with fruit and vegetable intake in the lowest quartile. Mean number of cigarettes smoked per day among the smokers was 16. Venous blood was collected after overnight fasting.

2.2. Chemical analysis

Samples were analysed for cholesterol oxides and for carotenoids and tocopherols. Cholesterol oxides were determined by GC essentially as described by Granelli et al.[4], and by GC/MS. The carotenoids and tocopherols were analysed by HPLC with diode-array detection after extraction without hydrolysis or saponification. The extraction was based on a method published by Franke et al. with some modifications[5], but echinenone was chosen as internal standard for the carotenoids. For the tocopherols, δ-tocopherol was used as internal standard. The carotenoids determined were α-carotene, ß-carotene, lycopene, ß-cryptoxanthin and lutein + zeaxanthin. The tocopherols determined were α- and γ-tocopherol.

3 RESULTS

Mean values and ranges for α- and γ-tocopherol, lutein + zeaxanthine, β-cryptoxanthin, lycopene, α-carotene and for β-carotene are shown in Table 1 and are comparable to levels published before.[5]

Table 1

Mean values and ranges of antioxidants (μg/ml plasma).

	Mean	**Range**
Alfa-tocopherol	11.03	6.73-15.87
Gamma-tocopherol	1.11	0.00-3.25
Lutein+ zeaxanthin	0.27	0.03-0.67
Beta-cryptoxanthine	0.08	0.01-0.30
Lycopene	0.19	0.01-0.38
Alfa-carotene	0.06	0.00-0.28
Beta-carotene	0.26	0.02-1.14

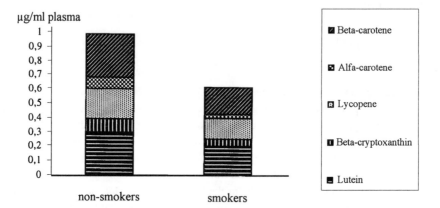

Figure 1

Mean carotenoid concentrations in plasma of smoking and non-smoking women.

Compared to smokers, non-smokers had significally higher levels of α-carotene, β-carotene, lutein, β-cryptoxanthin (P < 0.05) and slightly higher for lycopene (P= 0.07). For α- and γ-tocopherol no significant difference between the smokers and non-smokers was observed. Women with a high intake of fruit and vegetables surprisingly had a significantly lower level of γ-tocopherol (P< 0.05) than women with a low intake, and no differenses in the concentrations of α-tocopherol or the carotenoids were observed.

Cholesterol oxides were analyzed in the plasma from the 39 women and identified by GC/MS. The following cholesterol oxides were detected: 7α-hydroxycholesterol, 7ß-hydroxycholesterol, 5α, 6α-epoxycholestanol, 5ß, 6ß-epoxycholestanol, 20α-hydroxy-cholesterol, 7-ketocholesterol, 25-hydroxycholesterol and cholestanetriol. Preliminary calculations showed that the concentrations of each of α- and ß-epoxides, the most interesting cholesterol oxides from their association with cancer,[6] are in the range of 3-15 ng/ml plasma. These figures are substantially higher than those reported by Kucuk et al.,[6] but considerably lower than those reported by Björkhem et al.[7] It is known that levels of cholesterol oxides reported for a single food sample can vary by a factor 100 or more between laboratories using different methods.[8]

4 CONCLUSIONS

There was a wide range of variation between all subjects in the levels of α- and γ-tocopherols and the five carotenoids studied.

There were significally higher levels of α-carotene, ß-carotene, lutein and ß-cryptoxanthin in the nonsmoking group of women compared to the smoking women. There was no significant difference between the levels of tocopherols in the smokers compared to the nonsmokers.

References

1.   Morin R.J., Hu B., Peng S.K. and Sevanian A. in Peng S.K. and Morin R. *J. Biological Effects of Cholesterol Oxides.* CRC Press, Boca Raton, FL, USA, 1992, 191-202.
2.   Block G., Patterson B., Subar A. *Nutrition and Cancer.* 1992, **18**, 1-29.
3.   Brown A.J. *Journal of Nutritional Biochemistry.* 1996, **7**, 29-39.
4.   Granelli K., Fäldt P., Appelqvist L-Å. and Bergenståhl B. *J. Sci. Food Agric.* 1996, **71**, 75-82.
5.   Franke A., Custer L and Cooney R. *Journal of Chromatography.* 1993, **614**, 43-57.
6.   Kucuk O., Churley M., Goodman M. T., Franke A., Custer L., Wilkens L.R., and Pyrek J.S. *Cancer Epidemiology, Biomarkers and Prevention.* 1994, **3**, 571-574.
7.   Björkhem I., Breuer O., Angelin B. and Wikström S-Å. *Journal of Lipid Research.* 1988, **29**, 1031- 1038.
8.   Appelqvist L-Å. *Bulletin of the IDF.* 1996, **315**, 52-58.

IMPORTANCE OF *IN VITRO* STABILITY FOR *IN VIVO* EFFECTS OF FISH OILS

Tom Saldeen, Karin Engström, Ritva Jokela and Rolf Wallin.

Dept of Surgical Sciences/Forensic Medicine, University of Uppsala, Dag Hammarskjölds väg 17, 752 37 Uppsala, Sweden.

1 INTRODUCTION

There is epidemiological, clinical and experimental support for the belief that fish oils containing omega-3 fatty acids have protective qualities against both cardiovascular disease and inflammatory complaints.[1] Omega-3 fatty acids in fish oils are highly unsaturated and potentially unstable compounds. Consequently, exposure of most fish oils to air quickly leads to their deterioration with increase in indices of rancidity such as peroxide and anisidine values.[2]

In the present study the importance of the *in vitro* stability of fish oils for their *in vivo* effect was determined in humans and rats.

2 MATERIALS AND METHODS

For studies of *in vitro* stability 14 different commercially available fish oils were used. 50 g of each fish oil was stored in a 100 ml beaker with a diameter of 50 mm. The beakers were stored open in room temperature and darkness. Samples were analyzed every day for peroxide value (American Oil Chemical Society (A.O.C.S) Official Method Cd 8-53. Vitamin E was determined by HPLC (A.O.C.S. Official Method Ce 8-89).

In the human studies 15 subjects with joint stiffness due to osteoarthritis participated in a randomized, double blind study. Seven of them were given 15 ml daily for 4 weeks of a fish oil (Eskimo-3®, Cardinova, Uppsala, Sweden) containing a mixture of natural antioxidants (Pufanox®) and with an *in vitro* stability of 200 days and eight subjects the same fish oil containing vitamin E alone as antioxidant and with an *in vitro* stability of 14 days. At the end of the experiment blood samples for determination of LDL- and HDL-cholesterol were taken and changes in joint stiffness were registered.

In one rat experiment six rats were given Eskimo-3®, six rats the same fish oil without Pufanox® and containing vitamin E alone as antioxidant whereas six rats were controls. After 18 days plasma was taken for measurement of malondialdehyde (MDA) in plasma and myocardium by HPLC[3] and 6-keto-prostaglandin F1a and thromboxane B_2 in serum and myocardium by radioimmunoassay. [4]

In another experiment six rats were given fish oil with Pufanox® and six rats the same oil without Pufanox®. After ten days DHA and nitric oxide synthase (NOS) was determined in the brain.

In the third experiment four rats were given 10 IU vitamin E daily, four rats 30 IU vitamin E daily and four rats were controls. After 14 days MDA was determined in myocardium and 6-keto PGF1a and thromboxane B2 were determined in myocardium and serum and cholesterol and triglycerides in serum.

3 RESULTS

The *in vitro* stability varied between one day and 200 days among the 14 different fish oils. (Table I). Eskimo-3® (Natural stable fish oil) containing a mixture of natural antioxidants (Pufanox®) had by far the best stability. There was no correlation between the amount of vitamin E in the fish oil and the *in vitro* stability.

Table 1
Stability of different fish oils°.

	Stability (days)	Vitamin E (IU/g)		Stability (days)	Vitamin E (IU/g)
Fish oil 1*	1	6.8	Fish oil 8	14	1.0
Fish oil 2*	3	20.0	Fish oil 9	14	1.5
Fish oil 3*	4	4.4	Fish oil 10	14	8.5
Fish oil 4*	4	5.0	Fish oil 11	14	3.7
Fish oil 5	6	4.4	Fish oil 12	16	0.3
Fish oil 6	10	1.4	Fish oil 13	21	1.5
Fish oil 7	13	1.0	Natural stable fish oil	200	4.5

° Stability= time to rancidity (peroxide value 20) after exposure of the oil to air at room temperature. *chemically modified fish oils. Other fish oils are natural.*

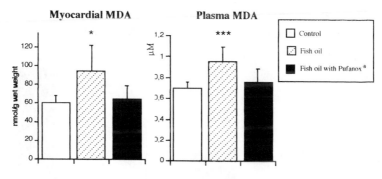

Figure 1
*Effects of 18 days of dietary supplementation with fish oil with or without Pufanox® on malondialdehyde concentration in the rat heart (nmol/g wet weight) and in plasma (μM). Mean ±SEM, n=6 in each group. *p<0.05, **p<0.01 and ***p<0.001 compared with controls.*

Unstable fish oil when given to rats produced increased lipid peroxidation (MDA) in plasma and myocardium whereas intake of the fish oil with high *in vitro* stability resulted in no increase in lipid peroxidation (Fig.1). Stable fish oil also increased prostacyclin/thromboxane ratio more than unstable fish oil (Fig.2). A high prostacyclin/thromboxane ratio is regarded beneficial, especially in coronary artery disease.

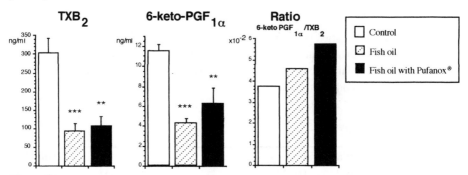

Figure 2
Effects of 18 days of dietary supplementation with fish oil with or without Pufanox® on in vivo production of TxB_2 *and 6-keto* PGF_{1a}/TxB_2 *ratio in the serum. Mean $\pm$SEM, n=6 in each group. *p<0.05, **p<0.01 and ***p<0.001 compared to control.*

Intake of the stable fish oil resulted in significantly higher amounts of DHA and significantly higher NOS-activity in the brain compared to rats given ordinary fish oil (Fig.3). NOS is a mediator of neurotransmission in the brain and is regarded important for learning ability and memory.

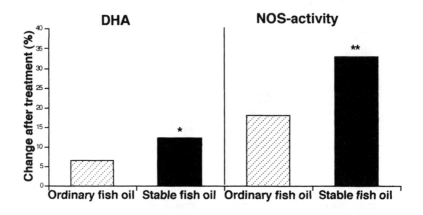

Figure 3
*Effects of ten days of dietary supplementation with ordinary and stable fish oil on DHA and Nitric Oxide Synthase (NOS) activity in the brain. * p<0.05, ** p<0.01 compared with ordinary fish oil.*

In humans stable fish oil decreased LDL-cholesterol (Fig.4) and joint stiffness (Fig.5) whereas fish oil showed no such effect. Both fish oils increased HDL-cholesterol. Later studies showed that stable fish oil increased HDL-cholesterol more than ordinary fish oil. Later studies also showed that joint stiffness is further improved after intake of the fish oil for longer periods of time.

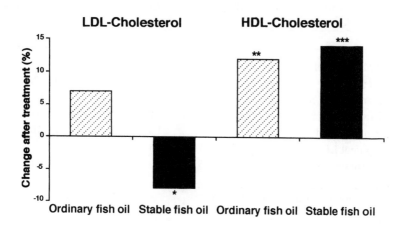

Figure 4

*Effects of four weeks of dietary supplementation with ordinary and stable fish oil on LDL-ad HDL-cholesterol in blood. *p<0.05, **p<0.01, ***p<0.001.*

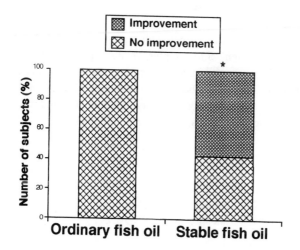

Figure 5

*Effects of four weeks of dietary supplementation with ordinary and stable fish oil on joint stiffness. * p<0.05 compared with ordinary fish oil.*

Vitamin E given alone had no effect on the prostacyclin/thromboxane ratio or other parameters studied, indicating that the major effect of fish oil is due to the fish oil itself and not to the vitamin E added. (Table II).

Table 2

Heart and serum metabolite concentrations in rats given drinking water supplemented with 10 IU or 30 IU vitamin E daily for 14 days1. Control rats received only normal tap water.

	Control	**10 IU vit.E**	**30 IU vit.E**
Myocardial MDA (nmol/g)	35.7±4.2	37.6±5.8	39.0±6.1
TxB2, serum (mg/L)	136.2±46.8	145.9±78.9	140.9±76.5
TxB2, myocardium (ng/mg)	14.9±6.2	14.1±4.3	21.2±14.6
6-keto-PGF1a, serum (mg/L)	9.1±2.3	8.8±3.2	8.9±3.5
6-keto-PGF1a, myocardium (ng/mg)	93±28	113±134	204±170
Cholesterol (mmol/L)	2.1±0.1	1.9±0.4	2.0±0.3
Triglycerides (mmol/L)	1.9±0.5	1.4±0.3	1.5±0.2

[1] *Values are means ± SEM (n=4 in all groups).*

3 CONCLUSION

In summary, the *in vitro* stability of fish oils has major importance for the *in vivo* effects.

References
1. A. Leaf and PC Weber. Cardiovascular effects of n-3 fatty acids. *N. Engl. J. Med.* 1988, **318**, 549.
2. T. Saldeen. Fish oil and health. SwedeHealth Press, Uppsala, Sweden. 1997. pp 1-64.
3. S.H.Y. Wong, J.A. Knight, S.M. Hopfer, O. Zacharia, C.N. Leach Jr and F.W. Sunderman Jr. Lipoperoxides in plasma as measured by liquidchromatographic separation of malondiadehydethiobarbituric acid adduct. *Clin. Chem.* 1987, **33**, 214.
4. P. Saldeen, Esquivel, C. Björck, D. Bergqvist and T. Saldeen. Thromboxane production in umbilical vein grafts. *Thromb. Res.* 1984, **33**, 259.

ENHANCEMENT OF ABSORPTION OF VITAMIN E BY SESAMINOL
- AN ACTIVE PRINCIPLE OF SESAME SEED -

Kanae Yamashita*, Yoshie Iizuka,* and Ikuo Ikeda**.

*Department of Food and Nutrition, Sugiyama Jogakuen University, Nagoya 464-8662, Japan**Laboratory of Nutrition Chemistry, Kyusyu University School of Agriculture 46-09 Fukuoka 812-8581 Japan.

1 INTRODUCTION

We Have shown previously that sesaminol, a sesame lignan, caused significant elevation of α-tocopherol concentration in the plasma and tissues of rats fed diets containing α-tocopherol and γ-tocopherol, but significant elevation of γ-tocopherol was limited when α-tocopherol intake was low.[1,2] On the other hand, K. Nesaretanam et al.[3] reported that the tocotrienol-rich fraction of palm oil inhibited the growth of human breast cancer cell line in culture. Therefore, we attempted to determine whether sesaminol caused an enhanced tocotrienol concentration using a similar fraction of palm oil containing of tocotrienols and some α-tocopherol. Our results showed that sesaminol caused remarkably elevated α-tocopherol and very slight α-tocotrienol concentrations in rats fed a diet containing α-tocopherol and a mixture of tocotrienols, but we could hardly detect any tocotrienols without sesaminol.[4] These studies showed the superiority of α-tocopherol in comparison to γ-tocopherol and tocotrienols in animal models. In this study we investigated whether the marked elevation of α-tocopherol by sesaminol is caused by an enhanced absorption of α-tocopherol using the same tocotrienol-rich fraction of palm oil.

2 EFFECT OF ALTERNATING DAY SUPPLIES OF TOCOTRIENOL(TOC-3) -RICH FRACTION OF PALM OIL (T-MIX) AND SESAMINOL ON α-TOCOPHEROL (TOC) CONCENTRATION IN RATS

The tocotrienol-rich fraction of palm oil (T-mix) contains α-Toc; 23 %, α-Toc-3; 22 %, β-Toc-3; 5 %, γ-Toc-3; 34 %, δ-Toc-3; 10 %. Thirty rats (3 weeks old Wistar male) were divided into 5 groups. Group 1 (control) received a daily supply of a vitamin-E free diet, Group 2 a daily supply of a 20 mg/kg T-mix (vitamin E-deficient) diet, Group 3 a daily supply of a 20 mg/kg T-mix with 0.1 % sesaminol diet, Group 4 alternating day supplies of either a 40 mg/kg T-mix diet or a vitamin E-free diet, and Group 5 alternating day supplies a 40 mg/kg T-mix diet or a vitamin E-free with 0.2 % sesaminol diet. After the rats were raised for 8 weeks and fasted overnight, the animals were sacrificed and tocopherol and tocotrienol contents in the plasma and tissues were determined as well as red blood cell hemolysis, plasma pyruvate kinase activity and TBARS in the plasma and tissues as indexes of vitamin E status. The results of vitamin E status and the concentration of vitamin E in the plasma, liver and kidney are shown in Tables 1 and 2.

Table 1

Vitamin E status in rats supplied daily T-mix with sesaminol and alternating diet of T-mix and sesaminol.

	Hemolysis	Pyruvate kinase	Liver	Kidney	Plasma
				TBARS	
	%	U/L		nmol MDA/g	
VE-free	91.6 ± 1.8 [d]	150 ± 21[c]	335 15[c]	495±33[c]	13.2±1.8 [b]
20 mgT-mix	70.9 ± 2.2 [c]	85 ± 20[b]	3l4 23b[c]	429±26[b]	10.2±0.5 [ab]
20 mgT-mix + 0.1 % sesaminol	0.6 ± 0.5 [a]	34 ± 9 [a]	165 ±10[a]	305±13[a]	7.4±0.9 [a]
Alternating days					
40 mgT-mix or VE-free	12.1 ± 1.0[b]	83 ± 6 [b]	277±12[b]	420±15[b]	12.1±1.15
Alternating days					
40 mg T-mix or 0.2 % sesaminol	0.6 ±0.2 [a]	41 ±7 [a]	194±19 [a]	277±18 [a]	11.3±0.95

As shown in Table 1 Group 2 (daily supply of 20 mg/kg T-mix) and Group 4 (alternating supplies of 40 mg/kg T-mix or vitamin E-free) showed almost the same degree of vitamin E deficient status, but Group 3 (daily supply of 20 mg/kg T-mix with 0.1 % sesaminol) and Group 5 (alternating suppies of 40 mg/kg T-mix or vitamin E-free with 0.2 % sesaminol) showed almost the same degree of improvement in vitamin E status. Furthermore, the results showed α-tocopherol concentrations in the plasma and tissues of rats fed sesaminol were significantly higher than those in rats without sesaminol. Among the tocotrienols ingested, only α-tocotrienol was detected in very low concentrations and the other tocotrienols were not detected in any rats. Sesaminol caused a significant increase of α-tocotrienol but the concentrations were very low. No difference in the α-tocopherol and α-tocotrienol raising effects of sesaminol were shown between Group 3 and Group 5. Therefore, the effect of sesaminol was observed to the same extent whether the group was fed vitamin E and sesaminol at the same time or at alternating times.

3 EFFECT OF SESAMINOL ON α-TOCOPHEROL CONCENTRATION IN LYMPH FLUID

In another experiment in which α-tocopherol was supplied into the stomach with and without sesaminol and was determined the α-tocopherol content in the lymph fluid, sesaminol did not enhance the absorption of α-tocopherol.

These results indicated that the enhancing effect of α-tocopherol by sesaminol was not caused by the enhnced absorption of α-tocopherol.

Table 2

The plasma, liver and kidney concentrations of α-tocopherol and α-tocotrienol in rats supplied daily T-mix with sesaminol and alternating diet of T-mix and sesaminol

	Plasma		Liver		Kidney	
	α-Toc	α-Toc-3	α-Toc	α-Toc-3	α-Toc	α-Toc-3
	µmol/L		nmol/g wet liver		nmol/g wetkidney	
daily VE-free	0.95±0.13 [a]	ND	1.60±0.05 [a]	ND	2.02±0.14 [a]	ND
daily 20 mg T-mix	3.58±0.21 [b]	ND	7.28±0.49 [b]	0.12±0.05 [ab]	6.90±0.21 [b]	0.28±0.02 [a]
daily 20 mg T-mix + 0.1 % sesaminol	6.97±0.91 c	0.12±0.05	12.02±0.69	0.35±0.14 [bc]	13.05±0.74 [d]	1.57±0.16 [c]
Alternating days 40 mg T-mixor VE - free	3.95±0.17[b]	0.02±0.01	7.05±0.05 [b]	0.05±0.02 a	7.04±0.16 [b]	0.35±0.02 [a]
Alternating days 40 mg T-mix or 0.2 % sesaminol	6.53±0.30[c]	0.12±0.05	11.90±0.72 [c]	0.47±0.05 [c]	10.88±0.60 [c]	1.05±0.19 [b]

4 CONCLUSION

A significant amount of α-tocopherol and a negligibly low level of α-tocotrienol was detected in the plasma and tissues of rats fed α-tocopherol and tocotrienols mixture from palm oil, but the other tocotrienols were not detected in any rats. Significant elevation in α-tocopherol concentrations caused by sesaminol were almost the same in the plasma and tissues of rats fed sesaminol and T-mix at the same time and in those of rats fed them at alternating times. α-tocopherol concentrations in the lymph fluid of the group given α-tocopherol with sesaminol were not different from the concentrations in the group without sesaminol. These results indicated that the elevation of plasma and tissue α-tocopherol by sesaminol was not caused by the enhanced absorption of α-tocopherol. Now we are considering other mechanisms than absorption, for example, the participation of α-tocopherol transfer protein in the liver.

References

1.   K. Yamashita, Y. Nohara, K. Katayama and M. Namiki. *J. Nutr*. 1992, **122**, 2440.
2.   K. Yamashita, Y. Iizuka, T. Imai and M. Namiki. *Lipids*. 1995, **30**, 1019.
3.   K. Nesaretnam, N. Guthrie, A.F. Chambers and K.K. Carrol. *Lipids*. 1995, **30**, 1139.
4.   Y. Iizuka, M. Namiki and K. Yamashita. *J. Home Econ*. 1997, **48**, 575.

MODEL *IN VITRO* STUDIES ON THE PROTECTIVE ACTIVITY OF TOCOCHROMANOLS WITH RESPECT TO β-CAROTENE

Malgorzata Nogala-Kalucka and Jan Zabielski.

Department of Food Biochemistry and Analysis, Faculty of Food Technology, Agricultural University, Mazowiecka 48, 60 -623 Poznaò, Poland.

1 INTRODUCTION

Tocopherols, being natural antioxidants, play crucial role in stabilization of the quality of food. Natural dye-stuffs present in food are relatively little resistant to the influence of various physical and chemical factors. Visual evaluation is the first step in assessing the quality of a foodstuff. Colouring matters have been used in foodstuffs for a very long time already, and some of them, apart from dyeing, also have some physiological importance.

One of such substances is β-carotene which displays significant sensitivity to light, oxygen and oxidants. Its presence in human diet is of particular importance due to the possibility of conversion into retinole, and the rate of the carotene absorption varies depending upon many factors.[1] Protecting foodstuffs against oxidation becomes of significance because products of the oxidation of carotenes are biologically inactive. Because of the increasing degree of processing and refining the food which is the only source of exogenic antioxidants supplies less and less of these naturally occuring substances. An example here can be refining of plant oils-mainly their deodorization-depriving them of natural antioxidants.[2] During this process many undesirable compounds, imparting unpleasant flavour and scent to oils, are being distilled off but so are the compounds which are nutritionally indispensable: EFA, phytosterols and tocopherols. The latter ones are being collected as a by-product called the postdeodorization condensate (oil scum). Tocopherols, being natural antioxidants, are of particular significance as the vitamin E-active compounds, and they tend to protect β-carotene.

The aim of the study was to obtain the tocopherol concentrates out of the postdeodorization condensate and then to determine their protective properties with respect to β-carotene in the model system.

2 MATERIALS AND METHODS

Tocopherol concentrates were obtained by elimination of acylglycerols and fatty acids from postdeodorization condensates of rapeseed oil by their crystallization from the acetone solution at -70 °C and elimination of sterols from methanol solution after 24 h at -20 °C.[3,4] The content of homologous tocopherols in concentrates was determined by the HPLC.[5,6]

In order to determine the protective properties of tocopherols with respect to β-carotene the Bickoff test[7] in which to every 0.02 mM of β-carotene (Fluka AG) 0.01 mM of the tocopherol standards/alpha-T (α-T), gamma-T(γ-T) and delta-T (δ-T) as well as the concentrate obtained and Mix-Toc (Eisai Ltd.) were added and dissolved in 100 ml of extraction naphta. The solution was stored in closed measuring cylinders at 25 °C ± 1 °C. The contents of β-carotene in the samples was determined by measuring the absorption at the wave-lenght 450 nm.

3 RESULTS AND DISCUSSION

In the investigation the antioxidant activity of the studied compounds was determined by comparing disintegration of β-carotene, the tocopherol standards added and their concentrates in individual samples. The parameter was evaluation of the dynamics of changes of the β-carotene disintegration in the time function "t" for individual experimental systems.

This relationship can be expressed at the linear equation:

y = A - b •t
y - β-carotene disintegration (%)
A - initial tocopherol content (%)
b - factor characterizing the dynamics of changes (%/d)
t - storage time (days)

Fig. 1-4 present the course of the above relationship. It was stated-on the basis of the calculations carried out-that degree of the influence of the storage time on the β-carotene decomposition does not exceed 90.7 % in experimental systems with no tocopherols added. Dynamics of the rate of the decomposition of β-carotene is defined by the directional coefficient of linear regression, and for this system it amounts to 5.5 % /d (Fig.1, pattern A).

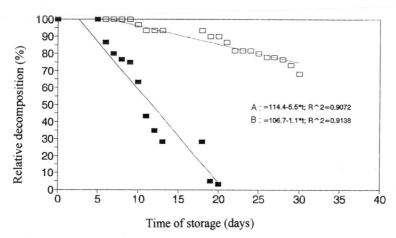

A : =114.4-5.5*t; R^2=0.9072
B : =106.7-1.1*t; R^2=0.9138

Figure 1
Dynamics of the decomposition of β-carotene with addition of α-T ■ without α-T addition; experimental data ☐ with α-T addition; experimental data ——— without α-T addition; calculated data --------with α-T addition; calculated data.

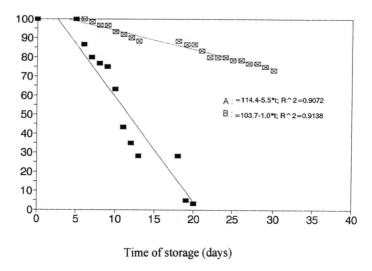

Time of storage (days)

Figure 2
Dynamics of the decomposition of β-carotene with addition of γ-T ■ *without γ-T addition;*
experimental data ⊠ *with γ-T addition; experimental data* ——— *without γ-T addition;*
calculated data -------- *with γ-T addition; calculated data.*

Laboratory tests showed that most efficient antioxidants, as far as β-carotene was concerned, were individually added tocopherol standards: α-T, γ-T and δ-T. In the samples with addition of α-T and γ-T decomposition was over five times decreased (Fig. 1,2; pattern B,B), because the value of the directional coefficient was -1.1 %/d. In their studies Rahmani and Saad[8] also observed the greatest protective properties of α-T with respect to β-carotene. Lower dynamics of decomposition -4.5 % times -was noticed when δ-T was added (Fig. 3, pattern B). Addition of the tocopherol concentrates obtained and Mix-Toc caused decrease of the rate of the β-carotene decomposition by 2 and 2.3 times (Fig.4, pattern B,C).

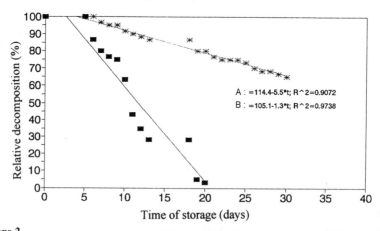

Figure 3
Dynamics of the decomposition of β-carotene with addition of δ-T ■ *without δ-T addition;*
experimental data ✳ *with δ-T addition; experimental data* ——— *without δ-T addition;*
calculated data -------- *with δ-T addition; calculated data.*

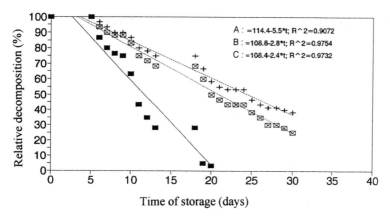

Time of storage (days)

Figure 4

Dynamics of the decomposition of β-carotene with addition of tocopherol concentrates ■ with addition of rapeseed conc.; calculated data ----- without tocopherol conc.; experimental data —— with addition of Mix-Toc conc.; experimental data + without tocopherol conc.; calculated data ⊠ with addition of Mix-Toc conc.; calculated data ······· with addition of rapeseed conc.; calculated data.

When compared to the tocopherol standards, concentrates of tocopherols display slightly more than twofold lower efficiency in inhibiting decomposition of β-carotene. In comparison to the sample containing only β-carotene addition of tocopherol concentrates also caused twofold decrease of its decompostion. By complementing fodder with tocopherol concentrates obtained after deodorization of plant oils one can not only prolong the period when no quality changes occur by decreasing the rate of the β-carotene decomposition, but also increase the nutritional value of the product by increasing the content of vitamin E -active compounds.

Feeding poultry with such fodder prior to slaughtering one can positively influence further stability of lipids in the stored meat and decrease the rate of formation of harmful oxidation products such as e.g. oxysterols. Therefore utilization of tocopherols obtained from postdeodorization condensates as mixtures of full-value natural antioxidants seems purposeful.

References

1. D.L. Madhavi, S.S Deshpande and D.K.Salunke. Food Antioxidants. Marcel Dekker Inc., New York, Basel, Hong Kong, 1995.
2. S. Ghosh and D.K. Bhattacharyya. *JAOCS*. 1996. 73, 1271.
3. C.K. Chow, H.H. Draper and A.S. Csallany. *Anal. Biochem.* 1969. 32, 81.
4. M. Gogolewski, M. Nogala-Kalucka and A. Luczyòski. *Ann. of Poznaò Agricultural Univ.* 1976. 89, 59.
5. C.Gertz and K. Herrmann. Z. Lebensm. Unter. Forsch. 1982. 174, 390.
6. H. Schulz, K. Müller K. and W. Feldheim. *Helgolander Meersuntersuchungen.* 1984, **38**, 75.
7. M. Rahmani and L.Saad. *Rev. Franc. des Corps Gras.* 1989, **36**, 355.

THE EFFECT OF PROCESSING ON TOTAL ANTIOXIDATIVE CAPACITY IN STRAWBERRIES

Uno Viberg[1], Charlotte Alklint[1], Björn Åkesson[2], Gunilla Önning[2] and Ingegerd Sjöholm[1].

[1]Departments of Food Engineering, and [2]Applied Nutrition and Food Chemistry, Chemical Center, University of Lund, P.O. Box 124, S-221 00 Lund, Sweden.

1 INTRODUCTION

There is an increasing interest in foods rich in antioxidants, since they may reduce the risk of contracting cardiovascular disease and cancer. Little is known of the antioxidant potential in foods and the effects of food processing. Recently it was reported that strawberries had significantly greater antioxidant capacity than other fruits investigated.[1] Apart from the common antioxidants ascorbic acid, ß-carotene, and tocopherols, strawberries contain a number of phenolic substances with antioxidative properties, such as anthocyanins, quercetin, kaempferol and ellagic acid, which are important for the total antioxidative capacity (TAC). In this study the changes in TAC, ascorbic acid and anthocyanins during processing were studied. It forms part of a larger project on processing of strawberries.[2]

2 METHODS

TAC, ascorbic acid and anthocyanins were measured in juice extract from the pulp after freezing, processing and accelerated storage for 7 days at 40 °C. The pulp was stored with and without headspace in the presence of 50 µmol/kg Fe^{+3} or Cu^{+2}. Half of the samples were fortified with ascorbic acid (5.68 mmol/kg pulp). TAC was measured as the capacity to inhibit formation of the ABTS radical cation according to the principle described.[3]

3 RESULTS

TAC in fresh strawberry pulp was 13.1±1.3 µmol Trolox equivalents per gram and the content of ascorbic acid was 2.1 mmol/kg. The content of anthocyanins was set to 100 %. Losses of ascorbic acid and anthocyanins after freezing and processing were 30 % and 10 %, respectively. After processing and accelerated storage without headspace, the TAC was similar to that in fresh pulp (Fig. 1). The presence of headspace caused a clear reduction in the TAC for all combinations (5-23 %). The difference in TAC between samples stored with and without headspace was greater in jars with added ascorbic acid (15-23 %) than in those without (5-15 %). Addition of Fe^{+3} and Cu^{+2} had no consistent effect on TAC (Fig. 1).

Ascorbic acid in pulp decreased by 57 % after storage without headspace. In jars withheadspace only traces of ascorbic acid could be detected. In pulp fortified with ascorbic acid and stored without headspace the amount of ascorbic acid left after one week was

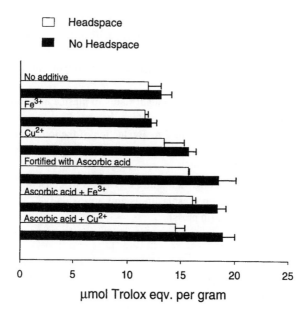

Figure 1
TAC in strawberry pulp after prosessing and storage (7 days at 40 °C).

approx. 5.7 mmol/kg both in the presence and absence of metal ions. With headspace and metal ions less ascorbic acid remained in jars stored with copper ions, 1.6 mmol/kg, compared to 2.4 mmol/kg for pulp without additives and pulp stored with iron ions. In all cases headspace had a more degenerating effect than metal ions. Also in a previous study losses of ascorbic acid during storage were clearly reflected in TAC measurements.[4]

During storage the anthocyanin content decreased more in samples that were fortified with ascorbic acid. The lowest amount of anthocyanins was found in pulp fortified with ascorbic acid and stored with headspace (14 %), while the differences between all other combinations were small (20-39 %). Also for the anthocyanin content headspace had a negative effect in all cases.

4 CONCLUSIONS

It can be concluded that TAC, ascorbic acid and anthocyanins in strawberry pulp were affected differently after a treatment involving freezing, heating and accelerated storage. TAC seems to be a valuable new method to follow changes during storage and processing. Previously the method has been used mainly on blood plasma.[3,5] The data also indicate that strawberry pulp contain important antioxidants other than ascorbic acid. Also in apple juice significant antioxidant capacity was demonstrated after disappearance of ascorbic acid upon storage.[4]

References

1.   Wang H, Cao G, Prior RL. Total antioxidant capacity of fruits. *J. Agric. Food Chem.* 1996, **44**, 701-705.

2.   Viberg U. Studies of the effects of industrial processing on fruit. *Doctoral thesis, Lund University.* 1998.

3.   Rice-Evans C, Miller NJ. Total antioxidant status in plasma and body fluids. *Meth. Enz.* 1994, **234**, 279-293.

4.   Miller NJ, Diplock AT, Rice-Evans CA. Evaluation of the total antioxidant activity as a marker of the deterioration of apple juice on storage, *J. Agric. Food Chem.* 1995, **43**, 1794-1801.

5.   Önning G, Åkesson B, Öste R, Lundquist I. Effects of consumption of oat milk, soya milk, or cow's milk on plasma lipids and antioxidative capacity in healthy subjects. *Ann. Nutr. Metab.* (in press).

QUERCETIN CONTENT IN BERRY PRODUCTS

Sari Häkkinen,[1,2] Pirjo Saarnia[1] and Riitta Törrönen.[1,2]

[1]Department of Clinical Nutrition and [2]Department of Physiology, University of Kuopio, P.O. Box 1627, FIN-70211 Kuopio, Finland.

1 INTRODUCTION

Only a small proportion of berries in Finland is consumed fresh, while most of them are frozen or processed to various products such as jams, jellies, juices, desserts etc. Flavonoids, the natural phenolic compounds in berries and other foods of plant origin, are suggested to have health-promoting effects. The flavonoid quercetin is a potential anticarcinogen,[1] possessing antioxidative[2] capacity. Precise knowledge of the effects of processing on the quercetin content of berries is lacking.

The aim of this study was to investigate the effects of household processing methods on quercetin in the four most common berries growing and consumed in Finland (strawberry *Fragaria x ananassa* 'Senga Sengana', black currant *Ribes nigrum* 'Öjebyn', bilberry *Vaccinium myrtillus* and lingonberry *Vaccinium vitis-idaea*). The berry products studied were strawberry jam, bilberry soup, black currant steam juice, crushed lingonberries and lingonberry raw juice.

2 MATERIALS AND METHODS

Three sets of the berry products were prepared and analyzed within 24 h. Strawberry jam was made by cooking 1500 g of fresh strawberries with 750 g of sugar for 30 min. For black currant steam juice, 3000 g of black currants and 900 g of sugar were cooked in a steamer insert in alternate layers and the juice was extracted for 60 min. To make a bilberry soup, frozen bilberries (appr. 175 g), 6 dl water and 0.5 dl sugar were cooked for 10 min and thickened with potato starch (1 tablespoonful in 0.5 dl water). Lingonberries were crushed partly so that all the berries were covered with juice when poured in glass containers. To make lingonberry raw juice, frozen berries (1500 g) were crushed in a food processor and mixed with cool water (2 l) and citric acid (0.5 teaspoonful). After 48 h at $+4\pm6$ °C, the juice was filtered and sugar was added.

For quercetin assays, homogenized samples of berries and berry products were extracted and hydrolyzed in aqueous methanol plus HCl[3]. The concentrated extract was injected to an HPLC system equipped with a variable wavelength detector and a LiChroCART 125-3 column.[3]

Quercetin was quantified with UV detection at 360 nm. Quercetin was identified by comparing the retention time and UV/VIS spectrum of the peak with that of the quercetin aglycone standard. Recoveries were measured in each berry and berry product[3] and they have been taken into account in the calculation of the final results.

3 RESULTS AND DISCUSSION

Quercetin content in strawberry jam was half of that in fresh strawberries (Table 1). During the cooking process, the quercetin content was reduced by 20 % in jam. Cooking with sugar did not affect quercetin content of strawberry as much as cooking with water and sugar did for that of bilberry. In bilberry soup, 60 % of the quercetin of the berries was present after processing.

The quercetin content in black currant steam juice and lingonberry raw juice (8 and 9 mg/kg, respectively) compared well with the levels found in fruit juices (2.5-13 mg/l)[4] or red wines (4-16 mg/l)[4]. It should, however, be noticed that these berry juices are diluted (1/4-1/5 v/v) before drinking. Our results show that in traditional juice making most of the quercetin of black currant and lingonberry remains in the solid waste and only small portion (15 %) is extracted to the juice, respectively.

In lingonberries, the quercetin content was reduced by 40 % when the berries were crushed and kept at +4±6 °C overnight. This could be explained by chemical (antioxidative) and enzymatic reactions that start when the the cell walls in berries are broken.

Table 1

Quercetin content in berries and berry products.

	Quercetin (mg/kg)[1]	*Recovery*[2]
Strawberry (fresh)	6	88
•Jam	3	88
Bilberry (frozen)	41	86
• Soup	6	86
Black currant (fresh)	53	50
• Steam juice	8	75
Lingonberry (fresh)	169	54
• Raw juice	9	60
• Crushed	100	67

[1] *Mean of triplicate determinations.* [2] *Recoveries were taken into account in the results.*

References

1.   M. Strube, L.O. Dragsted and J.C. Larsen. Naturally occuring antitumourigens. I Plant phenols. The Nordic Council of Ministers, Copenhagen, 1993.
2.   C.A. Rice-Evans, N.J. Miller and G. Paganga. *Free Rad. Biol. Med.* 1996, **20**, 933
3.   S. Häkkinen, H. Mykkänen, S. Kärenlampi and R. Törrönen, in this issue.
4.   M.G.L. Hertog, P.C.H. Hollman and B. van de Putte. *J. Agric. Food Chem.* 1993, **41**, 1242.

THE ANTIPEROXIDATIVE EFFECT OF DUNALIELLA β-CAROTENE ISOMERS

Moshe, J. Werman, Michal Yeshurun, Ami Ben-Amotz* and Shoshana Mokady.

Technion, Dept. of Food Engineering & Biotechnology, Haifa 32000, Israel, and *The National Institute of Oceanography, Haifa 35471, Israel.

1 INTRODUCTION

In recent years the consumption of foods rich in β-carotene was shown to be associated with reduced risk of several pathological events such as malignant and cardiovascular diseases.[1,2] The properties of β-carotene as a potent free radical quencher, singlet oxygen scavenger and an antioxidant rather than its activity as pro-vitamin A has been implicated as paramount in this protective role.[3] Recently two human studies point to the possible carcinogenicty of synthetic all-trans β-carotene.[4,5] Therefore, attention was drawn to β-carotene from natural sources, such as that present in fruits and vegetables which also contain small amounts of other β-carotene isomers.[6] The unicellular algae *Dunaliella bardawil* has drawn much attention in the last years as a rich source of β-carotene composed of the *all-trans* and *9-cis* stereoisomers in approximately equal amounts[7]. This unique isomer mixture was shown to accumulate to a higher extent in livers of rats and chicks[8,9] and to cope more efficiently with oxidative stress, as compared to synthetic all-trans β-carotene.[10,11]

In the present study we determined the ability of the natural β-carotene isomer mixture to efficiently prevent peroxidation induced in vivo by chronic alcohol consumption, or by dietary oxidized oil.

2 MATERIALS AND METHODS

Weanling female rats were divided into 4 groups of 8 animals each, and fed AIN[12] diets. In the ethanol study, two groups were gradually adapted to ethanol consumption by increasing the alcohol levels in their drinking water, from 5 to 30 % at 5 % intervals per week. The remaining rats continued to drink water. One water or ethanol drinking group were supplemented with *Dunaliella* powder to contain 0.1 % of β-carotene mixture and the feeding period lasted 3 months. In the Oxidized oil study rats were fed for 7 weeks diets containing 10 % fresh or oxidized soybean oil with or without *Dunaliella* powder.

Analytical procedures: β-carotene, retinol and conjugated dienes analyses[10] and level of red blood cells (RBC) fluorescence peroxidation products[13] analysis were carried out. The activity of alanine amino-transferase (ALT) and aspartate amino-transferase (AST) were determined using Sigma kits No. 505-OP.

3 RESULTS

Ethanol drinking (Table 1) was shown to elevate total hepatic β–carotene stores and the levels of the *all-trans* and *9-cis* isomers and to reduce liver vitamin A levels. *Dunaliella* feeding reduced the amount of hepatic and plasma conjugated dienes in the water drinking rats, and counteracted ethanol-induced oxidative stress.

Table 1

The effect of b-carotene isomer mixture of Dunaliella bardawill on growth parameters and oxidative stress (CD) of rats consuming alcohol.

	Control-Water	Control-EtOH	*Dunal*-Water	*Dunal*-EtOH
Body weight, g	244±16[ab]	224±22[c]	263±15[a]	233±12[bc]
Liver β-carotene (μg/g dry weight)				
Total	ND	ND	98.29±12.42[b]	135.87±26.06[a]
all-trans	ND	ND	62.15±7.51	85.19±16.65
9-cis	ND	ND	18.56±2.3[b]	28.60±3.94[a]
Liver Vitamin-A (μg/g dry weight)	2077.9±111.0[c]	946.5±161.3[d]	7137.1±249.7[a]	6059.9±882.9[b]
Hepatic CD	1.000	1.633	0.497	1.104
Plasma CD	1.000	4.257	2.598	3.473

BW- body weight; CD- conjugated dienes; ND- non-detectable. The results are means (SD, n=8. Within a raw, values with different superscript are significantly different (P<0.05).

Dietary oxidized oil (Table 2) reduced hepatic total β–carotene levels. Rats fed oxidized oil supplemented with *Dunaliella*, exhibited a significantly lower activity of ALT, and a reduced amount of RBC fluorescent peroxidative products. The effect on the activity of AST was less pronounced.

Table 2

The effect of dietary oxidized oil on hepatic b-carotene, erythrocytes' peroxidation, and the activity of plasma AST and ALT in rats.

Experimental groups	Liver β-carotene* (mg/g dry weight)	RBC fluorescent peroxidative products** (U/g Hb)	AST acivity** (U/dL plasma)	ALT activity** (U/dL plasma)
Control				
Fresh oil	ND	901±245[ab]	77.2±7.52[c]	30.8±3.9[c]
Oxidized oil	ND	1287±620[a]	131.8±41.7[a]	52.8±2.5[a]
Dunaliella				
Fresh oil	87.2±8.7[a]	830±167[b]	104.3±13.7[bc]	30.8±1.7[c]
Oxidized oil	40.6±9.3[b]	780±128[b]	119.5±23.2[ab]	44.3±3.9[b]

*AST- aspartate aminotransferase; ALT- alanine aminotransferase; Hβ– hemoglobin; ND- non- detectable; RBC- red blood cells. The results are means (SD, n=8. Within a column, values with different superscript are significantly different (*P<0.01, ** P<0.05).*

4 DISCUSSION

Several studies suggest that β–carotene has the therapeutic potential to reduce the manifestation of liver injury associated with peroxidation processes. Thus, patients with liver cirrhosis caused by chronic alcohol consumption, which is known to be responsible for the generation of free radicals[14] were shown to benefit from supplementation of synthetic β–carotene.[15] Indeed, *Dunaliella* feeding was found to counteract oxidative stress as expressed by the amount of hepatic conjugated dienes in the rats fed alcohol, and by the lower levels of RBC fluorescent products caused by dietary oxidized oil. A restored liver function as expressed by the lower amount of plasma AST and ALT was also shown in the oxidized oil fed rats. These results suggest that the consumption of *Dunaliella bardawil*, a natural rich source of β–carotene may ameliorate *in-vivo* peroxidation processes. The way by which *Dunaliella* feeding led to its antioxidative effects is most likely due to the high levels and the unique pattern of the β–carotene it contains.

References
1.   G. van Poppel. *Eur. J. Cancer*, 1993, **29A**, 1335.
2.   K. F. Gey, U. K. Moser, P. Jordan, H. B. Stahelin, M. Eichholzer and E. Ludin *Am.J.Clin Nutr.* 1933, **57**, 787s.
3.   N. I. Krinsky. *Free Radical Biol. Med.* 1989, **7**, 617.
4.   O. P. Heinonen and D. Albanes. *New Engl. J. Med.*, 1994, **331**, 316.
5.   E. R. Greenberg, J. A. Baron and T. D. Tosteson. *New Engl. J. Med.*, 1994, **331**, 147.
6.   L. A. Chandler and S. J. Schwartz. *J. Food Sci.*, 1987, **52**, 669.
7.   A. Ben-Amotz and M. Avron. *Trends in Biotechnology*, 1991, **8**, 121.
8.   A. Ben-Amotz, S. Mokady, S. Edelstein and M. Avron. *J. Nutr.* 1989, **119**, 1013.
9.   S. Mokady, M. Avron and A. Ben-Amotz. *J. Nutr.* 1990, **120**, 889.
10.   A. Ben-Amotz and Y. Levy. *Am J. of Clin. Nutr.* 1996, **63**, 729.
11.   G. Levin, M. Yeshurun and S. Mokady. *Nutr. and Cancer*, 1994, **27**, 293.
12.   American Institute of Nutrition. Report of the AIN Ad-hoc Committee on Standards for Nutritional Studies. *J. Nutr.* 1977, **107**, 1340.
13.   J. M. Gutteridge, L. F. Taffs, C. M. Hawkey and C. Rice-Evans. *Lab. Anim.* 1986, **20**, 140.
14.   L. H. Chen, S. Xi and D.A. Cohen. *Alcohol*, 1995, **12**, 453.
15.   S. Ahmed, M. A. Leo and C. S. Lieber. *Am. J. Clin. Nutr.*, 1994, **60**, 430.

Dietary Intakes and Modes of Action of Potentially Anticarcinogenic Dietary Compounds

DIETARY PHYTOESTROGENS - MECHANISMS OF ACTION AND POSSIBLE ROLE IN THE DEVELOPMENT OF HORMONALLY DEPENDENT DISEASES

Sari Mäkelä[1,2], Leena Strauss[1], Niina Saarinen[1], Saija Salmi[1], Tomi Streng[1], Suresh Joshi[1] and Risto Santti[1].

[1]Institute of Biomedicine and Medicity Research Laboratory, University of Turku, Finland, [2]Unit for Preventive Nutrition, Department of Medical Nutrition, Karolinska Institute, Stockholm, Sweden.

1 INTRODUCTION

Phytoestrogens are nonsteroidal plant-derived estrogenic compounds. Based on their chemical structure, phytoestrogens are divided in four main groups, isoflavonoids, flavonoids, coumestans and mammalian lignans. Any plant food may contain phytoestrogens, although the amounts and the combinations of the various compounds vary considerably. For isoflavones (e.g. genistein and daidzein), the main dietary source for humans is soybean, while flavonoids (e.g. apigenin, naringenin, luteolin) are found in several different vegetables, fruits, berries, herbs and green tea. For coumestans (e.g. coumestrol) the main sources are sprouts of alfalfa and various beans. Mammalian lignans (e.g. enterolactone and enterodiol) are not present in our diets as such, but as precursors, plant lignans (e.g. matairesinol and secoisolariciresinol), which are converted to mammalian lignans by the gut microflora. Mammalian lignan precursors are present in fibre-rich food; such as flaxseed, unrefined grain products, particularly rye, and some berries.

Epidemiological studies suggest that diets rich in phytoestrogens, particularly soy and unrefined grain products, may be associated with low risk of breast and prostate cancer.[1] It has also been proposed that dietary phytoestrogens could play a role in the prevention of other estrogen-related conditions, namely cardiovascular disease, menopausal symptoms and postmenopausal osteoporosis. However, there is very little, if any, direct evidence for the beneficial effects of phytoestrogens in humans. It has not been unequivocally established that the beneficial effects associated with the consumption of phytoestrogen-rich diets would be explicitly due to phytoestrogens, rather than other coinciding dietary components. Therefore, the causal relationship and the mechanisms of phytoestrogen action in humans still remain to be demonstrated. In addition, the possible adverse effects of phytoestrogens in humans have not been evaluated. It is plausible that phytoestrogens, as any exogenous hormonally active agent, might also cause adverse effects in the endocrine system, i.e. act as endocrine disrupters. The endocrine disrupting activity of phytoestrogens has been clearly demonstrated in domestic and experimental animals, but there is no evidence to show that phytoestrogens would have induced any harmful effects in humans.

2 MULTIPLE MECHANISMS OF PHYTOESTROGEN ACTION

Phytoestrogens are a complex group of bioactive compounds, and they exert multiple effects in several target tissues. Only part of the effects of phytoestrogens are related to their hormonal activity, while others are mediated via other mechanisms. The effects may be divided in three major categories: **1)** estrogen receptor (ER) -related actions resulting in estrogenic or antiestrogenic effects; **2)** modulation of endogenous sex steroid production or bioavailability (via inhibition of the key enzymes in sex steroid production, and/or stimulation of the production of sex steroid binding proteins), leading to hormone- or antihormone-like actions without direct interaction with sex steroid receptors; and **3)** other, non-hormonal actions. It is apparent that the individual phytoestrogens differ considerably from each other with regard to their hormonal potencies, as well as in terms of their mechanisms of actions.

2.1. Estrogen receptor -mediated actions

Numerous *in vitro* studies show that phytoestrogens exert many of the typical estrogen actions: **1)** they bind to both known ER subtypes (ERα and ERß), **2)** they induce estrogen-responsive genes, **3)** they stimulate the growth of estrogen-sensitive cancer cells, and **4)** their actions are blocked by pure antiestrogens.[2-4] Some phytoestrogens, such as genistein and coumestrol are very good ligands for both ERα and ER-ß, while some other isoflavones or flavonoids bind with much lesser activity.[2] For example, genistein shows significant estrogenic activity *in vitro* at 10-1000 nM concentrations. This is well in the range of serum genistein concentrations of soy-consuming individuals, suggesting that effects *in vivo* are likely. This is further supported by studies showing that genistein, when given in doses relevant to dietary exposure, produces the typical estrogen-like actions in female and male experimental animals.[5-7]

In general, phytoestrogens are weaker than the endogenous steroidal estrogens, and, therefore, it has been postulated that phytoestrogens might compete with the more potent endogenous estrogens for the binding to ERs, i.e. act as antiestrogens. However, antiestrogenic activity has not been demonstrated for isoflavones or coumestrol, and only few phytoestrogens with very weak intrinsic estrogenic activity, such as naringenin (a flavanone), have been shown to counteract the effects of more potent steroidal estrogens.[8]

The discovery of the novel estrogen receptor subtype, ER-ß, and its intriguing interaction with phytoestrogens, has brought up novel theories about phytoestrogen action. The two ER subtypes are expressed differently in estrogen target tissues,[9] and are likely to mediate different estrogen actions. Isoflavone, coumestan and flavone type phytoestrogens all prefer ERß for binding, while endogenous steroidal estrogens do not show such difference.[5] This may suggest that phytoestrogens could exert their actions preferentially through ER-ß, resulting in unique, tissue- or organ-specific actions, different from those of endogenous hormones.

2.2. Modulation of endogenous steroid production and bioavailability

Several, but not all, phytoestrogens have been shown to inhibit some of the key enzymes of sex steroid biosynthesis *in vitro*. The target enzymes for phytoestrogen action include 17ßhydroxysteroid oxidoreductase type 1 and aromatase, both required for estrogen production, as well as 5α-reductase, which converts testosterone to the more active androgen, dihydrotestosterone.[10-12] Furthermore, some phytoestrogens stimulate the production of sex hormone binding globulin *in vitro*.[13] In theory, this suggests that phytoestrogens may modulate the production and bioavailability of endogenous steroidal estrogens and/or androgens, and could reduce the exposure to these hormones, resulting in antihormone-like acitivity. However, none of these actions have been confirmed *in vivo*, and the biological significance of these findings remains to be established.

2.3. Other effects

Several phytoestrogens have been reported to exert multiple non-hormonal, non-ER-mediated effects *in vitro*.[14-18] These include: **1)** antiproliferative activity, **2)** inhibition of tyrosine kinase, **3)** inhibition of protein kinase C, **4)** inhibition of DNA topoisomerase II, **5)** antioxidant activity, **6)** inhibition of angiogenesis, and **7)** inhibition of prostaglandin synthase. All of these actions require high concentrations, typically over 10 µM, i.e. well above concentrations reported to produce ER-mediated actions. In theory, the non-ER-mediated actions could all play a role in cancer prevention, but it is still questionable whether sufficient concentrations are ever reached *in vivo*, and these effects need to be confirmed *in vivo* before any conclusions on their biological relevance can be made.

3 EFFECTS OF PHYTOESTROGENS IN HUMANS, AND THEIR ROLE OF IN THE PREVENTION OF HORMONE-RELATED DISEASES

Very little is known about the possible effects of phytoestrogens in humans. Epidemiological studies suggest that there is inverse correlation between the dietary intake of phytoestrogens and the risk of certain hormone-related diseases and conditions, but the causal relationship has not yet been shown. Only few short-term clinical trials have been conducted so far, and in all of them food items containing combinations of phytoestrogens (e.g. soy protein or flaxseed), rather than purified compounds, have been used. Detailed conclusions about the role of individual phytoestrogens in the prevention of human diseases are thus not yet possible.

In women, daily soy consumption has been shown to have slight, estrogen-like effects, such as stimulation of vaginal cell maturation, reduction in menopausal hot flushes, and stimulatory effect on breast cell epithelium.[19-23] This indicates that dietary isoflavones are capable of inducing typical estrogen actions also in humans, although it is not clear which compounds are responsible for the effects. In addition, soy and flax diets have also been reported to lengthen the menstrual cycle and to induce changes in gonadotrophin and sex hormone concentrations,[24-26] but the mechanisms by which phytoestrogens might exert these actions are not yet known.

3.1. Phytoestrogens and prevention of menopausal symptoms, postmenopausal osteoporosis and cardiovascular disease

Epidemiological studies and preliminary clinical trials suggest that soy isoflavones may prevent or alleviate diseases or conditions associated with estrogen deficiency, such as menopausal vasomotor symptoms[19,20] and cardiovascular disease.[27,28] Experimental animal studies suggest that soy protein and/or isoflavones may prevent cardiovascular diseases by multiple mechanisms,[29,30] and inhibit ovariectomy-induced bone loss.[31] Several plant-based preparations (natural products and health or dietary supplements) claimed to have high isoflavone content, and marketed as a "natural and safe" alternative to traditional postmenopausal hormone replacement are already available. These marketing arguments are based on very preliminary, short-term human trials, or on animal experiments, and their validity is still questionable. Well controlled long-term clinical trials are required to confirm whether phytoestrogens could be used as an effective, safe and reliable form of menopausal hormone replacement therapy.

3.2. Phytoestrogens and prevention of breast and prostate cancer

Recently it has been shown that urinary phytoestrogen excretion and serum entrolactone concentration correlate inversely with breast cancer risk,[32,33] which is well in accordance with the findings in epidemiological studies. The chemopreventive effect of phytoestrogen-rich diets (soy and flax), as well as that of a dietary mammalian lignan precursor (secoisolariciresinol) has been also demonstrated in experimental mammary cancer models.[34-36] However, the direct causal relationship between phytoestrogen exposure and reduced cancer risk, as well as the plausible mechanisms of phytoestrogen action in humans, still remain to be demonstrated. In theory, the protective effect of phytoestrogens could be explained by their antiestrogenic activity. However, for soy isoflavonoids, there is no evidence for ER-mediated antiestrogenicity. On the contrary, all the data from *in vitro* studies, animal experiments and human trials indicate that isoflavones are estrogen agonists also in breast epithelium,[4,7,23,24] and it is thus not easily conceivable how they could prevent the development or growth of breast cancer. A more logical explanation for phytoestrogen action would be reduced cumulative exposure to endogenous estrogens, either due to the lengthening of menstrual cycle, or the decrease in endogenous estrogen synthesis in ovaries or breast tissue itself via aromatase or 17ß-hydroxysteroid oxidoreductase type 1 enzyme inhibition.

The protective effect of phytoestrogen-rich diets (soy and rye) on experimental prostate cancer models has been well demonstrated,[37-39] again in accordance with the epidemiological observations, although the mechanisms of phytoestrogen action are still unknown. Based on findings in experimental studies, the protective effect could be related to reduction in testicular androgen production due to the estrogenic action on central nervous system -gonadal axis (i.e. decrease in pituitary gonadtrophin secretion), or reduction in the local synthesis of dihydrotestosterone the most active androgen in prostate through inhibition of prostatic 5α-reductase. Another attractive hypothesis would be the direct effect of phytoestrogens on prostate epithelium via ERß, which is the predominant ER subtype in prostate. So far, ERß-specific functions in human prostate are not known, and there is no evidence for the role of ERß in prevention of prostate carcinogenesis, and the latter theory remains purely hypothetical.

It has also been suggested that some of the non-ER-mediated effects, such as the antioxidant activity, antiproliferative effects or inhibition of angiogenesis could play a role in cancer prevention. Via these mechanisms phytoestrogens would prevent tumor growth generally, and not only in estrogen target tissues. In accordance with this idea, dietary soy has been shown to prevent cancer development and growth in multiple organ sites. These actions, however, require high concentrations, and it is not yet known if such concentrations are ever reached *in vivo*. Furthermore, it should be noted that, in addition to isoflavones, soy contains several other bioactive, potentially chemopreventive compounds, which could be responsible for the beneficial actions.

4 PHYTOESTROGENS -POTENTIAL ENDOCRINE DISRUPTORS?

It is well known that excessive exposure to estrogens causes adverse effects in the reproductive system in several animal species, including humans. At present, there is no logical explanation why plant estrogens would exert only beneficial effects, and would be devoid of the adverse effects of estrogens. Studies in experimental and domestic animals demonstrate that isoflavones and coumestrol, if given in doses high enough or during critical stages of development, cause severe reproductive tract disorders, including impaired fertility.[9,40,41] Fortunately, there is no evidence for any adverse effects of dietary phytoestrogens in humans. Phytoestrogens have been a part of human diets for hundreds of years, and there is no data on any obvious adverse effects, and it is quite unlikely that normal plant based diet would contain phytoestrogens in amounts sufficient to cause reproductive abnormalities.

However, there may be groups that are exposed to very high phytoestrogen doses, and in whom adverse effects could occur. One of the two most likely groups at high risk would be those taking phytoestrogens at high doses during critical stages of reproductive tract development, namely infants consuming soy milk as their major source of nutrition. Soy formulas have been used in many countries for decades without any signs of adverse effects occurring during or shortly after the use of soy milk. On the other hand, there are no long-term follow up studies, and it is not known if the soy formula-fed individuals would display reproductive tract disorders later in life or have altered risk of breast or prostate cancer. Another potential high risk group would be adults consuming plant extracts or concentrates with very high phytoestrogen content. Such preparations are sold in several countries for various purposes, such as an alternative for postmenopausal hormone replacement therapy. At present, these preparations have not been sufficiently tested in well controlled clinical trials, and there is no evidence for their long-term safety.

5 CONCLUSIONS

Phytoestrogens are an extremely complex group of bioactive compounds capable of exerting both beneficial and adverse effects. Information on their effects in humans is very limited, and it is not yet possible to predict the actions of the complex phytoestrogen mixtures present in different human diets. Currently there is not enough firm evidence to justify any recommendations about the optimal intake of phytoestrogens. Many of the arguments used for promoting the use of phytoestrogen-enriched products are based on circumstantial or very preliminary findings, or are not scientifically relevant at all. Furthermore, it is not justified to claim that individual phytoetrogens, such as genistein, would account for all the possible benefits of complex phytoestrogens-rich diets, and that these benefits would be achieved by simply adding isolated phytoestrogens to diet.

More information on the cell-, and tissue-, and organ-specific actions of the individual phytoestrogens is required, in order to understand how they may act in the various target tissues. Long-term animals studies are needed to confirm the activity and safety *in vivo*. Systematic pharmacokinetic and dose-response studies are required to determine the bioavailability of the individual compounds, and to estimate the amounts in diet likely to exert biological effects. Finally, only after the possible adverse effects have been ruled out, long-term clinical trials with well characterised phytoestrogen preparations are necessary to confirm the beneficial health effects in humans.

References

1. H. Adlercreutz, H. Honjo, A. Higash, T. Fotsis, E. Hämäläinen, T. Hagesawa, H. Okada, *Am. J. Clin. Nutr.* 1991, **54**, 1093.
2. G.G.J.M. Kuiper, J.G. Lemmen, B. Carlsson, J.C. Corton, S.S. Safe, P.T. Van der Saag, B. Van der Burg, J.-Å. Gustafsson, *Endocrinology.* 1998, **139**, 4252.
3. R.J. Miksicek, *Proc. Soc. Exp. Biol. Med.* 1995, **208**, 44.
4. S. Mäkelä, V.L. Davis, W.C. Tally, J. Korkman, L. Salo, R. Vihko, R. Santti, K. Korach, *Environ. Health Perspect.* 1994, **102**, 572.
5. S.R. Milligan, A.V. Balasubramanian, J.C. Kalita, *Environ. Health Perspect.* 1998, **106**, 23.
6. R.C. Santell, Y.C. Chang, M.G. Nair, W.G. Helferich, *J. Nutr.* 1997, **127**, 263.
7. L. Strauss, S. Mäkelä, S. Joshi, I. Huhtaniemi, R. Santti, *Mol. Cell Endo.* 1998, **144**, 83.
8. M.F. Ruh, T. Zacharewski, K. Connor, J. Howell, I. Chen, S. Safe, *Biochem. Pharmacol.* 1995, **50**, 1485.
9. G.G. Kuiper, B. Carlsson, K. Grandien, E. Enmark, J. Haggblad, S. Nilsson, J.-A. Gustafsson, *Endocrinology.* 1997, **138**, 863.
10. S. Mäkelä, M. Poutanen, M.-L. Kostian, N. Lehtimäki, L. Strauss, R. Santti, R. Vihko, *Proc. Soc. Exp. Biol. Med.* 1998, **217**, 310.
11. Y.C. Kao, C. Zhou, M. Sherman, C.A. Laughton, S. Chen, *Environ. Health Perspect.* 1998, **106**, 85.
12. B.A. Evans, K. Griffiths, M.S. Morton, *J. Endocrinol.* 1995 **147**, 295.
13. H. Adlercreutz, K. Hockerstedt, C. Bannwart, S. Bloigu, E. Hamalainen, T. Fotsis, A.J. Ollus, *Steroid Biochem..* 1987 **27**, 1135.
14. T. Akiyama, J. Ishida, S. Nakagawa, H. Ogawara, S. Watanabe, N. Itoh, M. Shibuya, Y. Fukami, *J. Biol. Chem.* 1987, **262**, 5592.
15. J. Markovits, S. Junqua, F. Goldwasser, A.M. Venuat, C. Luccioni, J. Beaumatin, J.M. Saucier, A. Bernheim, Jacquemin-Sablon, *Biochem. Pharmacol.* 1995, **50**, 177.
16. A. Arora, M.G. Nair, G.M. Strasburg, *Free Radic. Biol. Med.* 1998, **24**, 1355.
17. T. Fotsis, M. Pepper, H. Adlercreutz, G. Fleischmann, T. Hase, R. Montesano, L. Schweigerer, *Proc. Natl. Acad. Sci.* 1993 **90**, 2690.
18. G.H. Degen, *J. Steroid Biochem.* 1990, **35**, 473.
19. P. Albertazzi, F. Pansini, G. Bonaccorsi, L. Zanotti, E. Forini, D. De Aloysio,

Obstet. Gynecol. 1998, **91**, 6.
20. A.L. Murkies, C. Lombard, B.J. Strauss, G. Wilcox, H.G. Burger, M.S. Morton, *Maturitas.* 1995, **21**, 189.
21. D.D. Baird, D.M. Umbach, L. Lansdell, C.L. Hughes, K.D. Setchell, C.R. Weinberg, A.F. Haney, A.J. Wilcox, J.A. McLachlan, *J. Clin. Endocrinol. Metab.* 1995, **80**, 1685.
22. G. Wilcox, M.L. Wahlqvist, H.G. Burger, G. Medley, *Br. Med. J.* 1990, **301**, 905.
23. N.L. Petrakis, S. Barnes, E.B. King, J. Lowenstein, J. Wiencke, M.M. Lee, R. Miike, M. Kirk, L. Coward, *Cancer Epidemiol. Biomarkers Prev.* 1996, **5**, 785.
24. A. Cassidy, S. Bingham, K.D. Setchell, *Am. J. Clin. Nutr.* 1994, **60**, 333.
25. L.J. Lu, K.E. Anderson, J.J. Grady, M. Nagamani, *Cancer Epidemiol. Biomarkers Prev.* 1996, **5**, 63.
26. W.R. Phipps, M.C. Martini, J.W. Lampe, J.L. Slavin, M.S. Kurzer, *J. Clin. Endocrinol. Metab.* 1993, **77**, 1215.
27. P.J. Nestel, T. Yamashita, T. Sasahara, S. Pomeroy, A. Dart, P. Komesaroff, A. Owen, M. Abbey, *Arterioscler. Thromb. Vasc. Biol.* 1997, **17**, 3392.
28. M.J. Tikkanen, K. Wahala, S. Ojala, V. Vihma, H. Adlercreutz, *Proc. Natl. Acad. Sci. U.S.A.* 1998, **95**, 3106.
29. M.S. Anthony, T.B. Clarkson, C.L. Hughes Jr, T.M. Morgan, G.L. Burke, *J. Nutr.* 1996, **126**, 43.
30. M.S. Anthony, T.B. Clarkson, B.C. Bullock, J.D. Wagner, *Arterioscler. Thromb. Vasc. Biol.* 1997, **17**, 2524.
31. K.E. Anderson, W.W. Ambrose, S.C. Garner, *Proc. Soc. Exp. Biol. Med.* 1998, **217**, 345.
32. D. Ingram, K. Sanders, M. Kolybaba, D. Lopez, *Lancet.* 1997, **350**, 990.
33. K. Hulten, H. Adlercreutz, A. Winkvist, P. Lenner, G. Hallmans, A. Ågren, *COST 916 Workshop.* 1998, p. 34.
34. S. Barnes, *Breast Cancer Res. Treat.* 1997, **46**, 169.
35. M. Serraino, L.U. Thompson, *Cancer Letters.* 1991, **60**, 135.
36. L.U. Thompson, M.M. Seidl, S.E. Rickard, L.J. Orcheson, H.H. Fong, *Nutr. Cancer.* 1996, **26**, 159.
37. M. Pollard, P.H. Luckert PH, *Nutr. Cancer.* 1997, **28**, 41.
38. J.X. Zhang, G. Hallmans, M. Landström, A. Bergh, J.E. Damber, P. Åman, H. Adlercreutz, *Cancer Letters.* 1997, **114**, 313.
39. M. Landström, J.X. Zhang, G. Hallmans, P. Åman, A. Bergh, J.E. Damber, W. Mazur, K. Wähälä, H. Adlercreutz, *Prostate.* 1998, **36**, 151.
40. C.D. Burroughs, K.T. Mills, H.A. Bern, *J. Toxicol. Environ. Health.* 1990, **30**, 105.
41. N.R. Adams, *Proc. Soc. Exp. Biol. Med.* 1995, **208**, 87.

DIETARY INTAKES AND LEVELS IN BODY FLUIDS OF LIGNANS AND ISOFLAVONOIDS IN VARIOUS POPULATIONS

Witold Mazur and Herman Adlercreutz.

Department of Clinical Chemistry, University of Helsinki and Folkhälsan Research Center, P.O. Box 60, FIN-00014 Helsinki, Finland.

1 INTRODUCTION

Epidemiology has established connection between semivegetarian diet in some Asian countries and a reduced incidence of many chronic and degenerative diseases (i.e. the major hormonedependent cancers, colon cancer, atherosclerosis and coronary heart disease), indicating that some unknown substances in this diet may contribute to homeostasis and thus play a role in the maintenance of health. It is now clear that the link between our health and the food we eat cannot be adequately explained by nutrient composition alone. These findings triggered research to identify nutritional biomarkers that could be responsible for the international variations in the incidence of these chronic diseases. A special interest is focused on non-nutrient phytoestrogens, the plant chemicals that mimic the biological effects of estrogens because of their ability to bind to and activate estrogen receptors.

The detection and identification in human body fluids of two groups of active principles, lignans and isoflavonoids, both of plant origin with molecular weights and structures similar to those of steroids, suggested that they could be important modulators of the human hormonal system and hormone action.[1-5] The plant lignan and isoflavonoid glycosides are transformed by intestinal bacteria to hormone-like compounds.[6-9] The mechanism(s), through which the phytoestrogens may influence sex hormone production, metabolism and biological activity and exert anticancer, cancer-protective, antiatherogenic, cardioprotective, bone-maintaining effects, seems to depend, at least in part, on their mixed estrogen agonist-antagonist properties. Furthermore, these weakly estrogenic molecules were demonstrated to affect intracellular enzymes, protein synthesis, growth factor action, malignant cell proliferation, cell differentiation, cell adhesion, angiogenesis, and apoptosis. A heavy experimental evidence and results of animal studies[10,11] suggest that both lignans and isoflavonoids are among the dietary factors affording protection against cancer and atherosclerosis.

Isoflavonoids and lignans (Fig. I) have phenolic groups at both ends of the molecule, making them unique from the chemical, biochemical and biological point of view and rendering them exceptionally stable. The mammalian lignans[12-15] were previously unknown compounds, whereas the isoflavonoids, the most important group of the so-called phytoestrogens, have long been known in the veterinary field,[16] although there were no studies in human subjects before 1982.

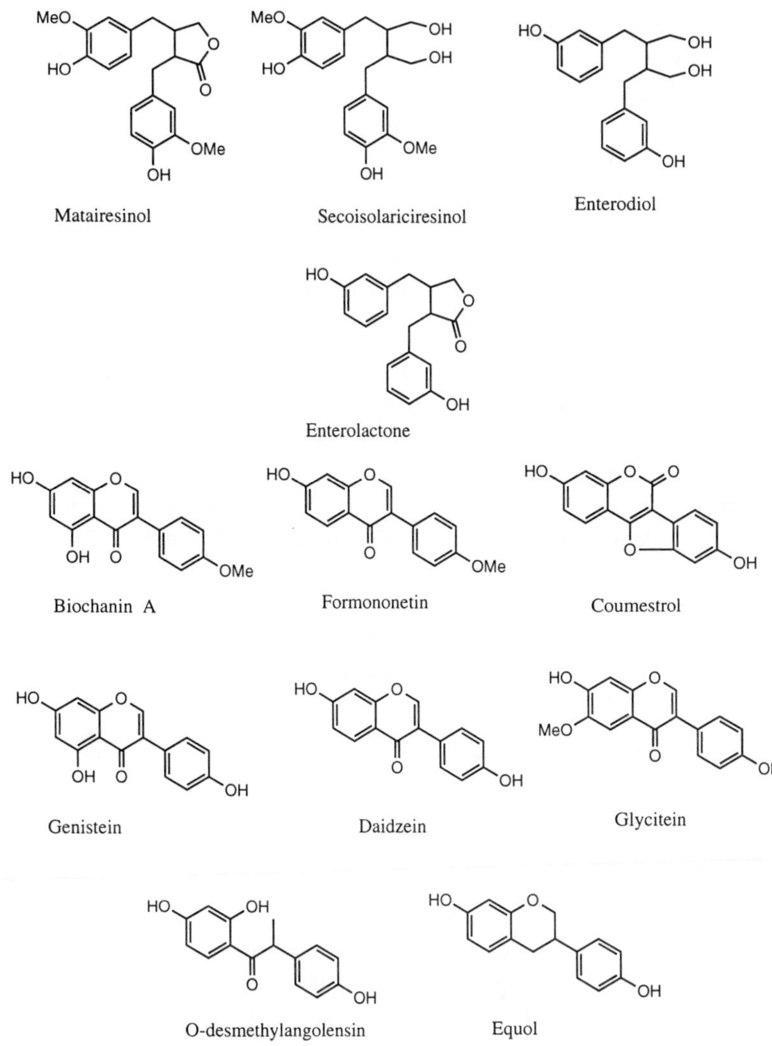

Figure 1
Structures of plant and mammalian lignans and isoflavonoids, and coumestrol.

Herein, we review the presence of naturally occurring estrogens in plant foods, their dietary intake and concentrations in the human biological fluids. Data on metabolism of phytoestrogens in mammalian systems are not sufficient, we summarize metabolic pathways of isoflavonoids and lignans described in literature till now.

2 DIETARY SOURCES OF PHYTOESTROGENS

Although many studies have been published on the relation of food groups such as fruits and vegetables and chronic diseases (i.e., cancer and coronary heart disease), the lack of reliable food composition data has led to few studies on the active ingredients in foods. Food composition tables do not list lignan and isoflavonoid content, making it difficult to calculate their intake. Thus we developed a method for the quantitative determination of the phytoestrogens biochanin A, formononetin daidzein, genistein, and coumestrol and the lignans matairesinol (MAT) and secoisolariciresinol (SECO) in plant-derived foods.[17] This method -the isotope dilution gas chromatographic-mass spectrometric method (ID-GC-MS)- permits foods that contain phytoestrogens to be identified and enables the intake of all biologically important isoflavonoids and lignans to be assessed.

Flaxseed (linseed) is the most abundant source of lignans in foods, the main component being SECO, with MAT present in minor amount (Fig. 1). Values in the literature range from 800 µg/g[18] to 3700 µg/g.[17] Other sources of lignans are various grains, other seeds, fruits and berries, some vegetables and beverages such as tea and coffee.[19,20] The main components identified have been MAT and SECO.[17] The analysis of green and black teas[21] yielded relatively high levels of SECO (5.61-28.9 mg/kg) and MAT (0.56-4.13 mg/kg) but only low levels of isoflavonoids. Surprisingly low levels of the isoflavonoids and lignans have been measured in fruits whereas vegetables contain mostly lignans (0.32-38.7 mg/kg). No or only trace amounts of isoflavonoids were detected in nuts, cereals and berries, however, these foods contain high quantities of the lignans (dry weights, 0.96-2.98 mg/kg, 0.25-2.8 mg/kg and 1.16-15.1 mg/kg respectively) (unpublished observation).

Table I shows the concentrations of SECO and MAT in selected food items. It has been well known since 1931 that soybeans contain rather high amounts (up to 1000-3000 µg/g) of the glycosides of the two isoflavones daidzein and genistein.[22] Much later a third major compound, glycitein, was found also mainly as a glycoside (glycitin)[23] (Fig. 1). Small amounts of these three compounds occur in the free form. Coumestrol is present in soybean sprouts[24] and in relatively large amounts in alfalfa and clover sprouts.[25] The richest source of coumestrol in human food that we have found is mung bean sprouts which contain 20 times as much coumestrol (about 1 mg/100 g) as do alfalfa sprouts. In addition they contain large amounts of daidzein (about 700 µg/100 g dry weight) and genistein (about 2000 µg/100 g dry weight). Soy sauce does not contain any isoflavones[24] but does contain the lignan precursor coniferyl alcohol.[26] The total concentrations of genistein and daidzein measured in soybean meal after hydrolysis of the glycosides ranges from 1000 to 2000, and 20 to 1200 µg/g, respectively.[22,27] In soy flour we measured concentrations of 969 µg of genistein and 674 µg/g of daidzein.[17]

Another rich source of isoflavonoids is clover, clover seeds, sprouts and leaves. Some beans other than soybeans contain isoflavonoids, but the concentrations are much lower. Biochanin A has been detected in bourbon[28] and genistein and daidzein in beer.[29] After the development of radioimmunoassays for daidzein, genistein, formononetin and biochanin A in a collaborative study, all four isoflavonoids were chromatographically identified in beer and their concentrations determined.[30,31] The concentrations of daidzein and genistein in selected food products are shown in Table 1.

Table 1

Phytoestrogen content of selected foods (mg/100 g dry weight, ID-GC-MS-SIM method).

Product analyzed	Genistein	Daidzein	SECO*	Matairesinol
Seeds				
Flaxseed	0	0	369,900	1,087
Clover seed***	322.5	178.3	13.2	3.8
Sunflower seed	13.9	8.00	609.5	0
Poppy seed	6.7	17.9	14.0	12.1
Caraway seed	8.00	0.14	220.7	5.7
Soybean products				
Soybean flour (Soyolk)**	96,900	67,400	130	0
Hatcho Miso	14,500	13,700	0	0
Soy drink, (First Alternative)	2,100	700	0	0
Vegemil Adult soy milk	9,900	5,130	12.0	0
Soy oil	0.40	0.02	1.20	0.20
Cereals				
Wheatwhite meal	tr	tr	8.1	0
Oat meal	0	0	13.4	0.3
Barley (whole grain)	7.7	14.0	58.0	0
Rye meal (Amando, whole grain)	0	0	47.1	65.0
Rye bran (Amando)	0	0	132	167
Triticale meal (mean of four different brands)	1.1	1.9	21.4	10.7
Legumes				
Groundnut peanut	82.6	49.7	333.4	tr
Chick peas ****	76.3	11.4	8.4	0
Mung bean °	365	9.7	172	0.25
Mung bean sprouts °°	1,902	745	468	0.87
Urid dahl bean °°°	60.3	30.3	240	79.4
Fruits				
Apple	0	12.4	tr	0
Banana	0	0	10.0	0
Plum	0	0	5.1	0
Vegetables				
Beetroot	0	0	99.5	tr
Carrots	1.7	1.6	192	2.86
Garlic	1.45	2.08	379	3.62
Broccoli	6.6	4.7	414.2	23.1
Cranberry	0	0	1,510	0
Nuts				
Walnut	tr	5.3	162.9	4.6
Almond	0	4	107.0	tr

Product analyzed	Genistein	Daidzein	SECO*	Matairesinol
Beverages				
Earl Grey black tea	0	29.0	1,590	197.0
Japanese green tea	nd	tr	2,460	186
Arabica instant coffee	0	0	716.0	nd

nd- Not determined due to low concentration and interferences by other compounds.
tr- Present in trace amounts.
I *In addition to the compounds shown we also measured formononetin, biochanin A and coumestrol in all samples, but the amounts were usually very low. * Secoisolariciresinol.*
*** In addition, 30 μg/100 g of formononetin and 70 μg/100 g of biochanin A were found in this flour. *** In addition, 1,272 μg/100 g of formononetin, 380.6 μg/100 g of biochanin A and 5.3 μg/100 g of coumestrol. **** In addition, 215 μg/100 g of formononetin, 838 μg/100 g of biochanin A and 5.0 μg/100 g of coumestrol. ⁷In addition, 7.5 μg/100 g of formononetin, 14.1 μg/100 g of biochanin A and 1.8 μg/100 g of coumestrol. °° In addition, 1,032 μg/100 g of coumestrol. °°° In addition, 81.1 μg/I00 g of biochanin A and 9.5 μg/100 g of coumestrol.*

3 PHYTOESTROGENS - CHEMISTRY AND METABOLISM

About seventeen years ago two cyclically occurring unknown compounds, now known as enterolactone and enterodiol, were detected by us in the urine of the female vervet monkey, and women[12,32,33] and subsequently identified separately and independently by two groups.[13,32] Furthermore, small amounts of four plant lignans -MAT (immediate precursor of enterolactone), lariciresinol, isolariciresinol and SECO (immediate precursor of enterodiol)- were identified along with some other metabolites.[34,35] Enterolactone, enterodiol, and later MAT and SECO were measured by ID-GC-MS-SIM in human plasma, urine and faces.[36-39] Enterolactone, alone or together with enterodiol, has also been measured in cow milk,[19] human breast cyst fluid, saliva and prostatic fluid.[10,40]

The following isoflavonoid phytoestrogens have been identified or detected in human urine: formononetin methylequol, daidzein, dihydrodaidzein, *O*-desmethylangolensin (*O*-DMA), genistein and 3', 7-dihydroxyisoflavan and some other metabolites.[3,15,34,41-45] Daidzein, genistein, equol, *O*-DMA have been measured[36,37,46,47] by ID-GC-MS-SIM in human plasma and faces.

Studies on the origin, formation and metabolism of the phytoestrogens in animals [16] and in man[48-50] have been reviewed. When consumed, the plant isoflavonoids and lignans undergo metabolic conversions in the gut resulting in the formation of hormone-like compounds with the ability to bind with low affinity to estrogen receptors and with weak estrogen activity.[16,50] However, many if not most, of the effects of these compounds may be unrelated to their estrogenic effects.

The mammalian lignans enterolactone and enterodiol are formed from the plant precursors MAT and SECO by the action of intestinal bacteria.[6-9] Recent results from our laboratory indicate, however, that most of the mammalian lignan precursors are still unidentified (unpublished observation). The compounds occur mainly in the glycosidic form in the plants and these glycosides are hydrolyzed in the proximal colon. The colonic microflora convert MAT to enterolactone and SECO to enterodiol, and the latter is readily oxidized to enterolactone.[8,9,50] At least in rats the plant lignans undergo an enterohepatic circulation and it is most likely that this is the case in man also as shown for phenolic estrogenic steroids.[51,52] Lignans are excreted in both urine and faces, and in human subjects the fecal excretion pathway seems to be more important than it is for estrogens,[51,53] since the amounts are similar or only slightly lower in faces than in urine.[47] This could be explained by assuming that some of the formed mammalian lignans escape absorption.

Administration of antibiotics almost completely eliminates the formation of enterolactone and enterodiol from plant precursors in the gut[6,54] and later leads, after initial rapid lowering of the lignan levels in urine, to a relative increase in the enterodiol/enterolactone ratio.

Urinary lignan excretion in Finnish women correlates with the intake of total vegetable, berry and fruit, and legume fibre intake.[6,54-57] The best correlation was found when fibre intake was calculated per kg body weight.

The metabolism of formononetin daidzein and biochanin A and genistein has been studied particularly in sheep.[58,59] Biochanin A is converted to genistein, which is further metabolized to p-ethylphenol and dihydrogenistein. Formononetin is converted to daidzein and daidzein to *O*-DMA, equol and some other metabolites. The metabolism varies with the animal and metabolism in man may therefore not be identical with that found in sheep or cattle.[60] The hydrolysis of the flavonoid glycosides takes place in the proximal colon and several bacteria have been found which produce the necessary enzyme.[60-62] It is also likely that the lignan glycosides are hydrolyzed in the proximal colon.

In man, equol[15,63] and *o*-desmethylangolensin are most likely formed, as in sheep, by intestinal bacterial action trom formononetin and daidzein present in foods such as soy products.[50] Some people are unable to produce equol or they excrete this isoflavon in very low amounts. Equol excretion also depends on diet; a high fat and meat diet increases equol production (5, 50, 63, 64, and unpublished results).

As estrogens[51,52] and flavonoids,[65] the isoflavonoids seem to undergo an enterohepatic circulation, at least in the rat.[7] There are no data on isoflavonoid metabolites in human bile.

4 PHYTOESTROGEN CONCENTRATIONS IN HUMAN

In our laboratory we have developed ID-GC-MS-SIM methods for the identification and quantitative determination of lignans and isoflavonoids in human urine,[66] plasma[36,38] and feces.[47] We have also collaborated in the development of analytical methodologies for the identification and measurement of phytoestrogens in human saliva, breast aspirate or cyst fluid and prostatic fluid.[40] A summary of all our results regarding urinary lignan and isoflavonoid excretion in various populations (Finnish women, American women, Asian immigrant women in Hawaii, breast cancer patients, and Japanese women and men) and dietary groups (omnivores, vegetarians, lactovegetarians, adherents of macrobiotics, and those who consume the traditional Japanese diet) has recently been presented.[11,67] None of the subjects investigated had been treated with antibiotics during the last 3 months as antibiotics negatively affect lignan metabolism.

American macrobiotic subjects, lacto-vegeterians, and Japanese men excreted the greatest quantities of urinary isoflavones, with levels at 3412-8770, 885-2188, and 1820-3630 nmol/d, respectively. Japanese women exhibited the greatest variation (347-6610 nmol/d), and Finnish breast cancer patients showed the lowest urinary isoflavon concentrations (67.5-324 nmol/d). With regards to urinary lignans, American macrobiotics excreted the greatest quantities (15,228-35,363 nmol/d), while Finnish women suffering from breast cancer had the lowest excretion of 1302-2835 nmol/d. What is of interest, Japanese male and female subjects-despite their high urinary isoflavone excretion-had very low concentrations of lignans in urine (840-1792 nmol/d).

We studied fecal excretion of lignans and isoflavones in 10 omnivore and 10 vegetarian women. The quantitative analysis yielded the following results (mean values) for the omnivore and vegetarian women, daidzein 45.4 and 259.1 nmol/d, genistein 11.6 and 189.6 nmol/d, equol 14.9 and 257.8 nmol/d, *O*-DMA 5.67 and 114.7 nmol/d, enterolactone 1510 and 3280 nmol/d, enterodiol 147.7 and 479.2 nmol/d, and MAT 22.3 and 89.3 nmol/d, respectively.

One of us has studied Japanese men and assayed isoflavonoids concentrations in their plasma.[37] The plasma concentration of daidzein, genistein *O*-DMA and equol were generally very high, but varied greatly. The geometric mean values for total individual isoflavonoids were 7-110 times higher than those in the Finnish men. It was observed, based on our recent measurements that Japanese men as well as Finnish men have relatively high plasma levels of enterolactone and enterodiol sulfates. About 20 % of the main mammalian lignan enterolactone in plasma in Finnish men occurs in the free+sulfate-conjugated fraction. This fraction is probably biologically active on the basis of the natural estrogens, estrone sulphate, being a source of biologically active estrogens, because phenol sulphates are abundant in the organism.[37,38] The Japanese men, however, excrete in urine relatively low amounts of the lignans. This could be due to very low levels of lignan glucuronides in plasma. The explanation for this finding is unknown. Racial differences in glucuronyl transferase activity could be consideled or that the amount of isoflavonoids and lignans is high exceeding the capacity of the enzyme.

In another study[38] we characterized the concentrations of isoflavones and lignans in plasma of Finnish omnivorous and vegetarian women and Japanese men who consumed a traditional Japanese diet. Two fractions -free+sulfate and glucuronide- of isoflavones and lignans were measured. Among the Finnish women generally all the plasma isoflavones and lignans (except for both fractions of MAT and equol and the glucuronide fraction of genistein) were significantly greater in the vegetarian subjects whose mean values of total daidzein, genistein, equol, *O*-DMA, enterolactone, enterodiol, and MAT were 18.5, 17.1, 0.7, 0.8, 89.1, 5.4, and 0.06 nM, respectively. For the Finnish omnivore women, mean values of total daidzein, genistein, equol, *O*-DMA, enterolactone, enterodiol, and MAT were 4.2, 4.9, 0.8, 0.07, 28.5, 1.4, and 0.02 nM, respectively. For the Japanese men, plasma levels of isoflavones were significantly higner than those of the Finnish subjects but the concentrations varied greatly ranging from 60 to 924, 90 to 1204, 0.54 to 24.6, and from 0.98 to 223 nM for total daidzein, genistein, equol and *O*-DMA, respectively. However, these Japanese men concentrations of total enterolactone, enterodiol, and MAT (varied from 2.87-17.8, 0.68-1.61, and 0.21-3.5, respectively) in plasma were much lower than their concentrations of isoflavones and similar to those of the Finnish female omnivore subjects plasma lignan levels.

5 CONCLUDING REMARKS AND FUTURE PERSPECTIVES

Recent studies show that several single compounds, like the isoflavonoids genistein and daidzein, and perhaps equol and the lignan enterolactone, reach plasma levels that would be compatible with biological activity in man. On the other hand, practically nothing is known about their tissue concentrations. This problem could be solved soon, as we have developed new time resolved fluorescence immunoassay based (TR-FIA) methods for genistein, daidzein, enterolactone and equol.[30,68,69] Preparations for tissue procedures would be expedited.

Our studies have shown that in addition to a large diversity of chemicals with a broad spectrum of biological properties, plants contain biologically active phytoestrogens-precursors of hormone-like compounds in mammalian systems. The lignan content of plants may be underestimated, however, or most of the mammalian lignan precursors may still not have been identified, as shown by our recent experiments on rye-derived lignans and their metabolism in pigs (unpublished observations). Cell walls (which comprise most of the dietary fiber in cereals) or other components of the food matrix may protect dietary lignans from analytical hydrolysis. They may also be underestimated because of the presence of lignan structures (e.g., lignan polymers[70]) that may escape analytical determination and still be degraded and available in the body. This discrepancy between lignan intake and later excretion in urine calls for further investigation of their precursors in food and their absorption and metabolism in humans.

The different activities of dietary phytoestrogens suggest that phytoestrogens may offer protection against a wide range of human conditions, including cardiovascular disease, hormonedependent breast, prostate, bowel, and other cancers, osteoporosis, and menopausal symptoms. Although the mechanisms for this protection are not yet known, a number of different mechanisms and interactions witn endogenous factors such as steroid hormones seem to be involved.

Acknowledgments

The research, which has been carried out in these laboratories since 1979, was supported in the begin g by the Medical Research and Natural Science Councils of the Academy of Finland and the Sigrid Juselius Foundation, Helsinki. Later funding has been by NIH grants I R01 CA56289-01 and R01 CA56289-04, a grant from the Nordic Industrial Foundation, EC contract FAIR-CT95-0894, and grants from the King Gustav Vth and Queen Victorias Foundation, Sweden, and the Finnish Cancer Foundations. One author (W. M.) thanks the

Graduate School of Steroid Research for scholarship.

References

1. Adlercreutz H. Western diet and Western diseases, some hormonal and biochemical mechanisms and associations. *Scand. J. Clin. Lab. Invest.* 1990 b, 50, Suppl. **201**, 3-23.

2. Adlercreutz H. Does fiber-rich food containing animal lignan precursors protect against both colon and breast cancer? An extension of the "fiber hypothesis". *Gastroenterology.* 1984, **86**, 761-764.

3. Bannwart C, Fotsis T, Heikkinen R, Adlercreutz H. Identification of the isoflavonic phytoestrogen daidzein in human urine. *Clinica. Chimica. Acta.* 1984 a, **136**, 165-172.

4. Adlercreutz H, Musey Pl, Fotsis T, et al. Identification of lignans and phytoestrogens in urine of chimpanzes. *Clin. Chem. Acta.* 1986 c, **158**, 147-154.

5. Setchell KDR, Borriello SP, Hulme P, Axelson M. Nonsteroidal estrogens of dietary origin, possible roles in hormone-dependent disease. *American Journal of Clinical Nutrition.* 1984, **40**, 569-578.

6. Setchell KDR, Lawson AM, Borriello SP, et al. Lignan formation in man-microbialinvolvement and possible roles in relation to cancer. *Lancet.* 1981, **2**, 4-7.

7. Axelson M, Setchell KDR. The excretion of lignans in rats -evidence for an intestinalbacterial source for this new group of compounds. *FEBS Letters.* 1981, **123**, 337-342.

8. Borriello SP, Setchell KDR, Axelson M, Lawson AM. Production and metabolism of lignans by the human faecal flora. *J. Appl. Bacteriol.* 1985, **58**, 37-43.

9. Setchell KDR, Lawson AM, Borriello SP, Adlercreutz H, Axelson M. Formation of lignans by intestinal microflora. In, Malt RA, Williamson RCN, eds. Colonic Carcinogenesis, Falk Symposium 31. *Lancaster, MTP Press,* 1982, 93-97.

10. Griffiths K, Adlercreutz H, Boyle P, Denis L, Nicholson Rl, Morton MS. Nutrition and Cancer. Oxford, *ISIS Medical Media,* 1996.

11. Adlercreutz H, Mazur W. Phyto-oestrogens and Western diseases. *Ann. Med* .1997, **9**, 95-10.

12. Setchell KDR, Adlercreutz H. The excretion of two new phenolic compounds (180/442 and 180/410) during the human menstrual cycle and in pregnancy. *Journal of Steroid Biochemistry.* 1979, **1** I , xv-xvi.

13. Setchell KDR, Lawson AM, Mitchell FL, Adlercreutz H, Kirk DN, Axelson M. Lignans in man and in animal species. *Nature.* 1980 a, **287**, 740-742.

14. Setchell KDR, Lawson AM, Axelson M, Adlercreutz H. The excretion of two new phenolic compounds during the menstrual cycle and in pregnancy. In, Adlercreutz H, Bulbrook R, van der Molen H, Vermeulen A, Sciarra F, eds. Endocrinological Cancer, Ovarian Function and Disease. *Amsterdam, Excerpta Medica,* 1980 c, 297-315. International Congress Series No. 515,

15. Adlercreutz H, Fotsis T, Heikkinen R, et al. Excretion of the lignans enterolactone and enterodiol and of equol in omnivorous and vegetarian women and in women with breast cancer. *Lancet.* 1982 a, **2**, 1295-1299.

16. Price KR, Fenwick GR. Naturally occurring oestrogens in foods - A review. *Food Additions and Contaminants.* 1985, **2**, 73-106.
17. Mazur W, Fotsis T, Wähälä K, Ojala S, Salakka A, Adlercreutz H. Isotope dilution gas chromatographic-mass spectrometric method for the determination of isoflavonoids, coumestrol, and lignans in food samples. *Anal. Biochem.* 1996, **233**, 169-180.
18. Obermeyer WR, Warner C, Casey RE, Musser S. Flaxseed lignans. Isolation, metabolism and biological effects. *FASEB Journal.* 1993, **7** (3 Pt II), Abstract 4985.
19. Adlercreutz H, Fotsis T, Bannwart C, et al. Assay of lignans and phytoestrogens in urine of women and in cow milk by GC/MS (SIM). In, Todd JFJ, ed. Advances in Mass Spectrometry-85. Proceedings of the 10th International Mass Spectrometry Conference. *Chichester, Sussex, John Wiley,* 1986 b, 661-662 .
20. Hutchins AM, Lampe JW, Martinit MC, Campbell DR, Slavin JL. Vegetables, fruits, and legumes, Elfect on urinary isoflavonoid phytoestrogen and lignan excretion. *J. Am. Diet. Ass.* 1995 , **95** , 769-774.
21. Mazur W, Wähälä K, Rasku S, Salakka A, Hase T, Adlercreutz H. Lignan and isoflavonoid concentrations in tea and coffee. *British Journal of Nutrition.* 1997, **79**(1), 37-45.
22. Eldridge A, Kwolek WF. Soybean Isoflavones, effect of environment and variety on composition. *Journal of Agriculture and Food Chemistry.* 1983, **31**, 394-396.
23. Naim M, Gestetner B, Kirson 1, Birk Y, Bondi A. A new isoflavone from soya beans. Phytochemistry 1973, **22**, 237-239.
24. Murphy PA. Phytoestrogen content of processed soybean products. *Food Technology.* 1982, **4**, 60-64.
25. Franke AA, Custer LJ, Cerna CM, Narala KK. Quantitation of phytoestrogens in legumes by HPLC. *J. Agr. Food Cllem.* 1994 a, **42**(9), 1905-1913.
26. Yokotsuka T. Soy sauce biochemistry. *Advances in Food Research.* 1986, **30**, 195-329.
27. Pettersson H, Kiessling K-H. Liquid chromatographic determination of the plant coumestrol and isoflavones in animal feed. *J. Assoc. Off Anal. Chem.* 1984, **67**, 503-506.
28. Rosenblum ER, Van Thiel DH, Campbell IM, Eagon PK, Gavaler JS. Separation and identifcation of phytoestrogenic compounds isolated from bourbon. *Alcohol.* 1987, Suppl. I, 55 1-555.
29. Rosenblum ER, Campbell IM, Vanthiel DH, Gavaler JS. Isolation and identifcation of phytoestrogens from beer. *Alcohol. Clin. Exp. Res.* 1992, **16**(5), 843-845.
30. Lapcik O, Hampl R, Al-Maharik N, Salakka A, Wähälä K, Adlercreutz H. A novel radioimmunoassay for daidzein. *Steroids.* 1997 a, **62**(3), 315-320.
31. Lapcik O, Hill M, Hampl R, Wähälä K, Adlercreutz H. Chromatographic identifcation and radioimmunoassay of isoflavonoids in beer. *Steroids.* 1997 b, 00-00, in press.
32. Stitch SR, Toumba JK, Groen MB, et al. Excretion, isolation and structure of a phenolic constituent of female urine. *Nature.* 1980 b, **287**, 738-740.

33.	Setchell KDR, Bull R, Adlercreutz H. Steroid excretion during the reproductive cycle and in pregnancy of the vervet monkey (Ceropithecus aethiopus pygerethus). *Journal of Steroid Biochemistry.* 1980 b, **1**, 375-384.

34.	Bannwart C, Adlercreutz H, Fotsis T, Wähälä K, Hase T, Brunow G. Identification of *O*-desmethylangolensin, a metabolite of daidzein, and of matairesinol, one likely plant precursor of the animal lignan anterolactone, in human urine. *Finn. Chem. Lett.* 1984b, **4-5**, 120-125.

35.	Bannwart C, Adlercreutz H, Wähälä K, Brunow G, Hase T. Detection and identification of the plant lignans lariciresinol, isolariciresinol and secoisolariciresinol in human urine. *Clin. Chem. Acta.* 1989, **180**, 293-302.

36.	Adlercreutz H, Fotsis T, Lampe J, et al. Quantitative determination of lignans and isoflavonoids in plasma of omnivorous and vegetarian women by isotope dilution gas-chromatography mass-spectrometry. *Scand. J. Clin. Lab. Invest.* 1993 b, **53** (Suppl. 215), 5-18.

37.	Adlercreutz H, Markkanen H, Watanabe S. Plasma concentrations of phyto-oestrogens in Japanese men. *Lancet.* 1993 c, **342**(888 1), 1209-1210.

38.	Adlercreutz H, Fotsis T, Watanabe S, et al. Determination of lignans and isoflavonoids in plasma by isotope dilution gas chromatography-mass spectrometry. *Cancer Detect. Prev.* 1994 a, **18**, 259-271.

39.	Adlercreutz H, Fotsis T, Kurzer MS, Wähälä K, Mäkelä T, Hase T. Isotope dilution gas chromatographic mass spectrometric method for the determination of unconjugated lignans and isoflavonoids in human feces, with preliminary results in omnivorous and vegetarian women. *Anal. Biochem.* 1995g, **225**(1), 101-108.

40.	Finlay EMH, Wilson DW, Adlercreutz H, Griffiths K. The identification and measurement of phyto-oestrogens in human saliva, plasma, breast aspirate or cyst fluid, and prostatic fluid using gas chromatography-mass spectrometry. *Journal of Endocrinology.* 1991, **129**(suppl), abstract no. 49.

41.	Axelson M, Kirk DN, Farrant RD, Cooley G, Lawson AM, Setchell KDR. The identification of the weak oestrogen equol [7-hydroxy-3-(4'-hydroxyphenyl)chroman] in human urine. *Biochemical Journal.* 1982 b, **201**, 353-357.

42.	Bannwart C, Adlercreutz H, Fotsis T, Wähälä K, Hase T, Brunow G. Identification of isoflavonic phytoestrogens and of lignans in human urine and in cow milk by GC/MS. In, Todd JFJ, ed. Advances in Mass Spectrometry-85. Proceedings of the 10th International Mass Spectrometry Conference. *Chicllester, Sussex, John Wiley*, 1986, 661-662.

43.	Bannwart C, Adlercreutz H, Wähälä K, Brunow G, Hase T. Identification of the isoflavonic phytoestrogens formononetin and dihydrodaidzein in human urine. International Symposium on Applied Mass Spectrometry in the Health Sciences. *Abstracts. Barcelona, Fira de Barcelona, Palau de Congressos*, 1987, 169.

44.	Bannwart C, Adlercreutz H, Wähälä K, et al. Identification of the phytoestrogen 3',7-dihydroxyisoflavan, an isomer of equol, in human urine and cow milk. *Biomedical and Environmental Mass Spectrometry.* 1988 c, **17**,1-6.

45. Joannou GE, Kelly GE, Reeder AY, Waring M, Nelson C. A urinary profile study of dictary phytoestrogens. The identifcation and mode of metabolism of new isoflavonoids. *J. Steroid. Biochem. Molec. Biol.* 1995, **54**(3-4), 167-184.

46. Adlercreutz H, Gorbach SL, Goldin BR, Woods MN, Dwyer JT, Hämäläinen E. Estrogen metabolism and excretion in oriental and caucasian women. *J. Nat. Cancer Inst.* 1994b, **86**(14), 1076-1 082.

47. Adlercreutz H, Fotsis T, Kurzer MS, Wähälä K, Mäkelä T, Hase T. Isotope dilution gas chromatographic-mass spectrometric method fore the determination of unconjugated lignans and isoflavonoids in human feces, with preliminary results in omnivorous and vegetarian women. *Anal. Biochem.* 1995 b, **225**, 101-108.

48. Adlercreutz H. Lignans and phytoestrogens. Possible preventive role in cancer. In, Horwitz C, Rozen P, eds. Progress in Diet and Nutrition. Basel, S. Karger, 1988a, 165-176. Rozen P, ed. Frontiers of Gastrointestinal Research 14.

49. Adlercreutz H, Mousavi Y, Loukovaara M, Hämäläinen E. Lignans, isoflavones, sex hormone metabolism and breast cancer. In, Hochberg R, Naftolin F, eds. The New Biology of Steroid Hormones. *New York, Raven Press,* 1991 d, Serono Symposia Publications from Raven Press. Vol. 74.

50. Setchell KDR, Adlercreutz H. Mammalian lignans and phyto-oestrogens. Recent studies on their formation, metabolism and biological role in health and disease. In, Rowland 1, ed. Role of the Gut Flora in Toxicity and Cancer. *Londoll, Academic Press*, 1988, 315-345.

51. Adlercreutz H, Martin F, Järvenpää P. Steroid absorption and enterohepatic *Contraceptioll.* 1979, **20**, 20 1-224.

52. Adlercreutz H, Martin F. Review. Biliary excretion and intestinal metabolism of progesterone and estrogens in man. J. Steroid Biochem. 1980, **13**, 231-244.

53. Adlercreutz H, Järvenpää P. Assay of estrogens in human feces. *J. Steroid.Biochem.* 1982, **17**, 639-645.

54. Adlercreutz H, Fotsis T, Bannwart C, et al. Determination of urinary lignans and phytoestrogen metabolites, potential antiestrogens and anticarcinogens, in urine of women on various habitual diets. *J. Steroid. Biochem.* 1986 d, **5**, 791-797.

55. Adlercreutz H, Fotsis T, Heikkinen R, et al. Diet and urinary excretion of lignans in female subjects. *Medical Biology.* 1981, **59**, 259-261.

56. Adlercreutz H, Höckerstedt K, Bannwart C, et al. Effect of dietary components, including lignans and phytoestrogens, on enterohepatic circulation and liver metabolism of estrogens, and on sex hormone binding globulin (SHBG). *Journal of Steroid. Biochemistry.* 1987, **27**, 1135-1144.

57. Adlercreutz H, Höckerstedt K, Bannwart C, Hämäläinen E, Fotsis T, Bloigu S. Association between dietary fiber, urinary excretion of lignans and isoflavonic phytoestrogens, and plasma non-protein bound sex hormones in relation to breast cancer. In, Bresciani F, King RJB, Lippman ME, Raynaud J-P, eds. Progress in Cancer Research and Therapy, Vol. 35, Hormones and Cancer 3. New York, Raven Press, 1988 b, 409-412.

58. Nilsson A, Hill JL, Davies HL. An *in vitro* study of formononetin and biochanin A in rumen fluid from sheep. *Biochem. Biophys. Acta.* 1967, **148**, 92-98.

59. Shutt DA, Weston RH, Hogan JP. Quantitative aspects of phyto-oestrogen metabolism in sheep fed on subterranean clover (Trifolium subterraneum cultivar clare) or red clover (Trifolito pralense). *Austr. J. Agric. Res.* 1970, **21**, 713-722.

60. Braden AWH, Thaun Rl, Shutt DA. Comparison of plasma phyto-oestrogen levels in sheep and cattle after feeding on fresh clover. *Austr. J. Agric. Res.* 1971, **22**, 663-70.

61. Bokkenheuser VD, Shackleton CHL, Winter J. Hydrolysis of dietary flavonoid glycosides by strains of intestinal Baclerioicles from humans. *Biochem. J.* 1987, **248**, 933-956.

62. Bokkenheuser VD, Winter J. Hydrolysis of flavonoids by human intestinal bacteria. Plant Flavonoids in Biology and Medicine 11, Biochemical, Cellular, and Medicinal Properties. Alan R. Liss, Inc, 1988, 143-145. *Prog Clin Biol Res*, vol 280.

63. Axelson M, Sjövall J, Gustafsson BE, Setchell KDR. Soya - a dietary source of the non-steroidal oestrogen equol in man and animals. *Journal of Endocrinology.* 1984, **102**, 49-56.

64. Adlercreutz H, Honjo H, Higashi A, et al. Urinary excretion of lignans and isoflavonoid phytoestrogens in Japanese men and women consuming traditional Japanese diet. *Am. J. Clin. Nutr.* 1991 e, **54**, 1093-1 100.

65. Hackett AM. The metabolism of flavonoid compounds in mammals. In, Cody V, Middleton Jr CV, Harborne JB, eds. Plant Flavonoids in Biology and Medicine, Biochemical, Pharmacological, and Structure-Activity Relationships. New York, Alan R. Liss, Inc., 1986, 125-140. *Prog Clin Biol Res,* vol 213.

66. Adlercreutz H, Fotsis T, Bannwart C, Wähälä K, Brunow G, Hase T. Isotope dilution gas chromatographic-mass spectrometric method for the determination of lignans and isoflavonoids in human urine, including identification of genistein. *Clin. Chem. Acta.* 1991 c, **199**, 263-278.

67. Adlercreutz CHT, Goldin BR, Gorbach SL, et al. Soybean phytoestrogen intake and cancer risk. *J. Nutr.* 1995 a, **125**, 757S-770S.

68. Adlercreutz H, Wang G-J, Uehara M, et al. Immunoassays of phytoestrogens in human plasma. The Cost 916 Workshop on Phytoestrogens, exposure, bioavailability, health benefits and safety concers. April 17.18, 1998, Doorwerth, The Netherlands 1998, in press.

69. Adlercreutz H, Lapcik O, Hampl R, et al. Immunoassay of phytoestrogens in human plasma. Symposium on Phytoestrogen Research Methods, Tucson, Arizona, USA 1997, Abstract.

70. Anderegg RJ, Rowe JW. Lignans, the major component of resin from Araucaria angustifolia knots. *Holzforschung.* 1974, **28**, 171-175.

INHIBITION OF AROMATASE BY FLAVONOIDS IN CULTURED JEG-3 CELLS

S.C. Joshi[1], M. Ahotupa[1], M.L. Kostian[1], S.I. Mäkelä[1,2] and R.S.S. Santti[1].

[1]Institute of Biomedicine, University of Turku, Kiinamyllynkatu 10; FIN-20520 Turku, Finland. [2]Unit for Preventive Nutrition, Department of Medical Nutrition, Karolinska Institute, S-14157 Huddinge, Sweden.

1 INTRODUCTION

The risk of human breast cancer (BC) is related to cumulative exposure of breast tissue to endogenous estrogens. Experimental evidence also strongly favors the role of estrogens in the development and growth of breast cancer. In males, the estrogen concentration and the ratio of the concentrations of estrogens and androgens in serum, which increase with age, have been postulated to be risk factors for the lower urinary tract symptoms (LUTS) and benign prostatic hyperplasia (BPH). Estrogens may also play a role in prostate carcinogenesis as suggested by studies involving animal models, but their role in human prostate cancer (PC) has not been fully elucidated yet.

The diet may have a crucial role in the development of BC, LUTS/BPH and PC. Their incidence rates are high in populations consuming high-fat, low-fiber diets. Instead, risks are lower in populations having diets rich in fruits and vegetables. Consumption of high-fiber diets may be associated with lower levels of estrogens and androgens in serum. In addition to the fiber effect, the possible beneficial effects of low-fat, high-fiber diets have been suggested to be related to contents of weakly estrogenic, plant-derived compounds called phytoestrogens.

Phytoestrogens are polyphenolic nonsteroidal compounds which may be classified into three structurally different groups: isoflavonoids, lignans and coumestans. The intake and serum concentrations of isoflavonoids (such as daidzein and genistein) are higher in Asian countries where the incidences of BC, BPH and PC are low.[1] In addition to isoflavonoids derived from soy, edible plants contain a wide variety of flavonoids which have been shown to exert similar biological actions with phytoestrogens. They may also contribute to the chemopreventive actions of vegetable- and fruit-rich diets.

It is not easily conceivable how phytoestrogens and structurally related compounds as estrogens could prevent the development of BC, BPH or PC. Besides binding to estrogen receptors (ERα and ERß) some phytoestrogens and structurally related compounds may compete with endogenous substrates for active sites of estrogen biosynthesizing and metabolizing enzymes, such as aromatase[8,9] and 17ß-hydroxysteroid oxidoreductase, type 1.[12,13] Inhibition of these two enzymes would lower estrogen concentrations in target tissues which could consequently decrease the risks of BC, BPH and perhaps PC.[10] Moreover, isoflavonoids and flavonoids have been reported to posses strong antioxidant activity[14] which may also play a role in chemoprevention of these cancers.

As a part of our efforts to identify the targets of isoflavonoids and flavonoids as well as to understand the mode of their chemopreventive action, we have studied the inhibition of [3]H-17ß-estradiol formation from ^{3}H-androstenedione in cultured JEG-3 cells (a human choriocarcinoma cell line). The JGG-3 choriocarcinoma cells[11] are a useful aromatase model enabling the study of aromatase inhibition *in vitro*. The aim of this presentation is to discuss the aromatase inhibitory and antioxidant properties of isoflavonoids and flavonoids and the structure/activity relationships.

2 MATERIALS AND METHODS

Various isoflavonoids and flavonoids were tested to determine the structural requirement for inhibition of aromatase and to evaluate the antioxidative properties. The test compounds were purchased from the following sources: apigenin, chrysin, kaempferol and quercetin from Sigma Chemical Co. (St. Louis, MO), genistein from Gibco Life Technologies Inc. (Gaithersburg, MD); daidzein from Research Biochemicals International (Natic, MA); 7-hydroxyflavone, fisetin, kaempferide, luteolin, galangin, naringenin and pinostrobin from Carl Roth GmbH (Karlsruhe, Germany). Their systematic names are given in table 1.

Table 1.
Aromatase inhibitory and antioxidant properties of isoflavonoids and flavonoids in vitro.

Compounds	Inhibition (%) of 17ß-estradiol formation in JEG-3 cells (1µM)	Antioxidative capacity (t-BuOOH-LP) IC_{50} (µM)
Isoflavones		
Genistein (4',5,7-trihydroxyisoflavone)	22	15
Daidzein (4',7-dihydroxyisoflavone)	30	164
Biochanin A (5,7-dihydroxy-4'-methoxyisoflavone)	46	12
Flavones		
7-hydroxyflavone	82	170
chrysin (5,7-dihydroxyflavone)	68	18
apigenin (4',5,7-trihydroxyflavone)	62	7
luteolin (3',4',5,7-tetrahydroxyflavone)	35	0.25
Flavonols		
Galangin (3,5,7-trihydroxyflavone)	20	0.44
kaempferol (3,4',5,7-tetrahydroxyflavone)	11	0.9
fisetin (3,3',4',7-tetrahydroxyflavone)	11	0.3
quercetin (3,3',4',5,7-pentahydroxyflavone)	0	0.4
Flavanone		
naringenin (4',5,7-trihydroxyflavanone)	45	9
Methoxyflavones		
pinostrobin (5-hydroxy-7-methoxyflavone)	69	29
acacetin (5,7-dihydroxy-4'-methoxyflavone)	13	7
kaempferide (3,5,7-trihydroxy-4'-methoxyflavone)	18	0.5

The aromatase activity in JEG-3, human choriocarcinoma cells was assayed by determining the capability of the intact cellular monolayer to convert added ^{3}H-androstenedione to ³H-17ß-estradiol.[8] Cells were maintained DMEN containing 10 % fetal calf serum. Experiments were carried out by using confluent monolayers (approximately 5 days after subculturing). Before the experiment, the cells were washed with EDTA and treated with EDTA and trypsin mixture. Experiments were carried out in serum free medium. The incubation mixture contains 50 µl ³H-androst-4-ene, 3,17-dione (0.5 nM), 50 µl unlabelled androstenedione (0.5 nM), 100 µl test compounds (10 µM) and 800 µl cells (1 million).

After the incubation for 4 h, unlabelled carriers (androstenedione, testosterone, estradiol and estrone) were added. The steroids were extracted twice with 3 ml dichloromethane. The combined dichloromethane solutions were evaporatd to dryness under nitrogen. Extracted steroids were dissolved in 35 % acetonitrile before HPLC run. HPLC was used for separation and quantification of the radiolabelled products. The column system consisted of a guard column followed by a C18. The mobile phase was acetonitrile/water (35/65). For in-line detection of the radioactive metabolites, the eluent of the HPCL column was continuously mixed with liquid scintillant and then monitored with in-line radioactivity detector. Methodological details for the quantifications of ³H-labelled products have been given earlier.[13] The aromatase activity (formation of ³H-17ß-estradiol from ³H-androstenedione) was calculated as percentage of ³H-androstenedione converted to ³H-17ß-estradiol. The number of the cells in each assay was adjusted so that the conversion of the substrate to products was 5-30 % during the incubation.

The antioxidative capacity of isoflavonoids and flavonoids was estimated by their potency to inhibit tert-butylhydroperoxide-induced lipid perioxidation (t-BuOOH-LP) in rat liver microsomes *in vitro*.[2] The test for the t-BuOOH-LP was performed as follows: The buffer (50 mM sodium carbonate, pH 10.2, with 0.1 mM EDTA) was pipetted in a volume of 0.8 ml in the luminometer cuvette. Twenty microliters of diluted liver microsomes, final concentration 1.5 µg protein/ml, were added, followed by 6 µl of luminol (0.5 mg/ml) and test chemicals. Test compounds were added to incubation mixtures in a small volume diluted in ethanol or dimethylsulphoxide (2 % of incubation volume), and the lipid peroxidation potency was compared to that of the vehicle (ethanol or dimethyl sulphoxide). The reaction was initiated by 0.05 ml of 0.9 mM t-BuOOH at 33 °C. The chemiluminescence was measured for about 45 min at 1 min cycles and the area under curve (integral) was calculated. Chemiluminescence measurements were performed with Bio-Orbit 1251 Luminometer connected to a personal computer with special programs for the assays (Bio-Orbit, Turku, Finland). The protocol for each assay included automatic pipetting of the appropriate reagent that initiated the radical reaction in all cuvettes. Six or twelve samples were analyzed simultaneously, depending on the assay. Disposable plastic cuvettes were used throughout.

3 RESULTS

The data showing the inhibition of ³H-17ß-estradiol formation from ³H-androstenedione in cultured JEG-3 cells by isoflavonoids and flavonoids at the final concentration of 1µM are given in Table 1. Several isoflavonoids and flavonoids such as genistein, daidzein, biochanin A, flavone, 7-hydroxyflavone, chrysin, apigenin, luteolin, galangin, naringenin and pinostrobin caused at least 20 % inhibition of aromatization at the concentration of 1 μM while kaempferol, kaempferide, acacetin, quercetin and fisetin did not show any inhibition at this concentration. However, the noninhibitory compounds: kaempferol, kaempferide, quercetin and fisetin appeared to be potent antioxidants as supported by the data from the t-BuOOH-LP assay shown in table 1. Luteolin showed both inhibitory and antioxidant capacities.

4 DISCUSSION

Several isoflavonoids and flavonoids were found to inhibit the formation of ^{3}H-17ß-estradiol from ^{3}H-andostenedione in JEG-3 cells. The number and location of hydroxyl groups appeared to be important for the inhibition. The hydroxyl groups of flavones at C-5 (7-OH-flavone versus chrysin) and at C-4' (apigenin versus chrysin) are permissible for the inhibition. C-2, C-3 double bond in the C ring (apigenin versus naringenin) is not critical for the inhibition, either. The hydroxyl group in position 3 of flavones (flavonol structure) (galangin versus chrysin; kaempferol versus apigenin; quercetin versus luteolin) and C-3', C-4'-dihydroxy (luteolin versus apigenin) reduce the inhibitory activity of 17ß-estradiol formation. The inhibitory effect of O-methylation on 17ß-estradiol formation was related to the position of the hydroxy group. Methylation of the 7-OH group has no effect (pinostrobin versus chrysin) whereas the methoxy group in position 4' eliminates the inhibitory capacity (acacetin versus apigenin). The contribution of the position of the B ring (in the 3 position adjacent to the 4 keto group in the isoflavonoids compared to the 2 position in the flavonoids) depend on the substitutions on the molecules (apigenin stronger than genistein but biochanin A stronger than acacetin). In general, our findings on inhibition of ^{3}H-17ß-estradiol formation from ^{3}H-androstenedione are in agreement with earlier observations made on inhibition of placental and preadipocyte aromatases with the assay quantitating the production of ^{3}H$_2$O released from ^{3}H-andostenedione after aromatization.[4,6,7,9,17]

The structural requirements for inhibition of 17ß-estradiol formation do not correlate with structures known for participation in antioxidative capacity of isoflavonoids and flavonoids. In general, the more hydroxyl subtitutions the stronger the antioxidant activities. Isoflavonoids were weaker antioxidants in comparison to some flavonoids (apigenin versus genistein; acacetin versus biochanin A). The specific structural criteria defining the antioxidant acitivities of flavonoids include: **1.** 2,3-double bond with the 4'-hydroxy group; **2.** 3-hydroxyl group in the C ring; **3.** 5,7-dihydoxy structure in the A ring and **4.** orthodihydroxy structure in the B ring. This is in agreement with the findings from other laboratories.[15,16] The contribution from the 5,7-dihydroxy group in the A ring in the isoflavonoids is approximately similar to the total antioxidant activities as in the flavonoids (daidzein versus genistein; chrysin versus 7-OH-flavone). A single hydroxyl group in the 4' position of the B ring in the flavonoids does not contribute to the antioxidant activity (apigenin versus chrysin; kaempferol versus galangin). This is in contrast to the findings on the isoflavone structure.[15] O-methylation of hydroxyl substitutions inactivates antioxidant activities of flavonoids confirming earlier findings.[3,5]

In conclusion, several isoflavonoids and flavonoids such as genistein, daidzein, biochanin A, apigenin, naringenin and luteolin known to be present in human diets were found to inhibit the formation of 17 ß-estradiol in JEG-3, human choriocarcinoma cells. These compounds do not have an optimal structure to act as antioxidants with luteolin as a notable exception. In theory, these compounds could decrease endogenous estrogen formation and consequently, the risks of BC, BPH and PC. However, their effects *in vivo* cannot be predicted on the *in vitro* results alone because both bioavailability and metabolism may influence the biological actions *in vivo*. *In vivo* experiments are needed to confirm their aromatase inhibitory potencies.

References

1.   H. Adlercreutz, H. Markkanen and S. Watanabe. Plasma concentrations of phyto-oestrogens in Japanese men. *Lancet*. 1993, **342**, 1209.

2.   M. Ahotupa, E. Mäntylä and L. Kangas. Antioxidant properties of the triphenylethylene antiestrogen drug toremifene. Naunyn-Schmiedeberg's Arch. *Pharmacol*. 1997, **356**, 297.

3.   A. Arora, M.G. Nair and G.M. Strasburg. Structure-activity relationships for antioxidant activities of a series of flavonoids in liposomal system. *Free Radical. Biol. Med*. 1998, **24**, 1355.

4.   D.R. Campbell and M.S. Kurzer. Flavonoid inhibition of aromatase enzyme activity in human preadipocytes. *J. Steroid Biochem. Molec. Biol*. 1993, **46**, 381.

5.   G. Cao, E. Sofice and R.L. Prior. Antioxidant and prooxidant behaviour of flavonoids: structure-activity relationships. *Free Radic. Biol. Med*. 1997, **22**, 749.

6.   F.H. de Jong, K. Oishi, R.B. Hayes, J.F.A.T. Bogdanowicz, A.R. Ibrahim and Y.J. Abul Hajj. Aromatase inhibition by flavonoids. *J. Steroid Biochem. Molec. Biol*. 1990, **37**, 257.

7.   Y.-C. Kao, C. Zhou, M. Sherman, C.A. Laughton and S. Chen. Molecular basis of the inhibition of human aromatase (estrogen synthetase) by flavone and isoflavone phytoestrogens: A site-directed mutagenesis study. *Environ. Health Perspect*. 1998, **106**, 85.

8.   J.T. Kellis, Jr. and L.E. Vickery. Inhibition of estrogen synthetase (aromatase) by 4-Cyclohexylaine. *Endocrinology*. 1984 a, **114**, 2128.

9.   J.T. Kellis, Jr. and L.E. Vickery. Inhibition Of Human Estrogen Synthetase (Aromatase) By Flavones. *Science*. 1984b, **225**, 1032.

10.   G.J. Kelloff, R.A. Luber, R. Lieberman, K. Eisenhauer, V.E. Steele, J.A. Crowell, E.T. Hawk, C.W. Boone and C.C. Sigman. Aromatase inhibitors as potential cancer chemopreventives. *Cancer Epid. Biomarkers & Prevention*. 1998, **7**, 65.

11.   M.D. Krekels, W. Wouters, R. Decoster, R. Van Ginckel, A. Leonares and P.A. Janssen. Aromatase in the human choriocarcinoma JEG-3: inhibition by R76713 in cultured cells and in tumors grown in nude mice. *J. Steroid Biochem. Mol. Biol*. 1991, **38**, 415.

12.   S. Mäkelä, M. Poutanen, M.-L. Kostian, N. Lehtimäki, L. Strauss, R. Santti and R. Vihko. Inhibition of 17ß-hydroxysteroid oxidoreductase by flavonoids in breast and prostate cancer cells. *P.S.E.B.M*. 1998, **217**, 310.

13.   S. Mäkelä, M. Poutanen, J. Lehtimäki, M.-L. Kostian, R. Santti and R. Vihko.

Estrogen-specific 17ß-hydroxysteroid oxidoreductase type I (E.C. 1.1.1.62) as a possible target for the action of phytoestrogens. *Proc. Soc. Exp. Biol. Med*. 1995, **208**, 51.

14.   C. Rice-Evans, N.J. Miller and G. Paganga. Structure-antioxidant activity relationships of flavonoids and phenolic acids. *Free. Radic. Biol. Med*. 1996, **20**, 933.

15.   C. Rice-Evans. Antioxidant properties of phytoestrogens and comparision with model compounds. Proceedings of symposium on phytoestrogen research methods. Chemistry, analysis & Biological properties. Sept. 21-24, 1997. Tucson, Arizona, USA.

16.   S.A. van Acker, D.J. van den Berg, M.N. Tromp, D.H. Griffionen, W.P. van Bennekom, W.J. van der Vijg and A. Bast. Structural aspects of antioxidant activity of flavonoids. *Free Radic. Biol. Med*. 1996, **20**, 331.

17.   C. Wang, T. Mäkelä, T. Hase, H. Adlercreutz and M.S. Kurzer. Lignans and flavonoids inhibit aromatase enzyme in human adipocytes. *J. Steroid Biochem. Molec. Biol*. 1994, **50**, 205.

THE EFFECT OF PHYTOESTROGEN - RICH FOODS ON URINARY OUTPUT OF PHYTOESTROGEN METABOLITES - A PILOT STUDY

J.V. Woodside, *M.S. Morton and A.J.C. Leathem.

Department of Surgery, UCL, 67-73 Riding House Street, London, W1P 7LD and
*Tenovus Cancer Research Centre, University of Wales College of Medicine, Heath Park, Cardiff CF4 4X.

1 INTRODUCTION

There is growing interest in the reported health benefits of the isoflavonoid and lignan phytoestrogens, which are oestrogen-like plant compounds. Epidemiological, *in vitro* and animal studies provide evidence that phytoestrogens may reduce cancer risk.[1-3] The weak antiestrogenic effect of phytoestrogens has been proposed as one mechanism by which these compounds reduce the risk of hormone-dependent cancers.[4] However, phytoestrogens also have antiproliferative, antioxidant and angiogenesis-inhibiting properties, which may help reduce the risk of many types of cancer.[3]

Cross-sectional studies show that urinary isoflavone (daidzein, genistein and equol) excretion is substantially greater in Japanese volunteers consuming a traditional diet compared to levels in adults living in Boston and Helsinki.[5] Soy products are rich in isoflavonic phytoestrogens, while the lignans are found in a number of fruits, vegetables and grains. British isoflavone intake is estimated at <1 mg/d.[6]

We have conducted a small-scale feeding study in healthy volunteers to assess to what extent urinary output of phytoestrogens can be affected by a short-term isoflavonoid- and lignan-rich diet.

2 MATERIALS AND METODS

Seven healthy volunteers, three male and four female, followed a special diet over a period of three days. Each were given 100 g of soya chunks, 150 g lentils and 250 g of kidney beans daily to be incorporated into their normal diet. Twenty-four h urine samples were collected for 2 days prior to the initiation of the phytoestrogen-rich diet, during the three days of intervention, and for two days following intervention. Subjects filled in food diaries to assess habitual diet and also to ensure compliance with the test diet. The 24-h urine samples were analysed by GC-MS[6] for the isoflavones daidzein and equol and the lignan enterolactone.

3 RESULTS

Table 1 shows the distribution (mean and standard deviation) of the phytoestrogens daidzein, equol and enterolactone in urine over the 7-day period. Daidzein levels were significantly elevated on days 3-6, compared to days 1-2 and day 7, reaching over 4000 ng/ml on days 4-5. However, equol levels did not show a similar rise, indicating that at least some of these subjects were unable to produce equol from daidzein. It is known that there is considerable intra-individual variability in metabolic response to a known dose of isoflavone-rich food, and this may be due to variation in bacterial enzymes in gut microflora.[7] Enterolactone levels did not change significantly over the seven day period, although an increase may have been observed if urine collections had been carried out for a longer period post-intervention.

Table 1
Mean phytoestrogen levels in urine during study period.

Day	Daidzein (n=7) Mean	SE	Equol (n= 7) Mean	SE	Enterolactone (n= 7) Mean	SE
1 (baseline)	321	73	6.09	1.07	862	329
2 (baseline)	592	261	6.26	1.02	1094	365
3 (intervention)	2608	530	5.63	1.61	751	239
4 (intervention)	4151	692	7.71	2.69	910	277
5 (intervention)	4138	1284	4.57	0.40	774	328
6 (post-intervention)	2461	777	4.92	1.23	838	232
7 (post-intervention)	490	157	7.09	1.47	699	310

4 CONCLUSION

This study, in a small number of subjects, shows that phytoestrogen intake can be affected in the short-term by changes to the Western diet. This finding may have implications for the primary and secondary prevention of hormonally-dependent cancers.

Reference

1.   B.K. Armstrong, J.B. Brown, H.T. Clarke, et al. *J. Natl.Cancer Inst.* 1981, **67**, 761.
2.   H. Adlercreutz, R. Heikkinen, M. Woods, et al. *Lancet*. 1982, **2**, 1295.
3.   M. Messina, V. Persky, KDR Setchell and S. Barnes. *Nutr.Cancer.* 1994, **21**, 113.
4.   S.A. Bingham, C. Atkinson, J. Liggins, et al. *Br. J.Nutr.* 1998, **79**, 393.
5.   H. Adlercreutz, H. Honjo, T. Higashi, et al. *Am. J.Clin. Nutr.* 1991, **54**, 1093.
6.   A. Jones, K.R. Price and G.R. Fenwick. *J. Sci. Food .Agric.* 1989, **46**, 357.
7.   M.S. Morton, G. Wilcox, M. Wahlquist, et al. *J .Endocrinol* .1994, **142**, 251.
8.   A. Cassidy, S.A. Bingham, and K.D.R. Setchell. *Am. J.Clin. Nutr.* 1994, **60**, 333.

MECHANISMS OF ACTION OF THE ANTIOXIDANT LYCOPENE IN CANCER

Joseph Levy, Michael Danilenko, Michael Karas, Hadar Amir, Amit Nahum, Yudit Giat and Yoav Sharoni.

Departments of Clinical Biochemistry, Faculty of Health Sciences, Ben-Gurion University of the Negev and Soroka Medical Center of Kupat Holim, Beer-Sheva 84105, Israel.

1 INTRODUCTION

In the Western world the incidence of prostate and breast cancer is increasing at an accelerating pace. To fight this trend it is important to detect risk factors and to use appropriate preventive measures. Two recent prospective studies carried out by groups from Harvard and McGill Universities suggest that insulin-like growth factor-I (IGF-I) may be a risk factor for these cancers. One study was done on prospectively collected plasma from prostatic cancer cases and matched controls from participants in the Physicians' Health Study. A strong positive association between IGF-I plasma levels and prostate cancer risk was found so that healthy people with high IGF-I blood levels were four time more likely to develop prostate cancer than those with lower IGF-I levels.[1] An equally strong association between these growth factor levels and breast cancer risk was also reported in a case control study within the Nurses' Health Study cohort.[2] Thus, plasma IGF-I levels may be useful for identifying high risk breast and prostatic cancer patients and for administering preventative measures based on risk reduction strategies, either by lowering IGF-I levels or interference with its action.

A nutritional approach to the prevention of cancer has been prevalent for many years, including the anti cancer effect of various micronutrients present in the diet such as carotenoids. For example, the same group of researchers who reported an involvement of IGF-I in prostatic and breast cancer reported recently that intake of lycopene, the major tomato carotenoid - (but not other carotenoids), reduce the risk of prostate cancer,[3] see also paper by E. Giovannucci, this volume). In addition, in a case-control study higher breast adipose concentrations of lycopene was associated with decreased risk of breast cancer.[4]

These observations pose the question of whether the data showing that IGF-I is a risk factor for prostate and breast cancer can be linked to the cancer preventive effects of lycopene? The possibility that lycopene exerts its effect, at least partially, by interfering with the action of IGF is an alluring idea and will be discussed later in this review.

2 EPIDEMIOLOGICAL STUDIES INVOLVING LYCOPENE AND CANCER

In addition to the epidemiological studies mentioned above, several other investigations have examined data on lycopene intake (based on questionnaires) or lycopene plasma levels in relation to cancer risk. A study of cervical intra-epithelial neoplasia was designed to test the potential protective effects of various carotenoids.[5] In this study, both dietary and serum lycopene manifested a strong inverse association with this malignancy. No such association was detected in the same study in regard to β-carotene intake. A lower level of serum lycopene was also observed in patients who subsequently developed bladder[6] and pancreatic[7] cancers.

A protective effect of vitamin-A and β-carotene was found in squamous cell and small cell lung cancers.[8] In a later study, however, β-carotene-rich foods such as papaya, sweet potato, mango and yellow orange vegetables showed little influence on survival of lung cancer patients.[9] The authors concluded that β-carotene intake before diagnosis of lung cancer does not affect the progression of the disease. In contrast, a tomato-rich diet which contributes only small amounts of β-carotene to the total carotenoid intake had a strong positive relationship with survival, particularly in women.

A significant trend in risk reduction of gastric cancers by high tomato consumption was observed in a study which estimated dietary intake in low risk versus high risk areas in Italy.[10] It is interesting that a similar regional impact on stomach cancer risk was found also in Japan.[11] Out of several micronutrients including vitamins A, C and D and β-carotene in plasma, only lycopene was strongly-inversely associated with stomach cancer. A consistent pattern of protection for many sites of digestive tract cancer was associated with an increased intake of fresh tomatoes.[12] Nevertheless, the possibility exists that other benefits of a Mediterranean life style get confused with high lycopene intake from a tomato rich diet. No direct epidemiological data are available on the potential role of lycopene consumption in skin cancer. However, it was demonstrated that lycopene, but not β-carotene, levels decrease significantly in the area of UV irradiated skin.[13]

3 *IN VIVO* AND *IN VITRO* EVIDENCE FOR THE INHIBITION OF CANCER BY LYCOPENE

The anticancer activity of carotenoids has been reviewed by Krinsky[14,15] and more recent reviews by others have specifically addressed the anticancer activity of lycopene.[16-18]

The effect of several retinoids and carotenoids, including lycopene and β-carotene, was tested on rat C-6 glioma cells.[19] All the tested retinoids and carotenoids inhibited cellular growth at the 10 μM range. Lycopene effects were similar to those of β-carotene. The same group also demonstrated that lycopene is an effective inhibitor in an *in vivo* model of glioma cells transplanted in rats.[20] Lycopene and other carotenoids were found to reduce hepatocytes cell injury induced by carbon tetrachloride.[21] Several carotenoids including lycopene protect against liver tumor promoter microcystin-LR.[22] We compared the potential anticancer activity of lycopene to that of α- and β-carotene in human cancer cells in culture. Lycopene was observed to strongly inhibit the growth of mammary (MCF-7), endometrial (Ishikawa) and lung (H226) cancer cells. The inhibition was dose-dependent- half maximal inhibition was at ~2 μM. α- and β-carotene were far less effective. For example, in Ishikawa endometrial cancer cells, a four-fold higher concentration of α-carotene or a ten-fold concentration of β-carotene were needed for growth inhibition of the same order. Inhibition of cell growth by lycopene was detected after 24 h of incubation and was maintained for at least three days. In contrast to cancer cells, human fibroblasts were less sensitive to inhibition by lycopene, and the cells gradually escaped inhibition.[23,24]

Using other *in vivo* models Nagasawa and colleagues[25] found that lycopene inhibits spontaneous mammary tumor development in SHN virgin mice probably by modulating the immune system in tumor-bearing mice.[26] The inhibitory effect of four carotenoids found in human blood and tissues which are effective against the formation of colonic aberrant crypt foci induced by N-methylnitrosourea was examined in Sprague-Dawley rats.[27] Lycopene, lutein, α-carotene and palm carotenes (a mixture of α-carotene, β-carotene and lycopene), but not β-carotene, inhibited the development of aberrant crypt foci. The same research team also compared the cancer preventive effect of five kinds of carotenoids on mouse lung carcinogenesis with similar results.[28]

We have also tested the effects of lycopene and β-carotene in a well-known animal model for hormone-dependent mammary cancer DMBA-induced rat mammary tumors.[29] In rats injected i.p. with lycopene-enriched tomato oleoresin or β-carotene, both carotenoids were absorbed into blood, liver, mammary gland and mammary tumors. Our results show that animals treated with tomato oleoresin, containing lycopene, developed significantly fewer tumors and a smaller tumor area than unsupplemented rats. β-carotene treatment, on the other hand, did not protect against tumor development.

4 POSSIBLE MECHANISMS FOR THE ANTICANCER ACTIVITY OF LYCOPENE

4.1. Antioxidant activity

The antioxidant activity of carotenoids is a major mechanism of action against cancer. As antioxidants, they may prevent cancer and degenerative diseases by preventing oxidative damage of proteins, lipids and DNA and by affecting signaling mechanisms initiated by oxidative stress. The antioxidant activity of carotenoids is versatile, as they prevent both singlet oxygen- and free radical-mediated damage. Among the carotenoids, lycopene exhibits the highest physical quenching rate constant with singlet oxygen[30] and is at least three-fold more effective than β-carotene in quenching NOO$^{\bullet}$ radicals which are present in tobacco smoke.[31] Smedman et al.[32] have reported that lycopene protects DNA damage to colon cells induced by 1-methyl 3-nitro-1-nitrosoguanidine (MNNG) and H_2O_2. The greatest protective effect was observed at a lycopene concentration approximating that found in blood of subjects consuming normal levels of tomatoes. The authors concluded that this effect cannot be explained simply by the antioxidant capacity of lycopene because the genotoxic effect of the alkylating agents, MNNG, was also inhibited. Support for additional actions is found also in a report which showed that carotenoids increase the expression of a gene encoding connexin-43- a gap junction protein.[33] This effect was independent of provitamin-A or the antioxidant properties of carotenoids (see also review by J. Bertram, this volume).

The above information suggests that other mechanisms, aside from the antioxidant activity of lycopene should be considered. Thus, we investigated several additional mechanisms which may explain the anticancer effect of lycopene and may link it to the importance of IGF-I as a risk factor for various types of cancer.

4.2. Inhibition of growth factor effects

One of the reasons for the uncontrolled growth of cancer cells is the augmented secretion of autocrine or paracrine growth factors which stimulate cell proliferation via different signaling pathways. The autocrine role of the IGF system in human mammary and endometrial cancer cells has been stressed in several studied including our own.[34,35] The strong association, discussed above, between these growth factor blood levels and cancer risk also substantiates their importance for the malignant process. We hypothesize that a possible mechanism for the inhibitory effect of lycopene may involve the modulation of the complex cellular IGF systems. We found that growth stimulation of mammary and endometrial cancer cells by IGF-I is markedly reduced by physiologic concentrations of lycopene (0.4-0.8 µM), while growth of unstimulated cells is inhibited only at higher concentrations of the carotenoid (3-4 µM).[36] Lycopene treatment did not affect the number or affinity of IGF-I receptors but significantly reduced IGF-induced tyrosine phosphorylation of insulin receptor substrate-1 and up-regulation of binding capacity of the AP-1 transcription complex. These effects of the carotenoid may be explained by the accompanied increase in membrane-associated IGF-binding proteins. Recently we demonstrated that IGFBPs specifically associated with the cell membrane inhibits IGF-I receptor signaling in an IGF-dependent manner as revealed by measurement of short- and middle-term receptor mediated responses.[37]

4.3. Intervention in cell cycle progression

The inhibitory effect of lycopene on cell proliferation was not accompanied by either necrotic or apoptotic cell death. Thus, we performed experiments to analyze the effect of lycopene on the cell cycle in mammary and endometrial cancer cells.[38] Lycopene slowed down cell cycle progression induced by IGF-I in serum starved synchronized cells. Moreover in cells synchronized in the G1-S boundary by mimosine treatment, lycopene delayed S-G2 cell cycle progression after release from the mimosine block. These results suggest interference by lycopene at several points of cell cycle progression. Such multiple changes at several steps of the cell cycles can be achieved by modulating the expression and/or activity of major components which regulate cell cycle progression. Preliminary studies along this line suggest that such changes do occur during lycopene treatment.

4.4. Synergism with other anticancer active compounds

Although lycopene has major effects on the growth of cancer cells, we do not consider it is a new "magic bullet" since the idea of using a single compound which has been successful in *in vitro* and *in vivo* models failed to show beneficial effects in several clinical studies.[39-42] Thus, one aim of our studies is to show that plant derived constituents, such as carotenoids, have the ability to synergize with other anti-cancer compounds such as various antioxidants (e.g. carnosic acid, polyphenols and garlic-derived sulfur compounds) or ligands of the nuclear receptor superfamily (e.g. 1,25-dihydroxyvitamin D_3 and retinoic-acid) which when used alone, are active only at high or toxic concentrations.

We found that in HL-60 leukemic cells lycopene alone inhibits cell proliferation, slows down cell cycle progression and induces cell differentiation as measured by phorbol ester-dependent reduction of nitroblue tetrazolium (NBT). The combination of low concentrations of lycopene and 1,25 $(OH)_2$ Vitamin-D_3 (the active metabolite of vitamin-D) exhibits a synergistic effect on cell proliferation and cell differentiation. Similar effects were found with retinoic acid. The synergistic inhibition of cell proliferation by lycopene and two members of the nuclear receptor super-family suggests a genomic effect for carotenoid action. Similar synergism was found in these cells between lycopene and carnosic acid, a polyphenolic compound extracted from rosmarine. When used alone, carnosic acid did not show any effect of HL-60 cell differentiation but it highly potentiated the effect of lycopene on this process. Another type of dietary derived active component is allicin, a garlic derived sulfur-containing compound. Allicin was tested in MCF-7 mammary cancer cells and was found to inhibit cell growth in a dose-dependent manner (4-32 µM). An additive effect with lycopene was found at the low doses of allicin (4-12 µM) suggesting that the two compound are inserting their effects by different mechanisms.

5. CONCLUDING REMARKS

Prevention of tumor induction and inhibition of tumor growth by natural food constituents is an intriguing idea which has promoted numerous studies. Among various plant constituents, carotenoids have been extensively studied and implicated as cancer preventive agents. In recent years lycopene is being recognized as one of the most important compounds among members of this family of phytochemicals.

Collectively, the gathered information suggests that the inhibitory effects of lycopene on cancer cell growth are not due to the toxicity of the carotenoid treatment but rather to interference in basic mechanisms pertaining to the proliferation of cells. It also should be noted that the activity of lycopene may reside in the carotenoid molecule or in a metabolic product which may be formed from it.

A strong positive correlation between plasma IGF-I level and risk of breast and prostate cancer[1,2] as well as interference by lycopene of IGF-I receptor activation (suggested here) may provide a putative explanation for the anti-cancer activity of lycopene found in numerous *in vivo* and *in vitro* studies. Lycopene is the major carotenoid in human plasma in some Western countries where the diet is rich in tomato and tomato products. However, its levels may be low and vary considerably in individuals on low-carotenoid diets due to lack of availability or to disease states which reduce the possibility to ingest or absorb carotenoids.

In summary, lycopene is a natural, non-toxic anticancer compound. Its wide availability and the elucidation of its mechanism of action will endorse its inclusion as an important component of dietary regimens which will act in concert with physiologic concentrations of other compounds in the prevention of cancer.

Acknowledgments

The authors gratefully acknowledge the various preparations of tomato lycopene received from LycoRed, Natural Products Industries, Ltd. (Beer-Sheva, Israel). The work presented here was supported in part by the Israel Cancer Association (to Y. S. and J. L.), the Chief Scientist, Israel Ministry of Health (to Y. S. and J. L. and to M. D.), and by the Israel Science Foundation, founded by the Israel Academy of Science and Humanities.

References

1.   J. M. Chan, M. J. Stampfer, E. Giovannucci, et al., Plasma insulin-like growth factor-I and prostate cancer risk: A prospective study. *Science*. 1998, **279**, 563.
2.   S. E. Hankinson, W. C. Willett, G. A. Colditz, et al., Circulating concentrations of insulin-like growth factor I and risk of breast cancer. *Lancet*. 1988, **351**, 1397.
3.   W. Giovannucci, A. Ascherio, E. B. Rimm, M. J. Stampfer, G. A. Colditz and W. C. Willett, Intake of carotenoids and retinol in relation to risk of prostate cancer. *J. Natl. Canc. Inst.* 1995, **87**, 1767.
4.   S. M. Zhang, G. W. Tang, R. M. Russell, K. A. Mayzel, M. J. Stampfer, W. C. Willett and D. J. Hunter, Measurement of retinoids and carotenoids in breast adipose tissue and a comparison of concentrations in breast cancer cases and control subjects. *Am. J. Clin. Nutr.* 1997, **66**, 626.
5.   J. VanEenwyk, F. G. Davis and P. E. Bowen, Dietary and serum carotenoids and cervical intraepithelial neoplasia. *Int. J. Cancer*. 1991, **48**, 34.
6.   K. J. Helzlsouer, G. W. Comstock and J. S. Morris, Selenium, lycopene, α–tocopherol, β-carotene, retinol, and subsequent bladder cancer. *Cancer Res.* 1989, **49**, 6144.
7.   P. G. Burney, G. W. Comstock and J. S. Morris, Serologic precursors of cancer: serum micronutrients and the subsequent risk of pancreatic cancer. *Am. J. Clin. Nutr.* 1989, **49**, 895.
8.   R. G. Ziegler, T. J. Mason and A. Stemhagen, Carotenoid intake, vegetables, and the risk of lung cancer among white men in New Jersey. *Am. J. Epidemiol.* 1986, **123**, 1080.
9.   M. T. Goodman, L. N. Kolonel, L. R. Wilkens, C. N. Yoshizawa, L. Lemarchand and J. H. Hankin, Dietary factors in lung cancer prognosis. *Eur. J. Cancer*.1992, **28**, 495.
10.   E. Buiatti, D. Palli, A. Decarli, et al., A case-control study of gastric cancer and diet in Italy: II. Association with nutrients. *Int. J. Cancer*. 1990, **45**, 896.
11.   S. Tsugane, M. Tsuda, F. Gey and S. Watanabe, Cross-sectional study with multiple measurements of biological markers for assessing stomach cancer risks at the population level. *Environ Health Perspect.* 1992, **98**, 207.
12.   S. Franceschi, E. Bidoli, C. La Vecchia, R. Talamini, B. D'Avanzo and E. Negri, Tomatoes and risk of digestive-tract cancers. *Int. J. Cancer.* 1994, **59**, 181.
13.   J. D. Ribaya Mercado, M. Garmyn, B. A. Gilchrest and R. M. Russell, Skin lycopene is destroyed preferentially over β-carotene during ultraviolet irradiation in humans. *J. Nutr.* 1995, **125**, 1854.

14. N. I. Krinsky, Effects of carotenoids in cellular and animal systems. *Am. J. Clin. Nutr.* 1991, **53**, 238S.

15. N. I. Krinsky, Anticarcinogenic activities of carotenoids in animals and cellular systems. in "Free Radicals and Aging" (eds. I. Emers and B. Chance) *Verlag, Basel,* 1992, 227.

16. H. Gerster, The potential role of lycopene for human health. *J. Am.. Coll. Nutr.* 1997, **16**, 109.

17. W. Stahl and H. Sies, Lycopene: A biologically important carotenoid for humans? *Arch. Biochem. Biophys.* 1996, **336**, 1.

18. S. K. Clinton, Lycopene: Chemistry, biology, and implications for human health and disease. *Nutr. Rev.* 1998, **56**, 35.

19. C.-J. Wang and J.-K. Lin, Inhibitory effects of carotenoids and retinoids on the *in vitro* growth of rat C-6 glioma cells. *Proc. Natl. Sci. Counc. B. ROC* . 1989, **13**, 176.

20. C. J. Wang, M. Y. Chou and J. K. Lin, Inhibition of growth and development of the transplantable C-6 glioma cells inoculated in rats by retinoids and carotenoids. *Cancer Letters.* 1989, **48**, 135.

21. H. Kim, Carotenoids protect cultured rat hepatocytes from injury caused by carbon tetrachloride. *Int. J. Biochem. Cell. Biol.* 1995, **27**, 1303.

22. R. Matsushima-Nishiwaki, Y. Shidoji, S. Nishiwaki, T. Yamada, H. Moriwaki and Y. Muto, Suppression by carotenoids of microcystin-induced morphological changes in mouse hepatocytes. *Lipids.* 1995, **30**, 1029.

23. J. Levy, E. Bosin, B. Feldman, Y. Giat, A. Miinster, M. Danilenko and Y. Sharoni, Lycopene is a more potent inhibitor of human cancer cell proliferation than either α–carotene or β–carotene. *Nutr. & Cancer.* 1995, **24**, 257.

24. Y. Sharoni and J. Levy, Lycopene, the major tomato carotenoid, inhibits endometrial and lung cancer cell growth. in "Proceedings of the XVI international cancer congress" (eds. R. S. Rao, M. G. Deo and L. D. Sanghvi) *Monduzzi Editore, Bologna.* 1994, **1** 641.

25. H. Nagasawa, T. Mitamura, S. Sakamoto and K. Yamamoto, Effect of lycopene on spontanous mammary tumor development in SHN virgin mice. *Anticancer Res.* 1995, **15**, 1173.

26. T. Kobayashi, K. Itjima, T. Mitamura, K. Torilzuka, J. Cyong and H. Nagasawa, Effect of lycopene, a carotenoid, on intrathymic T cell differentiation and peripheral CD4/CD8 ratio in a high mammary tumor strain of SHN retired mice. *Anticancer drugs.* 1996, **7**, 195.

27. T. Narisawa, Y. Fukaura, M. Hasebe, et al., Inhibitory effects of natural carotenoids, α–carotene, β-carotene, lycopene and lutein, on colonic aberrant crypt foci formation in rats. *Cancer Lett.* 1996, **107**, 137.

28. D. J. Kim, N. Takasuka, J. M. Kim, et al., Chemoprevention by lycopene of mouse lung neoplasia after combined initiation treatment with DEN, MNU and DMH. *Cancer Lett.* 1997, **120**, 15.

29. Y. Sharoni, E. Giron, M. Rise and J. Levy, Effects of lycopene enriched tomato oleoresin on 7,12-dimethyl-benz[a]anthracene-induced rat mammary tumors. *Cancer Detect. Prevent.* 1997, **21**, 118.

30.   P. Di Mascio, S. Kaiser and H. Sies, Lycopene as the most efficient biological carotenoid singlet oxygen quencher. *Arch. Biochem. Biophys.* 1989, **274**, 532.

31.   F. Bohm, J. H. Tinkler and T. G. Truscott, Carotenoids protect against cell membrane damage by the nitrogen dioxide radical. *Nature Medicine.* 1995, **1**, 98.

32.   A. E. M. Smedman, C. Smith, I. R. Davison and I. R. Rowland, Antigenotoxic effects of lycopene of human colon cells *in vitro. Anticancer Res.* Abs. 82. 1995, **15**, 1656.

33.   L. X. Zhang, R. V. Cooney and J. S. Bertram, Carotenoids up-regulate connexin-43 gene expression independent of their provitamin-A or antioxidant properties. *Cancer Res.* 1992, **52**, 5707.

34.   D. Kleinman, C. T. Roberts Jr., D. LeRoith, A. V. Schally, J. Levy and Y. Sharoni, Growth regulation of endometrial cancer cells by insulin-like growth factors and by the luteinizing hormone-releasing hormone antagonist SB–75. *Regulatory Peptides.* 1993, **48**, 91.

35.   E. Hershkovitz, M. Marbach, E. Bosin, et al., Luteinizing hormone-releasing hormone antagonists interfere with autocrine and paracrine growth stimulation of MCF-7 mammary cancer cells by insulin-like growth factors. *J. Clin. Endo. Metab.* 1993, **77**, 963.

36.   Y. Sharoni and J. Levy, Anticarcinogenic properties of lycopene. in "Natural antioxidants and food quality in atherosclerosis and cancer prevention" (eds. J. T. Kumpulainen and J. K. Salonen) *The Royal Society of Chemistry., Cambridge.* 1996, 378.

37.   M. Karas, M. Danilenko, D. Fishman, D. LeRoith, J. Levy and Y. Sharoni, Membrane associated IGFBP-3 inhibits insulin-like growth factor-I (IGF-I)-induced IGF-I receptor signaling in Ishikawa endometrial cancer cells. *J. Biol.Chem.* 1997, **272**, 16514.

38.   J. Levy, M. Karas, H. Amir, J. Giat, M. Danilenko and Y. Sharoni, Lycopene, the major tomato carotenoid, delays cell cycle in breast, lung and endometrial cancer cells. *Anticancer Res.* Abs. 80. 1995, **15**, 1655.

39.   C. H. Hennekens, J. E. Buring, J. E. Manson, et al., Lack of effect of long-term supplementation with β-carotene on the incidence of malignant neoplasms and cardiovascular disease. *New Engl. J. Med.* 1996, **334**, 1145.

40.   E. R. Greenberg, J. A. Baron, T. D. Tosteson, et al., Clinical trial of antioxidant vitamins to prevent colorectal adenoma. *New Engl. J. Med.* 1994, **331**, 141.

41.   O. P. Heinonen, J. K. Huttunen, D. Albanes, et al., Effect of vitamin E and beta carotene on the incidence of lung cancer and other cancers in male smokers. *New Engl. J. Med.* 1994, **330**, 1029.

42.   G. S. Omenn, G. E. Goodman, M. D. Thornquist, et al., Effects of a combination of beta carotene and vitamin A on lung cancer and cardiovascular disease. *New Engl. J. Med.* 1996, **334**, 1150.

INTAKES AND MODES OF ACTION OF OTHER ANTICARCINOGENIC DIETARY COMPOUNDS

L. O. Dragsted.

Institute of Food Safety and Toxicology, Danish Veterinary and Food Administration
Mørkhøj Bygade 19, DK-2860 Søborg, Denmark.

1 INTRODUCTION

Fruits and vetetables in the diet have been identified through epidemiological studies as highly important cancer protective factors in human diets. Foods of plant origin have also been found to contain several non-vitamin compounds with cancer preventive effects in animal studies, see figure 1 for examples. Animal studies have often been performed with only a single, high dose level of the preventive compound. Information on their modes of action can therefore be useful for the evaluation of potential low-dose effects. More detailed information on their modes of action is emerging for some of the preventive compounds, notably for some compounds formed from the glucosinolates in cruciferous vegetables and also for some simple terpenes, but biological actions of potential significance in cancer prevention are also known for various phenolic compounds and for allyl sulphides. Information on quantitaties of these compounds in human foods are quite scattered so that actual intake levels in various populations can be difficult to estimate. In the present mini-review an estimate of Danish intake levels for some of these compounds will be made and compared with effective levels in animal studies in the light of existing mechanistic information.

2 DIETARY LEVELS OF ANTICARCINOGENS

Intake levels of a simple phenol, chlorogenic acid, of a simple terpene, limonene, of allyl sulphides, and of indolyl-, aromatic- and aliphatic glucosinolates were estimated in Denmark based on a recent dietary survey and reported analytical data from major food sources. The survey registered all food items ingested by a group of more than 1800 Danish men and women matched to the general population with respect to age, sex distribution, and social class (N. Lyhne Andersen, personal communication). Available analytical data from international journals are quite limited and may not be representative of levels in Danish foods. When more than one analytical report was identified data from countries geographically close to Denmark were selected. The calculated intakes may therefore be taken only as an estimate of the order of magnitude of human intakes in a Northern European population like the Danish.

2.1 Glucosinolates

Glucosinolates are mainly found in cruciferous vegetables and may belong to one of several classes, including indolylglucosinolates, other aromatic glucosinolates or aliphatic

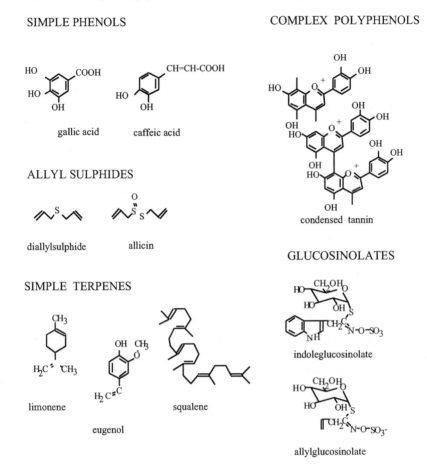

SIMPLE PHENOLS

gallic acid   caffeic acid

ALLYL SULPHIDES

diallylsulphide   allicin

SIMPLE TERPENES

limonene   squalene

eugenol

COMPLEX POLYPHENOLS

condensed tannin

GLUCOSINOLATES

indoleglucosinolate

allylglucosinolate

Figure 1
Some naturally occurring anticarcinogens.

glucosinolates. The intakes of aromatic glucosinolates from water cess and garden cress are exceedingly low, so only the other two groups of glucosinolates were included in table 1. Indolylglucosinolates break down to indole-3-carbinol and analogs upon ingestion, whereas aliphatic glucosinolates are degraded into isothiocyanates such as allyl isothiocyanate or sulphoraphane. It may be estimated that a fraction of about 50 % of the glucosinolates from ingested vegetables end up as indoles, including the vitamin C adduct, ascorbigen,[1] whereas less than 30 % by weight is transformed into isothiocyanates.[2]

Table 1

Calculated intakes of indolylglucosinolates(IGs)) and aliphatic glucosinolates (AGs) in Denmark.

Source	Average Intakes	Contents of IGs[3]	Daily dose	Contents of AGs[4]	Daily dose
	g/d	µg/g	mg/d	µg/g	mg/d
cauliflower	3.9	500	2	250	1
broccoli	2.3	1500	3.5	3400	7.8
white cabbage	4.4	1800	8	2900	12.8
others	~3	1500	4.5	2000	6
Total			~18		~28

2.2 Simple phenolic acids

Simple phenolic acids include ferulic, caffeic, and chlorogenic acid among others. They are quite widespread in fruits, vegetables and grain, but contents vary.[5] Some of the major sources of chlorogenic acid in the Danish diet are shown in table 2 together with a calculation of the average daily dose. As it may be seen, intakes of simple phenols may be quite large compared with the other compounds dealt with here, and taken together average intakes of simple phenolic acids may reach a level of several grams per day. Important major sources contributing to very large intakes of simple phenolic acids are wines (gallic acid),[6] coffee (chlorogenic acid),[7] apples (several),[8] grains and vegetables (caffeic acid).[9] A systematic quantitative survey of the contents of these phenolics in foods is, however, lacking, and such a study might lead to corrections of this present picture. A large fraction of the phenolic acids are present in foods in the form of esters with other phenols, and the fraction which is liberated in the gut is not well known.

Table 2

Calculated intakes of chlorogenic acid in Denmark.

Source	Contents	Average Intakes	Daily dose
	µg/g	g/d	mg/d
boiled potatoes	30-70	125	3.8-8.8
apples and juices	100-800	60	6-48
coffee	1000	750	750

2.3. Simple terpenes

Simple terpenes are mainly found in herbs and in citrus fruits, but again, systematic quantitative surveys are lacking from the literature. The major simple terpene in the diet is probably limonene which is found in all citrus products. Orange products are quantitatively the largest contributor (Table 3). Since terpenes are not very polar and the highest levels are found in the peels, we seem to get most of this compound from flavourings containing orange oil, which is composed of about 80 % limonene. The figures on orange oil intakes in table 3 are from a recently published database on natural toxicants.[10]

Table 3

Calculated intakes of limonene in Denmark.

Source	Contents	Average Intakes[a]	Daily dose
	µg/g	g/d	mg/d
orange pulp	50	16	0.8
orange juice	100	24	2.4
orange oil + flavourings	800,000	0.027	21.6
Total			~25

2.4. Allyl sulphides

Little quantitative data exist on the contents of allyl sulphides in onions and garlic, which seem to be their major dietary sources. It has been published that one liter of onion oil is produced from 1000 kg of onions.[50] Onion oil has a very high content of allyl sulphides but their concentration in onions must still be below 1ppm.

3 MODES OF ACTION

3.1. Anticarcinogens from glucosinolates

Glucosinolates break down to a range of degradation products upon chewing and cooking (Fig. 2). Indolylglucosinolates break down to indole-3-carbinol and related compounds, which can dimerize or oligomerize into several compounds with potent biological actions. One of the dimers, indolo[3,2-*b*]carbazole (ICZ), is the most potent natural *Ah*-receptor agonist known.[11,12] Induction of a range of enzymes involved in biotransformation of xenobiotic compounds, including CYP1A, glucoronosyltransferase type 1 (UDPGT1), sulfotransferases and glutathione-S-transferases are mediated through this receptor. Since CYP1A can both activate and detoxify procarcinigens, it is not entirely clear whether this is positive or negative. Like other known *Ah*-receptor agonists indole-3-carbinol oligomers may promote cancer at high dose levels, whereas low levels seem to be protective. This dual action is caused by the induction of liver hypertrophy in rodents at very high doses and seems irrelevant at human dietary levels, which are smaller by about 3-4 orders of magnitude. Indole-3-carbinol at doses of 50mg/kg bw/day has been found to significantly reduce tumour multiplicity or incidence in two studies. It decreased the number of liver tumours in ACI/N rats when given before and simultaneously with N-nitrosodiethylamine,[13] and it decreased the number of tumours of the tongue when applied either before or after a dose of 4-Nitroquinoline-N-oxide.[14] Doses at a similar level were also found to reduce the incidence of dimethylbenz[*a*]anthracene (DMBA)-induced breast tumours in SD rats and the incidence of Benzo[*a*]pyrene induced forestomach tumours in ICR/Ha mice.[15] This level is also similar to the effective dose levels observed to induce CYP1A related enzyme activities in rats,[16] whereas somewhat lower doses have been found to induce such activities in humans.[17-18] In a study on the ability of I3C to inhibit spontaneous endometrial tumours in Donryo rats, a dose-dependent decrease in response was observed in the dose range 10-50 mg/kg bw/day, but the decrease in incidence was only significantly reduced (by 63 %) at the highest dose level.[19] The result of this study indicates that given a study with a large enough number of animals per group an antitumourigenic response of indole-3-carbinol might be evident at a somewhat lower dose level, possibly in the range of 1-10 mg/kg bw/day, which may also be close to the dose threshold for enzyme induction.

Other glucosinolates like sinigrin and glucoiberin form stable isothiocyanates as part of their degradation products which are formed when the vegetables are chewed. These isothiocyanates are very potent inducers of phase II enzymes such as glutathione-S-transferases and quinone reductase (QR) and the induction of QR has been used to rank their inducing potency.[20] Isothiocyanates containing sulphinyl groups are particularly potent inducers of these enzymes[21] and may be active *in vivo* at doses of a few mg/kg bw/day. This is on the same order of magnitude as the lowest reported anticarcinogenic dose levels of isothiocyanates.

3.2. Simple phenolic acids

Simple phenolic acids are potent antioxidants[22-23] and scavengers of electrophilic activated forms of polycyclic aromatic hydrocarbons (PAHs) *in vitro.*[24] They may also inhibit CYP1A like activities of mouse liver and skin *in vivo,*[25] and these effects may be involved in their ability to decrease cytogenetic effects of PAH in rodents.[26-27] Several studies have been performed to test the anticarcinogenic properties of simple phenolic acids. Very low doses, 8 mg/kg bw/day, of topically applied chlorogenic acid or ferulic acid were unable to inhibit skin papilloma formation in mice treated with an initiating dose of DMBA and subsequently with weekly applications of 12-*O*-tetradecanoylphorbol-13-acetate (TPA),[28] whereas significantly reductions in animals with tumours were observed at 20-30 times higher doses of caffeic acid or chlorogenic acid applied only during tumour promotion.[29] Dietary chlorogenic acid at 30 mg/kg bw/day have also been reported to inhibit completely colon adenocarcinomas and reduce adenomas by 50 % in methyl azoxymethanol treated SG hamsters.[30] The mechanism behind these anticarcinogenic actions has not been elucidated but may be related to inhibition of carcinogen activation or antioxidant and scavenging actions.

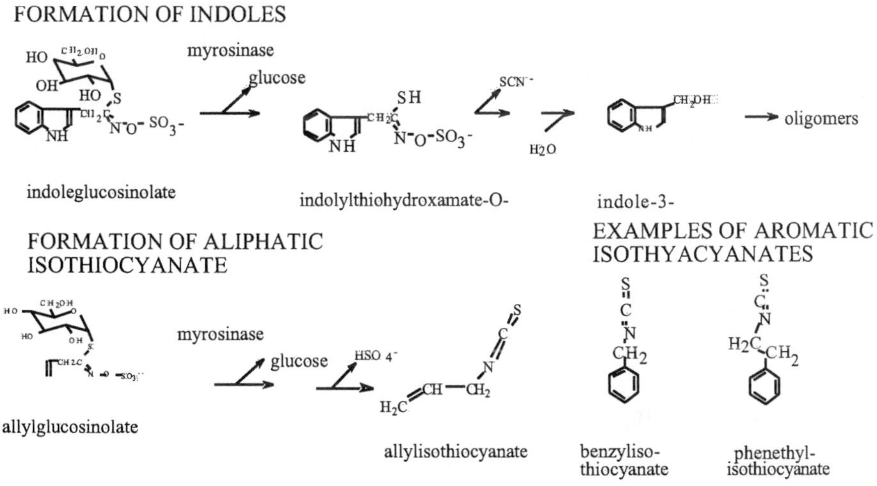

Figure 2. *The degradation of glucosinolates following disruption of the cells from cruciferous vegetables.*

3.3. Simple terpenes

The simple terpenes are potent antioxidants and have several biological effects related to anti-initiation and antipromotion of tumours.[31] The most studied compound in terms of antitumourigenic activities is limonene, which is a weak anti-tumour promoter against mammary tumours in the rat[32] and an inhibitor of aberrant crypt foci in the rat colon.[33] Limonene has also anti-initiating activities. Limonene is also by itself a tumour promoter at very high dose levels in the male rat kidney due to its ability to induce alpha-2 microglobulin,[34] but this specific male rat cancer induction mechanism is not relevant to humans. Limonene is a very weak inhibitor of farnesyl transferase which is necessary to transfer a farnesyl moiety to the *ras* p21 protein for its insertion into the plasma membrane. Some of the biotransformation products of limonene, perillyl alcohol and particularly the methyl ester of perillic acid, are much stronger inhibitors of p21 farnesylation[35] and proper insertion of the *ras* protein which according to the Vogelstein model of colon carcinogenesis is necessary at the adenoma II stage of tumour development. Perillyl alcohol is also a much stronger inhibitor of tumour promotion than limonene and it seems plausible that biotransformation of limonene into more potent antipromoters is a prerequisite for its action as a antitumourigen.[36]

3.4. Allyl sulphides

There may be several mechanisms behind the anticarcinogenic action of diallyl sulphide and analogs from *Allium* species. They are potent antioxidants,[37-38] they can inhibit some CYP isoenzymes,[39-40] induce glutathione dependent phase II enzymes[41-42] and antioxidant enzymes,[43] and the analog, diallyl disulphide can influence the processing and membrane insertion of *ras* p21.[44] Diallyl sulfide and analogs have been observed to inhibit dimethyl hydrazine induced colon cancer in mice,[45] nitrosamine-induced esophageal[46] and mammary cancers[47] in rats. Their effects seem to be most pronounced at the promotion stage in carcinogenesis, but anti-initiation due to inhibition of CYP2E1-mediated nitrosamine activation has also been reported.[48] Some *allium* organosulphur compounds have also been observed to increase liver cell proliferation and tumour promotion in the rat liver.[49] Such dual actions of antioxidants which act on the promotion stage of carcinogenesis are known for a range of structurally different compounds, e.g. butylated hydroxytoluene in the mouse lung, beta-carotene in human lung, caffeic acid in mouse forestomach etc., but there are no definitive explanations to the phenomenon. Early confidence in their antitumourigenic actions came from anti-tumour promotion studies with onion and garlic extracts corresponding to doses around 150 mg/kg bw/day of allyl sulphides,[50] but the contents of organoselenium compounds have later been observed also to be very important for the anticarcinogenic actions of these extracts.[51]

4 COMPARISON OF DIETARY LEVELS AND EFFECTIVE LEVELS IN ANIMAL STUDIES

The Danish median intake levels and the lowest active antitumourigenic dose levels reported from animal studies have been extracted from the text and tables above into table 4. As it can be seen there are many orders of magnitude between potentially antitumourigenic doses of allyl sulphides and human intakes and one or two orders of magnitude for glucosinolate degradation products and for limonene, whereas there is only a factor of about three for chlorogenic acid. Since intake levels vary greatly within the population, individuals with high intakes of certain foods or beverages might attain intake levels similar to those leading to significant tumour prevention in rodents.

In the case of glucosinolates the prediction would be that high intakes of cruciferous vegetables should be sufficient for enzyme induction in humans and this has also been observed by several authors.[52-54] Consequently, if xenobiotic metabolising enzyme induction can reduce cancer risk in humans, individuals with a very high regular intake levels of cruciferous vegetables might be at a decreased risk due to their glucosinolate contents.

In the case of chlorogenic acid which comes mainly from coffee and which seems to be particularly active against colon cancer, the prediction would be that high coffee consumption might decrease the risk of this disease. This is in contrast to the general belief which ranks coffee as a colon cancer risk factor. Interestingly, however, coffee was found to be inversely related to colon cancer risk in a prospective cohort study from Norway, where the habit is to prepare the coffee rather strong compared with northern American habits.[55]

Table 4

Comparison between lowest reported effective doses in animal studies and calculated dietary intakes in Denmark.

Compound	Dietary intake, mg/kg/d	Min. effective dose, mg/kg/d
limonene	~0.4	2.0
chlorogenic acid	~13	30
indole-3-carbinol	~0.3	50
Isothiocyanates	<0.47	7.5
allyl sulphides	<0.0017	150

5 CONCLUSIONS

Dietary anticarcinogenic compounds have several mechanisms of action by which they might decrease cancer risks, e.g. by induction of xenobiotic metabolizing enzymes or conjugating enzymes, through antioxidant effects, by decreasing carcinogen-DNA binding, by decreasing inflammatory responses or by increasing intercellular communication. Some of these effects have in fact been observed in both animals and man after intakes of crucifers or after intake of specific glucosinolate breakdown products. However, most glucosinolate breakdown products have toxic side effects which may be related to their cancer preventive effects, and the overall evaluation of these compounds will therefore have to take potential toxicity into account. For other potential cancer preventive compounds, such as the phenolic acids from coffee and the polyphenols from tea, the evidence in favor of anticarcinogenic actions in animals are fairly convincing, even at low doses, whereas human evidence from observational studies seems to indicate that such compounds have only limited cancer preventive activity at physiological intake levels. Taken together, there is some promise that cancer preventive compounds of relevance to humans can be identified experimentally and that the impact of dietary intake levels can be assessed, but current animal and human study techniques may be inadequate to identify them and to assess their impact unambiguously.

In conclusion, non-vitamin anticarcinogens from various plant food sources seem to inhibit tumourigenesis in rodents by diverse mechanisms and at both early and later stages of carcinogenesis. The median dietary intake levels of these compounds in northern European countries are generally one or more orders of magnitude lower than reported active dose levels in rodents, but the combined effect of several of these compounds -such as they are found in diets rich in fruits and vegetables- may explain some of the beneficial effects of these diets on human cancer risk.

Acknowledgements

The author wishes to thank Niels Lyhne Andersen for retrieval of Danish food intakes. This work was supported by a Danish Food Technology (FØTEK2) grant to the author.

References

1.   Vang O. and Dragsted L. O. ,Naturally occurring antitumourigens III. Indoles', Nordic Council of Ministers, Copenhagen, 1996.

2.   Strube M. and Dragsted L.O. 'Naturally occurring antitumourigens II. Isothiocyanates'. Nordic Council of Ministers, Copenhagen, 1994.

3.   Otte J. Dissertation, Royal Veterinary and Agricultural University, Copenhagen, 1991.

4.   Sones K., Heaney R.K. & Fenwick G.R. *J.Sci.Food Agric.* 1984, **35**, 762-766.

5.   Strube M., Dragsted L.O., and Larsen J.C. 'Naturally occurring antitumourigens I. Plant Phenols', Nordic Council of Ministers, Copenhagen, 1993.

6.   Stich H.F. & Powrie D. (1983) in: *Carcinogens and mutagens in the environment, Vol. 1, Food products* (Stich H.F., ed.), CRC Press,Inc., Boca Raton, Florida., pp. 136-145.

7.   Challis B.C. & Bartlett C.D. *Nature.* 1975, **254**, 532-533.

8.   Mosel H.D. & Herrmann K., *Z.Lebensm.Unters.Forsch.* 1974, **154**, 6-11.

9.   Herrmann K., *Z.Lebensm.Unters.Forsch.* 1988, **186**, 1-5.

10.   Gry J., Kovatsis A., Jongen W., Møller A., Rhodes M., Rosa E., and Rosner H. 'Inherent Food Plant Toxicants (NETTOX)' Danish Veterinary and Food Administration, Søborg, 1998.

11.   Bjeldanes L.F., Kim J.Y., Grose K.R., Bartholomew J.C. & Bradfield C.A. *Proc.Natl.Acad.Sci.U.S.A.* 1991, **88**, 9543-9547.

12.   Gillner M., Bergman J., Cambillau C., Alexandersson M. & Fernström B. *Mol.Pharmacol.* 1993, **44**, 336-345.

13.   Tanaka T., Mori Y., Morishita Y., Hara A., Ohno T., Kojima T. & Mori H. *Carcinogenesis.* 1990, **11**, 1403-1406.

14.   Tanaka T., Kojima T., Morishita Y. & Mori H. *Jpn.J.Cancer Res.* 1992, **83**, 835-842.

15.   Wattenberg L.W. & Loub W.D. *Cancer Res.* 1978, **38**, 1410-1413.

16.   Bradfield C.A. & Bjeldanes L.F.*Food Chem.Toxicol.* 1984, **22**, 977-982.

17. Bradlow H.L., Michnovicz J.J., Halper M., Miller D.G., Wong G.Y. & Osborne M.P. *Cancer Epidemiol.Biomarkers.Prev.* 1994 , **3**, 591-595.
18. Michnovicz J.J. & Bradlow H.L.*J.Natl.Cancer Inst.* 1990, **82**, 947-949.
19. Kojima T., Tanaka T. & Mori H.*Cancer Res.* 1994, **54**, 1446-1449.
20. Talalay P., De Long M.J. & Prochaska H.J. *Proc.Natl.Acad.Sci.U.S.A.* 1988, **85**, 8261-8265.
21. Zhang Y., Talalay P., Cho C.G. & Posner G.H. *Proc.Natl.Acad.Sci.U.S.A.* 1992, **89**, 2399-2403.
22. Fraga C.G., Martino V.S., Ferraro G.E., Coussio J.D. & Boveris A., *Biochem. Pharmacol.* 1987, **36**, 717-720.
23. Miller N.J., Diplock A.T. & Rice E.C.A. *Journal of Agricultural and Food Chemistry.* 1995, **43**, 1794-1801.
24. Wood A.W., Huang M.T., Chang R.L., Newmark H.L., Lehr R.E., Yagi H., Sayer J.M., Jerina D.M. & Conney A.H. *Proc.Natl.Acad.Sci.USA*. 1982, **79**, 5513-5517.
25. Mukhtar H., Das M., Del T.-B.J., Jr. & Bickers D.R. *Xenobiotica.* 1984, **14**, 527-531.
26. Raj A.S., Heddle J.A., Newmark H.L. & Katz M. *Mutat.Res.* 1983, **124**, 247-253.
27. Wargovich M.J., Eng V.W. & Newmark H.L. *Food Chem.Toxicol.* 1985, **23**, 47-49.
28. Lesca P. *Carcinogenesis.* 1983, **4**, 1651-1653.
29. Huang M.T., Smart R.C., Wong C.Q. & Conney A.H. *Cancer Res.* 1988, **48**, 5941-5946.
30. Mori H., Tanaka T., Shima H., Kuniyasu T. & Takahashi M. *Cancer Lett.* 1986, **30**, 49-54.
31. Crowell P.L. *Breast Cancer Res.Treat.* 1997, **46**, 191-197.
32. Russin W.A., Hoesly J.D., Elson C.E., Tanner M.A. & Gould M.N. *Carcinogenesis.* 1989, **10**, 2161-2164.
33. Kawamori T., Tanaka T., Hirose Y., Ohnishi M. & Mori H. *Carcinogenesis.* 1996, **17**, 369-372.
34. Hard G.C. & Whysner J. *Crit.Rev.Toxicol.* 1994, **24**, 231-254.
35. Gelb M.H., Tamanoi F., Yokoyama K., Ghomashchi F., Esson K. & Gould M.N. *Cancer Lett.* 1995, **91**, 169-175.
36. Crowell P.L., Kennan W.S., Haag J.D., Ahmad S., Vedejs E. & Gould M.N.*Carcinogenesis.* 1992, **13**, 1261-1264.
37. Prasad K., Laxdal V.A., Yu M. & Raney B.L. *Mol.Cell Biochem.* 1995, **148**, 183-189.
38. Kourounakis P.N. & Rekka E.A. *Res.Commun.Chem.Pathol.Pharmacol.* 1991, **74**, 249-252.
39. Siess M.H., Le-Bon A.M., Canivenc L.M. & Suschetet M. *Cancer Lett.* 1997, **120**, 195-201.
40. Reicks M.M. & Crankshaw D.L. *Nutr.Cancer.* 1996, **25**, 241-248.
41. Singh S.V., Pan S.S., Srivastava S.K., Xia H., Hu X., Zaren H.A. & Orchard J.L. *Biochem.Biophys.Res.Commun.* 1998, **244**, 917-920.
42. Hu X., Benson P.J., Srivastava S.K., Xia H., Bleicher R.J., Zaren H.A., Awasthi S., Awasthi Y.C. & Singh S.V. *Int.J.Cancer.* 1997, **73**, 897-902.

43. Gudi V.A. & Singh S.V. *Biochem.Pharmacol.* 1991, **42**, 1261-1265.
44. Singh S.V., Mohan R.R., Agarwal R., Benson P.J., Hu X., Rudy M.A., Xia H., Katoh A., Srivastava S.K., Mukhtar H., Gupta V. & Zaren H.A. *Biochem. Biophys. Res. Commun.* 1996, **225**, 660-665.
45. Wargovich M.J.*Carcinogenesis.* 1987, **8**, 487-489.
46. Wargovich M.J., Imada O. & Stephens L.C. *Cancer Lett.* (1992) **64**, 39-42.
47. Schaffer E.M., Liu J.Z., Green J., Dangler C.A. & Milner J.A. *Cancer Lett.* 1996, **102**, 199-204.
48. Surh Y.J., Lee R.C., Park K.K., Mayne S.T., Liem A. & Miller J.A. *Carcinogenesis.* 1995, **16**, 2467-2471.
49. Takada N., Kitano M., Chen T., Yano Y., Otani S. & Fukushima S. *Jpn.J.Cancer Res.* 1994, **85**, 1067-1072.
50. Belman S. *Carcinogenesis.* 1983, **4**, 1063-1065.
51. Ip C., Lisk D.J. & Stoewsand G.S. *Nutr.Cancer.* 1992, **17**, 279-286.
52. Kall M., Vang O. & Clausen J. *Carcinogenesis.* 1996, **17**, 793-799.
53. Vistisen K., Loft S. & Poulsen H.E. *Adv.Exp.Med.Biol.* 1990, **283**, 407-411.
54. Vistisen K., Poulsen H.E. & Loft S. *Carcinogenesis,* 1992, **13**, 1561-1568.
55. Jacobsen B.K., Bjelke E., Kvale G. & Heuch I. *J.Natl.Cancer Inst.* 1986, **76**, 823-831.

INHIBITION OF CYP1A1 *IN VITRO* BY BERRIES WITH DIFFERENT QUERCETIN CONTENTS

Liisa Kansanen,[1,2] Hannu Mykkänen [2] and Riitta Törrönen[1,2].

1Department of Physiology and 2Department of Clinical Nutrition, University of Kuopio, P.O. BOX 1627, FIN-70211 Kuopio, Finland.

1 INTRODUCTION

Flavonoids, the potentially anticarcinogenic non-nutrients, are present in several wild and cultivated berries consumed in Finland.[1] In a previous study we showed that flavonol aglycones and acid-treated extracts of strawberry and black currant inhibit the enzymatic activity of CYP1A1 in Hepa-1 cell culture.[2] CYP1A1, an isozyme of cytochrome P450, is responsible for the metabolic activation of many carcinogens. In the present study we investigated the contribution of quercetin, the main flavonol in most berries,[3] to the inhibitory effect produced by berries.

2 MATERIALS AND METHODS

Six berries with different quercetin contents were studied: strawberry (*Fragaria x ananassa*), red raspberry (*Rubus idaeus*) and cloudberry (*Rubus chamaemorus*) of the family *Rosaceae*, bilberry (*Vaccinium myrtillus*) and lingonberry (*Vaccinium vitis-idaea*) of the family *Ericaceae*, and black currant (*Ribes nigrum*) of the family *Grossulariaceae*. The quercetin contents of these berries are given in Table 1.

Samples (5 g) of the frozen berries were extracted in 50 % (v/v) aqueous methanol and treated with 1.2 M HCl at 85 °C for 2 h with constant swirling to hydrolyze quercetin glycosides to aglycones,[3] or only extracted without the acid. Aliquots of 10 ml were evaporated to dryness, and the residues corresponding to 1 g of berry (fresh weight) were dissolved in 2 ml of DMSO (dimethyl sulphoxide).

The effects of the berry extracts and pure quercetin on the enzymatic activity of CYP1A1 were studied using the subclone Hepa-1c1c7 of the mouse hepatoma cell line Hepa-1. To increase the CYP1A1 level, the cells were pretreated with TCDD, a potent inducer of CYP1A1. The enzymatic activity of CYP1A1 was assayed as AHH (aryl hydrocarbon hydroxylase, i.e.hydroxylation of benzo(a)pyrene to its fluorescent 3-hydroxy metabolite).[4]

The berry extracts (corresponding to 1.25 to 12.5 mg of berry/assay) or pure quercetin (dissolved in DMSO) at final concentrations corresponding to those in the berry extracts (Table 1) were incubated with 100 μl of the cell suspension in the AHH reaction mixture. The concentration of DMSO did not exceed 2.5 %. DMSO alone was a solvent control. AHH activities detected in the presence of the berry extracts or pure quercetin were compared to that of the control (AHH activity produced by the TCDD-treated cells without the extracts or pure quercetin).

The possible interfering effect of the berry extracts on the fluoresence measurement was studied by adding them to the AHH reaction mixture after stopping the enzymatic reaction. No interference with the measurement of fluoresence of 3-hydroxybenzo(a)pyrene was observed.

3 RESULTS

A dose-dependent decrease of AHH activity was detected by all the berry extracts studied, while the solvent alone had no effect. The IC_{50} values (the inhibitor concentrations which produce 50 % inhibition) for the acid-treated berry extracts varied from 2.5 to 5.5 mg of berry/assay (Table 1). For the berry extracts not treated with the acid the IC_{50} values were somewhat higher, from 4.5 to 8.1 mg/assay.

Pure quercetin at the concentrations (7.5–1400 ng/assay) corresponding to those in the berries showed no inhibition.

Table 1

IC_{50} values (mg of berry/assay) for the inhibition of CYP1A1-dependent AHH activity by berries with different quercetin contents.

	Quercetin content[1]		IC_{50}	
	mg/100 g of berry (fresh weight)	ng/assay	extract treated with acid	extract not treated with acid
Cloudberry	0.6	7.5–75	3.6	7.4
Strawberry	0.7	8.7–87	3.5	4.5
Red raspberry	0.8	10–100	3.0	7.4
Bilberry	2.9	36–360	5.5	6.6
Black currant	4.4	55–550	2.5	6.4
Lingonberry	11	140–1400	5.5	8.1

[1] *analyzed by Häkkinen et al.[3]*

4 DISCUSSION

The results of this study show that the extracts of strawberry, red raspberry, cloudberry, lingonberry, bilberry and black currant contain inhibitors of the catalytic activity of the carcinogen-activating enzyme CYP1A1.

No apparent correlation was detected between the quercetin content in these berries and the inhibitory potency of the berry extracts. Although the quercetin contents of strawberry, red raspberry and cloudberry are low (<1 mg/100 g), after acid treatment they appeared to be more potent inhibitors than lingonberry or bilberry with up to 18-fold higher quercetin contents. Under these conditions (1.2 M HCl at 85 °C for 2 h) quercetin glycosides are hydrolyzed to aglycones.[3]

Previously we reported that acid-treated extracts of freeze-dried strawberry and black currant, in spite of their different phenolic profiles,[5] showed similar inhibition properties. A 50 % inhibition was produced by the extracts corresponding to 1 mg of berry (dry weight).[2] Also in this study, no clear difference was observed between the inhibitory potencies of strawberry and black currant. Despite the 6-fold difference in the quercetin content of strawberry and black currant, the half-maximal inhibition was produced by the acid-treated extracts corresponding to 3.5 and 2.5 mg of berry (fresh weight), respectively.

The concentrations of pure quercetin simulating the concentrations in the berry extracts (7.5–1400 ng/assay) produced no inhibition of AHH activity. Obviously, the quercetin concentrations present in the berry extracts were too low to produce the inhibition of AHH activity observed in this test system.

The results of this study suggest that the inhibition of CYP1A1-dependent AHH activity by these berries cannot be explained by their content of quercetin. At present, the inhibitor(s) remain unknown. However, the effect of the inhibitor(s) seems to be potentiated by acid treatment of the berry extracts.

References
1.   R. Törrönen, S. Häkkinen, S. Kärenlampi, and H. Mykkänen. *Cancer Letters.* 1997, **114**, 191.
2.   L. Kansanen, H. Mykkänen and R. Törrönen. Flavonoids and extracts of strawberry and black currant are inhibitors of the carcinogen-activated enzyme CYP1A1 in vitro. In: Natural Antioxidants and Food Quality in Atherosclerosis and Cancer Prevention, J.Kumpulainen and J.Salonen. *The Roayl Society of Chemistry, Cambridge.* 1996.
3.   S. Häkkinen, H. Mykkänen, S. Kärenlampi and R. Törrönen, in this issue.
4.   P. Kopponen, E. Mannila and S. Kärenlampi. *Chemosphere.* 1992, **24**, 201.
5.   S. Häkkinen, H. Mykkänen, S. Kärenlampi, M. Heinonen and R. Törrönen. HPLC method for screening of flavonoids and phenolic acids in berries: phenolic profiles of strawberry and black currant. In: Natural Antioxidants and Food Quality in Atherosclerosis and Cancer Prevention, J. Kumpulainen and J. Salonen. The Roayl Society of Chemistry, Cambridge, 1996.

CHEMOPROTECTIVE PROPERTIES OF COCOA AND ROSEMARY POLYPHENOLS

E. A. Offord, T. Huynh-Ba, O. Avanti and A. M.A. Pfeifer.

Nestlé Research Center ,P.O.Box 44, Vers-chez-les-Blanc, 1000 Lausanne 26.

1 INTRODUCTION

Polyphenolic compounds are widespread components of the human diet found in vegetables, fruits, cereals, spices and beverages such as wine, tea, coffee and cocoa.[1] Many of these compounds show both antioxidant and anticancer activities,[2,3] a typical example being green tea.[4,5] Cocoa is also a rich source of polyphenols which makes a major contribution to colour and flavour.[6] The polyphenolic composition of the cocoa bean includes anthocyanins, anthocyanidins, leucoanthocyanidins, flavonols, phenolic acids and flavonols.[6] The major flavonol in cocoa is (-)-epicatechin. Cocoa has antioxidant activity and modulates immune function *in vitro*.[7-9]

In this study, the chemoprotective potential of cocoa polyphenols was tested in *in vitro* liver and bronchial cell models with the procarcinogen aflatoxin B_1 (AFB_1). The immortalized liver epithelial cells (THLE) or bronchial cells (BEAS-2B) were previously stably transfected with individual cytochrome P450 (CYP450) enzymes to render them competent for metabolic activation of carcinogens.[10,11] Both the THLE and BEAS-2B cells express endogenous phase II enzymes such as quinone reductase (QR) and glutathione S-transferase (GST), although the isomers are not the same (GSTμ (M3) in THLE cells: GSTπ in BEAS-2B cells).[12,13] Cocoa extract was tested for its ability to inhibit AFB_1–induced DNA adduct formation by inhibiting the incorporation of tritiated metabolites of AFB_1 into DNA. The mechanism of action was further explored by testing the effect of cocoa extract on phase I and phase II enzymes.

These *in vitro* models were previously validated with respect to primary hepatocytes and to the well-characterized chemoprotective agent oltipraz.[14] Futhermore, a polyphenolic extract of the spice rosemary, was shown to act similarly to oltipraz in inhibiting the genotoxicity of carcinogens aflatoxin B_1 and benzo(a)pyrene through a dual mechanism: blocking of the metabolic activation stage through inhibition of phase I enzymes and induction of detoxifying, phase II enyzmes.[12-14] Here, the effects of cocoa and rosemary extracts are compared.

2 METHODS

2.1. Cells and Extracts

Generation, characterization and culture conditions of the CYP1A2-expressing THLE and BEAS-2B cell lines (T5-1A2, B.1A2) are fully described by Macé *et al.*[10,11] Cells (2 x 10⁶) were seeded in 6 cm dishes 24 h prior to treatment.

Cocoa extract was prepared from partially defatted cocoa powder by extraction with ethanol/water (8:2; v:v), heating, filtration, solvent concentration and lyophilization. The total polyphenol content of the cocoa extract was 131.4 mg/g in gallic acid equivalents (GAEs) as determined by the Folin-Ciocalteu method. The epicatechin content was 2.9 mg/g extract (dosed by HPLC analysis). For use in cell culture, the cocoa extract was made up as a stock solution at 100 mg/ml in DMSO and stored at -20 °C. The final concentration in the cell medium was 0.05-0.50 mg/ml which was equivalent to 0.5-5.0 μM epicatechin. Rosemary extract was prepared as previously described.[12,13]

2.2. Assays

Methods for measurement of DNA adducts, CYP450 enzymes and GST have all been described previously.[10,12,13,15] Regulation of the antioxidant-responsive element (ARE) was tested in BEAS-2B bronchial cells transiently transfected by the lipofection technique with 10 μg plasmid ARE-CAT containing the ARE element from the upstream promoter region of the human quinone reductase gene linked to the reporter gene chloramphenicol acetyl transferase[16] (kindly donated by G. Williamson, Inst Food Research, Norwich). After 24 h, the cells were treated or not with cocoa or rosemary extracts. Incubation continued for a further 24 h before harvesting and assay of chloramphenicol acetyl transferase (CAT).

3 RESULTS

3.1. Inhibition of AFB$_1$-DNA adduct formation

DNA adduct formation was assessed by the incorporation of ^{3}H-AFB$_1$ metabolites into DNA. Cells were preincubated for 1 h with cocoa extract followed by a coincubation with tritiated carcinogen overnight. Control cells were incubated with ^{3}H-AFB$_1$ alone. In untreated liver (T5-1A2) or bronchial (B-1A2) cells the level of DNA adduct formation was 4.9 and 4.2 pmol/mg DNA, respectively. In the presence of cocoa extract, DNA adduct formation was inhibited in a dose-dependent manner in T5-1A2 cells reaching 70 % inhibition with 0.5 mg/ml extract (see Fig. 1). In B-1A2 cells, the inhibition of DNA adduct formation was more marked than in the T5-1A2 cells and reached 90 % inhibition with 0.5 mg/ml extract (see Fig. 1). Rosemary extract (RE-S) at 0.005 mg/ml inhibited DNA adducts by 85 % and 30 % in B-1A2 and T5-1A2 cells, respectively.[13]

3.2. Effect of Cocoa Extract on Phase I and Phase II enzymes

Inhibition of DNA adduct formation may result from at least two mechanisms: (i) inhibition of the CYP450 enzymes (ii) induction of the detoxification enzymes such as GST. The CYP1A2-expressing liver or bronchial cells were incubated with or without cocoa extract for 18 h and methoxyresorufin O-de-ethylase (MROD) activity measured in intact cells. In control cells, MROD activity was 4.8 and 4.2 pmol/mg/min in T5.1A2 and B.1A2 cells, respectively. Treatment of cells with cocoa extract had no effect on MROD activity. RE-S, on the other hand, inhibited CYP1A2 and CYP3A4 activities.[13]

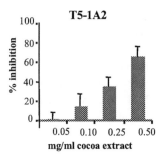
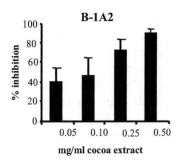

Figure 1

Inhibition of AFB_1-DNA adducts by cocoa extract in CYP1A2 expressing liver (T5-1A2) and bronchial (B-1A2) cells. The control level of DNA adducts is given in the text. The error bars represent the standard deviation calculated from at least three independent experiments. The percentage inhibition of DNA adduct formation by cocoa extract was calculated as follows: **%** inhibition = 100 − [(dpm/mg DNA test cells) x 100]
[(dpm/mg DNA control cells)

The effect of cocoa extract on phase II enzymes was tested in two ways: (i) expression of GST mRNA in BEAS-2B cells (ii) modulation of the antioxidant responsive element (ARE). To test the effect of cocoa extract on GST expression, BEAS-2B cells were incubated for 24 h with 0.1-0.5 mg/ml cocoa extract, RNA extracted and analysed by Northern blotting. As shown in Figure 2, the mRNA for GSTπ was induced by cocoa extract reaching 2.4 fold induction with 0.5 mg/ml extract (lanes 3-5) compared to untreated, control cells (lanes 1, 2). RE-S at 0.005 mg/ml induced GST by 1.5 fold (lane 6). Equivalent concentrations (1 μg/ml or 3 μM) of carnosol or carnosic acid, the principal active components of RE-S, induced GSTπ by 2-fold.[12]

Many phase II enzymes are transcriptionally regulated through the ARE in the upstream promoter sequence.[17,18] To test if cocoa extract had an effect on the ARE, BEAS-2B cells were transiently transfected with 10 μg of the reporter plasmid, ARE-CAT and subsequently incubated with cocoa extract for 24 h. The relative CAT activity was compared (see Fig. 3). Cocoa extract (0.5 mg/ml) induced the CAT activity by 2-fold (lane 5). In comparison, incubation with rosemary extract (0.005 mg/ml) led to a 4-fold induction of ARE-CAT (lane 6).

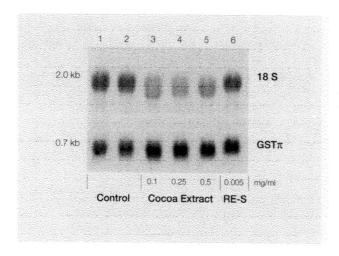

Figure 2

Northern blot analysis of RNA from bronchial cells treated as follows: control, untreated cells (lanes 1,2), 0.1-0.5 mg/ml cocoa extract (lanes 3-5), 0.005 mg/ml rosemary extract (RE-S), (lane 6). The blot was first probed for the GSTπ mRNA (0.7 kb, lower panel), then stripped and probed for the control ribosomal 18S mRNA (2.0 kb, top panel).

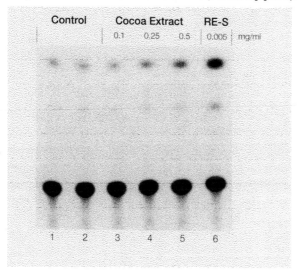

Figure 3

Activation of the antioxidant responsive element (ARE) by cocoa and rosemary extracts. Bronchial cells were transfected with 10 μg of the reporter plasmid ARE-CAT and subsequently incubated with control medium (lanes 1,2), 0.1-0.5 mg/ml cocoa extract (lanes 3-5) or rosemary extract (lane 6). The figure shows a higher intensity of acetylated chloramphenicol forms in the presence of the cocoa or rosemary extracts compared to the control.

4 SUMMARY AND DISCUSSION

Cocoa extract (0.5 mg/ml) significantly inhibited (70-90 %) AFB_1-induced genotoxicity in CYP1A2-expressing human liver (T5-1A2) or bronchial (B-1A2) cells. Equivalent concentrations of epicatechin (5 μM) had no effect on DNA adduct formation in this assay, suggesting that the active compound was another of the polyphenols. The cocoa extract had no effect on the metabolic activation step catalyzed by CYP1A2 but did induce phase II detoxifying enzymes shown through induction of GST and transcriptional activation of the antioxidant responsive element. A further mechanism of action of cocoa polyphenols may be direct scavenging of reactive intermediates in aflatoxin metabolism.

The potency of the cocoa extract was approximately 100-fold less than the rosemary extract on a gram weight basis in both the inhibition of DNA adducts and the induction of phase II enzymes. However, the level of consumption of these two products in the diet would also be considerably different. The greater inhibition of DNA adducts by both extracts in the bronchial cells compared to the liver cells may be due to the greater expression and inducibility of phase II enzymes such as GST in the bronchial cells. An important difference between the two extracts was the inhibition of phase I enzymes by rosemary but not cocoa extract. Although blocking of metabolic activation of a procarcinogen through inhibition of CYP450 enzymes is a powerful mechanism of chemoprotection, these same enzymes are also important for detoxification reactions in the liver. Therefore, it would be worthwhile to identify the agents in cocoa which can specifically induce the phase II enzymes without affecting phase I.

References

1. D. E. Pratt, "Phenolic compounds in food and their effects on health II: Antioxidants & cancer prevention" M.-T. Huang, C.-T Ho and C.Y.Lee, eds., Washington DC. 1992, ACS Symposium Series Vol. 507, p54.

2. C.A. Rice-Evans, N.J. Miller and G. Paganga. *Trends in plant science.* 1997, **2(4)**, 152–159.

3. H.L. Newmark, "Dietary phytochemicals in cancer prevention and treatment", Plenum Press, New York and London, 1996. Advances in experimental medicine and biology Vol. 401, chapter 3, p. 25.

4. S.A Wiseman, D.A. Balentine, and B. Frei. *Critical Reviews in Food Science and Nutrition.* 1997, **37(8)**, 705–718.

5. I.E. Dreosti, M.J. Wargovich and S.S. Yang. *Critical Reviews in Food Science and Nutrition.* 1997, **37(8)**, 761–769.

6. J.A. Williams. *J. Biol. Appl. Chem.* 1971, **14(1)**, 10-19.

7. A.L. Waterhouse, J.R. Shirley, J.L. Donovan. *Lancet.* 1996, **348**, 834.

8. K. Kondo, R. Hirano, A. Matsumoto, O. Igarashi and H. Itakura. *Lancet.* 1996, **349**, 1414.

9. C. Sanbongi, N. Suzuki and T. Sakane. *Cell. Immunol.* 1997, **177**, 129–136.

10. K. Macé, F.J. Gonzales, I.R. McConnell, R.C. Garner, O. Avanti, C.C. Harris, and A.M.A. Pfeifer. *Mol. Carcinogenesis.* 1994, **11**, 65–73.

11. K. Macé, F. Aguilar, J.-S. Wang, P. Vautravers, M. Gomez-Lechon, F.J. Gonzalez, J. Groopman, C.C. Harris and A.M.A. Pfeifer. *Carcinogenesis.* 1997, **18**(7), 1291–1297.

12. E.A. Offord, K. Macé, C. Ruffieux, A. Malnoe and A.M.A. Pfeifer, *Carcinogenesis,* 1995, **16**(9), 2057–2062.

13. E. A. Offord, K. Macé, O. Avanti and A M. A. Pfeifer. *Cancer Lett.* 1997, **114**, 275–281.

14. E. A. Offord, M. Gomez-Léchon, K. Macé, O. Avanti and A. M. A. Pfeifer. *Proceedings of the American Association for Cancer Research.* 1997, **38**, abst 2473.

15. S. Sharma, J. D. Stutzman, G. J. Kelloff and V. E. Steele. *Cancer Res.* 1994, **54**, 5848-5855.

16. B. Wang and G. Williamson. *Biochim. Biophys. Acta.* 1994, **1219**, 645-652.

17. T.H. Rushmore, M.R. Morton and C.B. Pickett. *J. Biol. Chem.* 1991, **266**, 11632-11639.

18. W.W. Wasserman and W.E. Fahl. *Proc. Natl. Acad. Sci.* 1997, **94**, 5361-5366.

ANTIOXIDANT COMPOUND 4-NEROLIDYLCATECHOL INHIBITS *IN VITRO* KB CELLS GROWTH AND TOPOISOMERASE I ACTIVITY

E. Mongelli,[1] A. Romano,[2] C. Desmarchelier,[1] J. Coussio,[2] G. Ciccia.[1]

[1]Cátedra de Microbiología Industrial y Biotecnología, [2]Cátedra de Farmacognosia (IQUIMEFA-CONICET). Facultad de Farmacia y Bioquímica, Universidad de Buenos Aires, Junín 956 (1113) Buenos Aires, Argentina.

1 INTRODUCTION

Leaves of *Pothomorphe peltata* (L.) Miq. (Piperaceae) are widely used in the treatment of liver diseases and other inflammatory disorders in tropical South America.[1] Different extracts of this plant have shown to reduce oxidative stress when tested in *in vitro* free radical scavenging, lipid peroxidation and DNA damage models.[2-4] This activity has been attributed to the presence of catechol derivative 4-nerolidylcatechol (4-NC).

As part of a program to study bioactive antitumour compounds from plants, the cytotoxicity of the 4-NC was studied. Herein we report the results of the brine shrimp lethality test, the crown gall tumour and KB cells growth inhibition assays. The use of these assays has proved to be effective in order to detect cytotoxicity in plant extracts and isolated natural products.[5] Since interaction with DNA and inhibition of topoisomerase I activity have been described as pivotal mechanisms by which antitumour compounds exert their cytotoxic effects,[7,8] and in order to determine a possible mechanism of action, the methyl green-DNA bioassay and the DNA- topoisomerase I inhibition test were carried out.

2 MATERIALS AND METHODS

2.1. Plant material

Leaves of *P. peltata* were collected by one of the authors (C. Desmarchelier) in the district of Pozuzo, Cerro de Pasco, Peru, in August 1996.

2.2. Extraction and isolation

Extracts were prepared by extracting 5 g of dry powdered plant material during 24 h at room temperature in a stoppered flask with 50 ml MeOH/H$_2$O (1:1). The extract was filtered, concentrated under reduced pressure at 43 °C in a Savant Speed Vac Plus SC210A concentrator and finally freeze-dried in a Gamma A lyophilizer (Chriss, Germany). DMSO was used to pre-solubilize the extracts (final concentration = 1 % V/V). After concentration under reduced pressure, the extract was submitted to a chromatographic column (Silicagel 6OH, 1.5 x 30 cm) (Bäckström SEPARO AB, Sweden) under vacuum (5 x 5 cm) eluted with a gradient of CHCl$_3$/EtOAc (1 to 50). Based on the brine shrimp lethality bioassay (BST), the CHCl$_3$/EtOAc (95:5) fraction was identified as the most bioactive. The presence of 4-NC was detected by thin-layer chromatography (TLC)[9] using previously isolated 4-NC as a standard[4] and further submitted to the same chromatographic system described above. Pooled fractions containing 4-NC were concentrated under reduced pressure and the pure compound was detected by TLC.

2.3. Brine Shrimp Test (BST)

The BST was performed according to standard protocols.[10] The LC_{50} values in μg/ml (ppm) were determined using Finney probit analysis computer program.[11]

2.4. Potato disc bioassay

The *Agrobacterium tumefaciens* potato disc assay for tumour/antitumour induction was performed following procedure previously described.[12]

2.5. KB cells cytotoxicity

KB cells cytotoxicity was determined using human oral epidermoid carcinoma cells (KB)[13] provided by the University of Chicago at Illinois.

2.6. DNA-methyl green bioassay (DNA-MG)

Interaction with DNA was determined using the DNA-methyl green reagent.[14]

2.7. Inhibition of Topoisomerase I activity

Inhibition of topoisomerase I activity was determined using a relaxation assay,[15] with modifications.

3 RESULTS AND DISCUSSION

The *in vitro* cytotoxic activity in extracts and fractions of *P. Peltata* were studied using different bioassays. The methanolic extract was tested using the brine shrimp microwell cytotoxicity assay, showing an LC_{50} = 89 μg/ml. The extract was further submitted to bioassay guided fractionation using a silica gel column eluted with chloroform: ethanol. Thin layer chromatography (TLC) confirmed the presence of 4-NC in the active fraction (LC_{50} = 9.6 μg/ml). Further studies on the cytotoxic activity demonstrated a 22 % crown gall tumour inhibition for the crude extract, and an EC_{50} = 1,3 μg/ml in KB cells growth for the purified compound (EC_{50} = 123 μg/ml for the crude extract).

In order to see if the cytotoxicity observed was due to interaction of 4-NC with DNA, a colorimetric microassay was carried out using the Methyl Green-DNA (MG-DNA) complex. No intercalation with DNA could be observed. An inhibition in the activity of Topoisomerase I using agarose gel electrophoresis was detected in the presence of the purified compound, when compared to camptothecin, suggesting that this could be a possible mechanism for the cytotoxicity observed in KB cells.

Acknowledgements

Research supported by the International Foundation for Science (IFS), in Stockholm, Sweden (Grant agreement No. F/2628-1), BID-CONICET (grant PMT-SID0370) and the University of Buenos Aires (grant UBACYT FA002).

Reference

1.   C. Desmarchelier, A. Gurni, G. Ciccia and A.M. Giulietti. *J. Ethnopharmacol*. 1996, **52,** 45.
2.   C. Desmarchelier, E. Mongelli, J. Coussio and G. Ciccia. *Braz. J. Med. Biol.Res.* 1997, **30,** 85.
3.   S. Barros, D. Teixeira, A. Aznar, J. Moreira Jr, I. Ishii and C. Freitas. *Cienc.Cult. São Paulo*. 1996, **48,** 114.
4.   C. Desmarchelier, S. Barros, M. Repetto, L. Ribeiro Latorre, M. Kato, J. Coussio and G. Ciccia. *Planta Med.* 1997, **63,** 561.
5.   B. Falch, G. König, A. Wright, O. Sticher, C. Angerhofer, J. Pezzuto and H. Bachmann. *Planta Med*. 1995, **61,** 321.
6.   K. He, L. Zeng, G. Shi, G. Zhao, J. Kozlowski, and J. McLaughlin. *J. Nat. Prod.* 1997, **60,** 38.
7.   K. Bonjean, M. De Pauw-Gillet, R. Bassleer, J. Quentin-Leclercq, L. Angenot, and W. Wright. *Phytother. Res*. 1996, **10,** 159.
8.   K. Yagi. *Chem. Phys. Lipids*. 1987, **45,** 337.
9.   K. Duve and P. White. *J. Am. Oil Chem. Soc*. 1991, **68,** 365.
10. P. Solis, C. Wright, M. Anderson and J.D. Phillipson. *Planta Med*, 1993, **59,** 250.
11. J. McLaughlin. in: Methods in Plant Biochemistry (Harborne, J. B., ed.) pp. 1-32. Academic Press, New York, 1991.
12. N. Ferrigni, J. Putman, B. Andserson, L. Jacobsen, D. Nichols, D. Moore, J. McLaughlin, R. Powell, and C. Smith Jr. *J. Nat. Prod*. 1984, **45,** 679.
13. K. Likhitwitayawuid, C. Argenhoffer, G. Cordell, J. Pezzuto and N. Ruangrungsi. *J. Nat. Prod*. 1993, **56,** 30.
14. N. Burres, A. Frigo, R. Rasmussen and J. McAlpine. *J. Nat. Prod*. 1992, **55,** 1582.
15. L. Liu and K. Miller. *Proc. Natl. Acad. Sci. USA*. 1981, **78,** 3487.

PHYTOESTROGEN PROFILING FROM BIOLOGICAL SAMPLES USING HPLC WITH COULOMETRIC ELECTRODE ARRAY DETECTION

Tarja Nurmi*, Philip Lewis**, Kristiina Wähälä** and Herman Adlercreutz*.

* Department of Clinical Chemistry, University of Helsinki and Folkhälsan Research Center
** Department of Chemistry, University of Helsinki.

1 INTRODUCTION

Several methods for profiling biological samples using HPLC with coulometric electrode array detector (CEAD) have been published, but not any allowing quantitation close to the detection limits.[1,2,3,4] We have developed a reliable and precise method for profiling 11 trace level phytoestrogens in plasma. Altogether we have been able to quantify 13 phytoestrogens in biological samples of the following groups: isoflavonoids, their glycosides, plant and mammalian lignans and their metabolites. These compounds possess a number of biological activities making them e.g. possible factors lowering the risk of CHD and cancer. Regarding these beneficial properties of the phytoestrogens, the profiles among the people following their habitual Western diets are of particular interest. The HPLC-CEAD method is far more simple than other techniques with similar sensitivity, and also the selectivity is high. The HPLC-CEAD methods for urine and food samples, based on the original plasma method, are also presented.

2 MATERIALS AND METHODS

2.1. HPLC-CEAD analysis

The thirteen compounds, quantified from biological samples, were daidzein, genistein, dihydrodaidzein, dihydrogenistein, equol, O-desmethylangolensin, daidzein-7-O-glucoside, genistein-7-O-glucoside, secoisolariciresinol (seco), matairesinol, anhydrosecoisolariciresinol (seco is converted to anhydroseco during hydrolysis), enterodiol and enterolactone. The compounds were dissolved in MeOH and a standard mixture was prepared from the separate standard stock solutions. The standard mixture was diluted with mobile phase containing 20 % eluent B. Mobile phase consisted of two eluents: A: 50 mM sodium acetate-buffer pH 5: MeOH (80:20, v/v) B: 50 mM sodium acetate-buffer pH 5: MeOH:ACN (40:40:20, v/v/v). Elution was performed at the flow rate of 0.3 ml/min with the gradient beginning from 20 % eluent B and rising up to 100 % during 50 minutes. Total run time, including stabilizing time, was 85 minutes. Column was GL Sciences Inertsil ODS-3 C18, with dimensions 150 x 3 mm and 3 μm particles, made from endcapped material. Instrument consisted of two ESA solvent pumps model 580, ESA thermostated autosampler model 540, thermal chamber for column and detector cells, ESA 8 channel coulometric electrode array, system control module and computer. Detection potentials varied from 200 mV at Ch 1 to 720 mV at Ch 8. Thermal chamber for the column and detector cells were maintained at 37 ° C and the samples were stored at 10 ° C in autosampler.

2.2. Samples

Plasma samples from 22 Finnish women, omnivores (8), vegetarians (8) and breast cancer patients (6), following their habitual diets, were analyzed. Four samples: spring, summer, autumn and winter, from all the persons were taken, if available. (total number of the plasma samples was 82). Urine samples of the vegetarians were also analyzed. The method is shortly as follows. Tritiated estrone-glucuronide was added to the samples and after mixing were the samples kept at room temperature for 20 minutes. ß-Glucuronidase and sulphatase in sodium acetate buffer were added and the samples were incubated overnight at 37 °C. The samples were extracted twice with diethyl ether. The ether extracts were combined and evaporated to dryness under N_2 flow. Plasma samples were then dissolved in MeOH using 1/5 of the initial sample volume, and urine samples were diluted. 1/10 of each sample was taken and the recovery was counted with a ß-counter. Using the method for food, different soy products, flaxseed, cereal products, vegetables and exotic fruits were analyzed. Food samples can be analyzed with, or directly without, the hydrolysis of phytoestrogen conjugates. Dry samples were weighed and dissolved in water. Proteins were precipitated with EtOH. The conjugates were converted to their aglycones with enzyme and/or acid hydrolysis and then the samples were extracted twice with diethyl ether. Finally the samples were dissolved in MeOH and submitted to the analysis.

3 RESULTS AND DISCUSSION

The HPLC-CEAD plasma method reliability was evaluated by determining the intra- and interassay precision values using control plasma samples. Coeffiecient of variations (CVS) varied from 5.11 to 22.3 % for intra-assay precisions and from 7.35 to 32.2 % for interassay precisions (n=5). The lowest quantitation limits for the compounds, usually lignans, were 1.5 times the detection limit values with the interassay precisions. Identification was done by using retention time window of 10-30 s, and additionally by comparing the oxidation patterns of the analytes to the patterns of the standard compounds. Some cases were also confirmed with standard additions. Signal linearities were determined using on column amounts from the detection limits up to 5.1-7.5 ng, corresponding to the concentration range from 0.4 to 500-700 ng/ml. Upper limits of the linearities were not determined. On column amounts were presented as a function of the peak height by using the least square method. Correlation factors for signal linearities varied from 0.994 (dihydrodaidzein) to 1.000 (secoisolariciresinol and enterodiol). Detection limits were determined using a signal to noise ratio of 3. Values varied from 4.29 pg (anhydrosecoisolariciresinol) to 25.2 pg (genistein) on column. Equol and enterolactone are chromatographically very similar and they are detected on the same channel. These compounds form a critical peak pair in chromatogram, when selectivity of the method is evaluated. Resolution for equol (rt 44.3 min.) and enterolactone (rt 45.1min.) is 1.3 when 2.0 is equal for baseline separation. In multicomponent analysis resolution between 1.0-1.5 is satisfactory, and errors (due to the peak overlapping) are avoidable if the peak height for quantitation is used.[5] Method is also selective enough to separate glucosides and malonylglucosides of daidzein and genistein in food samples. Identification of the conjugates were confirmed with HPLC-MS (ESI) at VTT Biotechnology and Food Research. Repeatability of the separation was evaluated comparing the retention times from 23 separate runs. The CVS for the retention times varied from 0.47 % to 0.78 %. When changes in retention times are 0.5 % or less, column is assumed to be completely equilibrated.[5]

Table 1
Phytoestrogen profile in plasma of the 22 Finnish women consuming their habitual diets. (MV; nmol/l).

Isoflavonoids	omnivores (8)	vegetarians (8)	breast cancer patients (6)
Daidzein	4.33	11.7	7.96
Genistein	17.0	49.7	50.1
Dihydrodaidzein	2.64	3.30	4.19
Dihydrogenistein	0.97	0.85	nd +
Equol	2.78	2.78	1.25
O-Desmethylangolensin	0.65	1.02	nd +
Lignans			
Enterolactone	17.8	33.0	11.7
Enterodiol	1.56	8.71	1.71
Matairesinol	1.51	1.23	0.66
Secoisolariciresinol*	3.60	2.66	3.06

* *Anhydrosecoisolariciresinol included*
+ *nd = not determined, values below the quantitation limits.*

The values presented were corrected using the tritiated estrone-glucuronide recoveries. Mean recovery was 76 %, ranged 68-91 %. Phytoestrogen levels and profiles in plasma were similar to results reported earlier.[6] The highest concentrations and the most variable phytoestrogen profiles were found for vegetarians. Dihydrodaidzein and dihydrogenistein were detected in some plasma samples and more frequently in urine samples of the vegetarians (data not shown). Our methods for urine and food need minor modifications in order to increase the sensitivity.

Acknoledgement

This work was supported by NIH grant and S. Juselius Foundation.

References:
1. P. H. Gamache and I. N. Ackworth. *Proc. Soc. Exp. Biol. Med.* 1998, **217**, 274.
2. V. Rizzo, G. M. D'eril, G. Achilli, G. P. Cellerino. *J. Chrom.* 1991, **536**, 229.
3. P. H. Gamache, E. Ryan, I. N. Ackworth. *J. Chrom.* 1993, **635**, 143.
4. G. Achilli, G. P. Cellerino, P. H. Gamache, G. M. D'eril. *J. Chrom.* 1993, **632**, 111.
5. L.R. Snyder, J. J. Kirkland and J. L. Glajch. Practical HPLC Method Developement. John Wiley & Sons, 1997, 2nd ed., pp. 13-17, 22-27, 652-6536. H. Adlercreutz, T. Fotsis, S. Watanabe, J. Lampe, K. Wähälä, T. Mäkelä and T. Hase.*Cancer Detection and Prevention* 1994, **18**, 259.

Antioxidants, Oxidative Damage and Cancer

DIETARY CANCER PREVENTION: CAVEATS SEEN BY A TOXICOLOGIST

Hans Verhagen.

TNO Nutrition and Food Research Institute, Zeist, The Netherlands.

1 INTRODUCTION

Diet contains thousands of chemicals, nutrients as well as non-nutrients. Nutrients can be subdivided into macronutrients (protein, fat, carbohydrate) and micronutrients (vitamins, minerals, trace elements). All nutrients are necessary to support life: 'necessity', and some may have an added value beyond the delivery of essential nutrients: 'beneficity'. However, an excess consumption of nutrients may result in adverse effects: 'toxicity'. Non-nutritive dietary constituents can be of natural or of synthetic origin. Synthetic non-nutrients in the diet are either contaminants or food additives, and subject to officially set limits. Natural occurring non-nutrients can comprise compounds with mere adverse potential ('natural toxicants') and compounds with beneficial potential ('dietary chemoprevention').

In this paper the current interest in bioactive dietary constituents will be briefly discussed in the light of the established science of toxicology and several caveats are presented to take into account when claiming a dietary constituent to have chemopreventive potential.[1,2]

2 HEALTH RISK ASSESSMENT = DOSE *VERSUS* EFFECT

In toxicology, health risk assessment is essentially based on only two aspects:[3] dose (amount per weight or volume) and effect (response, including underlying mechanisms). All substances are toxic. Any chemical, either synthetic or natural, has a threshold dose at and below which no effect will occur: the 'no-observed-adverse-effect-level' (NOAEL), which is commonly generated in studies with experimental animals. In order to account for possible intra- and interspecies differences, a potentially safe level for human exposure (e.g. the acceptable daily intake, ADI) is calculated by dividing the NOAEL by a 'safety factor' (SF), e.g. 10 x 10 = 100 (ADI = NOAEL/SF). This NOAEL-SF approach is applicable to all chemicals except genotoxic carcinogens and perhaps allergens.

Hence 'dose' is the factor that discriminates between the presence or absence of an effect. This holds also true in pharmacology and nutrition. As such, for nutrients 'dietary reference values are set:[4] 'lower reference nutrient intake' (LRNI), 'estimated average require ment' (EAR), 'reference nutrient intake' (RNI), and 'safe intake', whereas for medicines or putative beneficial dietary constituents a desired effect this may be called 'lowest effect level' (LEL).[1,2]

3 DIETARY CHEMOPREVENTION = BEING BEWARE OF CAVEATS

The basic concept of toxicology is that health risk assessment is a function of dose and effect: health risk assessment = f (dose, effect). In the area of dietary chemoprevention one should beware that there are several caveats when claiming "beneficity" for a food (ingredient).[1, 2]

3.1. Caveats in relation to dose

3.1.1. Caveat 1: the threshold. Chemoprevention is a non-stochastical event: in theory one molecule cannot prevent cancer. Also for beneficial effect levels the threshold principle applies. Thus, there will be a LEL for an chemopreventive effect to become manifest, and exposure to beneficial substances below the LEL is necessarily without effect. This is far from new: also for medicines a high enough dose is needed to have the desired effect (e.g. to cure a disease).

3.1.2. Caveat 2: toxicity! For chemopreventive substances the toxicological (NOAEL) and beneficial dose levels (LEL) should be considered together in one evaluation. A beneficial effect is only valuable in the absence of toxicity: the LEL should be well below the safe human dose. It should be noted that for nutrients RNI and NOAEL levels may be very close and a proper SF may not be feasible; rather a 'margin of safety' applies. However, with medicines toxic side effects may be unavoidable. In such cases the necessity of therapy outweighs the concomitant toxicity, but for dietary chemopreventive agents it is not acceptable to have toxicity at beneficial dose levels.

3.1.3. Caveat 3: the matrix. Dietary constituents are not consumed as pure substances. Rather diet comprises thousands of chemical substances and these are consumed simultaneously. In toxicology there is uncertainty as to how the combined toxicity of chemicals should be assessed and how combined toxicity should be taken into account in setting standards for individual compounds.[5] This problem equally well applies to chemoprevention. Indeed, in general it is unlikely that single compounds may be consumed in sufficient quantity to elicit the desired effects. However, a combination of beneficial substances in a matrix may yet result in "beneficity" in humans through additivity, synergism and/or potentiation. Moreover, by spreading "beneficity" over a number of substances non-desirable compound-specific toxic side effects may be overcome.

3.2. Caveats in relation to effect

3.2.1. Caveat 4: assessment of potential. Genotoxicity of a compound is generally tested in a tiered approach ranging from short-term *in vitro* tests, and short-term *in vivo* tests in experimental animals, to long-term *in vivo* studies in which the carcinogenic potential of a compound is assessed by life-time exposure to the test compound up to toxic dose levels. Occasionally also human data are available from epidemiological studies and/or studies applying biomarkers.

This approach can equally well be applied to determine the chemopreventive potential of compounds, bearing in mind that an antimutagenic response *in vitro* should be verified *in vivo*[6], and that antimutagenic effects should be verified for anticarcinogenicity in long-term animal studies and ideally also in human studies.

3.2.2. Caveat 5: the mechanism. A carcinogen is not a carcinogen merely based on functionality: is rather a poor way of doing toxicology. Data should be available on the mechanism underlying the carcinogenicity. Therefore, in Europe a discrimination is made between genotoxic and non-genotoxic carcinogens and this is of major importance for health risk assessment. The same concept is true for denominating an anticarcinogen. If the underlying mechanism is not known a claim of anticarcinogenic potential has a weak basis. The mechanisms of chemopreventive agents are multiple:

- antioxidant activity
- inhibition or induction of biotransformation enzymes
- modifying effect on the activities of other enzymes
- prevention of formation/uptake of carcinogens
- scavenging effect on the (activated) carcinogens
- shielding of nucleophilic sites in DNA
- inhibition of DNA/carcinogen complex

3.2.3. Caveat 6: (anti)mutagens are not always (anti)carcinogens and vice versa. Initially toxicologists thought that carcinogens could be assessed by performing short/term genotoxicity tests *in vitro* and *in vivo*. Indeed, carcinogens are sometimes mutagens and vice versa, but this is only partially correct and one of the main reasons for this is that nowadays the rodent carcinogenicity assays are overly sensitive by the necessity to test at a 'maximum tolerated dose', thereby rendering almost every second compound a carcinogen.[7] In this way many a 'carcinogen' is a non-genotoxic carcinogen (and thus in fact a non-carcinogen). By giving a similar consideration anticarcinogens may not always be antimutagens and vice versa.

3.2.4. Caveat 7: weight of the evidence. In toxicology generally a whole series of data is necessary to make a complete judgement of the toxicological status of a compound. For synthetic chemicals dozens of study reports performed according to officially recognized (eg. OECD) guidelines and conducted under Good Laboratory Practice are obligatory prior to the market introduction of a compound. On the basis of the entire dossier, which can take years to compile, an expert judgement can be made taking into account all data from regular toxicity studies as well as information on the mechanism of toxicity. By using similar argumentation the chemopreventive potential of compounds cannot be claimed on just one or a few studies, rather many data from many studies should be considered in concert.

4 THE PROOF OF THE PUDDING = IN THE EATING!

Is dietary chemoprevention still feasible to occur in humans? Indeed, despite all caveats given above, there are many indications available now.[8,9,10] These may come from epidemiology and be based on dietary questionnaires or result from experimental studies using biomarkers. In fact such epidemiological findings have triggered the onset of experimental chemoprevention studies. Also experimental biomarker studies in human volunteers have indicated that it is indeed feasible in humans to have a potential beneficial effect in the absence of adverse effects.

References

1. Verhagen H., Rompelberg C.J.M., Strube M., van Poppel G. and van Bladeren P.J. *J.Environm.Pathol.Toxicol.Oncol.* 1997, **16**, 343-360.
2. Verhagen H, 1998, In: "Functional Foods; the consumer, the products and the evidence", M.J. Sadler & M. Saltmarch (Eds.), The Royal Society of Chemistry, Cambridge UK, ISBN 0-85404-792-1, pp 87-93.
3. C.D. Klaassen, 1986, 'Casarett and Doull's Toxicology', 3rd Ed. (C.D. Klaassen *et al.* eds.), Macmillan Publ. Co., Chapter 2.
4. P. Mason, 'Handbook of dietary supplements', Blackwell Science, 1995.
5. Feron,V.J., Groten,J.P., van Zorge, J.A., Cassee,F.R., Jonker,D., and P.J. van Bladeren, *Toxicology Letters* 1995, **105**, 415-427.
6. L.R. Ferguson, *Mutat. Res.* 1994, **307**, 395-410.
7. B. N. Ames and L.S. Gold, *Proc. Natl. Acad. Sci. USA.* 1990, **87**, 7772-7776.
8. L.W. Wattenberg, *Cancer Res.* 1992, **(Suppl.) 52**, 2085 s-2091 s.
9. National Research Council, 'Carcinogens and anticarcinogens in the human diet', Natl. Acad. Press, 1996.
10. Verhagen H. and van Poppel G., 1997, In: "Food Ingredients Europe: Conference Proceedings 1996, Miller Freeman Plc, Maarssen, The Netherlands, pp 65-68.

OXIDATIVE DAMAGE TO DNA: A LIKELY CAUSE OF CANCER?

Andrew R. Collins.

Rowett Research Institute, Greenburn Road, Bucksburn, Aberdeen AB21 9S.

1 INTRODUCTION

The endogenous oxidation of cellular DNA by reactive oxygen released during respiration is generally accepted as a contributory cause of cancer, for three main reasons. First, when methods developed for the identification of oxidation products in DNA were applied to normal cells, very high concentrations of oxidised bases (notably 8-oxoguanine) were detected. Second, epidemiological studies have shown fruit and vegetable consumption to be associated with decreased risk of various cancers. Since fruit and vegetables contain antioxidants (vitamin C, vitamin E, carotenoids etc.), which can scavenge free radicals, we have a convincing explanation; fruit and vegetables protect because the antioxidants limit the amount of potentially mutagenic DNA damage. Third, it has been demonstrated, *in vivo*, that 8-oxoguanine is indeed a mutagenic lesion, as it tends to base-pair with adenine rather than cytosine in replication, resulting in G:C T:A transversions. Thus, if one in a thousand guanines are oxidised, surely the damage must have serious consequences.

In this paper, I challenge some of the elements of this rather too comfortable explanation. I look at the latest estimates of the baseline level of oxidative DNA damage in normal cells; examine the hypothesis that dietary antioxidants limit DNA oxidation; and seek direct evidence that tumorigenic transformation can result from oxidative damage.

2 HOW MUCH OXIDATIVE BASE DAMAGE IS THERE IN NORMAL HUMAN CELLS?

Three kinds of analysis of DNA oxidation are most commonly applied to lymphocytes or other human cells. *Gas chromatography-mass spectrometry (GC-MS)* can unambiguously identify oxidation products; however, during the essential derivatisation step, oxidation of dG occurs, leading to serious over-estimation of the background level of 8-oxoguanine.[1] If the unoxidised guanine is removed before derivatisation, the level of 8-oxoguanine measured is similar to that detected by *high performance liquid chromatography with electrochemical detection (HPLC-EC)* -the second analytical method. Even with HPLC, estimates cover a wide range -from 8 down to 0.2 8-oxo-dG per 10^5 dG (reviewed in reference 2). It is now clear that spurious oxidation of dG can occur during the isolation and hydrolysis of DNA for HPLC analysis.[3] The lowest estimates so far reported are from an anaerobic preparation method.[4] The third approach is based on the use of bacterial repair enzymes with specific glycosylase/endonuclease activities, recognising oxidised pyrimidines (endonuclease III), or, in the case of formamidopyrimidine glycosylase (fpg), 8-oxo-dG as well as ring-opened purines. The lesions are converted to strand breaks, which can be measured by several *DNA breakage assays* -alkaline elution,[5] alkaline unwinding[6] and the comet assay.[7] These enzyme-based assays consistently indicate a basal level of 8-oxoguanine in lymphocytes of around 0.05 per 10^5 guanines -several times lower than the lowest estimate by HPLC. It is possible that the enzyme-based assays are under-estimating DNA damage -if some lesions are not accessible to enzymes, or if several lesions occur very close together in a cluster which registers as a single strand break. It is also possible that the oxidation artefacts have not yet been totally eliminated from the HPLC method. At the moment, then, the most likely estimate of the amount of 8-oxo-dG is around 0.1 per 10^5 dG, or about 2000 per cell -a far smaller burden of genetic damage than was previously supposed.

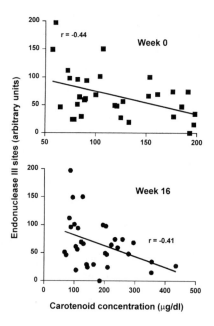

Figure 1

Correlations of lymphocyte DNA damage (oxidised pyrimidines) with serum carotenoid concentrations. Carotenoids were measured before (week 0) and after (week 16) dietary supplementation with carotenoids. (Redrawn from reference 10.)

3. DO DIETARY ANTIOXIDANTS PROTECT AGAINST DNA OXIDATION *IN VIVO*?

There is persuasive evidence to sustain this element of the antioxidant hypothesis. For instance, we carried out a 20-week supplementation trial in which smokers and non-smokers (male, 50-59) received either a placebo or a mixture of vitamin C (100 mg/d), β-carotene (25 mg/d) and vitamin E (280 mg/d). Base oxidation was measured as endonuclease III-sensitive sites with the comet assay; a decrease of about 40 % was seen in the lymphocytes of subjects receiving supplement compared with those from placebo groups.[8] The lymphocytes were also more resistant to *in vitro* attack by H_2O_2. A similar protection was seen in experiments where lymphocytes were isolated before or up to 24 h after a single large dose of vitamin C, vitamin E or β-carotene.[9]

Supplements may be effective; what about normal dietary sources of antioxidants? We have looked for correlations between oxidative DNA damage and blood antioxidant concentrations. Subjects taking part in a carotenoid supplementation trial showed a negative correlation between endonuclease III-sensitive sites (measured at the end of the supplementation period) and several carotenoids.[10] Curiously, the correlation was at least as strong with carotenoid levels at week 0 (before supplementation began) as with carotenoids at week 16 (Fig. 1). The individual carotenoid supplements increased blood carotenoid concentrations but had no effect on endogenous DNA damage. It seems that the natural dietary balance of carotenoids (or of other phytochemicals associated with them) is most effective at protecting against DNA oxidation.

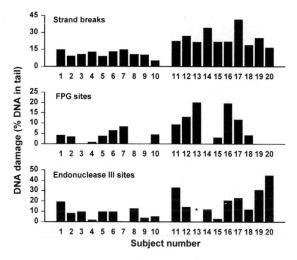

Figure 2

DNA damage (strand breaks and oxidised bases) measured with the comet assay in normal and diabetic subjects. Subjects 1-10 are normal healthy males, matched with male IDDM patients (subjects 11-20). The asterisk indicates a missing sample. (Redrawn from reference 11.)

4 ARE THERE BIOLOGICALLY SIGNIFICANT DIFFERENCES IN LEVELS OF OXIDATIVE DNA DAMAGE BETWEEN INDIVIDUALS OR BETWEEN POPULATION GROUPS?

The comet assay has been used to investigate levels of endogenous DNA damage in patients with diseases associated with oxidative stress. DNA strand breaks and oxidised pyrimidines were significantly elevated in patients with insulin-dependent diabetes mellitus compared with normal controls (Fig. 2), and fpg-sensitive sites (altered purines) showed a strong positive correlation with blood glucose level.[11] In a similar study of patients with ankylosing spondylitis, a rheumatoid condition, strand breaks and fpg-sensitive sites were significantly higher than normal.[12]

8-oxo-dG was measured (by HPLC) in lymphocyte DNA of subjects (aged 25-45) from 5 European countries; Spain, France, the Netherlands, Ireland and UK (Ulster).[13] Mean levels of damage were similar for women in all 5 countries, and for men in Spain and France. However, the samples from men in northern Europe, and particularly Ireland, had levels of DNA base oxidation that were up to 3 times as high (Fig. 3). The differences did not reflect different levels of dietary antioxidants measured in the blood; vitamin C, vitamin D and carotenoids were slightly higher in men than in women in all countries, and between countries the differences were surprisingly small. If lymphocytes are in any way representative of other cells, tissues and organs in which cancer might arise, and if oxidative DNA damage is a contributory cause of cancer, these results (8-oxo-dG levels) might be reflected in variations in the incidence of cancer in the countries concerned. Using WHO statistics for mortality from common cancers, we found a significant positive

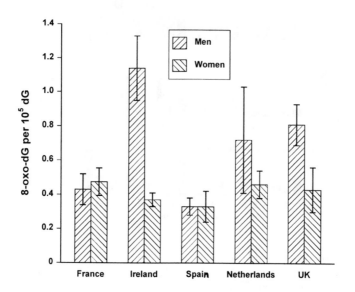

Figure 3

8-oxo-dG measured by HPLC in lymphocytes from volunteers in different European countries. (Data derived from study described in reference 13.)

correlation with 8-oxo-dG levels for colorectal cancer in men but a significant negative correlation for gastric cancer in women. Overall, there was no correlation of oxidative DNA damage with total cancer mortality. The strongest correlation (r=0.95, P<0.01) was seen between 8-oxo-dG levels and premature mortality from coronary heart disease in men. DNA base oxidation in lymphocytes may be a useful biomarker of oxidative stress (which is considered a serious risk factor in heart disease); it may even have predictive value for CHD risk when measured in individuals, but prospective studies and comparisons with other CHD risk factors are needed to test this.

5 OXIDATIVE DAMAGE AND CANCER

I have discussed evidence that antioxidants, given as supplements or taken in as part of the normal diet, can decrease the level of DNA oxidation in lymphocytes. When we look at attempts to influence the incidence of cancer by long-term supplementation, however, the evidence is discouraging. -carotene, at least when given at high dose to high-risk groups (smokers/asbestos workers) actually increased the incidence of lung cancer in two studies.[14,15]

Is there, in fact, any evidence for involvement of oxidative damage in the aetiology of human cancer? Different kinds of DNA damage leave a 'signature' in the kinds of mutation that they lead to, and it is, in theory, possible to analyse the mutations in cancer genes from human tumours and suggest what agent was responsible. The spectrum of mutations resulting from DNA oxidation is generally similar to that of other damaging agents, but a minor product, a tandem mutation of CC TT, is distinctive. This mutation also results from UV irradiation and has been linked to skin cancer; but CC TT transitions in internal tumours should be of oxidative origin. It has not been identified in the internal tumours studied.[16]

In cell culture, H_2O_2 can be used to introduce oxidative DNA damage. Analysis of the resulting mutations according to type and pattern (base substitutions at different sites) did not show marked similarity to mutations occurring spontaneously (i.e. from endogenous causes) in the same cells.[17] Also in cultured cells, the spontaneous mutation frequency is much less than would be predicted from the number of base oxidation events.[18]

Lack of evidence for a link between oxidative DNA damage and cancer is not conclusive. However, it should perhaps cause us to think again about the biological factors involved in genetic homeostasis. DNA is very prone to damage, and effective defence mechanisms have evolved. DNA base excision repair or nucleotide excision repair pathways can deal with most DNA lesions before they are fixed as mutations during replication. Reactive oxygen species are released in quantitity during respiration and might constitute a threat to genetic stability; but very powerful antioxidants -notably glutathione -and antioxidant enzymes (catalase, superoxide dismutase, etc.) are present, and consequently very little of the predicted oxidation of DNA actually takes place. The best guess (from measurements of rates of urinary excretion of 8-oxo-dG, presumed to be DNA repair products) is that a few hundred guanines are oxidised, and repaired, per cell per day.[19] If the steady state level of 8-oxo-dG in DNA is 2000 per cell, as suggested above, then we have a biologically reasonable dynamic situation with turnover of damage in a matter of days. The argument, in short, is that a balance is struck (in evolutionay terms) between the risk of a certain level of DNA damage (unavoidable because a by-product of respiration) and the resources needed to reduce that risk. Perhaps phytochemical protectants, derived from a primaeval human diet rich in fruit and vegetables, play an important part in the equation - not just as antioxidants, but, for example, as inducers of phase II metabolising enzymes that are important in eliminating potential genotoxins.

Acknowledgments
I gratefully acknowledge the support of the Scottish Office Agriculture, Environment and Fisheries Department, the Ministry of Agriculture, Fisheries and Food, and the European Commission.

Reference

1. J.-L. Ravanat, R.J. Turesky, E. Gremaud, L.J. Trudel and R.H. Stadler, *Chem. Res.Toxicol.* 1995, **8**, 1039-1045.

2. A. Collins, J. Cadet, B. Epe and C. Gedik, *Carcinogenesis.* 1997, **18**, 1833-1836.

3. C.M. Gedik, S.G. Wood and A.R. Collins, *Free Radical Res.* in press.

4. M. Nakajima, T. Takeuchi, T. Takeshita and K. Morimoto, *Environ. Health Perspect.* 1996, **104**, 1336-1338.

5. M. Pflaum, O. Will and B. Epe, *Carcinogenesis.* 1997, **18**, 2225-2231.

6. A. Hartwig, H. Dally and R. Schlepegrell, *Toxicol. Lett.* 1996, **88**, 85-90.

7. A.R. Collins, M. Dusinská, C.M. Gedik and R. Stìtina, *Environ. Health Perspect.* 1996, **104**, suppl. 3, 465-469.

8. S.J. Duthie, A. Ma, M.A. Ross and A.R. Collins, *Cancer Res.* 1996, **56**, 1291-1295.

9. M. Panayiotidis and A.R. Collins, *Free Radical Res.* 1997, **27**, 533-537.

10. A.R. Collins, B. Olmedilla, S. Southon, F. Granado and S.J. Duthie, *Carcinogenesis.* in press.

11. A.R. Collins, K. Raslová, M. Somorovská, H. Petrovská, A. Ondrusová, B. Vohnout, R. Fábry and M. Dusinská, *Free Radical Biol. Med.* 1998, **25**, 373-377.

12. M. Dusinská, J. Lietava, B. Olmedilla, K. Raslová, S. Southon and A.R. Collins, 'Antioxidants and Disease', CAB International, Oxford, in press.

13. A.R. Collins, C.M. Gedik, B. Olmedilla, S. Southon and M. Bellizzi, *FASEB J.* 1998, **12**, 1397-1400.

14. α-tocopherol, β-carotene cancer prevention study group, *New Engl. J. Med.* 1994, **330**, 1029-1035.

15. G.S. Omenn, G.E. Goodman, M.D. Thornquist *et al.*, *New Engl. J. Med.* 1996, **334**, 1150-1155.

16. T.M. Reid and L.A. Loeb, *Cancer Res.* 1992, **52**, 1082-1086.

17. A.R. Oller and W.G. Thilly, *J. Molec. Biol.* 1992, **228**, 813-826.

18. J.F. Ward, *Radiat. Res.* 1995, **142**, 362-368.

19. S. Loft, K. Vistisen, M. Ewertz, A. Tjänneland, K. Overvad and H.E. Poulsen, *Carcinogenesis.* 1992, **13**, 2241-2247.

INCREASED FRUIT AND VEGETABLE CONSUMPTION REDUCES INDICES OF OXIDATIVE DNA DAMAGE IN LYMPHOCYTES AND URINE

Albert D. Haegele and Henry J. Thompson.

Division of Laboratory Research, AMC Cancer Research Center, Denver, CO 80214. Send correspondence and proofs to:Albert D. Haegele, Division of Laboratory Research, AMC Cancer Research Center, 1600 Pierce Street, Denver, CO 80214.

1 INTRODUCTION

Epidemiological studies have shown a consistent inverse relationship between fruit and vegetable consumption and cancer occurrenc.[1-3] Evidence for the effectiveness of whole foods in preventing cancer is stronger than for supplements, suggesting that substances other than those generally used as dietary supplements are contributing to this effect . Because fruits and vegetables are rich in substances with antioxidant activity, it has been proposed that the antioxidant properties of fruits and vegetables are responsible for their cancer preventive activity.[7-9] We sought to test the hypothesis that increased consumption of fruit and vegetables confers protection against oxidative damage that may play a role in carcinogenesis; namely oxidative DNA damage and lipid peroxidation (LP).

Oxidative DNA damage has been implicated in a number of disease syndromes including cancer[12-17] and 8-OHdG is regarded as a reliable indicator of such damage.[15-17] We elected to measure 8-OHdG excreted in urine and in DNA isolated from peripheral lymphocytes as indices of oxidative DNA damage rate and steady state levels, respectively.

LP causes an increase in reactive species that can react with DNA, and some LP byproducts have mitogenic activity. Thus, LP is potentially pro-carcinogenic by both genetic and epigenetic mechanisms.[18-23] 8-iso-prostaglandin $F_{2\alpha}$ (8-EPG), the 8-epimer of prostaglandin F2α, is an isomer of a class of prostanoids, referred to collectively as F2 isoprostanes, that are produced *in vivo* by free radical-induced LP. Numerous studies have shown production of F2 isoprostanes to be induced in rats by treatments that are known to cause LP, such as CCl_4 and diquat exposure. Evidence indicates that they are generated primarily by cyclooxygenase (COX) independent pathways, as their production is not modulated by COX inhibitors such as ibuprofen, aspirin, and indomethacin.[24-29] They have been shown to arise in situ esterified to phospholipids[27] and to be excreted in urine.[24-26,28,29] Consequently, urinary F2 isoprostanes are proving to be excellent time integrated markers for *in vivo* free radical-induced LP, and measurement of urinary 8-EPG by competitive binding assay is a facile method of estimating *in vivo* LP.[24] Malondialdehyde is amenable to sensitive and selective HPLC measurement of its thiobarbituric acid (TBA) derivative, and has been used extensively as an index of *in vivo* LP.[19,30] We measured urinary 8-EPG and both plasma and urinary MDA as indices of subjects overall oxidative status with respect to

LP.

2 MATERIALS AND METHODS

2.1. Study design

Fifteen free-living adult females participated in a 14-day intervention during which they consumed a completely defined recipe-based diet. A pre-intervention assessment of dietary intake was obtained via three-day food records that provided an estimate of pre-intervention fruit and vegetable consumption. The subjectís calorie requirements and energy intake data were used to adjust portion size of each recipe so that subjects would maintain their pre-intervention body weights during the intervention. Pre-intervention and post-intervention blood samples and first void of the morning urine specimens were obtained from each subject.

2.2. Analytical methods

Lymphocytes were isolated using 8 mL cell preparation tubes (Becton Dickinson) and frozen for subsequent isolation of DNA at -80 °C in phosphate buffered saline containing 10 % DMSO. The contents of two tubes (~16 mL blood) were used for each analysis, and analyses were generally performed in duplicate. Nuclei were isolated from lymphocytes by use of non-ionic detergent, and DNA isolated from nuclei by a method utilizing proteinase K and organic extraction. 8-OHdG and dG in DNA from lymphocytes were measured by reverse phase HPLC with electrochemical and spectrophotometric detection for 8-OHdG and dG, respectively. We elected to employ phenol in our DNA isolation despite the controversy surrounding its use. Vigilance against the artificial generation of 8-OHdG during processing was exercised throughout our procedure, as described in a recent communication.[31] 8-OHdG in urine was measured with a commercially available ELISA kit (Genox Inc.). The assay is based on a primary antibody whose specificity is unique in that it has 100X greater affinity for 8-OHdG than for the corresponding ribonucleoside, 8-hydroxyguanosine (data not shown). MDA in both plasma and urine was measured as its thiobarbituric derivative by HPLC with fluorimetric detection. 8-EPG was assayed in urine by use of a commercially available ELISA kit (Assay Designs, Inc.). All urine indices (8-OHdG, 8-EPG, and MDA) were expressed per unit creatinine. Plasma carotenoids and tocopherols were measured by HPLC as biochemical indices of fruit and vegetable consumption.[32]

3 RESULTS

The described intervention resulted in a statistically significant increase in mean fruit and vegetable consumption from 5.2 to 12.5 servings per day. Pre- and post-intervention mean plasma carotenoids and tocopherols are summarized in Table I. Dramatic increases in α- and ß-carotene (+85 % and 47 % respectively) and decreases in γ- and α-tocopherols (-42 % and -11 % respectively) are particularly noteworthy. Plasma carotenoids, especially α-carotene, are regarded as reliable indicators of fruit and vegetable consumption.[33]

Table 1

Pre- and post-intervention plasma levels of selected carotenoids and tocopherols.

Compound	Mean plasma concentration ± SEM (ng/mL) Pre-Intervention	Post-Intervention	%Change	*p*-value (paired two tailed *t*-test)
α-carotene	92.3 ± 16.2	171 ± 15.4	85	0.000
ß-carotene	291 ± 64.7	430 ± 65.3	47	0.000
cryptoxanthine	130 ± 21.4	154 ± 18.9	18	0.065
lutein	158 ± 17.8	242 ± 29.9	53	0.000
lycopene	382 ± 40.3	394 ± 35.5	3.3	0.364
retinol	506 ± 21.3	496 ± 21.5	-1.9	0.495
γ-tocopherol	1601 ± 228	925 ± 151	-42	0.000
α-tocopherol	13805 ± 1146	12228 ± 863	-11	0.007

8-OHdG concentration in both urine and lymphocytes decreased significantly over the course of the intervention, as indicated in Fig. 1 and 2 respectively. Mean urinary 8-OHdG decreased 59 % (p=0.03, paired Studentís paired two-tailed t-test) and mean lymphocyte 8-OHdG decreased 18 % (p = 0.02, repeated measures analysis of variance) in response to the intervention. There was no correlation between levels of 8-OHdG in urine and in lymphocytes.

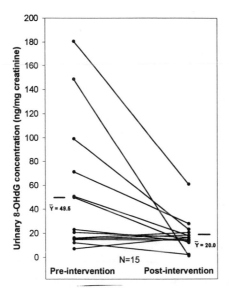

Figure 1

Mean urinary 8-OHdG concentration decreased 59 % over the course of the intervention. The decrease was statistically significant (p = .03, Student's paired two-tailed t-test).

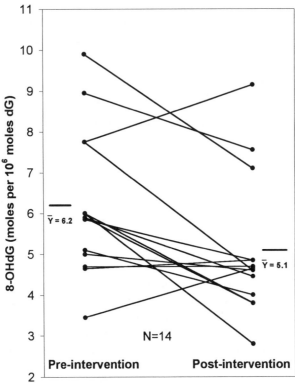

Figure 2

*Mean 8-OHdG concentration in DNA isolated from peripheral lymphocytes decreased 18 %
in response to the diet-based interevention. The reduction was statistically significant
(p=0.02) by repeated measures analysis of variance.*

Table 2

Noteworthy correlations between urinary indices of oxidative damage. N= 15.

Comparison	N	Spearman Correlation Coefficient	p-value
Change in urinary 8-EPG and change in urinary 8-OHdG	14	0.613	0.020
Change in urinary MDA and change in urinary 8-OHdG	15	0.482	0.069
Change in urinary MDA and change in urinary 8-EPG	14	0.587	0.027
Urinary MDA and urinary 8-OHdG (pre-intervention)	15	0.710	0.003
Urinary 8-EPG and urinary 8-OHdG (pre-intervention)	14	0.543	0.045
Polyunsaturated fat in pre-intervention diet (% of fat) and pre-intervention urinary MDA	15	0.542	0.037

Mean urinary 8-EPG decreased 33 % (P = 0.001, Studentís two tailed paired t-test)
over the course of the intervention as shown in Fig. 3. There was no change in
malondialdehyde concentration in either urine or plasma during the study. However, both
change in urinary MDA and change in urinary 8-EPG showed a positive correlation with
change in urinary 8-OHdG. The effect was statistically significant in the case of 8-EPG (p =

0.020) and approached statistical significance with MDA (p = 0.069). Both urinary MDA and urinary 8-EPG were initially positively correlated with urinary 8-OHdG, and the relationship was particularly strong in the case of MDA. The initial correlations were lost entirely at post-intervention. Dietary polyunsaturated fat intake was also positively correlated with urinary MDA, but not urinary 8-EPG, prior to the intervention. Table II summarizes the aforementioned noteworthy correlations.

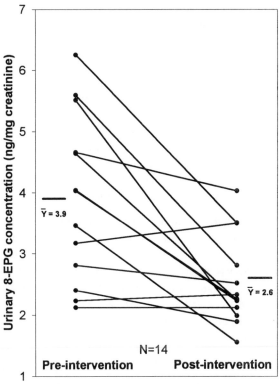

Figure 3

Mean urinary 8-EPG concentration decreased 33 % in response to the diet-based intervention. The decrease was statistically significant (p = 0.001, Student's two-tailed t-test).

4 DISCUSSION

Both urinary and lymphocyte 8-OHdG levels decreased over the course of the intervention, a finding that is consistent with other reports of oxidative DNA damage modulation by consumption of food or food extracts.[9,34] To our knowledge, this is the first report of a simultaneous reduction in both rate of DNA oxidation (reflected in urinary 8-OHdG) and equilibrium levels of oxidative DNA damage (in lymphocyte DNA). The observed decrease in urinary 8-OHdG was much greater than that in lymphocyte DNA, consistent with the concept that 8-OHdG concentration in DNA reflects the dynamic equilibrium between formation and repair, and is subject to homeostatic regulation. Previous accounts of the relationship between urinary 8-OHdG and oxidative stress have employed complex chromatographic methods, whereas our measurements were performed with a commercially available ELISA kit on untreated urine. Although the kit manufacturer claims an impressive degree of specificity for genuine 8-OHdG for this kit, it is possible that the ELISA assay detects other, presumably structurally related, compounds in addition to 8-OHdG, and indeed the levels of urinary 8-OHdG we observe are considerably higher than values reported by those using HPLC.[34-40] At this juncture, the reliability of the ELISA kit has not been verified by our laboratory, but efforts are underway to determine unequivocally the identity of compound(s) detected by the ELISA assay.

The observed decrease in 8-EPG indicates that LP was lowered by this intervention, and the significant positive correlation between change in urinary 8-EPG and change in urinary 8-OHdG supports the concept that LP and oxidative DNA damage are related, and that overall systemic oxidative stress was reduced by this intervention. The same may be said of the marginal positive correlation between change in urinary MDA and change in urinary 8-OHdG. Subjects with high initial levels of urinary 8-OHdG tended to have the greatest decrease associated with the intervention, and the same was true of subjects with high initial urinary MDA. Moreover, they tended to be the same subjects (i.e. urinary 8-OHdG and urinary MDA were initially positively correlated).

However, considering that mean urinary MDA remained virtually unchanged over the course of the intervention, the observed correlation between changes in urinary MDA and 8-OHdG is puzzling. MDA present in the diet may explain this apparent confounding observation. In animal models, ingested MDA from diets high in polyunsaturated fat has been shown both in our laboratory (unpublished data) and elsewhere[41,42] to contribute substantially to urinary MDA excretion. The positive correlation between polyunsaturated fatty acid consumption and urinary MDA observed in this study prior to the intervention suggests that this may be the case in humans as well. Thus urinary MDA presumably reflects both consumed MDA and *in vivo* LP. We hypothesize that *in vivo* LP was diminished as a result of increased fruit and vegetable consumption and that this reduction is reflected by reduced urinary 8-EPG and, less directly, urinary 8-OHdG. By virtue of the abundance of polyunsaturated fatty acids present in fruits and vegetables however, the dietary contribution to urinary MDA may have increased, offsetting decreased urinary MDA from *in vivo* LP in the group of subjects as a whole. The intervention diet employed in this study was in fact higher in polyunsaturated fat than most of the subjects pre-intervention diets. The utility of MDA as a marker for LP has been debated extensively,[42-46] and the complexity of our interpretation of the MDA data underscores its shortcomings. Our data suggest that urinary 8-EPG may offer considerable advantage over urinary MDA as an *in vivo* LP marker. To our knowledge however, the contribution of dietary sources to urinary 8-EPG has not been investigated. Such an investigation is important to validate the utility of 8-EPG as an index of *in vivo* LP.

The reduced levels of urinary 8-OHdG, lymphocyte 8-OHdG and urinary 8-EPG associated with this diet-based intervention are remarkable and statistically robust. Several correlations between indices are also noteworthy, and lend credence to our proposition that *in vivo* oxidative damage was modulated by the intervention. Collectively, our results support the hypothesis that increased consumption of fruits and vegetables can reduce *in vivo* oxidative damage, presumably by virtue of compounds with antioxidant activity contained therein. The results reported here are based on a small number of subjects however, and must be regarded as preliminary. Efforts to substantiate these findings are ongoing.

Acknowledgement

This work was supported by grant 97-A106 from the American Institute for Cancer Research. The authors thank the subjects who volunteered to participate in this study for their commitment and adherence to the dietary intervention.

References

1. Steinmetz, K.A. and Potter, J.D. Vegetables, fruit, and cancer. I. *Epidemiology. Cancer Causes.Control*. 1991, **2**, 325-357.
2. Block, G., Patterson, B. and Subar, A. Fruit, vegetables, and cancer prevention: a review of the epidemiological evidence. *Nutr.Cancer*. 1992, **18**, 1-29.
3. Ames, B.N., Gold, L.S. and Willett, W.C. The causes and prevention of cancer. *Proc.Natl.Acad.Sci.U.S.A.* 1995, **92**, 5258-5265.
4. Wattenberg, L.W. Inhibition of carcinogenesis by minor dietary constituents. *Cancer Res*. 1992, **52**, 2085 s-2091 s.
5. Fullerton, F.R., Greenman, D.L., McCarty, C.C. and Bucci, T.J. Increased incidence of spontaneous and 2-acetylaminofluorene-induced liver and bladder tumors in B6C3F1 mice fed AIN-76A diet versus NIH-07 diet. *Fundam.Appl.Toxicol*. 1991, **16**, 51-60.
6. Fullerton, F.R., Greenman, D.L. and Bucci, T.J. Effects of diet type on incidence of spontaneous and 2- acetylaminofluorene-induced liver and bladder tumors in BALB/c mice fed AIN-76A diet versus NIH-07 diet. *Fundam.Appl.Toxicol*. 1992, **18**, 193-199.
7. Aruoma, O.I. Nutrition and health aspects of free radicals and antioxidants [published erratum appears in Food Chem Toxicol 1994 Dec, **32(12)**, 1185]. *Food Chem.Toxicol*. 1994, **32**, 671-683.
8. Verhagen, H. Cancer prevention by natural food constituents. *Int.Food Ingred*. 1993, **1**, 22-29.
9. Pool-Zobel, B.L., Bub, A., Muller, H., Wollowski, I. and Rechkemmer, G. Consumption of vegetables reduces genetic damage in humans: first results of a human intervention trial with carotenoid-rich foods. *Carcinogenesis*. 1997, **18**, 1847-1850.
10. Bonorden, W.R. and Pariza, M.W., 1994. Antioxidant nutrients and protection from free radicals. In Kotonsis, F.N., Mackey, M. and Hielle, J. (eds.) *Nutritional Toxicology, Raven Press Ltd., New York*, pp. 19-48.
11. Dragsted, L.O., Strube, M. and Larsen, J.C. Cancer-protective factors in fruits and vegetables: biochemical and biological background. *Pharmacol.Toxicol.*, 1993, **72** Suppl. 1, 116-35, 116-135.

12.   Ames, B.N.,. Dietary carcinogens and anticarcinogens. Oxygen radicals and degenerative diseases. *Science.* 1983, **221**, 1256-1264.

13.   Cerutti, P.A. Prooxidant states and tumor promotion. *Science.* 1985, **227**, 375-381.

14.   Ames, B.N. Endogenous oxidative DNA damage, aging, and cancer. *Free Radic.Res.Commun.*1989, **7**, 121-128.

15.   Ames, B.N., Shigenaga, M.K. and Hagen, T.M. Oxidants, antioxidants, and the degenerative diseases of aging. *Proc. Natl. Acad. Sci U.S A.* 1993, **90**, 7915-7922.

16.   Shigenaga, M.K., Aboujaoude, E.N., Chen, Q. and Ames, B.N. Assays of oxidative DNA damage biomarkers 8-oxo-2'-deoxyguanosine and 8- oxoguanine in nuclear DNA and biological fluids by high-performance liquid chromatography with electrochemical detection. *Methods Enzymol.* 1994 **234:16-33**, 16-33.

17.   Floyd, R.A. The role of 8-hydroxyguanine in carcinogenesis. *Carcinogenesis.* 1990, **11**, 1447-1450.

18.   de Kok, T.M., ten Vaarwerk, F., Zwingman, I., van Maanen, J.M. and Kleinjans, J.C. Peroxidation of linoleic, arachidonic and oleic acid in relation to the induction of oxidative DNA damage and cytogenetic effects. *Carcinogenesis.* 1994, **15**, 1399-1404.

19.   Park, J.W. and Floyd, R.A. Lipid peroxidation products mediate the formation of 8-hydroxydeoxyguanosine in DNA. *Free Radic.Biol.Med.* 1992, **12**, 245-250.

20.   Halliwell, B. and Gutteridge, J.M., 1989. Lipid Peroxidation: a radical chain reaction. In AnonymousFree Radicals in Biology and Medicine, Clarendon Press, Oxford, pp. 188-276.

21.   Bull, A.W., Nigro, N.D., Golembieski, W.A., Crissman, J.D. and Marnett, L.J., 1984. *in vivo* stimulation of DNA synthesis and induction of ornithine decarboxylase in rat colon by fatty acid hydroperoxides, autoxidation products of unsaturated fatty acids. *Cancer Res.*, **44**, 4924-4928.

22.   Bull, A.W., Nigro, N.D. and Marnett, L.J., 1988. Structural requirements for stimulation of colonic cell proliferation by oxidized fatty acids. *Cancer Res.*, **48**, 1771-1776.

23.   Bull, A.W., Earles, S.M. and Bronstein, J.C., 1991. Metabolism of oxidized linoleic acid: distribution of activity for the enzymatic oxidation of 13-hydroxyoctadecadienoic acid to 13- oxooctadecadienoic acid in rat tissues. *Prostaglandins*, **41**, 43-50.

24.   Wang, Z., Ciabattoni, G., Creminon, C., Lawson, J., FitzGerald, G.A., Patrono, C. and Maclouf, J., 1995. Immunological characterization of urinary 8-epi-prostaglandin F2 alpha excretion in man. *J.Pharmacol.Exp.Ther.*, **275**, 94-100.

25.   Awad, J.A., Morrow, J.D., Takahashi, K. and Roberts, L.J., 1993. Identification of non-cyclooxygenase-derived prostanoid (F2- isoprostane) metabolites in human urine and plasma. *J.Biol.Chem.*, **268**, 4161-4169.

26.   Morrow, J.D., Hill, K.E., Burk, R.F., Nammour, T.M., Badr, K.F. and Roberts, L.J. , 1990. A series of prostaglandin F2-like compounds are produced *in vivo* in humans by a non-cyclooxygenase, free radical-catalyzed mechanism. *Proc. Natl. Acad. Sci. U.S.A.*, **87**, 9383-9387.

27. Morrow, J.D., Awad, J.A., Boss, H.J., Blair, I.A. and Roberts, L.J., 1992. Non-cyclooxygenase-derived prostanoids (F2-isoprostanes) are formed in situ on phospholipids. *Proc. Natl. Acad. Sci. U.S.A.*, **89**, 10721-10725.

28. Roberts, L.J. and Morrow, J.D., 1994. Isoprostanes. Novel markers of endogenous lipid peroxidation and potential mediators of oxidant injury. *Ann. N.Y. Acad. Sci.*, **744:237-42**, 237-242.

29. Roberts, L.J., Moore, K.P., Zackert, W.E., Oates, J.A. and Morrow, J.D., 1996. Identification of the major urinary metabolite of the F2-isoprostane 8- iso-prostaglandin F2alpha in humans. *J.Biol.Chem.*, **271**, 20617-20620.

30. Draper, H.H. and Hadley, M., 1990. Malondialdehyde determination as index of lipid peroxidation. Methods Enzymol., **186**, 421-431.

31. Haegele, A.D., Wolfe, P. and Thompson, H.J., 1998. X-radiation induces 8-hydroxy-2'-deoxyguanosine formation *in vivo* in rat mammary gland DNA. *Carcinogenesis*, **19**, 1319-1321.

32. Peng, Y.M., Peng, Y.S., Lin, Y., Moon, T., Roe, D.J. and Ritenbaugh, C., 1995. Concentrations and plasma-tissue-diet relationships of carotenoids, retinoids, and tocopherols in humans. *Nutr. Cancer*, **23**, 233-246.

33. Polsinelli, M., Rock, C.L., Henderson, S.A. and Drewnowski, A., 1998. Plasma carotenoids as biomarkers of fruit and vegetable servings in women. *J. Am. Diet. Assoc.*, **98**, 194-196.

34. Verhagen, H., Poulsen, H.E., Loft, S., van Poppel, G., Willems, M.I. and van Bladeren, P.J., 1995. Reduction of oxidative DNA-damage in humans by brussels sprouts. *Carcinogenesis*, **16**, 969-970.

35. Fraga, C.G., Shigenaga, M.K., Park, J.W., Degan, P. and Ames, B.N. (1990) Oxidative damage to DNA during aging: 8-hydroxy-2'-deoxyguanosine in rat organ DNA and urine. *Proc. Natl. Acad. Sci U.S A.*, **87**, 4533-4537.

36. Tagesson, C., Kallberg, M., Klintenberg, C. and Starkhammar, H., 1995. Determination of urinary 8-hydroxydeoxyguanosine by automated coupled- column high performance liquid chromatography: a powerful technique for assaying *in vivo* oxidative DNA damage in cancer patients. *Eur. J. Cancer*, **31A**, 934-940.

37. Park, E.M., Shigenaga, M.K., Degan, P., Korn, T.S., Kitzler, J.W., Wehr, C.M., Kolachana, P. and Ames, B.N., 1992. Assay of excised oxidative DNA lesions: isolation of 8-oxoguanine and its nucleoside derivatives from biological fluids with a monoclonal antibody column. *Proc. Natl. Acad. Sci U.S A.*, **89**, 3375-3379.

38. Shigenaga, M.K., Gimeno, C.J. and Ames, B.N., 1989. Urinary 8-hydroxy-2'-deoxyguanosine as a biological marker of *in vivo* oxidative DNA damage. *Proc. Natl. Acad. Sci U.S A.*, **86**, 9697-9701.

39. Loft, S., Vistisen, K., Ewertz, M., Tjonneland, A., Overvad, K. and Poulsen, H.E. , 1992. Oxidative DNA damage estimated by 8-hydroxydeoxyguanosine excretion in humans: influence of smoking, gender and body mass index. *Carcinogenesis*, **13**, 2241-2247.

40. Lagorio, S., Tagesson, C., Forastiere, F., Iavarone, I., Axelson, O. and Carere, A. , 1994. Exposure to benzene and urinary concentrations of 8-hydroxydeoxyguanosine, a biological marker of oxidative damage to DNA. *Occup. Environ. Med.*, **51**, 739-743.

432 *Natural Antioxidants and Anticarcinogens in Nutrition, Health and Disease*

41. Draper, H.H., Polensek, L., Hadley, M. and McGirr, L.G., 1984. Urinary malondialdehyde as an indicator of lipid peroxidation in the diet and in the tissues. *Lipids*, **19**, 836-843.
42. McGirr, L.G., Hadley, M. and Draper, H.H., 1985. Identification of N α-acetyl-epsilon-(2-propenal)lysine as a urinary metabolite of malondialdehyde. *J. Biol. Chem.*, **260**, 15427-15431.
43. Knight, J.A., Pieper, R.K. and McClellan, L. (1988) Specificity of the thiobarbituric acid reaction: its use in studies of lipid peroxidation. *Clin. Chem.*, **34**, 2433-2438.
44. Lazzarino, G., Tavazzi, B., Di Pierro, D., Vagnozzi, R., Penco, M. and Giardina, B. , 1995) The relevance of malondialdehyde as a biochemical index of lipid peroxidation of postischemic tissues in the rat and human beings. *Biol. Trace Elem. Res.*, **47**, 165-170.
45. Liu, J., Yeo, H.C., Doniger, S.J. and Ames, B.N., 1997. Assay of aldehydes from lipid peroxidation: gas chromatography-mass spectrometry compared to thiobarbituric acid. *Anal. Biochem.*, **245**, 161-166.
46. Duthie, G.G., Morrice, P.C., Ventresca, P.G. and McLay, J.S., 1992. Effects of storage, iron and time of day on indices of lipid peroxidation in plasma from healthy volunteers. *Clin. Chim. Acta*, **206**, 207-213.

ELEVATED DNA DAMAGE IN LYMPHOCYTES FROM ANKYLOSING SPONDYLITIS PATIENTS

Maria Dušinská[1], Katarína Raölová[1], Jan Lietava[2], Martina Somorovská[1], Helena Petrovská[1], Petra Dobríková[1], and Andrew Collins.[3]

[1]Institute of Preventive and Clinical Medicine, Bratislava, Slovak Republic. [2]2nd Department of Internal Medicine, Comenius University, Bratislava, Slovak Republic. [3]Rowett Research Institute, Aberdeen, Scotland, UK.

Keywords: Ankylosing spondylitis, hyperglycemia, hyperlipidemia, oxidative stress, DNA damage, comet assay, oxidized bases.

1 INTRODUCTION

Oxidative stress is a factor in many diseases, either as cause or effect. We suppose that oxidative damage to the DNA of lymphocytes may serve as an objective indicator of the level of oxidative stress. We therefore examined this parameter in a sample of patients with the rheumatoid disease ankylosing spondylitis (AS), in patients suffering from insulin dependent diabetes (IDDM) as well as in patients with primary hypertriglyceridemia/hypo-alphalipoproteinemia (HLP).

The comet assay, a sensitive method for measuring strand breaks (and alkali-labile sites) in individual cells was used. We have modified it for the detection of particular kinds of damage, by incubating nucleoids with lesion-specific repair endonucleases.[1,2] Here we use 2 enzymes -endonuclease III, which converts oxidized pyrimidines to strand breaks; and formamidopyrimidine glycosylase (FPG), to detect altered purines including 8-oxo-guanine. Over the range of damage we observe, the % DNA in the tail is linearly related to the frequency of strand breaks.[3] It is therefore valid to estimate the net endonuclease III sensitive or net FPG-sensitive sites as the difference between values of % DNA in the tail obtained with and without the respective enzyme.

2 MATERIALS AND METHODS

2.1. Subjects

Subjects were 8 male patients with AS, 10 patients with IDDM, 7 patients with primary HLP (hypertriglyceridemia/hypoalphalipoproteinemia) and 8 apparently healthy male controls. All subjects gave their informed consent, and the study was approved by the Ethical Committee of the Institute of Preventive and Clinical Medicine, Bratislava.

Table 1

Characteristics of patients and control group.

subjects	AS	IDDM	HLP	control
number	8	10	7	8
sex	M	M	M	M
average age	42.9 ± 6.7	48.4 ± 7.9	52.2 ± 8.1	44.4 ± 5.5
range	35 - 56	33 - 59	40 - 60	36 - 53
BMI (kg/m^2)	24.5 ± 3.6	27.5 ± 4.1	28.9 ± 3.7	24.7 ± 3.5

2.2. Measurement of DNA damage

The assay has been fully described.[3] Briefly, lymphocytes isolated by centrifugation on a Ficoll-based density gradient are embedded in agarose on a microscope slide and lysed in asolution containing Triton X-100 and 2.5 M NaCI. The resulting nucleoids are electrophoresed under alkaline conditions; the presence of breaks in the DNA allows it to extend towards the anode, forming a comet-like image when viewed by fluorescence microscopy. If, after lysis, the nucleoids are incubated with a repair endonuclease specific for certain kinds of damage, additional breaks are formed at sites of such lesions, and the relative amount of DNA in the tail of the comet is increased. In this study, endonuclease III was used to detect oxidized pyrimidines, and formamidopyrimidine glycosylase (FPG) to detect damaged purines including 8-oxo-guanine. Comets in each gel were analyzed (blind) using a CCD camera and Komet 3.0 image analysis programme (Kinetic Imaging Ltd, Liverpool, UK).

3 RESULTS AND DISCUSSION

Three groups of patients with chronic degenerative disease (AS, IDDM and HLP) were assessed for DNA damage in peripheral blood lymphocytes and compared with normal healthy subjects. AS patients were selected from the cohort already described.[4] All patients had verified AS (stage IV-V with functional classification b-c); average duration of symptoms was 11.2 years (range 1-18) and time since diagnosis 7.2 years (range 1-18). All subjects were without history of cardiovascular disease, without diabetes mellitus or hyperlipidemia, had normal resting blood pressure, gave normal results in a symptom-limited maximal exercise test, and showed no ECG, vectocardiographic or echocardiographic signs of myocardial infarction. IDDM patients were defined according to WHO criteria. The duration of diabetes ranged from 6-30 years, mean 15.6. The mean fasting glucose level (measured enzymically using a Hitachi 911 autoanalyzer) in the IDDM group was 10.2 ± 4.8 mmol/l, range 3.8-17.8. Controls had apparently normal glucose metabolism. Patients with primary HLP were from the Lipid Clinic of the IPCM, Bratislava. None of the patients had any disorder causing secondary hyperlipoproteinemia, and none had recent acute myocardial infarction, stroke, surgery, unstable angina pectoris or gastroduodenal ulcer. Table 1 summarizes the demographic characteristics of the subjects.

Table 2 gives mean values of % DNA in tail for strand breaks, FPG-sensitive, and endonuclease III-sensitive sites. Subtracting the % DNA in tail without enzyme incubation (i.e. strand breaks) from the % DNA in tail with enzyme incubation gives the net amount of damage represented by FPG- or endonuclease III-sensitive sites. 10 % of DNA in the tail corresponds to about 1000 breaks per cell. Our results indicate that strand breaks are significantly elevated in AS (P<0.01), IDDM (P<0.0001) and HLP patients' lymphocytes (P=0.0003) compared with control. We found DNA strand breaks to be significantly correlated with BMI, in the diabetic group. FPG-sensitive sites are significantly higher in AS Iymphocytes than in the controls (P<O.OS). There is no significant difference between diabetics and controls in FPG- sensitive sites (P=0.28). Endonuclease III-sensitive sites, an index of overall oxidative damage to DNA, are significantly elevated in diabetics (P<0.01) but not in AS (P=0.25). The levels of FPG- and Endonuclease III-sites in HLP patients is not higher than in controls. All 8 normal subjects had, as expected, low levels of strand breaks. The levels of FPG- and endonuclease III-sensitive sites were more variable than strand breaks, being as low in some patients as in normal controls (data not shown).

Table 2

Oxidative DNA damage associated with ankylosing spondylitis (AS), diabetics (IDDM) and hyperlipidemia (HLP).

	% of DNA in tail		
	Strand breaks	**FPG sites**	**Endo III sites**
Control	11.8±2.4	3.3±3.0	8.8±6.0
AS	34.4±10.7	14.5±7.6	12.6±6.7
IDDM	25.1±7.5	8.1±7.8	21.3±12.9
HLP	29.8±6.3	1.75±1.7	3.1±3.0

DNA damage was estimated by computer image analysis. The mean % of DNA in comet tails (mean of 50 comets per gel, 2 gels per sample) was calculated for each sample. The mean % DNA in tail was then calculated for each group of samples, and this value is shown here ± SD. The % of DNA in the tail is a measure of DNA break frequency.

It is clear that IDDM and AS as well as HLP lymphocytes are distinguished from normal lymphocytes by a higher level of DNA damage as measured by the comet assay. In the case of AS, the most likely cause of damage is the reactive oxygen released as a result of chronic inflammation of connective tissue. Diabetes and HLP are major risks factors for atherosclerosis. Higher oxidative DNA damage in IDDM patients was found also by Dandona et al.,[5] We found that serum glucose concentrations in IDDM are strongly correlated with oxidised purines.[6] In HLP patients the high level of DNA breaks could be caused via low density proteins (LDL) which were shown to have decreased resistance to in vitro oxidation probably as a result of changes in the LDL subfraction profile.[7, 8]

In conclusion, our results indicate that measurement of DNA damage as an indicator of oxidative stress may be a useful aid to diagnosis of certain diseases.

Acknowledgement

This work was supported by EC contract CIPA-CT94-0129, by EC contract IClS-CT96-1012, by the Ministry of Health of the Slovak Republic and the Scottish Office Agriculture, Environment and Fisheries Department.

References

1. A.R. Collins, S.J. Duthie. and V.L. Dobson, *Carcinogenesis*. 1993, **14**, 1733.
2. M. Dušinská and A. R. Collins, *ATLA*. 1996, **24**, 405.
3. A.R. Collins, M. Dušinská, C.M. Gedik, and R. Ätïtina, *Environ. Health Perspect.* 1996, **104**, suppl. 3, 465.
4. J. Lietava, M. Sitár, J. Lukáë, J. Zim Uová, P. Lukáë, P. Váûny, J. Šelko and A. Dukát, Proceeding of 14 th International Conference of International Society of Noninvasive Cardiology. 1996, Cambridge.
5. P. Dandona, K. Thusu, S. Cook, B. Snyder, J. Makowski, D. Armstrong and T. Nicotera, *Lancet*. 1996, **347**, 444.
6. A.R. Collins, K. Raölová, M. Somorovská, H. Petrovská, A. Ondruöová, B. Vohnout, R. Fá bry and M. Dusinska, *Free Radical Biology and Medicine*. 1998, **25**, 373.
7. K. Raölová, M. Dobiáöová, A. Nagyová, R. Fábry and M. Dušinská, *Eur. J. Clin. Pharmacology*, in press.
8. R. Kinscherf, R. Claus, M. Wagner, C. Gehrke, H. Kamencic, D. Hou, O. Nauen, W. Schmiedt, G. Kovacs, J. Pill, J. Metz and H-P. Diegner, *FASEB J*, 1998, 12, in press.

INHIBITORY EFFECT OF CITRUS EXTRACTS ON THE GROWTH OF BREAST CANCER CELLS *IN VITRO*

Najla Guthrie and Kenneth Carroll.

Centre for Human Nutrition, Department of Biochemistry, The University of Western Ontario, London, Ontario, N6A 5C1, Canada.

1 INTRODUCTION

Breast cancer is the most prevalent cancer in women of developed countries, and its incidence has been increasing world wide.[1] Attempts to improve survival and reduce the risk of relapse following diagnosis have shown limited success and there is still substantial room for improvement. Although researchers are evaluating new drugs, another promising approach is the investigation of dietary components as anti-cancer agents.

There is general agreement that plant-based diets, rich in whole grains, legumes, fruits and vegetables, reduce the risk of various types of cancer, including breast cancer, and a variety of compounds produced by plants have been investigated for their anti-cancer activity.[2-4] These include the flavonoids, which are widely distributed in plants.[5] They are present in citrus and other fruits and also in vegetables, grains, seeds, nuts, tea and wine.[6]

Previous studies have shown that diets containing a high level of palm oil do not promote chemically-induced mammary carcinogenesis in rats.[7] Palm oil stripped of its vitamin E fraction promoted mammary carcinogenesis like other unsaturated fats and oils.[8] The vitamin E of palm oil consists mainly of tocotrienols (~70 %) with the remainder being γ-tocopherol.[9]

2 CELL CULTURE STUDIES

We have previously reported that citrus flavonoids inhibited proliferation of human breast cancer cells in culture.[10,11] These include hesperetin (from oranges), naringenin (from grapefruit), tangeretin and nobiletin (from tangerines). In ER- cells the IC_{50}s for hesperetin and naringenin were 18 μg/mL, whereas for tangeretin and nobiletin, the IC_{50}s were 0.5 μg/mL. The citrus flavonoids were subsequently tested on ER+ human breast cancer cells and were found to inhibit proliferation and growth of these cells (IC_{50}s: naringenin 18, hesperetin 12, nobiletin 0.8 and tangeretin 0.4 μg/mL).[11]

We have also tested the tocotrienol-rich fraction (TRF) and the individual tocotrienols (α -, ß-, γ-) on the proliferation of both ER- and ER+ human breast cancer cells in culture. Tocotrienols were effective inhibitors of proliferation of both cell types, with IC_{50}s of 30-180 μg/mL in ER- cells and 2-6 μg/mL in ER+ cells, whereas γ-tocopherol was ineffective.[12]

In other studies with ER- and ER+ human breast cancer cells *in vitro*, we have observed that 1:1 combinations of flavonoids with tocotrienols or tamoxifen and 1:1 combinations of tocotrienols with tamoxifen inhibit proliferation of the cells more effectively than the individual compounds by themselves. The most effective combination with ER- cells was tangeretin and γ-tocotrienol (IC_{50} 0.05 μg/mL).[11] With ER+ cells, the best results were obtained with tangeretin and γ-tocotrienol (IC_{50} 0.02 μg/mL), nobiletin + tamoxifen (IC_{50} 0.004 μg/mL) and γ-tocotrienol + tamoxifen (IC_{50} 0.003 μg/mL).[11] When combined in 1:1:1 combinations of flavonoids, tocotrienols and tamoxifen, tangeretin + γ-tocotrienol + tamoxifen was the most effective in ER- cells (IC_{50} 0.01 μg/mL) and hesperetin + γ-tocotrienol + tamoxifen was the most effective in ER+ cells (IC_{50} 0.0005 μg/mL).[11]

3 ANIMAL STUDIES

We have previously reported that orange juice and naringin (glycoside form of the flavonoid naringenin from grapefruit) delayed the development of mammary tumors induced by 7,12-dimethylbenz(a)anthracene (DMBA).[10] In a more recent experiment, we investigated the effects of orange juice, grapefruit juice, their constituent flavonoids and their glycosides on the growth of MDA-MB-435 ER- human breast cancer cells injected into the mammary fat pad of nude mice and their ability to metastasize to the lymph nodes and lungs. The incidence of mammary fat pad tumors was reduced by more than 50 % in the mice given orange juice, grapefruit juice or naringin.[13] Lymph node metastases and lung metastases were lowest in the orange juice and grapefruit juice groups, followed by the groups given naringin, hesperidin or naringenin.[13] Our results indicate that growth and metastasis of these tumors in nude mice are strongly inhibited by orange and grapefruit juice and this inhibition cannot be entirely attributed to their constituent flavonoids. We have also investigated another class of compounds present in citrus, the limonoids, which have anti-cancer activity.

Citrus limonoids are one of the two bitter principles found in citrus fruits, such as lemon, lime, orange and grapefruit.[14] They have been shown to have anti-cancer activity.[15] Nomilin reduced the incidence and number of chemically-induced forestomach tumors in mice when given by gavage. Addition of nomilin and limonin to the diet inhibited lung tumor formation in mice and topical application of the limonoids was found to inhibit both the initiation and the promotion phases of carcinogenesis in the skin of mice. We have recently tested the effect of nomilin, limonin and limonin glucoside on the proliferation and growth of ER- human breast cancer cells in culture. Nomilin was the most effective, having an IC_{50} of 0.4 μg/mL. We also tested a glucoside mixture and found it to have an even lower IC_{50} of 0.08 μg/mL.[16]

Acknowledgements

The support of the State of Florida Department of Citrus for the studies on citrus juices and flavonoids and the Palm Oil Research and Development Board for the studies on tocotrienols is gratefully acknowledged. The authors would like to thank Dr. Vince Morris for advice concerning the mammary fat pad injections, Debbie Friedrich, Hung Dao Tran,

Juliet and Josephine Ho for excellent technical assistance, and Charlotte Harman for typing this manuscript.

References

1. P. Pisani. *J. Environ. Pathol. Toxicol. Oncol.* 1992, **11**, 313.
2. J. Kuhnau. *World Rev. Nutr. Diet.* 1976, **24**, 117.
3. K.A. Steinmetz and J.D. Potter. *Cancer Causes Control.* 1991, **2**, 427.
4. L.W. Wattenburg. *Cancer Res.* 1992, **52**, 2085s.
5. J.A. Manthey and B.S. Buslig. Flavonoids in the Living System. Plenum Press, New York, 1998.
6. M.G.L. Hertog, P.C.H. Hollman, M.B. Katan and D. Kromhout. *Nutr. Cancer.* 1993, **20**, 21.
7. K. Sundram, H.T. Khor, A.S.H. Ong and R. Pathmanathan. *Cancer Res.* 1989, **49**, 1447.
8. K. Nesaretnam, H.T. Khor, J., Janeson, Y.H. Chong, K. Sundram and A. Gapor. *Nutr. Res.* 1992, **12**, 63.
9. K. Nesaretnam, N. Guthrie, A.F. Chambers and K.K. Carroll. *Lipids.* 1995, **30**, 1139.
10. F.V. So, N. Guthrie, A.F. Chambers, M. Moussa and K.K. Carroll. *Nutr. Cancer.* 1996, **26**, 167.
11. N. Guthrie, A. Gapor, A.F. Chambers and K.K. Carroll. *Asia Pacific J.Clin. Nutr.* 1997, **6**, 41.
12. N. Guthrie, A. Gapor, A.F. Chambers and K.K. Carroll. *J. Nutr.* 1997, **127**, 544S.
13. N. Guthrie and K.K. Carroll. Biological Oxidants and Antioxidants: Molecular Mechanisms and Health Effects. AOCS Press, Champaign, IL, 1998.
14. S. Hasegawa, M. Miyake and Y. Ozaki. Food Phytochemicals for Cancer Prevention I, Fruits and Vegetables. American Chemical Society, Washington, DC, 1994.
15. L.K.T. Lam, J. Zhang, S. Hasegawa and H.A.J. Schut. Food Phytochemicals for Cancer Prevention I, Fruits and Vegetables. American Chemical Society, Washington, DC, 1994.
16. N. Guthrie, A.F. Chambers and K.K. Carroll. *Proc. Am. Assoc.Cancer Res.* 1997, **38**, 113.

THE EFFECT OF PHYSICAL PROCESSING ON THE PROTECTIVE EFFECT OF BROCCOLI IN RELATION TO DAMAGE TO DNA IN COLONOCYTES

Brian Ratcliffe[1], Andrew R. Collins[2], Helen J. Glass[1], and Kevin Hillman[3].

The Boyd Orr Research Centre-Aberdeen Research Consortium at [1]The Robert Gordon University, Kepplestone, Aberdeen AB15 4PH, [2]The Rowett Research Institute, Bucksburn, Aberdeen AB21 9SB and [3]The Scottish Agricultural College, Craibstone, Aberdeen AB21 9YA.

1 INTRODUCTION

There is increasing evidence for negative associations between fruit and vegetable consumption and the occurrence of many forms of cancer.[1] Many authors have explored these relationships in terms of the associated intakes of antioxidant vitamins.

For cancer of the large bowel, the protective effect of vegetables is unlikely to be a simple correlation with vitamin intake since work with supplements has shown no beneficial effect.[2] Some evidence shows that the consumption of raw vegetables and salads may be important in protection against colorectal cancer.[3] Why raw plant foods seem to impart greater protection merits further study.

It has been hypothesized that intact plant cells act as "vehicles" for the delivery of antioxidants and anticancer agents to the colon where they are "unpacked" by the process of fermentation of cell walls and associated material.[4] The aim of this experiment was to examine the effect of simple physical processing on the protective effect of broccoli at the level of colonic cells. Oxidative DNA damage was used as an index of carcinogenic potential.

2 MATERIALS AND METHOD

Pigs were used as experimental models because of the problems with accessing tissue from the large bowels of healthy human subjects. Fifteen, male Landrace x Large White pigs were divided into three groups which were matched for age and starting weight. Each group received subsequently a standard, high quality, cereal-based diet (control group, C) or this same diet supplemented with whole, raw broccoli (treatment group, W) or with similar amounts of broccoli which had been physically processed (treatment group, H). The latter group's vegetable supplement was homogenized using a domestic food processor to disrupt the cellular structure. The pigs were 87 d of age at the start of the experiment and they were maintained on the diets for twelve days. The animals received their feed daily in two meals and the amounts were controlled to maintain similar dry matter intakes in all groups. Feed consumption was monitored on a daily basis.

At the end of the experimental period, the pigs were weighed then anaesthetized and killed. Post mortem, the colon was excised from caecum to rectum. In addition, samples of faeces, colonic contents, and blood were obtained for further analyses. Enumeration of selected microbial groups (coliforms, lactobacilli and bifidobacteria) was performed immediately on the samples of colonic contents and faeces. The anatomical mid-point of the colon was located and a section of 200 mm length was taken for the isolation of colonocytes by a modification of the method of Brendler-Schwaab et al.[5] Isolated colonic cells were suspended in freezing medium (90 % FCS, 10 % DMSO) at a density of 3 x 106 ml^{-1} and were stored at -80 °. The thawed colonocytes were assayed subsequently for DNA strand breaks using a modification of the "comet" method of Collins et al.[6] With this technique, "comets" arising from colonic cells were allocated a score from 0-4 indicating the increasing degree of damage. A total of 100 colonocytes was examined blind and at random for each pig.

Plasma vitamin C concentrations were determined by reversed phase HPLC using ion-pairing reagent with UV detection.[7]

3 RESULTS

There was no significant difference in food intake (mean 1 070 g dry matter d^{-1}) or final body weights (mean 32.9 kg) between the groups. Groups W and H consumed similar amounts of broccoli (approximately 600 g d^{-1}).

The mean pooled scores (arbitrary units) for "comets" were respectively 207 (SD 10.3) for C, 212 (SD 30.6) for H and 113 (SD 23.7) for W (P<0.05).

There was no significant difference between the treatment groups in the bacterial numbers of coliforms, lactobacilli and bifidobacteria examined within the colonic contents and faeces.

While both treatment groups had elevated levels (P<0.05) of plasma vitamin C, 30.45 µM (SD 9.45) for W and 28.35 µM (SD 10.02) for H, compared with the control group, 16.87 µM (SD 2.32), there was no significant difference between W and H in this regard.

4 DISCUSSION AND CONCLUSIONS

The fresh vegetable supplement produced elevated levels of plasma vitamin C regardless of the physical state of the broccoli. This was probably related to the vitamin C released by digestive processes (and physical processing prior to consumption for group H) in the upper alimentary tract. However, at the level of the colon, the physical nature of the vegetable was important. There was clearly a very marked protective effect against DNA strand breaks as measured by "comet" analysis from the consumption of whole, raw broccoli and this protection was removed if the broccoli was homogenized prior to consumption. The observed effects were not mediated by large scale changes in bacterial populations since the numbers of coliforms, lactobacilli and bifidobacteria in colonic contents and faeces were not affected by treatment.

It is accepted that DNA damage is not necessarily an indicator of carcinogenic processes. Nevertheless, it can act as an indicator of undesirable conditions which may be conducive to mutagenesis or carcinogenesis.

The results support the hypothesis that intact plant cells are important for the health of the large bowel. The concept requires further investigation but it could have important significance for the nature of dietary recommendations aimed at preventing colonic cancer.

Acknowledgement

This work was supported by a grant from the World Cancer Research Fund.

References

1. World Cancer Research Fund, American Institute for Cancer Research. Food, nutrition, and the prevention of cancer: a global perspective. *American Institute for Cancer Research, Washington DC*, 1997.
2. E.R. Greenberg, J.A. Baron, T.D. Tosteson, D.H. Freeman, G.J. Beck, J.H. Bond, T.A. Colacchio, J.A. Coller, H.D. Frankl, R.W. Haile, J.S. Mandel, D.W. Nierenberg, R. Rothstein, D.C. Snover, M. M. Stevens, R.W. Summers and R.U. van Stolk. *New England Journal of Medicine*. 1994, **331**, 141.
3. K. A. Steinmetz and J. D. Potter *JD. Cancer Causes and Control.* 1991, **2**, 325.
4. B. Ratcliffe, A.R. Collins, H.J. Glass, K. Hillman and R.J.T. Kemble. *Cancer Letters*. 1997, **114**, 57.
5. S.Y. Brendler-Schwaab, P. Schmezer, U. Liegibel, S. Weber, K Michalek, A. Tompa and B.L. Pool-Zobel. *Toxicology in Vitro*. 1994. **8**, 1285.
6. A.R. Collins, S.J. Duthie and V.L. Dobson. *Carcinogenesis*. 1993, **14**, 1733.
7. M. Ross. *Journal of Chromatography*. 1994, **675**, 197.

DIFFERENTIAL EFFECT OF A NOVEL GRAPE SEED PROANTHOCYANIDIN EXTRACT ON CULTURED HUMAN NORMAL AND MALIGNANT CELLS

R.L. Krohn,[1] X. Ye,[1] W. Liu,[1] S.S. Joshi,[2] M. Bagchi,[1] H.G. Preuss,[3] S.J. Stohs[1] and D. Bagchi[1],*

[1]Creighton University School of Pharmacy and Allied Health Professions, Omaha; [2]University of Nebraska Medical Center, Omaha; [3]Georgetown University Medical Center, Washington, USA. *Author for correspondence

1 INTRODUCTION

Proanthocyanidins are naturally occurring polyphenolic bioflavonoids diverse in chemical structure, pharmacology and characteristics, and widely available in fruits, vegetables, nuts, seeds, flowers and bark. Proanthocyanidins are natural antioxidants which are known to possess a broad spectrum of biological, pharmacological, chemoprotective and medicinal properties against free radicals and oxidative stress.[1] We have previously assessed the concentration- or dose-dependent free radical scavenging abilities of a novel IH636 grape seed proanthocyanidin extract (GSPE), in both *in vitro* and *in vivo* models, and compared the free radical scavenging ability of GSPE with vitamin C, vitamin E and β-carotene. These experiments demonstrated that GSPE is highly bioavailable and provides significantly greater protection against biochemically generated free radicals and free radical-induced lipid peroxidation and DNA damage than vitamin C, vitamin E, a combination of vitamins C plus E, and β-carotene.[2-5]

In the present study, the concentration- and time-dependent cytotoxicity of GSPE was assessed against selected human cancer cells including MCF-7 human breast cancer cells, A-427 human lung cancer cells, CRL 1739 human gastric adenocarcinoma cells and K562 chronic myelogenous leukemic cells using cytomorphology and the tetrazolium salt MTT cytotoxicity assay. In addition, we compared the effects of GSPE on normal human gastric mucosal cells, with the effects on the cancer cell lines.

2 MATERIALS AND METHODS

A commercially available dried, powdered IH636 grape seed proanthocyanidin extract (GSPE) (batch no. AV 609016) was obtained from InterHealth Nutraceuticals Incorporated (Concord, CA). MTT [3-(4,5-dimethylthiazol-2-yl)-2,5-diphenyltetrazolium bromide; thiazolyl blue] (Catalog #M-2128) was purchased from Sigma Chemical Co. (St. Louis, MO). RPMI-1640 medium, Dulbecco's Modified Eagle Medium (DMEM) and fetal bovine serum were obtained from GIBCO BRL (Gaithersburg, MD). Ham's F12 medium was purchased from Sigma Chemical Co. (St. Louis, MO). Keratinocyte growth medium (KGM) was obtained from Clonetics Corporation (LaJolla, CA). Unless otherwise stated all other chemicals used in this study were obtained from Sigma Chemical Co. and were of analytical grade or the highest grade available. Phase contrast microscopy or cytomorphology was performed using a Nikon TMS-F inverted microscope equipped with a Nikon N2000 camera.

2.1. Cell Cultures and Treatment

MCF-7 human breast cancer cells, A-427 human lung cancer cells, K562 human chronic myelogenous leukemic cells and CRL 1739 human gastric adenocarcinoma cells were obtained from American Type Culture Collection (ATCC, Rockville, MD). MCF-7 breast cancer cells, A-427 lung cancer cells and K562 chronic myelogenous leukemic cells were maintained and grown in culture flasks in RPMI 1640 medium supplemented with 10 % fetal bovine serum. CRL 1739 human gastric adenocarcinoma cells were maintained and grown in Ham's F12 medium containing 10 % fetal bovine serum. Normal human gastric mucosal cells were cultured from *Helicobacter pylori*-negative endoscopic biopsies from human subjects as described earlier.[6] Gastric cells were grown in KGM growth medium supplemented with 10 % heat-inactivated fetal calf serum in an atmosphere of 5 % CO_2 in 100 % humidity. Cell cultures were incubated at 37 °C in a humidified atmosphere containing 5 % CO_2 with regular passage. Trypsin solution was used to split cultures whenever they were grown to confluence. The number of cells was determined using a Coulter Counter. Viability was checked using the Trypan blue exclusion technique. Cultured cells were individually incubated with 25 mg/lit and 50 mg/lit concentrations of GSPE for 0, 24, 48 and 72 h at 37 °C. GSPE was freshly prepared by dissolving in culture media prior to each experiment.

2.2. Cytomorphology

Changes in cell morphology or cytomorphology following treatment with GSPE were detected using phase contrast microscopy, and the concentration- and time- dependent detachment of dead or dying cells following exposure to GSPE was monitored. Healthy cells remain elongated while dead or dying cells become rounded and loose their adhesion to the culture plate.

2.3. Colorimetric MTT (tetrazolium) Assay

The cleavage of the tetrazolium salt MTT into a blue colored formazan by the mitochondrial enzyme succinate dehydrogenase is potentially very useful for assaying cell survival and proliferation. The conversion takes place only in living cells and the amount of formazan produced is proportional to the number of cells present. Thus, the MTT assay detects living but not dead cells and the signal generated is dependent on the degree of activation of the cells. The MTT assay was conducted spectrophotometrically at 570 nm as described earlier by Mosmann[7] and us.[8]

2.4. Statistical Analyses

Significance between pairs of mean values was determined by Student's t test. A $p<0.05$ was considered significant for all analyses. Each value is the mean + standard deviation of 4-6 replicates.

3 RESULTS

3.1. GSPE-Induced Changes in Cell Morphology

Phase contrast microscopy or cytomorphology demonstrated that GSPE induces concentration-dependent changes in cell morphology in the human cancer and normal cells. With increasing concentration of GSPE and time, the MCF-7 human breast cancer cells, A-427 human lung cancer cells and CRL 1739 human gastric adenocarcinoma cells lost their adhesion to the culture plates and the cells became rounded in appearance. No concentration- and time-dependent cytotoxic effect of GSPE on cell morphology was observed on K562 chronic myelogenous leukemic cells and normal human gastric mucosal cells. In fact, GSPE enhanced the growth and viability of normal human gastric mucosal cells. Fig. 1 and 2 demonstrate the representative comparative concentration- and time-dependent effect of GSPE on CRL 1739 human gastric adenocarcinoma cells and on a primary culture of normal human gastric mucosal cells. Fig. 1 demonstrates the representative time-dependent cytotoxic effect of 25 mg/lit concentration of GSPE on CRL 1739 human gastric adenocarcinoma cells at 0, 24, 48 and 72 h of treatment based on the numbers of detached and rounded cells. Figure 2 demonstrates the effect of 25 mg/lit concentration of GSPE on normal human gastric mucosal cells at 0, 24, 48 and 72 h of treatment. Fig. 2 clearly showed that GSPE exhibits no cytotoxicity towards normal human gastric mucosal cells.

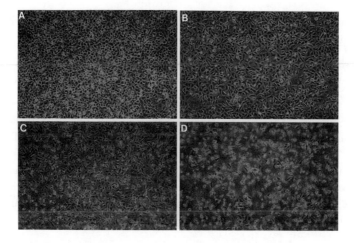

Figure 1
Time-dependent effect of GSPE on cultured CRL 1739 human gastric adenocarcinoma (GAC) cells as determined by cytomorphology. Cultured CRL 1739 GAC cells were incubated with 25 mg/lit concentration of GSPE for 0, 24, 48 and 72 h of treatment. A-Control; B- 24 h; C- 48 h and D- 72 h.

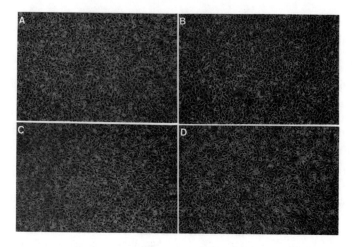

Figure 2
Time-dependent effect of GSPE on cultured human normal gastric mucosal cells as determined by cytomorphology. Cultured human normal gastric mucosal cells were incubated with 25 mg/lit concentration of GSPE for 0,24,48 and 72 h of treatment. A-Control; B- 24 h; C- 48 h and D- 72 h.

3.2. Concentration- and Time-Dependent Effects of GSPE on Selected Human Normal and Malignant Cells

The concentration- and time-dependent growth inhibitory effect of these cultured cells by GSPE *in vitro* was determined by MTT assay, a method for measuring cellular respiration that entails the conversion of MTT to colored formazan.

3.2.1. Cytotoxicity of GSPE against MCF-7 human breast cancer cells. Following incubation of the cultured MCF-7 human breast cancer cells at 37 °C without GSPE over a period of 0 to 72 h, a steady growth of these cells was observed. Following incubation of MCF-7 cells with 25 mg/lit of GSPE, approximately 7 %, 30 % and 43 % inhibition in cell growth was observed at 24, 48 and 72 h of incubation, respectively, while incubation of the breast cancer cells with 50 mg/lit of GSPE resulted in 11 %, 35 % and 47 % inhibition in cell growth at these same time points, respectively (Table 1).

3.2.2. Concentration- and time-dependent effect of GSPE against A-427 human lung cancer cells. A steady growth of A-427 lung cancer cells were observed following incubation without GSPE over a period of 0 to 72 h. Incubation of these A-427 cells with 25 mg/lit of GSPE, approximately 2 %, 20 % and 37 % inhibition in cell growth was observed at 24, 48 and 72 h of incubation, respectively, while incubation of the lung cancer cells with 50 mg/lit of GSPE resulted in 14 %, 32 % and 48 % inhibition in cell growth at these same time points, respectively (Table 1).

Table 1

Concentration and time-dependent cytotoxicity of GSPE against cultured MCF-7 human breast cancer and A-427 human lung cancer cells.

	Absorbance at 570 nm					
	MCF-7 Human Breast Cancer Cells			**A-427 Human Lung Cancer Cells**		
Time (h)	Control	GSPE 25 mg/lit	GSPE 50 mg/lit	Control	GSPE 25 mg/lit	GSPE 50 mg/lit
0	0.38±0.03	-	-	0.82±0.07	-	-
24	0.46±0.03	0.43±0.05	0.41±0.03	0.97±0.05	0.95±0.08	0.83±0.09
48	0.57±0.06	0.40±0.03*	0.37±0.07*	1.08±0.13	0.86±0.10*	0.73±0.12*
72	0.68±0.11	0.39±0.05*	0.32±0.04*	1.22±0.21	0.77±0.08*	0.64±0.09*

*Each value represents the mean ±S.D. of 4-6 experiments. *p<0.05 compared to the nontreated group.*

3.2.3. Cytotoxicity of GSPE against CRL 1739 human gastric adenocarcinoma cells. Following incubation of the cultured human gastric adenocarcinoma cells without GSPE over a period of 0 to 72 h, a steady growth of these cells was observed. Treatment of the human gastric adenocarcinoma cells with 25 mg/lit of GSPE, approximately 2 %, 18 % and 34 % inhibition in cell growth was observed at 24, 48 and 72 h of incubation, respectively, while incubation of the gastric adenocarcinoma cells with 50 mg/lit of GSPE resulted in 5 %, 22 % and 41 % inhibition in cell growth at these same time points, respectively (Table 2).

Table 2

Concentration and time-dependent cytotoxicity of GSPE against cultured CRL 1739 human gastric adenocarcinoma (GAC) and K562 human chronic myelogenous leukemic (CML) cells.

	Absorbance at 570 nm					
	CRL-1739 Human GAC Cells			**K562 Human CML Cells**		
Time (h)	Control	GSPE 25 mg/lit	GSPE 50 mg/lit	Control	GSPE 25 mg/lit	GSPE 50 mg/lit
0	0.72±0.05	-	-	0.77±0.07	-	-
24	0.83±0.06	0.81±0.10	0.79±0.09	0.92±0.08	1.01±0.08	0.96±0.09
48	0.98±0.07	0.80±0.06*	0.76±0.06*	1.13±0.13	1.13±0.20	1.08±0.10
72	1.16±0.14	0.76±0.05*	0.69±0.05*	1.28±0.14	1.29±0.17	1.33±0.28

*Each value represents the mean ±S.D. of 4-6 experiments. *p<0.05 compared to the nontreated group.*

3.2.4. Concentration- and time-dependent cytotoxic effect of GSPE on K562 human chronic myelogenous leukemic cells. Following incubation of the K562 leukemic cells over a period of 0 to 72 h, a steady growth of these cultured cells was observed. Following treatment of these cells with 25 or 50 mg/lit concentrations of GSPE, no significant changes in the growth of these cultured cells were observed at these same time points (Table 2).

3.2.5. Concentration- and time-dependent effect of GSPE on normal human gastric mucosal cells. Incubation of these gastric cells with 25 mg/lit of GSPE, approximately 8 %, 13 % and 9 % increases in cell growth were observed at 24, 48 and 72 h of incubation, respectively, as compared to the control cells, while under these same conditions incubation of these cells with 50 mg/lit of GSPE resulted in 32 %, 26 % and 18 % increases in cell growth at these same time points, respectively (Table 3).

Table 3

Concentration and time-dependent effects of GSPE on a primary culture of normal human gastric mucosal cells

<table>
<tr><th colspan="4" style="text-align:center">Absorbance at 570 nm
Human Normal Gastric Mucosal Cells</th></tr>
<tr><th>Time
(h)</th><th>Control</th><th>GSPE
25 mg/lit</th><th>GSPE
50 mg/lit</th></tr>
<tr><td>0</td><td>0.69±0.09</td><td>-</td><td>-</td></tr>
<tr><td>24</td><td>0.85±0.12</td><td>0.92±0.12</td><td>1.12±0.12</td></tr>
<tr><td>48</td><td>1.03±0.11</td><td>1.16±0.09</td><td>1.30±0.17*</td></tr>
<tr><td>72</td><td>1.27±0.23</td><td>1.38±0.14</td><td>1.50±0.19*</td></tr>
</table>

*Each value represents the mean ±S.D. of 4-6 experiments. *$p<0.05$ compared to the nontreated group.*

4 DISCUSSION

The increasing interest in proanthocyanidins is based on a variety of biological, pharmacological and therapeutic potential including inhibition of DNA topoisomerase II, modulation of protein kinase C, angiotensin-converting enzyme, and hyaluronidase enzyme activities.[9-15] The proanthocyanidins also known to possess potent antibacterial,[16] antiviral[17] and anti-HIV activities.[18] It has also been demonstrated that proanthocyanidins exhibit antihypertensive effects,[1] anti-peptic activity,[19] monocyte stimulating ability,[20] and anti-hepatotoxic activity.[21] The chemical properties of bioflavonoids, in terms of the availability of the phenolic hydrogens as hydrogen donating radical scavengers and singlet oxygen quenchers, predicts their antioxidant activity.[1,9,22] Bioflavonoids have also been demonstrated as potent inhibitors of of the enzymes phospholipase A_2, cyclooxygenase and lipooxygenase.[22-25] For a proanthocyanidin or a bioflavonoid to be defined as an antioxidant it must satisfy two basic conditions: (i) when present in low concentrations relative to the substrate to be oxidized it can delay, retard, or prevent autooxidation or free radical-mediated oxidative injury; and (ii) the resulting product formed after scavenging must be stable through intramolecular hydrogen bonding on further oxidation.[26]

The biological, pharmacological and medicinal properties of bioflavonoids have been extensively reviewed.[1,22] However, much less attention has been paid to the cytotoxic and antitumor activity, as well as the mechanistic pathways of cytoprotection of proanthocyanidins. The proanthocyanidins or polyphenolic bioflavonoids may act as antioxidants or by other mechanisms, contributing to anticarcinogenic and/or chemoprotective actions. However, no direct investigation has been conducted so far.

Based on these findings in conjunction with our previous studies on GSPE, we hypothesized that GSPE may possess selective cytotoxic activity towards human cancer cells. Furthermore, a major drawback of existing radiation or chemotherapy is the often unacceptable damage to normal cells at effective doses for eradicating cancer cells, and for this purpose we planned to assess the comprative cytotoxic effects of GSPE on selected human normal and malignant cells.

In our previous studies, we have demonstrated excellent concentration- or dose-dependent free radical scavenging abilities of GSPE, in both *in vitro* and *in vivo* models, and compared the free radical scavenging ability of GSPE with vitamin C, vitamin E and β–carotene.[2,3] Excellent concentration/dose-response inhibitions were demonstrated by GSPE in both *in vitro* and *in vivo* experiments.[2,3] These experiments demonstrated that GSPE is highly bioavailable and provides significantly greater protection against biochemically generated free radicals and free radical-induced lipid peroxidation and DNA damage than vitamin C, vitamin E, a combination of vitamins C plus E, and β-carotene[2,3] We have also demonstrated the protective ability of GSPE against smokeless tobacco extract (STE)-induced oxidative stress, DNA fragmentation and apoptopic cell death in a primary culture of human oral keratinocyte cells.[5] The results demonstrate that STE produces oxidative tissue damage and apoptosis, which can be attenuated by antioxidants. The GSPE demonstrated significantly better protection as compared to vitamins C and E, singly and in combination.[5]

The objective of the present study was to assess the concentration- and time-dependent cytotoxicity of GSPE on selected human cancer cells including MCF-7 human breast cancer cells, A-427 human lung cancer cells, CRL 1739 human gastric adenocarcinoma cells and K562 chronic myelogenous leukemic cells using cytomorphology and MTT cytotoxicity assay, and to compare the effects of GSPE on normal human gastric mucosal cells. Concentration- and time-dependent cytotoxicity was observed following incubation of the MCF-7 human breast cancer cells, A-427 human lung cancer cells and CRL 1739 human gastric adenocarcinoma cells with GSPE. However, no cytotoxic or growth inhibitory effects of GSPE were observed on human K562 chronic myelogenous leukemic cells and normal human gastric mucosal cells. In fact, GSPE enhanced the growth and viability of the normal human gastric mucosal cells and J774A.1 cells. These data demonstrate that GSPE exhibited cytotoxicity towards some cancer cells, while enhancing the growth and viability of the normal cells which were examined. These results indicate that GSPE may induce significant apoptopic cell death in human MCF-7, A-427 and CRL 1739 gastric adenocarcinoma cells, while enhancing the growth and viability in normal cells. The other possibility may include the modulation of bcl_2 family of genes, especially $bcl-X_L$ (death inhibitory) and $bcl-X_S$ (death promoter) genes, in these cultured human normal and malignant cells following incubation with GSPE. Further investigation is warranted to unveil the mechanism.

References

1. C.A. Rice-Evans and L. Packer, Flavonoids in Health and Disease, Marcel Dekker, Inc. New York, 1997.
2. D. Bagchi, A. Garg, R.L. Krohn, M. Bagchi, M.X. Tran and S.J. Stohs, *Res. Commun. Mol. Pathol. Pharmacol.* 1997, **95,** 179.
3. D. Bagchi, A. Garg, R.L. Krohn, M. Bagchi, D.J. Bagchi, J. Balmoori and S.J. Stohs,

Gen. Pharmacol. 1998, **30,** 771.

4. D. Bagchi, C. Kuszynski, J. Balmoori, M. Bagchi and S.J. Stohs, *Phytotherapy Res.* (in press).

5. M. Bagchi, J. Balmoori, D. Bagchi, S.D. Ray, C. Kuszynski and S.J. Stohs, *Free Rad. Biol. Med.* (submitted).

6. X. Ye, R.L. Krohn, W. Liu, S.S. Joshi, C.A. Kuszynski, T.R. McGinn, M. Bagchi, H.G. Preuss, S.J. Stohs and D. Bagchi, *Mol. Cell. Biochem.* (in press).

7. T. Mosmann, *J. Immunol. Meth.* 1983, **65,** 55.

8. A.K. Rao, C.A. Kuszynski, E. Benner, M.R. Bishop, J.D. Jackson, P.L. Iversen and S.S. Joshi, *Inter. J. Oncol.* 1997, **11,** 281.

9. Z.Y. Chen, P.T. Chan, K.Y. Ho, K.P. Fung and J. Wang, *Chem. Phys. Lipids.* 1996, **79,** 157.

10. M. Hanefeld and K. Herrmann, *Z. Lebensm. Unters. Forsch.* 1976, **161,** 243.

11. J. Masquelier, J. Michaud, J. Laparra and M.C. Dumon, *Int. J. Vitam. Nutr. Res.*1979, **49,** 307.

12. Y. Kashiwada, G. Nonaka, I. Nishioka, K.J. Lee, I. Bori, Y. Fukushima, K.F. Bastow and K.H. Lee, *J. Pharm. Sci.,* 1993, **82,** 487.

13. J. Inokuchi, H. Okabe, T. Yamauchi, A. Nagamatsu, G. Nonaka and I. Nishioka, *Life Sci.* 1986, **38,** 1375.

14. S. Uchida, N. Ikari, H. Ohta, M. Niwa, G. Nonaka, I. Nishioka and M. Ozaki, Ja*pan J. Pharmacol.*1987, **43,** 242.

15. H. Kakegawa, H. Matsumoto, K. Endo, T. Satoh, G. Nonaka and I. Nishioka, *Chem. Pharm. Bull.* 1985, **33,** 5079.

16. A.M. Balde, L. Van Hoof, L.A. Pieters, D.A. Vanden Berghe and A.J. Vlietinck, *Phytother. Res.* 1990, **4,** 182.

17. N. Kakiuchi, I.T. Kusumoto, M. Hattori, T. Namba, T. Hatano and T. Okuda, *Phytother. Res.*1991, **5,** 270.

18. H. Nakashima, T. Murakami, N. Yamamoto, H. Sakagami, S. Tanuma, T. Hatano, T. Yoshida and T. Okuda, *Antiviral Res.* 1992, **18,** 91.

19. N. Ezaki, M. Kato, N. Takizawa, S. Morimoto, G. Nonaka and I. Nishioka, *Planta Med.* 1985, **51,** 34.

20. H. Sakagami, K. Asano, S. Tanuma, T. Hatano, T. Yoshida and T. Okuda, *Anticancer Res.* 1992, **12,** 377.

21. H. Hikino, Y. Kiso, T. Hatano, T. Yoshida and T. Okuda, J. *Ethnopharmacol.* 1985, **14,** 19.

22. C.A. Rice-Evans, N.J. Miller and G. Paganda, *Free Rad. Biol. Med.,*1996, **20,** 933.

23. H. Kolodziej, C. Haberland, H.J. Woerdenbag and A.W.T. Konings, *Phytother. Res.* 1995, **9,** 410.

24. W. Bors and M. Saran, *Free Rad. Res. Commun.,*1987, **2,** 289.

25. N. Salah, N.J. Miller, G. Paganga, L. Tijburg, G.P. Bolwell and C. Rice-Evans, *Arch. Biochem. Biophys.*1995, **322,** 339.

26. F. Shahidi and P.K.J. Wanasundara, *Crit. Rev. Food Sci. Nutr.* 1992, **32,** 67.

REDUCTION OF THE INCIDENCE OF METACHRONOUS ADENOMAS OF THE LARGE BOWEL BY MEANS OF ANTIOXIDANTS: A DOUBLE BLIND RANDOMIZED TRIAL

Luigina Bonelli[1], Annalisa Camoriano[2], Paolo Ravelli[4], Guido Missale[4], Paolo Bruzzi[1] and Hugo Aste[2,3].

[1]Unit of Clinical Epidemiology and Trial, and [2]Unit. of Gastroenterology and Digestive Endoscopy, National Cancer Institute of Genova, and [3]Dept of Oncology, University of Genova, Italy, [4] Dept. of Surgical Endoscopy, University of Brescia, Italy.

1 INTRODUCTION

Most colorectal cancers arise from adenomatous polyps,[1-3] and 20 to 60 % of subjects who had an adenoma removed from the large bowel will develop further adenomas.[4-8] Several epidemiological studies suggest an association between a dietary intake rich in fat and poor in fiber and micronutrients, and the risk of developing adenomas and cancer of the large bowel.[9-13]

An increased risk of colorectal adenomas and cancer was observed in subjects with low serum concentrations of the microelements α-Tocopherol, vitamin A and selenium.[14-20]

In experimental models a dietary supplementation of selenium reduced the incidence of colorectal cancer.[21-24]

In healthy volunteers, oral supplementation with ascorbic acid and α-Tocopherol significantly reduced the fecal mutagenicity,[25] while long term administration of high doses of ascorbic acid significantly reduced the number of rectal adenomas in patients with Familial Adenomatous Poliposis who had had colectomy with ileorectal anastomosis.[26]

Aims of the present study were to evaluate: **a)** the efficacy of a combination of selenium, zinc and antioxidant vitamins in reducing the incidence of metachronous adenomas of the large bowel after endoscopic polypectomy, **b)** the compliance of the patients to both treatment and follow up procedures, **c)** the side effects associated to the treatment.

2 METHODS

This was a double blind, randomized clinical trial. Patients were randomized to receive daily, for five years, either the active compound, composed of 200 µg selenium (as l-selenomethionine), 30 mg zinc, 2 mg vitamin A, 180 mg vitamin C and 30 mg vitamin E, or a placebo. The intervention agent and the placebo were provided by Pharma Nord (Vejle, Denmark).

The study protocol was approved by the Ethics Committee of the National Cancer

Institute (NCI) of Genova, Italy. Patients informed consent was obtained before randomization.

The patients recruitment began in 1988 and was stopped at the end of 1995. Patients were eligible for the study if they were[1] aged 25-75 years,[2] had at least one histologically proven colorectal adenoma removed endoscopically from the large bowel within the last six months before randomization,[3] had a post polypectomy clean colon.

Patients were ineligible if they had a history of Familial Adenomatous Polyposis, inflammatory bowel diseases, ten or more adenomas, large sessile adenomas (3 cm. or larger in diameter), adenoma with invasive carcinoma, previous colorectal resection, invasive cancer at any site, life-threatening diseases, current use of vitamin or calcium supplements, a mental illness. Patients living outside the designated treatment area of the participating endoscopic centers were also excluded.

A clean colon was ascertained by means of a total colonoscopy. If a total colonoscopy was not feasible a double contrast barium enema was performed.

Follow-up colonoscopies were planned after one, three and five years from randomization. It was estimated that the expected risk of adenoma recurrence at five years in the control group was around 20 %, for a yearly recurrence rate of 4.5 %.[5] Sample size was estimated assuming a 50 % reduction in the 5-year cumulative risk of recurrence in the intervention group. With power=80 % and (=0.05, it was estimated that 650 patients had to be enrolled in 5 years with an additional follow up of three years.

By the end of 1995, when 304 patients had been enrolled, the accrual was stopped as during the recruitment period, many eligible patients refused to enter a study that required a long-term treatment and included a placebo arm. In addition, since the reported potential of antioxidants in terms of reduction of metachronous adenomas of the large bowel was not confirmed in other studies,[27-28] it was considered unethical to continue recruitment.

3 RESULTS

Of the 304 patients randomized into the study, 147 were assigned to the intervention agent group and 157 to the placebo group. Of the 304 randomized patients, 233 (76.6 %) had at least one follow-up colonoscopy (117 of 147, 79.6 %, in the intervention agent group and 116 of 157, 73.9 %, in the placebo group), and were evaluable for compliance and for adenoma recurrence.

The overall 5-year actuarial compliance to the treatment was 63.7 %. Compliance rates differed between the two centers, it was low in the group enrolled at the NCI in Genoa (40.5 %) at 5 years, while it was very high in the group enrolled at the University of Brescia (96.1 %).

The 233 patients provided 623 person-years of follow-up, with 51 patients developing a recurrence of metachronous adenomas, 20 in the intervention group and 31 in the placebo group. The observed incidence of metachronous adenomas was 5.9 % (20 cases/336 py) in the intervention group and 10.8 % (31 cases/287 py) in the placebo group (crude RR=0.52, c.i. 0.28-0.97, p = 0.04), the observed RRs of metachronous adenomas was 0.64 (c.i. 0.30-1.35) at the NCI and 0.23 (c.i. 0.06-0.92) at the University of Brescia.

The four-year actuarial adenoma recurrence-free survival was 76.9 % in the group assigned to the intervention agent and 64.2 % in the placebo group (P = 0.035).

The observed reduction in the risk of metachronous adenomas among patients assigned to the active compound was confirmed in multivariate analyses (RR=0.56, c.i. 0.32-0.99, p=0.04).

Minor side effects related to the treatment were reported by 12.5 % of the patients in the intervention agent group and by 2.9 % of the subjects in the placebo group.

4 CONCLUSIONS

The results of the present study suggest that supplementation with selenium, zinc, vitamins A, C and E, may reduce the risk of metachronous adenomas in subjects who previously had an adenoma removed endoscopically from the large bowel shortly before entering the study. The intervention agent was administered at a dosage that is within the range of a normal dietary intake in some populations as a consequence, the dose of nutrients used in this trial allows for a safe, long-term chemopreventive strategy.

It should be noted that in this study the efficacy of the antioxidant treatment was demonstrated using an intention to treat approach and it can be speculated that the observed effect could have been even more significant with a better compliance. In fact, the observed reduction in the recurrence of adenomas was 36 % at the NCI of Genova , where only one-third of the enrolled patients were compliant at 5-year treatment, as compared to 77 % at the University of Brescia, where the compliance to the treatment was over 90 %.

Limitations of the study are that only 304 patients were enrolled while 650 patients should have been randomized in five years. Despite the lower-than-planned power, the difference in recurrence rate between the intervention agent group and the placebo group are both clinically and statistically significant. In addition, due to the small sample size, estimates of the treatment effect show large confidence intervals and the exact magnitude of the treatment effect is uncertain.

Furthermore, a large number of patients discontinued the treatment in one of the two centers. The observed differences in compliance between the two study centers could be related to the availability of alternative health facilities in the two treatment areas. Compliance was apparently not affected by toxicity related to the treatment, since no major side effects were reported by subjects taking the intervention agent.

References
1.   Muto T, Bussey HJR, Morson BC. The evolution of cancer of the colon and rectum. *Cancer*. 1975, **36**, 2251-70.
2.   Eide JT. Risk of colorectal cancer in adenoma-bearing individuals within a defined population. *Int. J.Cancer*. 1986, **38**, 173-6.

3.   Striker SJ, Wolff BG, Culp CE, Libbe SD, Ilstrup DM, McCarty RL. Natural history of untreated colonic polyps. *Gastroenterology.* 1987, **93**, 1009-13.

4.   Matek W, Guggenmoos-Holzmann I, Demling L. Follow up of patients with colorectal adenomas. *Endoscopy.*1985, **17**, 175-181.

5.   Morson BC, Bussey HT. Magnitude of risk for cancer in patients with colorectal adenomas. *Br. J. Surg.* 1985, **72**, S23-25.

6.   Triantafyllou K, Papatheodoridis GV, Paspatis GA, Vasilakaki TH, Elemenoglou I, Karamanolis DG. Predictors of early development of advanced metachronous adenomas. *Hepatogastroenterology.* 1997, **44**, 533-38.

7.   Neugut AI, Jacobson JS, Ahasan H, Santos J, Garbowsky GC, Forde KA et al. Incidence of recurrence rates of colorectal adenomas, A prospective study. *Gastroenterology.* 1995, **108**, 402-8.

8.   Waye JD, Brawnfeld S. Surveillance intervals after colonoscopic polypectomy. *Endoscopy.* 1982, **14**, 79-81.

9.   Wynder EL, Reddy BS, Weisburg JH. Environmental dietary factors in colorectal cancer. *Cancer.* 1992, **70**, 1222-28.

10. Neugut AI, Jacobson JS, De Vivo I. Epidemiology of colorectal adenomatous polyps. *Cancer Epidemio.l Biomarkers Prev.* 1993, **2**, 159-76.

11. Giovannucci E, Willett WC. Dietary factors and risk of colon cancer. *Ann. Med.* 1994, **26,** 443-52,

12. Trock B, Lanza E, Greenwald P. Dietary fiber, vegetables, and colon cancer, critical review and meta-analyses of epidemiologic evidence. *J. Natl. Cancer Inst.* 1990, **82**, 650-61.

13. Haile RW, Witte JS, Longnecker MO, Probst-Hensch N, Chen MJ, Harper J, et al. A sigmoidoscopy-based case-control study of polyps, micronutrients, fiber and meat consumption. *Int. J. Cancer.* 1997, **73**, 497-502.

14. Ingles SA, Bird CL, Shikany JM, Frankl HD, Lee ER, Haile RW. Plasma Tocopherol and prevalence of colorectal adenomas in a multiethnic population. *Cancer Res.* 1998, **58**, 661-6.

15. Clark LC, Hixson LJ, Combs GFJr, Reid ME, Turnbull BW, Sampliner RE. Plasma selenium concentration predicts the prevalence of colorecctal adenomatous polyps. *Cancer Epidemiol. Biomarkers. Prev.* 1993, **2**, 41-6.

16. Russo MW, Murray SC, Wurzelmann LI, Woosley JT, Sandler RS. Plasma selenium levels and the risk of colorectal adenomas. *Nutr. Cancer.* 1997, **28**, 125-29.

17. Willett WC, Polk BF, Morris JS, Stampfer MJ, Pressel S, et al. Prediagnostic serum selenium and risk of cancer. *Lancet.* 1983, **ii,** 130-3.

18. Salonen JT, Salonen R, Lappetalainen R, Maenpaa P, Alftahan G, Puska P. Risk of cancer in relation to serum concentration of selenium and vitamin A and E, matched case-control analysis of prospective data. *BMJ.* 1985, **290**, 417-20.

19. Salonen JT, Alftahan G, Huttunen JK, Puska P. Association between serum selenium and the risk of cancer. *Am. J. Epidemiol.* 1985, **120**, 342-9.

20. Longnecker MP, Martin-Moreno JM, Knekt P, Nomura AMY, Shoeber SE, Stahelin HB et al. Serum a-Tocopherol concentration in relation to subsequent colorectal cancer, pooled data from five cohorts. *J. Natl. Cancer Inst.* 1992, **84**, 430-35.

21. Birt DF, Lawson TA, Julius AD, Runice CE, Salmasi S. Inhibition by dietary

selenium of colon cancer induced in the rat by bis (2-oxopropil)nitosamine. *Cancer Res.* 1982, **42**, 4455-59.

22.   Reddy BS, Sugie S, Maruyama S, El-Bayomy K, Marra P. Chemoprevention of colon carcinogenesis by dietary organoselenium, benzylselenocyanate, in F344 rats. *Cancer Res.* 1987, **47**, 5901-4.

23.   Salbe AD, Albanes D, Winick M, Taylor PR, Nixon DW, Levander OA. The effect of elevated selenium intake on colonic cellular growth in rats. *Nutr. Cancer.* 1990, **13**, 81-7.

24.   Reddy BS, Riverson A, El-Bayoumy K, Upadhayaya P, Pittman B, Rao CV. Chemoprevention of colon cancer by organoselenium compounds and impact of high- or low-fat diets. *J. Natl. Cancer Inst.* 1997, **89**, 506-12.

25.   Dion PW, Bright-See EB, Smith CC, Bruce WR. The effect of dietary ascorbic acid and a-Tocopherol on fecal mutagenicity. *Mutation Res.* 1982, **102**, 27-37.

26.   Bussey HJ, DeCosse JJ, Deschner EE, Eyers AA, Lesser ML, Morson BC et al. A randomized trial of ascorbic acid in Polyposis Coli. *Cancer.* 1982, **50**, 1434-39.

27.   McKeown-Eyssen G, Holloway C, Jazmaji V, Bright-See E, Dion P, Bruce P. A randomized trial of vitamin C and E in the prevention of recurrence of colorectal polyps. *Cancer Res.* 1988, **48**, 4701-5.

28.   Greenberg ER, Baron JA, Tosteson TD, Freeman DH, Beck GJ, Bond JH et al. A clinical trial of antioxidant vitamins to prevent colorectal adenomas. *N. Engl. J. Med.* 1994, **331**, 141-7.

INHIBITION OF COLORECTAL CARCINOMA DEVELOPMENT IN MIN MICE BY FLAVONOIDS

Barbara Raab, Andreas Salomon, Katrin Schmehl, Sabine Sander and Gisela Jacobasch.

German Institute of Human Nutrition Potsdam-Rehbruecke, D-14558 Bergholz-Rehbruecke, Germany.

1 INTRODUCTION

The potential health benefits of flavonoids, widely distributed plant secondary metabolites and substantial bioactive phytochemicals in the human diet, stimulated enormous research activities concerning their biological, pharmacological, and medical properties.

We hypothesised a causal connection between the antioxidative activity of the flavonoids and their anticarcinogenicity. Earlier results demonstrated high antioxidative activities of several flavonols prevalent in plant food and the reduction of oxidative DNA damages by the aglycon quercetin (Q) *in vitro*.[1] Here we report on the influence of Q and rutin (R) on the carcinogenesis in the Min mouse model and the human colon adenocarcinoma cell line HT-29. We wanted to clarify the following questions:

- Do Q and R inhibit the development of the intestinal tumorigenesis ?

- What mode of action can be deduced from histological and immunohistochemical findings and the course of the antioxidative capacity of plasma (AOCP)?

- What influence does Q exert on cell cycle control parameters and phases as well as processes like proliferation and apoptosis?

2 METHODS

C57BL/6J-Apc-Min-/+ and C57BL/6J-Apc-Min+/+ mice, a genetically defined model for inherited and sporadic forms of human colorectal tumorigenesis,[2,3] supplied by Bomholtgard Breeding & Research Center (Bomgard, Denmark), were fed with a semisynthetic diet as control or the same diet supplemented with 14 μg pure Q/g and 47 μg R/g, administered as dried and pulverised buckwheat leaves, respectively, and water *ad libitum* over 50 to 61 days.

Histological investigations were accomplished by counting of polyps after fixation with 5 % paraformaldehyde and staining with hematoxylin and the immunohistochemical detection of apoptosis in murine colon crypts by means of the *in situ* cell death detection kit, POD (TUNEL) from Boehringer Mannheim (Mannheim, Germany).

The AOCP of the Min mice was determined by a photochemiluminescence method described by I. N. Popov and G. Lewin.[4]

Human colon adenocarcinoma cells HT-29 were cultured according to B. Raab *et al.*[1] and treated with 50 or 100 µM Q over 3 days (daily change of culture media).

Cell cycle proteins were estimated by Western blot analysis. Cell lysates were subjected to sodium dodecyl sulfate electrophoresis on 14 % polyacrylamide gels. The proteins were transferred onto a polyvinylidene difluoride membrane by semi-dry blotting in a discontinuous buffer system and detected by incubation with the corresponding primary antibody followed by a peroxidase labeled secondary antibody.

Cell cycle analysis of trypsinised and ethanol fixed cells was carried out by measuring of the DNA content after staining with propidium iodine with a FACScan (Becton Dickinson, Heidelberg, Germany) using the software CellQuest version 3.1. Data were analysed with ModfFit LT software version 2.0.

3 RESULTS AND CONCLUSIONS

As shown in Fig. 1, Q and R significantly reduce the number of polyps in animals fed with standard diet from 72.6±16.4 to 10.8±8.1 and 26.3± 8.9, respectively, after supplementation. The AOCP of Min mice decreases with the severity of carcinogenesis. With Q and R in the diet the AOCP is maintained at a high level similar to control animals without mutation in the Apc gene (Fig. 2).

Determination of apoptosis in histological tissue sections from the colon crypts (not shown) mainly demonstrates apoptotic cells in the apical region with control animals but an increased number of such cells in the colon after Q treatment, especially in the depth of the crypts.

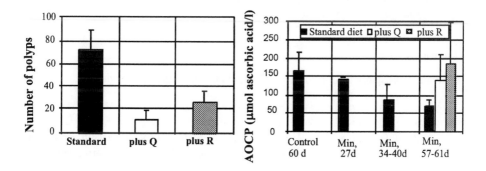

Standard diet vs. Q and R   *Control mice vs. Min-/+ mice at different days:*
supplementation: p < 0.001.   *p <0.02.*

Fig.1   **Fig. 2**
Influence of quercetin and rutin on the   *AOCP during tumorigenesis and influence*
development of intestinal polyps.   *of quercetin and rutin.*

Summarising the *in vivo* experiments we found an anticarcinogenic effect of the aglycon Q and its glycoside R, which may be hydrolysed by the gut microflora. The suppression of the intestinal tumorigenesis in the Min mouse is blood mediated. It runs parallel to an increase in apoptosis and the maintenance of AOCP.

Table 1

Changes in cell cycle phases of HT-29 cells after 3 days treatment with Q.
DMSO vs. Q treatment: p £ 0.005.

Sample	G0 / G1	S	G2 / M
DMSO	83.21 ± 1.16	11.16 ± 0.97	5.63 ± 0.32
50 μM Q	76.84 ± 1.18	15.57 ± 1.00	7.60 ± 0.47
DMSO	75.70 ± 1.61	17.54 ± 1.97	6.76 ± 0.36
100 μM Q	45.35 ± 3.22	42.25 ± 4.50	12.54 ± 0.53

In vitro Q exerts its influence in concentration dependent manner. It suppresses cell growth, documented by reduced cell numbers or protein content of cell lysates. Despite of an increase in the cyclin D1 we could not find a G1 arrest of the cell cycle as shown by F.O. Ranilletti.[5] Three days treatment with 50 and 100 μM concentrations of the flavonol with daily change of culture media lead to a significant decrease of cells in the G0/G1 phase and correspondingly to an increase in the S and G2/M phases (Table 1). An increased expression of different cyclins (A, B1, D1) and the cyclin dependent kinase cdc2 is already detectable with 10 to 25 μM Q as shown by Western blotting. The increase in the G2/M phase of the cell cycle by Q might be connected with a accumulation of cyclin B1 in the cytosol.[6]

We found only a weak influence of Q on apoptosis in HT-29 cells. That means that the reduction of proliferation might be more important than an increase of apoptosis to the normalisation of the balance between both processes by Q.

In accordance with literature our results allow to recommend food rich in Q to people high in risk of colorectal cancer.

Acknowledgement

The technical assistance of B. Kunkel and E. Chudoba is gratefully acknowledged.

References

1.   B. Raab, G. Jacobasch, L. Heinevetter and S. Maurer. Proc. 5th Chem. Congr. of North America, Symposium No. 503 "Phytochemicals and Phytopharmaceuticals", Nov. 11-15, Cancún, Mexico, in press. 1997.
2.   A. R. Moser, H. C. Pitot and W. F. Dove. *Sci.* 1990, **247**, 322.
3.   L.-K. Su, K. W. Kinzler, B. Vogelstein, A. C. Preisinger, A. R. Moser, C. Lugano, K. A. Gould and W. F. Dove. *Sci.* 1992, **256**, 668.
4.   I. N. Popov and G. Lewin. *Free Radic. Biol. Med.* 1994, **17**, 267.
5.   F. O. Ranilletti, R. Ricci, L. M: Larocca, N. Maggiano, A. Capelli, G. Scambia and P. Benedetti-Panici. *Int. J. Cancer.* 1992, **50**, 1992.
6.   P. Jin, S. Hardy and D. O. Morgan. *J. Cell Biol.* 1998, **141**, 875.

Subject Index

α -Carotene	221,376
α -Tocopherol	383
and diabetes	78,389
and antioxidant activity	221
α-Tocopherol acetate	3
Adenoma	451
Aging	
and glutathione	34
and oxygen free radicals	34
and peroxides	34
Allyl sulphides	385
Alzhaimer's disease	
and hippocampal norepinephrine	40
and melatonin	40
Amyloid beta-peptide (A,ß)	
Animal studies	
and citrus extracts	437
Anthocyanins	
and antioxidant activity	151
and bioavailability	151
and determination	151
and processing	332
and strawberries	332
Anthranilic acids	293
Anticarcinogenic compounds	395
and allyl sulphides	385
and chlorogenic acid	385
and dietary intake	385
and glucosinolates	385
and limonene	385
and melatonin	268
and phenolic acids	385
and terpenes	385
and flavonoids	395
and polyphenols	393
and quercetin	395
Antioxidant activity	
of anthocyanins	151
of avenanthramides	293
of flavonoids	166
Antioxidant capacity	
in non-smoking men	271
of foods	277
Antioxidant enzymes	
and depletion	52
and endothelial dysfunction	52
and nitrate	52
and nitrite	52
and superoxide dismutase	52
Antioxidant mechanism	
of atherosclerosis	46
and endothelium	46
and NO synthase	46
Antioxidants	
and antiatherogenity	9
and athero-thrombotic...	
...complications	20
and cancer	377
and cardiovascular disease	3
and carotenoids	9
and content in oats	293
and cytotoxicity	404
and dietary intake	271
and endothelial function	243
and LDL oxidation	9
and lipid peroxidation	3
and mechanisms	243
and oxidation af fatty acids	9
and oxidative modification	3
and plantago major	200
and plasma concentrations	274
and polyphenolic flavonoids	9
and selenium	9
and synergism	277
and ubiquinol Q 10	9

and urine concentrations 274
and vitamin C 9
and vitamin E 9
and 4-nerolidylcatechol 404
Apigenin 285
and contents in urine 188
Apoliprotein E
and atherosclerosis 9
Apoptosis
and cathepsin-D 57
and fas-activation 57
and fas/APO-1/CD95 receptor 57
and growth-factors 57
and iron-chelator desferrioxamine 57
and oxidative stress 57
and starvation 57
Apple juice 209
Aromatase 369
Ascorbic acid
and processing 332
and strawberries 332
Ascorbyl radical 251
Atherosclerosis
and aggregation 46
and hypercholesterolemia 46
and hypertension 46
and oxidative stress 46
and platelet adhesion 46
and risk factors 46
and smoking 46
Avenanthramides
and antioxidant activity 293
ß-carotene 178
and antiatherogenic effect 226
and antioxidant activity 221
and antiperoxidative effect
of dunaliella 337
and isomers 337
and protective activity of
vitamin E 328
and antioxidative effect 226
and diabetes 78
and LDL-oxidation 226
and oxidatitive damage 231
Berry products

and quercetin content 335
Broccoli 440
and antioxidant efficacy 285
Cancer 272,443
and antioxidants 451
and clinical trial 451
and DNA damage 417
and reactive oxygen 417
and cell culture 443
and grape seed extract 443
and DNA damage 456
and flavonoids 456
Cancer prevention
and adverse effects 413
and diet 413
and health risk assessment 413
Carboxymethyl-lysine derivates 74
Carcinogenesis
and free radicals 268
and melatonin 268
and reactive oxygen species 268
Carotenoids 377,423
and concentration in plasma 317
and intake of fruits and berries 317
and smoking 317
and actions 221
and antiatherogenic effect 226
and bioavailability 221
and cap junctional...
...communications 221
and cholesterol synthesis 226
and depletion 231
and free radicals 234
and functions 221
and intake 221
and LDL-oxidation 226
and supplementation 231
Catalase
and glycation 69
Cell morphology 443
Chemokines 46
Chlorogenic acid 385
Cholesterol oxides
and concentration in plasma 317
and determination in foods 303

and edible oils	314
and intake of fruits and berries	317
and interlaboratory comparison	
and smoking	317
Coenzyme Q	
and endogenous	238
and exogenous	238
and mitochondrial activities	238
and oxidative stress	238
and reactive oxygen species	238
Colorectal carcinoma	456
Comet assay	
and DNA oxidation	433
Coronary heart disease	
and antioxidants	20
and oxidative damage of LDL	20
and primary prevention	20
Coronary vasodilatation	
and acetylcholine	46
Coumestans	349
Cryptoxanthin	221
Cultured cells	
and citrus extracts	437
Cytochrome P-450	69,395,393
Daidzein	407
Daidzein-7-o-glucoside	407
Desferrioxamine	74
Diabetes	
and α-tocopherol	78
and body iron stores	80
and cardiovascular disease	69
and complications	69,74
and erythrocytes	83,69
and iron accumulation	74,80
and platelet aggregation	69
and selenium	86,78
and ß-carotene	78
and vitamin A	86
and vitamin C	74,83,86
and vitamin E	74,78,80,86
Dihydrogenistein	407
DNA damage	
and ankylosing spondylitis	433
and insulin dependent diabetes	433
and primary hypertriglyceridemia	433
and antioxidants	440
and diet	440,423
and lipid peroxidation	423
Edible oils	
and sterol oxides	314
and postdeodorization condensate	328
Eicosapentaenoic acid	20
Endoprotease	
and cathepsin-D	57
Endothelial cells	69,74
Endothelial dysfunction	
and cholesterol lowering drugs	46
and l-arginine	46
and vitamin C	46
and vitamin E	46
Endothelium-dependent vasodilatation	
and NO release	52
and oxygen free radics	52
and superoxide anion	52
Enterodiol	407
Enterolactone	407
Equol 407	
F2 isoprostanes	423
Ferritin	80
Fish oils	
and *in vitro* stability	320
and *in vivo* effects	320
Flavonoids	349,437,456,369
and absorption	93
and apigenin	349
and metabolism	137
and antioxidant activity	166,209
and antioxidant effect	137
and antioxidant potency	114
and antioxidative properties	106
and atherosclerosis	106
and berry products	395
and bioavailability	124,166
and biomarkers of intake	121
and contents in apple juice	209
and contents in herbs	206
and contents in vegetables	296
and contents of broccoli	285
and desirable vs. harmful levels	93
and determination	141,206

and diet	174	and oats	293
and dietary intake	395	Insulin-like growth factor	377
and effect of processing	209,297	Iron-chelator desferrioxamine	57
and induced toxicity	212	Isoflavans	161
and intake	93,141	Isoflavone	
and LDL oxidation	106,174	and daidzein	375
and luteolin	349	and equol	375
and modes of action	114	and genistein	375
and naringenin	349	Isoflavone fatty acid esters	215
and parsley intake	188	Isoflavones	200,215
and quercetin	395,174	Isoflavonoids	407
and Trolox Equivalent...		and acacetin	369
Antioxidant capacity (TEAC)	141	and apigenin	369
Flow cytofluorometry	57	and atherosclerosis	356
Free radicals	251	and bioavailability	121
and carcinogenesis	268	and biochanin	369
and gene expression	3	and chrysin	369
and reactive oxygen species	3	and colon cancer	356
and electron spin resonance...		and coronary heart disease	356
...(EPR) spectroscopy	251	and daidzein	349,369
and measurement	251	and dietary intake levels	356
Fructosamine	83	and fisetin	369
Fruit juice		and flavone	369
and antioxidative effect	193	and galangin	369
Genistein	407	and genistein	349,369
Genistein-7-0-glucoside	407	and hormone dependent cancers	356
Glabridin	9	and contents in body fluids	356
and modes of action	161	and isoflavans	161
Glucose autoxidation	69,74	and isoflavones	200,215
Glucosinolates	385	and kaempferide	369
and effects of food prosessing	297	and kaempferol	369
and endothelial cells	74	and luteolin	369
and lipoic acid supplementation	74	and naringenin	369
and oxidation	34	and pinostrobin	369
and oxidative damage	34	and plasma lipids	200
and retinal cells	74	and quercetin	369
and thiol-containing antioxidants	34	and 7-hydroxyflavone	369
Glutathione peroxidase	7 4,86	Isoprostanes	
Glycated end products	74	and measurement	260
Glycated proteins	69	Kaempferol	285
Glycosylated hemoglobin	74,83	Lag time	243
Grape seed proanthocyanidin extract	443	Lignans	369,325,407
Herbs	206	and atherosclerosis	356
Hydroxycinnamic acids		and colon cancer	356
and antioxidant activity	200	and coronary heart disease	356

and dietary intake	356	and tumour promoters	221
and flaxseed	356	and UV irradiated skin	377
and glycitein	356	in diet and serum	377
and hormonedependent cancers	356	Lysosomal proteinase	40
and levels in body fluids	356	Malondialdehyde (MDA)	3,272,423
and mammalian lignans	356	Mammalian lignans	
and matairesinol	356	and enterodiol	349
and secoisolariciresinol	356	and enterolactone	349
Linoleic acid	20	and matairesinol	349,407
Lipid peroxidation		and secoisolariciresinol	349
and arachidonic acid	260	Melatonin	268
and catalyst	3	and amyloid beta-peptide (A,ß)	40
and diabetes	69	and cholinergic neurons	40
and measurement	272	and dopamine relaese	40
and mercury	3	and free radicals	40
and *In Vivo*	260	and hydroxyl free radicals	40
Low density lipoprotein (LDL)		and lysosomes	40
and autoantibodies	3	and neuroblastoma cells	40
Lutein	221	and neuropsychiatric disoders	40
Luteolin	285	and peroxyl-free radicals	40
and antioxidant properties	166	and pineal gland	40
and bioavailability	166	and serotonin	40
Lycopene		Membrane polysaturated fatty acids	
and antiatherogenic effect	226	and oxidative damage	27
and anticancer activity	377	and peroxidation	27
and antioxidant activity	221	and vitamin E	27
and antioxidantive effect	226	Mitochondria	
and bioavailability	221	and gluconeogenesis	34
and cell cycle progression	377	and lactate	34
and free radical-mediated damage	377	and oxidative damage	34
and geometrical isomers	221	and reactive oxygen species	34
and IGF-1 receptor	377	Myricetin	285
and IGF-induced physpohrylation	377	NADPH	69
and insulin receptor	377	and active phagocytes	9
and intake	221	NAPDH oxidase	74
and LDL-oxidation	226	Nitrate	243
and occurrence	221	and urinary excretion	243
and oxidative damage	377	Nitric oxide (NO)	251
and prevention of singlet ...		and L-arginine	40
...oxygen damage	234	and L-citrulline	40
and singlet oxygen	377	and NO synthase	40
and skin cancer	377	NO synthase	
and stomach cancer	377	and coronary vasodilatation	46
and structure	221	and cyclic GMP	46
and tomato-rich diet	377	O-desmethylangolensin	407

Onions 124

Oxidative stress 407

    and arginine 243

    and atherosclerotic lesions 9

    and gluthatione...

    ... peroxidase (GPx) 9,27

    and glutathione (GSH) 9

    and Inflammation 260

    and insulin signaling 80

    and isoprostanes 260

    and nitric oxide 243

    and unsaturated fatty acids 260

    and vitamin E 243

    and metalloproteinase (MMP) 9

    and superoxide dismutase (SOD) 27

    and vitamin E 27

Oxidised proteins

    and oxidative stress 34

Oxidized pyrimidines 417,433

Oxygen free radicals

    and DNA 34

Parsley

    and antioxidative effect 188

Peroxynitrite 251

Phenolic acids 385

    and absorption and metabolism 93

    and desirable vs. harmful levels 93

    and hydroxycinnamic acids 200

    and intake 93

Phenolic compounds

    and antioxidant potency 114

    and mode of action 114

Phospholipids

    and fatty acid side chains 27

Phytochemicals 377,456

Phytoestrogens

    and antiestrogens 349

    and biochanin A 356

    and breast cancer 369,349

    and cardiovascular disease 349

    and coumestrol 356

    and diet 407

    and endogenous steroid...

    ... production 349

    and estrogen receptors 349

    and formononetin daidzein 356

    and genistein 356

    and lower urinary tract symptoms 369

    and measurement 407

    and menopausal symptoms 349

    and plasma 407

    and postmenopausal osteoporosis 349

    and prevention of hormone-

    related diseases 349

    and prostate carcinogenesis 369

    and prostatic hyperplasia 369

    and urine 407

Phytosterol oxides

    and determination 310

    and edible oils 314

    and pure phytosterol mixtures 310

    and supplements 310

Plantago major 200

Polyphenolic flavonoids

    and black tea 9

    and flavonol catechin 9

    and flavonol quercetin 9

    and red wine 9

Polyphenols 393

    and catechins 20

    and epigallocatechins 20

Proanthocyanidins

    and free radical scavenging ability 178

    and grape seeds 178

Prostaglandins

    and isoprostanes 260

Provitamin A carotenoids

    and ß-carotene 221

    and ß-cryptoxanthin 221

Quercetin

    and berry products 335,395,456

    and concentration in plasma 174

    and contents in urine 193

    and effect of processing 335

    and fruit juice intake 193

Reactive oxygen species 251,268

Redox balance

    and extracellular (SOD) 27

    and selenium 27

Redox imbalance	
and ferrous ion oxidising protein	27
and haptoglobins	27
and peroxidation	27
and polyunsaturated fatty acids	27
and reactive iron species	27
and 4-hydroxy-2-nonenal (HNE)	27
Rutin 456	
Secoisolariciresinol	407
Selenium	451
and diabetes	78,86
Serum paraoxonase (PON 1) activity	9
Sesaminol	325
Spin trapping	251
Sterol oxides	
and edible oils	314
Sterol oxides	
and heat teratment	314
Superoxide dismutase (SOD)	
and glycation	69,74
Superoxide radical	243
Tamoxifen	437
TBARS	272
Tea	
and flavonoids	124,274
Terpenes	385
Tocopherols	423
Tocotrienols	437
Total antioxidative capacity	
and edible oils	328
and effect of processing	332
and strawberries	332
Transferrin receptor	80
Transition metals	
and serum ferritin	3
and serum transferrin reseptors	3
Vascular oxidative stress	
and adhesion molecules	46
and cyclooxygenase	46
and lipoxygenase	46
and mononuclear cells	46
and NAPDH oxidase	46
and oxidation of LDL	46
and superoxide radicals	46
and xantine oxidase	46
Vitamin A	377,451
Vitamin C	440,451,178
and diabetes	74,83,86
and fruits	20
and pro-oxidants	34
and vegetables	20
Vitamin E	
and oxidative stress	243
and absoption	325
and all cause mortality	20
and concentration in plasma	317
and fish oils	320
and heat treatment	314
and intake of fruits and berries	317
and neutrophils	34
and protective properties	328
and smoking	317
and cardiovascular mortality	20
and diabetes	74,78,86
and lung cancer	20
Wine	
and flavonoids	243,451
Zeaxanthin	124,151
Zink	221
4-nerolidylcatechol	404
8-oxyguanosine	417